天津市科协资助出版

昆虫—植物生物学
第二版

［荷］路易斯·斯库霍温（Louis M. Schoonhoven）
约普·隆（Joop J. A. van Loon）
马塞尔·迪克（Marcel Dicke） 著
李　敏　朱耿平　刘　强　卜文俊　译

南开大学出版社
天　津

昆虫—植物生物学(第二版)
Insect-plant Biology（second edition） by Louis M. Schoonhoven, Joop J. A. van Loon, and Marcel Dicke
© 2005 Oxford University Press.
All Rights Reserved
"INSECT—PLANT BIOLOGY, SECOND EDITION" was originally published in English in 2005. This translation is published by arrangement with Oxford University Press.
本书英文版2005年出版。中文简体字版有牛津大学出版社有限公司授权南开大学出版社翻译出版。
出版所有，侵权必究。
天津市出版局著作权合同登记号：图字02-2016-67

图书在版编目（CIP）数据

昆虫—植物生物学：第二版/（荷）路易斯·斯库霍温，（荷）约普·隆，（荷）马赛尔·迪克著；李敏等译.—天津：南开大学出版社，2018.4（2019.7重印）
ISBN 978-7-310-05564-7

Ⅰ．①昆… Ⅱ.①路…②约…③马…④李… Ⅲ.①昆虫-关系-植物学 Ⅳ.①Q968.1

中国版本图书馆 CIP 数据核字(2018)第044307号

版权所有　侵权必究

南开大学出版社出版发行
出版人：刘运峰
地址：天津市南开区卫津路94号　邮政编码：300071
营销部电话：(022)23508339　23500755
营销部传真：(022)23508542　邮购部电话：(022)23502200

＊
天津市蓟县宏图印务有限公司印制
全国各地新华书店经销
＊
2018 年 4 月第 1 版　2019 年 7 月第 2 次印刷
260×185 毫米　16开本　27.75印张　684千字
定价：88.00元

如遇图书印装质量问题，请与本社营销部联系调换，电话：(022)23507125

第二版序

众所周知，昆虫与植物的种类在已知的多细胞生物中占半数左右。它们相互选择，相互影响，已经协同进化了近亿年。如今，它们之间的关系已涉及生物学的各个分支，从生物化学、遗传学到行为学和生态学等。

昆虫与植物有着互惠共生的一面，比如许多特化的花与传粉昆虫是对应存在的。早在两百多年前，传粉昆虫与花的关系作为昆虫—植物关系的范例已经引起了人们的普遍关注。但昆虫与植物也有着对抗的一面，特别是在农作物与植食性昆虫之间。在过去的几百年里，这种抗性作用为许多研究提供了灵感，使得人们对于植物化学、昆虫生理学、行为学和生态学有了更深的理解。分子系统学发展使人们重新认识了植物与昆虫的协同进化。在已有的研究中发现，对昆虫与植物关系的发展、变化和进化的影响可能是由多重因素相互作用造成的，如微生物、真菌、捕食者和昆虫寄生物。相比之下也有许多食草昆虫与植物无密切关系的例子被发现。

今天，昆虫—植物的关系已涉及生物学领域的各个方面。我们发现植食性昆虫对植物的选择绝不仅是取食没有防御能力的植物那么简单，而是由多重的历史因素共同作用影响了选择的过程，行为的表型可塑性、不同的营养级对于选择起到了一定的作用，而感觉与神经的反应是食草选择的主要因素。植物基因作为潜在的选择压力引导了单方或双方的进化方向。我们知道生态和生理因素会影响食性范围，同样了解昆虫种群限制在特定的植物取食也许是种化事件的第一步。我们认识到昆虫的进化跟随着植物的进化，然而关系密切的协同进化却未必呈现出来。

本书的三位作者发挥各自的专长与经验，充分展示了昆虫—植物关系广阔的领域。本书的作者们从大量的文献中整合出的实验及理论的经典案例可作为相关领域学生及研究者的综合指南。本书以一个对自然界现象的概述作为开端，有逻辑地转到了植物结构与化学，寄主植物寻找和选择，包括昆虫生理与变异。后面的章节包括了生态与进化、昆虫与花的关系，最后在昆虫和植物的关系中运用所学的知识来做结束。

自从1998年本书的第一版出版后，生物学在许多领域发生了很大的进步，尤其是在植物生物化学和进化生物学方面。而分子生物学技术的发展使得人们对于昆虫—植物关系有了进一步的了解。在本版书中，由于马塞尔·迪奇的参与，使得对一些新的工作和协同进化有了不同的解释。整本书比起上一版都有了更新并融入更为丰富的内容。对于对昆虫—植物关系各领域感兴趣的人来说，这本书是宝贵而可靠的资源。

伊丽沙白·伯奈斯
亚利桑那大学
图森
2005年5月

第二版前言

除了提供由过去七年迅速发展的文献所据知的资料的更新之外，本版和第一版主要有以下两点不同：第一，第一版的作者之一，昆虫与植物关系领域创始人蒂伯·杰米博士，不再负责本版的编写工作，但他丰富的观点与渊博的知识在第二版中仍有体现。我们谨以此书献给蒂伯·杰米，来表达我们对他深刻的洞察力和对昆虫—植物领域持久贡献的由衷尊敬。

第二，由于近年来昆虫—植物关系领域中多方面技术的进步，关于化学感应（chemoreception）的分子生物学和诱导植物防御及对更高层营养级的影响，提出了比7年前更新、更深刻的认识。分子生物学的发展，使得我们对昆虫—植物关系的理解更进一步。

感谢同行们对本书的评阅，感谢蒂伯·杰米、彼得·德·容、艾里克·斯得勒和弗莱迪·恰林基的鼓励和支持。感谢汉斯·斯米德提供的精美照片，和其他同行对其再版显微照片的允许。衷心感谢英国帝国学院种群生物研究委员会全国环境中心及中心主任查尔斯·戈弗雷，感谢马腾和艾丽（德国科隆）为我们提供了能启发灵感的编写此书的舒适环境。

最后，我们特别感谢牛津大学出版社的各位同仁，他们的激励与帮助使得这本书能够如我们所愿地顺利出版成一本有益并能激发灵感的作品！

瓦赫宁恩，2005年夏
路易斯·斯洪霍芬
乔普范隆
马塞尔·狄更斯

第一版前言

绿色植物占据了地球上绝大多数的陆地，而昆虫恰是植物的主要取食者。植物在长期的斗争中努力避免被昆虫取食，而昆虫努力取食植物最优化，二者的相互关系是本书的主题。这个主题包含的内容非常丰富，在过去的25年里，早期的文献数以指数级增长。它也是有史以来最具有挑战性的课题，尽管有许多实际事例，昆虫与植物间潜在关系的原则仍有大量的未解之谜。本书将对自然界及农业环境中的众多研究发现进行分类整理，试图发现其内在规律。希望本书能够成为相关领域学生的教材，同时生物学者们也可以从此书中得到对于理解复杂生物关系的新思路的启示。

现代农业生产需要在提高产量的同时尽量减少杀虫剂的使用，这使得农业昆虫学家需要学习自然界中的植物如何在昆虫的长期侵害下生存，以及植物自身的防御系统是否可以应用于农业生产中。因此本书或许可以帮助那些从事寻找保障日常食物产量新方法的应用昆虫学家。

由于近年来相关的文献信息太过丰富，以致我们无法将其全部收纳，因此我们只选择了其中我们认为最重要的部分。在这一过程中我们很有可能遗漏了一些同样重要甚至更为重要的观点和报道，对此我们表示由衷的歉意。为了对现存文献中阐述的事实和观点进行尽量客观的表述，我们不得不经过深思熟虑后给出一些个人观点。

我们谨以此书献给本领域三位伟大的奠基者：简·德·怀尔德、文森特·德蒂尔和约翰·肯尼迪。他们对我们的思想产生了深远的影响。正因为他们的远见、他们的热情、他们对于自然运转机制的洞察力，昆虫—植物关系领域才能获得如今的成就。

许多人在多个方面为我们提供了慷慨的帮助——积极的讨论、坦白的批评、插图材料的提供和已出版的数据信息的使用许可等。我们要特别感谢那些已经阅读过书稿的部分章节并给出有益建议的同行们：万·比克、毕斯马、迪克、哈瑞金、万·黑尔登、万·伦特恩、门肯、密斯陈德普、穆勒玛、罗斯汀、斯得勒、真蒂斯、恰林基和福休斯。最后，我们要感谢查普曼·霍尔公司的全体员工在本书生产过程中所作的努力。

瓦赫宁恩，布达佩斯
1996 年秋
路易斯·斯洪霍芬
文森特·德蒂尔
乔普·范隆

目 录

第1章 引言 ··· (1)
  1.1 提高关注:为什么要提高对昆虫—植物关系的关注度? ············· (1)
  1.2 昆虫与植物的关系 ·· (1)
  1.3 与农业的相关性 ·· (3)
  1.4 昆虫—植物研究所涉及的学科 ·· (3)
  1.5 参考文献 ··· (4)

第2章 植食性昆虫 ·· (5)
  2.1 寄主植物的专化性 ·· (6)
  2.2 被取食植物的范围和寄主植物的范围 ······································· (10)
  2.3 植物各部分的特化 ··· (12)
      2.3.1 取食植物地上部分的昆虫 ·· (12)
      2.3.2 取食植物地下部分的昆虫 ·· (13)
  2.4 每种植物上昆虫的数目 ··· (14)
  2.5 植食性昆虫:它们是植物的分类学家? ····································· (17)
  2.6 寄主植物多于被取食植物 ··· (17)
  2.7 植物周围的微气候 ··· (18)
  2.8 昆虫对自然生态系统和农业生态系统中的植物的危害程度 ····· (19)
  2.9 植物对植食性昆虫所造损伤的补偿 ··· (24)
  2.10 结论 ··· (25)
  2.11 参考文献 ·· (26)

第3章 植物结构:坚固的抗虫害防护 ··· (34)
  3.1 昆虫的取食系统 ··· (34)
  3.2 叶面 ·· (36)
      3.2.1 角质蜡层 ··· (36)
      3.2.2 毛状体 ·· (38)
  3.3 叶片韧性 ·· (40)
      3.3.1 上颚的磨损 ··· (40)
      3.3.2 C_3植物与C_4植物 ··· (43)
  3.4 互利共生关系中所包含的结构 ·· (44)
  3.5 植物虫瘿 ·· (45)
  3.6 植物的结构 ·· (46)
  3.7 结论 ·· (48)
  3.8 参考文献 ·· (49)

第4章 植物化学:无穷无尽的变化 ·· (56)

4.1 植物生物化学 ………………………………………………………… (57)
　4.1.1 植物的初生代谢物质 ………………………………………… (58)
　4.1.2 植物的次生代谢物质 ………………………………………… (58)
4.2 生物碱 ………………………………………………………………… (60)
4.3 萜类化合物和类固醇 ………………………………………………… (61)
4.4 酚类化合物(酚醛树脂) ……………………………………………… (62)
4.5 芥子油苷 ……………………………………………………………… (65)
4.6 生氰类化合物 ………………………………………………………… (65)
4.7 叶表面的化学物质 …………………………………………………… (66)
4.8 植物的挥发物质 ……………………………………………………… (67)
4.9 植物次生代谢物浓度 ………………………………………………… (71)
4.10 生产成本 …………………………………………………………… (74)
4.11 区室作用 …………………………………………………………… (76)
4.12 时间异质性 ………………………………………………………… (78)
　4.12.1 季节影响 ……………………………………………………… (78)
　4.12.2 昼夜影响 ……………………………………………………… (80)
　4.12.3 年间变化 ……………………………………………………… (81)
4.13 地理位置和肥料的影响 …………………………………………… (81)
　4.13.1 日晒和遮阴 …………………………………………………… (81)
　4.13.2 土壤因素 ……………………………………………………… (82)
4.14 诱导抗性 …………………………………………………………… (84)
　4.14.1 直接诱导 ……………………………………………………… (84)
　4.14.2 间接诱导 ……………………………………………………… (85)
　4.14.3 植食诱导变化引起的变异 …………………………………… (85)
　4.14.4 食草性昆虫引发的基因变异和代谢变化 …………………… (87)
　4.14.5 系统影响 ……………………………………………………… (88)
　4.14.6 长期反应 ……………………………………………………… (88)
　4.14.7 信号传导 ……………………………………………………… (90)
　4.14.8 食草动物诱导和病原体诱导的相互作用 …………………… (90)
　4.14.9 植物间的相互作用 …………………………………………… (91)
4.15 基因型差异 ………………………………………………………… (91)
　4.15.1 植物个体间的化学差异 ……………………………………… (91)
　4.15.2 植物个体内部的化学变化 …………………………………… (92)
　4.15.3 植物性别对昆虫的敏感性的影响 …………………………… (94)
4.16 结论 ………………………………………………………………… (95)
4.17 文献 ………………………………………………………………… (96)
4.18 参考文献 …………………………………………………………… (96)

第5章 植物并非昆虫最理想的食物来源 ………………………………… (116)
5.1 植物是次优的食物 …………………………………………………… (118)

5.1.1　氮 ………………………………………………………………………………… (119)
　　5.1.2　水 ………………………………………………………………………………… (121)
5.2　人工饲料 ………………………………………………………………………………… (122)
5.3　消耗和利用 ……………………………………………………………………………… (123)
　　5.3.1　被取食的食物数量 ………………………………………………………………… (123)
　　5.3.2　利用 ………………………………………………………………………………… (123)
　　5.3.3　次优食物和补偿性摄食行为 ……………………………………………………… (128)
　　5.3.4　化感物质及食物的利用率 ………………………………………………………… (130)
　　5.3.5　对植物化感物质的解毒作用 ……………………………………………………… (133)
5.4　共生体 …………………………………………………………………………………… (137)
　　5.4.1　食物利用与补充 …………………………………………………………………… (137)
　　5.4.2　植物化感物质的解毒作用 ………………………………………………………… (138)
5.5　微生物影响寄主植物特性 ……………………………………………………………… (138)
　　5.5.1　植物致病菌 ………………………………………………………………………… (138)
　　5.5.2　内生真菌 …………………………………………………………………………… (139)
5.6　寄主植物影响植食性昆虫对病原体和杀虫剂的敏感性 ……………………………… (140)
5.7　植物食物质量与环境因素的关系 ……………………………………………………… (142)
　　5.7.1　干旱 ………………………………………………………………………………… (142)
　　5.7.2　空气污染 …………………………………………………………………………… (142)
5.8　结论 ……………………………………………………………………………………… (144)
5.9　参考文献 ………………………………………………………………………………… (144)

第6章　选择寄主植物：如何发现寄主植物 ……………………………………………… (158)
6.1　术语 ……………………………………………………………………………………… (158)
6.2　寄主植物的选择：一个链过程 ………………………………………………………… (160)
6.3　搜寻机制 ………………………………………………………………………………… (162)
6.4　寄主植物的定位 ………………………………………………………………………… (166)
　　6.4.1　光学诱因与化学诱因 ……………………………………………………………… (166)
　　6.4.2　对寄主植物特性的视觉反应 ……………………………………………………… (167)
　　6.4.3　对寄主植物的嗅觉反应 …………………………………………………………… (172)
　　6.4.4　飞蛾和步甲：嗅觉定向的两个例子 ……………………………………………… (172)
6.5　对寄主的气味检测的化学感觉基础 …………………………………………………… (176)
　　6.5.1　嗅觉感受器的形态学研究 ………………………………………………………… (176)
　　6.5.2　嗅觉转导 …………………………………………………………………………… (177)
　　6.5.3　嗅觉电生理和敏感性 ……………………………………………………………… (177)
　　6.5.4　嗅觉特征与编码 …………………………………………………………………… (180)
6.6　自然中对寄主植物的搜寻 ……………………………………………………………… (182)
6.7　结论 ……………………………………………………………………………………… (184)
6.8　参考文献 ………………………………………………………………………………… (184)

第7章　选择寄主植物：何时接受植物 …………………………………………………… (197)

7.1 接触期:对植物特征的详细评价 …………………………………… (198)
7.2 接触期内自然界植物的物理特征 ………………………………… (198)
　7.2.1 毛状体 ……………………………………………………… (199)
　7.2.2 叶面结构 …………………………………………………… (199)
7.3 植物化合:接触化感作用评估 …………………………………… (201)
7.4 植物化学对寄主植物选择的重要性:一段历史性的间奏 ……… (201)
7.5 取食和产卵的刺激 ……………………………………………… (202)
　7.5.1 初级植物的代谢物 ………………………………………… (202)
　7.5.2 植物次级代谢物促进接受:信号刺激 …………………… (205)
　7.5.3 通常将产生的植物次生代谢产物作为刺激剂 …………… (209)
7.6 摄食和产卵的抑制 ……………………………………………… (211)
　7.6.1 抑制是寄主识别过程中的一个普遍原则 ………………… (211)
　7.6.2 寄主标记是避免植食性昆虫相互竞争的机制 …………… (211)
7.7 植物可接受性:刺激和抑制之间的平衡 ………………………… (212)
7.8 基于接触化学感受器的寄主植物的选择行为 …………………… (213)
　7.8.1 化学感受器 ………………………………………………… (213)
　7.8.2 味觉编码 …………………………………………………… (213)
　7.8.3 毛虫成为编码原则的模型 ………………………………… (215)
　7.8.4 信号刺激受体:卓越的专家 ……………………………… (216)
　7.8.5 糖和氨基酸的受体:养分的探测器 ……………………… (217)
　7.8.6 抑制受体:全面感受神经元 ……………………………… (218)
　7.8.7 外围化学物质的相互作用 ………………………………… (219)
　7.8.8 刺吸式昆虫的寄主植物选择 ……………………………… (222)
　7.8.9 产卵偏好 …………………………………………………… (224)
　7.8.10 寄主植物的选择:三级系统 ……………………………… (225)
7.9 化学感应系统与寄主植物偏好的演化 ………………………… (225)
7.10 结论 …………………………………………………………… (228)
7.11 参考文献 ……………………………………………………… (229)

第8章 寄主植物的选择:变化即规律 …………………………… (246)
8.1 地理变异 ………………………………………………………… (246)
8.2 相同地区的种群间差异 ………………………………………… (249)
8.3 个体间的差异 …………………………………………………… (249)
8.4 对寄主植物偏好引起改变的环境因素 …………………………… (250)
　8.4.1 季节性 ……………………………………………………… (250)
　8.4.2 温度 ………………………………………………………… (251)
　8.4.3 捕食风险 …………………………………………………… (252)
8.5 引起寄主植物偏好改变的内部因素 …………………………… (252)
　8.5.1 发育阶段 …………………………………………………… (252)
　8.5.2 昆虫性别影响食物选择 …………………………………… (254)

8.6 经验诱导寄主植物偏好的改变 …………………………………………………… (254)
   8.6.1 非联系式改变 …………………………………………………………… (254)
   8.6.2 联系式改变 ……………………………………………………………… (259)
8.7 幼虫和早期成虫的经验 …………………………………………………………… (262)
8.8 经验诱导寄主植物偏好改变的适应意义 ………………………………………… (263)
8.9 结论 ………………………………………………………………………………… (264)
8.10 参考文献 ………………………………………………………………………… (265)

第9章 植物对昆虫内分泌系统的影响 …………………………………………… (274)
9.1 发育 ………………………………………………………………………………… (274)
   9.1.1 形态 ……………………………………………………………………… (274)
   9.1.2 滞育 ……………………………………………………………………… (277)
9.2 繁殖 ………………………………………………………………………………… (278)
   9.2.1 成熟 ……………………………………………………………………… (278)
   9.2.2 交尾 ……………………………………………………………………… (281)
9.3 结论 ………………………………………………………………………………… (284)
9.4 参考文献 …………………………………………………………………………… (284)

第10章 生态学：共生却各自独立 ………………………………………………… (288)
10.1 植物对昆虫的影响 ……………………………………………………………… (289)
   10.1.1 植物物候学 …………………………………………………………… (290)
   10.1.2 植物化学 ……………………………………………………………… (291)
   10.1.3 植物形态学 …………………………………………………………… (293)
   10.1.4 替代食物 ……………………………………………………………… (293)
10.2 植食昆虫对植物的影响 ………………………………………………………… (294)
10.3 地上部分与地下部分的昆虫—植物交互作用 ………………………………… (296)
10.4 微生物和昆虫—植物的交互作用 ……………………………………………… (296)
10.5 脊椎动物和昆虫—植物的交互作用 …………………………………………… (297)
10.6 群落内间接的物种交互作用 …………………………………………………… (298)
   10.6.1 资源利用性竞争 ……………………………………………………… (299)
   10.6.2 似然竞争 ……………………………………………………………… (300)
   10.6.3 营养级联 ……………………………………………………………… (301)
10.7 物种交互与表型可塑性 ………………………………………………………… (302)
10.8 自上而下和自下而上的驱动力 ………………………………………………… (303)
10.9 食物网和化学信息网 …………………………………………………………… (304)
   10.9.1 食物网 ………………………………………………………………… (304)
   10.9.2 化学信息网 …………………………………………………………… (305)
10.10 群落 ……………………………………………………………………………… (307)
   10.10.1 为什么会有如此多的稀有植食昆虫存在？ ……………………… (308)
   10.10.2 群集性 ……………………………………………………………… (308)
   10.10.3 群落发展 …………………………………………………………… (309)

10.11	分子生态学	(310)
10.12	结论	(312)
10.13	参考文献	(313)

第11章 进化：昆虫与植物，永恒的斗争 (326)

11.1	昆虫与植物相互作用的化石记录	(327)
11.2	物种形成	(328)
11.2.1	生殖隔离	(330)
11.2.2	物种形成的速率	(333)
11.2.3	共生物种的形成	(333)
11.3	昆虫的基因变异	(334)
11.3.1	种间差异	(334)
11.3.2	种内变异	(335)
11.3.3	偏好与表型之间的关系	(336)
11.3.4	基因突变和寄主植物的适应性	(336)
11.4	植物抗性基因的突变	(337)
11.5	选择与适应	(337)
11.6	昆虫多样性的演变	(338)
11.7	专食性寄生昆虫的进化	(339)
11.7.1	应对植物次生代谢产物	(339)
11.7.2	竞争	(340)
11.7.3	降低由天敌引发的死亡率	(340)
11.7.4	系统发育关系	(340)
11.8	植食性昆虫与寄主间的进化	(341)
11.8.1	对协同进化理论的批判	(343)
11.8.2	对协同进化理论的支持	(344)
11.9	结论	(346)
11.10	参考文献	(346)

第12章 昆虫与植物：卓越的互利共生关系 (355)

12.1	互利共生	(357)
12.2	花的专一性	(359)
12.2.1	昆虫对花的识别	(361)
12.2.2	昆虫对花的采集	(363)
12.3	传粉的动力	(365)
12.3.1	距离	(365)
12.3.2	可接近性	(366)
12.3.3	温度	(367)
12.3.4	评价食物源	(368)
12.3.5	奖励策略	(368)
12.3.6	花蜜信号的状态	(370)

12.4	对多花序花的授粉运动	(371)
12.5	竞争	(372)
12.6	进化	(373)
12.7	自然保护	(377)
12.8	经济	(379)
12.9	结论	(380)
12.10	参考文献	(380)

第13章 昆虫和植物：如何运用我们的知识 (389)

- 13.1 植食性昆虫会演变成害虫及其原因 (390)
  - 13.1.1 害虫的特征 (390)
  - 13.1.2 引进农作物带来的影响 (390)
  - 13.1.3 农业活动促进了虫害的发生 (391)
- 13.2 寄主植物抗性 (392)
  - 13.2.1 寄主植物的抗性机制 (392)
  - 13.2.2 部分抗性 (393)
  - 13.2.3 植物特性与抗性 (394)
  - 13.2.4 培育植物抗性的方法 (395)
- 13.3 混养为何能减少病虫害 (398)
  - 13.3.1 破坏性作物假说 (400)
  - 13.3.2 天敌假说 (400)
  - 13.3.3 诱虫作物学说 (401)
  - 13.3.4 多样性原则 (402)
- 13.4 植物源杀虫剂和拒食剂 (402)
  - 13.4.1 拒食剂 (403)
  - 13.4.2 苦楝树：印楝素 (404)
  - 13.4.3 拒食素作为杀虫剂的前景 (405)
- 13.5 植食性昆虫可以控制杂草 (406)
  - 13.5.1 仙人掌和槐叶萍 (407)
  - 13.5.2 生物控制杂草项目的成功率 (407)
- 13.6 结论 (409)
- 13.7 参考文献 (409)

附录A：进一步阅读建议 (419)

附录B：植物次生代谢物的分子结构 (420)

附录C：本领域常用的实验方法 (422)

物种索引 (423)

作者索引 (425)

主题索引 (426)

译者说明 (428)

第1章 引言

1.1 提高关注:为什么要提高对昆虫—植物关系的关注度?
1.2 昆虫与植物的关系
1.3 与农业的相关性
1.4 昆虫—植物研究所涉及的学科
1.5 参考文献

近一个半世纪之前,柯伟林和威廉斯·彭斯在他们所撰写的《传奇昆虫学》[3]一书中描述了一只白色大蝴蝶寻觅合适植物产卵的过程:"她在一些植物中寻觅类似卷心菜的产卵地,在她比老练的植物学家更为精准的本能的引领下,一旦找到自己所需的植物后就会立刻接近它,并在上面完成产卵的使命。"紧接着,作者提出了一个至今仍困扰昆虫—植物研究者的基本问题:"她是如何从周围的蔬菜中识别出卷心菜的?"书中给出的答案是:"上帝赋予了她这项本能"[3]。这个答案显示,人们在很长一段时间内对植食性昆虫与寄主植物的关系难以进行因果分析。关于植食性昆虫对寄主植物的选择机制的科学性研究始于1900年[2,9],但在很长的一段时间内只引起了少数生物学家的关注。约半个世纪之前,动物学家开始对昆虫的行为进行了因果分析,如寄主植物的差别是否影响昆虫的行为,一些新的机制开始逐渐地被发现。

1.1 提高关注:为什么要提高对昆虫—植物关系的关注度?

昆虫—植物关系逐渐受到生物学家和农学家的关注。从地球生物圈基础知识角度,昆虫与植物关系显示了其重要性。首先是量的因素:不管是从种类还是从生物量上衡量,植物界与昆虫纲是两个非常大的生物范畴。绿色植物占有最大的生物量(图1.1),昆虫在种类数量上占据了绝对统治地位。

正如生态学家罗伯特·梅[6]所说:"大致来讲,如果忽略脊椎动物的体积,那么地球上所有的生物都是昆虫。"尽管它们体型微小,但它们的种类和总的体积却是巨大的。例如,温带陆生昆虫的生物量与陆生脊椎动物的生物量比例约为10:1(图1.2)[7]。

1.2 昆虫与植物的关系

植物与植食性昆虫的关系极其错综复杂。一方面包括昆虫在内的其他动物,均离不开为异养生物提供初级能量的绿色植物。另一方面,取食者对植物的影响正是造成植物多样性的重要原因。昆虫在形态及生活史上的巨大变化可能是促使植物分化的重要推动力[5]。正如埃利希和雷文[1]在一篇论文中提出,"植食性昆虫与植物的相互作用,造成了现今陆地生物多样性的产生。"由于昆虫取食植物,并由此达到了最大程度的分化,因此植物对昆虫意义非凡。如果单独研究昆虫或植物的特性会相对简单,但二者结合后的相互关系却极其复杂,很难再找到如昆虫—植物这样在类型与范围方面相匹敌的其他例子。因此,昆虫—植物关系成了在生物学研究中独特且成果丰硕的领域。

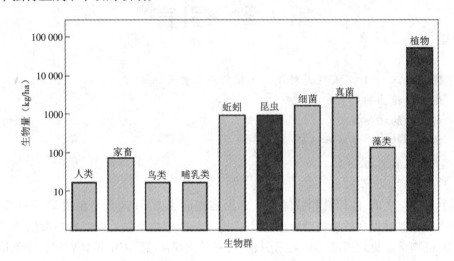

图1.1 美国每公顷土地人类、家畜的平均生物量以及其他主要的自然动植物
的估算生物量(昆虫包括非昆虫的其他节肢动物)

注:以对数比例记录(皮门特尔和安杜,1984)[7]。

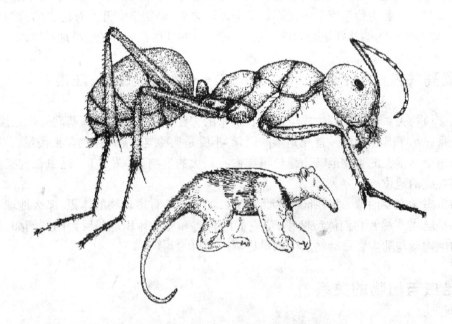

图1.2 蚂蚁与食蚁兽代表的生物量的对比

图中蚂蚁代表的是在巴西亚马孙区域蚂蚁的总生物量,而食蚁兽表示的是该地区所有陆生脊椎动物的总生物量,蚂蚁与脊椎动物的比例约为4∶1,如果将其他所有社会性和非社会性昆虫都包括在内,这个比例将会达到9∶1

1.3 与农业的相关性

从应用的角度来看,昆虫—植物相互作用关系起着至关重要的作用。尽管人类使用了昂贵并对环境造成了巨大危害的防控措施,但昆虫仍是农作物和仓储品的主要威胁,而且它们的破坏仍旧有增无减(图1.3)。

毋庸置疑,如果能够更好地掌握昆虫和植物相互作用的关系,或许可以帮助人类发现害虫增加的原因。基于这些知识,人们或许能够建立对害虫有效安全的控制策略。力普克和弗伦克尔[4]给了植物—昆虫关系研究一个极为恰当的评价——农业生态学的最核心环节。

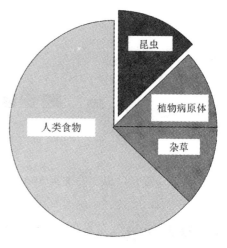

图1.3 估算美国谷物收获前由虫害(13%)、疾病(12%)和杂草(12%)造成潜在的损失(皮门尔特,1997)

1.4 昆虫—植物研究所涉及的学科

昆虫—植物关系包括不同层级的生物学问题,例如,从生物体水平方面提出"为什么卷心菜寄生虫吞食卷心菜叶而不吃马铃薯幼苗"的问题,而从生态学层级提出"为什么有些森林害虫比其他类的昆虫更容易发生大规模爆发"的问题。本书的重点是从生物体水平进行分析,包括生理学与分子生物学方面的研究。当然也包含生态学相关研究,因为从生物体水平得到的研究成果通常是生态模型的重要结论。另一个原因就是许多昆虫—植物行为或生理学方面的特征功能,只有从生态视角去看才会变得显而易见。

在其他生物学的亚学科中,研究昆虫—植物关系的学生可能会对直接问题(如何)或终极问题(为什么)感兴趣。像"昆虫如何识别它的寄主植物""植食性昆虫如何避免被食物中的有毒物质毒到"等这些问题属于第一类。"为什么沙漠植物比牧场的植物包含更多的萜类化合物""昆虫在何种程度上能够刺激有花植物的进化"等这些问题属于第二类。生理学家和分子生物学家最关心的是最直接的影响因素,而学习进化生物学的学生则更注重研究二者的最终原因。事实上,上述两种研究内容都是互补的,因此都被本书所采用,但许多例证并未被明确地归入其中的任何一类。

关于昆虫—植物相互作用关系的这类论题涉及范围非常广,无法在一本书中得到全面的阐述。大量的科学论文证明人们对于这一领域的关注在急速增长(图1.4),其中包括许多在过去二三十年间出版的期刊与书籍,可用信息的总量无法被统计,更不用说去单独研究了。由于物种与物种之间的变异在行为反应和生理适应方面是多样的,因此,本书的每个章节都提供了一定数量的参考文献,对昆虫—植物关系某一特例感兴趣的读者可选择相应的期刊或书籍进行考证。

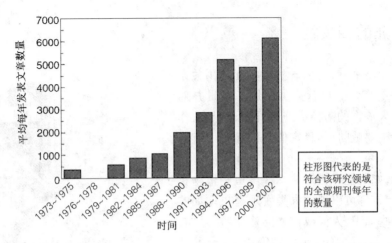

图1.4 从1973年至2002年7月来自联邦农业局(CAB)的主要参考数据

1.5 参考文献

1. Ehrlich, P. R. and Raven, P. H. (1964). Butterflies and plants: a study in coevolution. *Evolution*, 18, 586–608.
2. Errera, L. (1886). Un ordre de recherche trop négligé. L'efficacité des structures défensives des plantes. *Compets Rendus des Séances de la Société Royale de Botanique Belgique*, 25, 80–99.
3. Kirby, W. and Spence, W. (1863). *An introduction to entomology* (7th edn). Longman, Green, Longman, Roberts & Green, London.
4. Lipke, H. and Fraenkel, G. S. (1956). Insect nutrition. *Annual Review of Entomology*, 1, 17–44.
5. Marquis, R. J. (2004). Herbivores rule. *Science*, 305, 619–21.
6. May, R. M. (1988). How many species are there on earth? *Science*, 241, 1441–9.
7. Pimentel, D. and Andow, D. A. (1984). Pest management and pesticide impacts. *Insect Science and its Application*, 5, 141–9.
8. Pimentek, D. (ed.) (1997). *Techniques for reducing pesticides: euvironmental and economic benefits*. Wiley, Chichester.
9. Verschaffelt, E. (1910). The cause determining the selection of food in some herbivorous insects. *Proceedings Royal Academy, Amsterdam*, 13, 536–42. (Reprinted in *Proceedings Koninklijke Nederlandse Akademie van Wetenschappen*, 100, 362–68, 1997.)

第 2 章　植食性昆虫

2.1　寄主植物的专化性
2.2　被取食植物的范围和寄主植物的范围
2.3　植物各部分的特点
　　2.3.1　取食植物地上部分的昆虫
　　2.3.2　取食植物地下部分的昆虫
2.4　每种植物上昆虫的数目
2.5　植食性昆虫：它们是植物的分类学家？
2.6　寄主植物多于被取食植物
2.7　植物周围的微气候
2.8　昆虫对自然生态系统和农业生态系统中的植物的危害程度
2.9　植物对植食性昆虫所造成损伤的补偿
2.10　结论
2.11　参考文献

昆虫是地球上种类最多的一个纲,同时,绿色植物是地球上生物量最多的组成部分。所有现存的昆虫种类几乎一半都是以植物为食的。粗略统计证明,超过 400 000 种植食性昆虫以 300 000 种维管植物为食。(图 2.1)。

依据最近的一些研究材料估计,昆虫物种的整体数量比之前所想的要庞大得多,而且可能达到 400 万至 1 000 万种[82,85],如果这组数据反映的是现实情况的话,那么植食性动物的数量很可能需要调配比

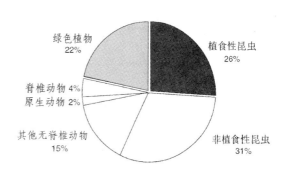

图 2.1　除去真菌、藻类和微生物之外,植物与动物主要分类单元所占比例(斯特朗等,1984)

例。有些研究利用 DNA 序列鉴定物种的同一性,表明目前估算物种数量具有相对性。举个例子来说,根据这样的研究方法,在新热带地区的蝴蝶(*Astraptes fulgerator*),这个物种是由至少十种不同的物种形成的拟态聚集体。它们的成体表现出的表现型差异极小,但是它们的幼虫通常随取食不同的植物而表现各异[53]。

植食性昆虫在类群的分布上并不是均等的。在一些目级分类单元的昆虫几乎完全是植食性的,而在另一些目中的植食性昆虫种类很少,甚至没有。在鳞翅目(蝴蝶和蛾)、半翅目(臭虫、叶蝉、蚜虫等)、直翅目(蚂蚱和蝗虫),还有一些较小的目,如缨翅目(蓟马)和竹节虫目(竹节虫)等目中,植食性昆虫所占的比例较大。在种类丰富的鞘翅目、半翅目和双翅目中,植食性昆虫占据了较大的比例,同时也包括许多肉食性昆虫和营寄生生活的昆虫种类(表 2.1)。

表 2.1　不同昆虫种类中植食性昆虫的数目

昆虫种类	整体数量	植食性昆虫	
		数量	%
鞘翅目	349 000	122 000	35
鳞翅目	119 000	119 000	100
双翅目	119 000	35 700	30
膜翅目	95 000	10 500	11
半翅目	59 000	53 000	90
直翅目	20 000	19 900	100
缨翅目	5 000	4 500	90
竹节虫目	2 000	2 000	100

由于陆生维管植物会庇护一些植食性昆虫，所以由植物滋生(plant-infesting)许多种类的昆虫并不稀奇。尽管过去人们认为在一些古老的植物上面都没有作为取食者的昆虫存在[125]，例如现在的"活化石"——银杏树(图 2.2)和蕨类植物。然而现在，众所周知，这类植物[125]如同其他残存种一样，像蕨类[54,93,94]、石松植物[61,104]、苔藓[66,108,110]、地衣[66,78]、蘑菇[22]都为一些昆虫提供食物。

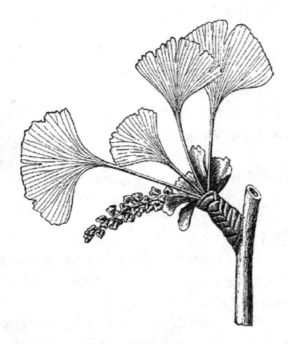

图 2.2　银杏嫩叶和雄蕊

2.1　寄主植物的专化性

在植物和昆虫相互作用的关系中，最显著的特征就是植食性昆虫对食物高度的专化性，这也是昆虫—植物关系的核心，以下章节的讨论大部分都是以此为中心扩展的。因此，对植食性昆虫摄食专化性和普遍性的了解是必要的。在自然界中，仅取食一种或少数几种有亲缘关系的同属植物的昆虫称为单食性昆虫。许多鳞翅目的幼虫、半翅目昆虫和鞘翅目昆虫均属单食性昆虫。寡食性昆虫只取食少数几种同属一科的植物,例如菜粉蝶(*Pieris brassicae*)和马铃薯甲虫(*Leptinotarsa decemlineata*)分别只取食十字花科和茄科植物。多食性昆虫可以接受许多不同科的植物。例如,绿桃蚜虫(*Myzus persicae*)在夏季时以50多科不同种的植物为食。但是,它在冬季时的寄主植物为桃树(*Prunus persica*)或是与其亲缘关系较近的蔷薇科植物。

然而,由此把昆虫分为三类是相当武断的,因为单食性和寡食性的精确定义在实际应用中很难证实。第一个问题在于在取食一种植物的昆虫和取食不同植物的昆虫之间,有一个完全的分级域。第二个问题是同种昆虫在不同地区表现出不同的植物食性偏好,甚至相同种群的不同个体,也能表现出特殊的偏好[13,57,87]。根据这些观察结果能够更容易地区分专食性昆虫(单食性和寡食性昆虫)和广食性昆虫(多食性昆虫)。

寄主植物的专化性与其说是特例,不如说是一种普遍原则。据统计,仅有不到10%的植食性昆虫以3科以上的植物为食[9]。与之相比单食性是一个更为普遍的习性。对英国大约5 000种植食性昆虫的研究统计表明,超过80%的昆虫被认为是专食性的昆虫。然而,不同的昆虫类群,其专食性昆虫所占的比例是不同的。在英国,25种直翅目昆虫中,51%是多食性昆虫,而有41%的昆虫取食仅限于草类和莎草类植物。相反的,所有蚜虫中有76%都是严格的单食性昆虫,18%是寡食性昆虫,而多食性昆虫只占6%。在叶蝉和潜叶蛾中,单食性是一个普遍的特征(图2.3)。总而言之,大部分目的植食性昆虫均为专食性昆虫。然而,也有一些例外,多种蝗虫以不同科的植物为食[16]。

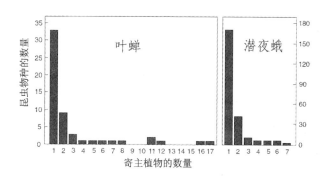

图2.3 在英国,大多数叶蝉与潜叶蛾都是严格的单食性昆虫(克劳利,1985)[25]

人们对植物的专化性及相关术语目前存在着一些争议,有人认为一些寡食性昆虫,甚至是多食性昆虫应该被认定为单食性,因为它们仅取食含某种特定化学物质的植物。菜粉蝶的幼虫,仅取食十字花科植物,偶尔也会在旱金莲(*Tropaeolum*)或是木樨草(*Reseda*)上取食。这两种植物隶属不同的科,但是这些植物都包含了芥子油甙(Glucosinolates),这是一种在十字花科中典型存在的化学成分。人们可以根据大菜粉蝶幼虫取食含芥子油甙这种化学物质的植物,把它定义为单食性昆虫;基于同样的论证,有人曾把多食性昆虫——棕尾毒蛾(*Euproctis chrysorrhoea*)的幼虫当作是专食性昆虫。因为像其他鳞翅目幼虫一样,它通常取食含单宁酸的植物叶片(图2.4)。但是,此定义却忽视了植物其他方面的特征,而这些特征在昆虫对寄主植物的选择上同样起着重要作用。然而,无论从实际情况还是理论研究出发,我们都更倾向将寄主植物专化性和自然状态下昆虫的寄主植物的范围(而非化学物质)联系起来。

昆虫的寄主植物的范围作为昆虫的主要生物特征受多重因素的影响,如形态、生理和生态等。探索食性宽度和植物或植食性昆虫的特征之间的关系有助于解释这些影响因素。目前,一些研究已经发现了一些有趣的结果。生活在草本植物上的昆虫比生活在灌木和乔木上的昆虫更具专化性(表2.2)[14,41]。这种情况可以解释为:与木本植物相比,草本植物在一些方面(如:生活史和化学组成方面)表现出更大程度的多样性。这些特点也就意味着,昆虫在适应各种变化时,专食性昆虫比杂食性昆虫更具有优势。

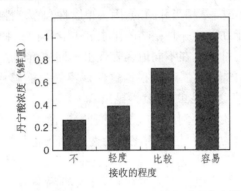

图2.4 杂食性昆虫黄毒蛾(*Euproctis chrysorrhoea*)幼虫更喜富含丹宁酸的植物,根据植物中的丹宁酸含量,61种植物种类可分成4个等级(格里弗柳斯,1905)

表2.2 鳞翅目对草本和木本寄主植物的专化性(福山,1976)

	物种数目	专化比(specialist)%
英国的蛾和蝴蝶		
在草本植物上	143	69
在木本植物上	229	54
北美的蝴蝶		
在草本植物上	110	88
在木本植物上	53	68

在寄主植物专化性方面,第二点值得注意的地方是,昆虫的取食范围与其体型大小有密切的联系:体型稍小的物种比稍大型的物种有更高程度的食物专化性(表2.3)[70,80],可能是因为对体型稍大的昆虫种类来说,对于食物选择的过程中因能量消耗大而风险较大,因而对寄主植物的选择度较小。

第三点值得注意的是,研究结果显示:食性较窄的植食性昆虫通常更喜欢取食正在发育的新鲜嫩叶。然而,杂食性昆虫的幼虫更喜欢取食不同寄主植物的成熟叶片。新鲜嫩叶通常含有更丰富的营养成分[107],然而,新叶也比老叶含更多有毒的植物次生物质[14]。为了补偿这个缺点,取食新叶的昆虫必须拥有更加专化的解毒系统。

上述三个方面可能并非偶然,而是客观反映了一些生物原则。也许这些直观的模式与看似完全不同的生物学特性是相关的,因为在某些昆虫类群中高度专化性的频率要远高于其他类群(表2.3)。

表2.3 昆虫取食仅一个属内植物,或一个科植物,或多于一个科植物的三种情况下各目昆虫的百分比(马特森等,1988)[76]

昆虫	物种数目	取食物种的百分比		
		仅取食一个属内的植物	仅取食一个科内的植物	取食多个科内的植物
木虱科(双翅目)	78	94	3	0
蚜科(同翅类)	445	91	7	2
木蠹科(鞘翅目)	NA	59	38	3
圆介壳虫科(同翅类)	64	58	8	34
缨翅目	88	56	15	29
蛱蝶科(鳞翅目)	88	56	11	33
小灰蝶科(鳞翅目)	89	55	14	31
粉蝶科(鳞翅目)	43	33	53	14
凤蝶科(鳞翅目)	89	25	21	54
其他(大鳞翅目)	430	17	23	60

注:表中前5组的昆虫和其他组相比,包括一些形体较小的类型。然而,昆虫的尺寸大小和寄主植物范围的相关性很低。已有的许多例子表明,同样大小的昆虫在寄主植物的范围宽度上有着很大的差异。

NA——数据未获得。

也许昆虫对于成为专食性或者广食性的进化选择是由诸多因素决定的,这个领域仍需要我们进一步的探索,例如在热带地区灰蝶科昆虫绝大多数种类表现为广食性,这与温带地区的混血种不同[35],而凤蝶科昆虫的情况却与灰蝶科大相径庭[102]。

广食性昆虫可食用多种不同种类的植物,然而,它们绝对不会不加选择地食用全部绿色植物,甚至爱好最广泛的捕食者所食用的食物种类都局限在几百种以内(表2.4),然而,对于这个范围之外的植物,昆虫即使在其他可替代食物资源匮乏的条件下也不会食用,甚至完全排斥它们。

表2.4 广食性昆虫寄主植物种类及数量

昆虫种类	寄主植物种类	植物所属科	参考文献
棉粉虱(*Bemisia tabaci*)	506	74	17
舞毒蛾(*Lymantria dispar*)	>500	>22	68
沙漠蝗(*Schidtocerca gregaria*)	>400	53	8
锈叶虫(*Lygus lineolaris*)	385	55	131
日本甲虫(*Poplilia japonica*)	>300	79	90
蛇纹石夜蛾(*Liriomyyza trifolii*)	>400	25	109

注:上述昆虫均为极端的广食性昆虫,隶属于5个不同的目。

这就不难想象,多食性昆虫和寡食性昆虫在它们所能接受的寄主植物范围内对寄主植物的选择是任意的,而它们其中一些也会表现出对某种植物有所偏好。即使是食性广泛的昆虫,如沙漠蝗(Schistocerca gregaria),也是只对一些植物情有独钟[15]。另外一种具有许多寄主植物的昆虫——舞毒蛾(Lymantria dispar)的幼虫,不仅对某些种类的栎树有特殊偏好[38],而且还显示出不凡的辨别力。在一次选择实验中,比起生长在阴面的栎木叶子来说,昆虫更喜欢食用同一棵树上阳面的叶子。另外,即使饲喂昆虫足量的嫩叶,当提供成熟叶片时,昆虫就会明显地趋向取食成熟的叶片(图2.5)。还有一些昆虫表现出这样的趋性,即喜欢树干的某些部分而不是树冠的一些部分[111]。

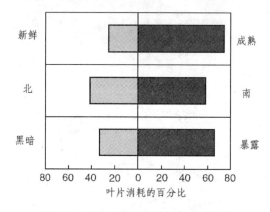

图2.5 不同状态下叶片消耗的百分比

在一次选择性测试中,桤木(Alnus glutinosa)叶子被提供给舞毒蛾(Lymantria dispar)的幼虫,这些幼虫吃掉的成熟叶片比新鲜的叶片多。同样地,它们更喜欢从阳面采集的叶片,而且与在黑暗中24小时生长的叶子来比较,它们更喜欢暴露在阳光中的叶子(斯洪霍芬,1977[99])。

起初,人们将广食性昆虫看作比专食性昆虫更有特权的昆虫,因为它们能取食更多的食物。然而伯奈斯[7]不同意这个观点,他认为从昆虫有限的神经容量来讲,广食性昆虫的发展并不占优势。专食性昆虫比广食性昆虫的选择空间更小,那么就会要求专食性昆虫必须具备更高效的消化吸收能力。

除了神经(行为)方面,其他生理和生态方面的因素也都会影响植食性昆虫对寄主植物选择[59]。在某些昆虫因为食物的增多而种族兴旺的时候,另一些昆虫反而选择了专一的发展模式,而这两种情况都有其各自的优点和缺点。

2.2 被取食植物的范围和寄主植物的范围

很多植食性昆虫在产卵时会选择将后代产在其后代可食用的植物上,那么由此就连带出一个问题,产卵成虫选择的寄主植物和幼虫阶段选择的食用植物是否相同呢?虽然,我们想象它们是一样的,但实际情况并非如此。通过观察证明,产卵雌虫对寄主植物的选择是受不同基因控制的,而非由幼虫阶段的取食选择行为决定[127]。有趣的是,幼虫的取食范围往往比产卵雌虫所选择的寄主植物的范围更广(图2.6)[45]。

显而易见,由于自然选择的作用,雌虫的产卵倾向和其后代取食偏好之间的差异不会过大。一些相关研究已经涉及这个问题:植食性昆虫的产卵倾向是否与它们的后代在特定植物上的生存以及繁殖能力相适应？由此产生了一个联想——雌虫也许更喜欢将卵产在最适合其幼虫生长发育的植株上[67]。这一点也促使了雌虫在选择不同植物种类的同时也考虑选择植物的不同部位（图2.7）[24]。但是总体来讲，凤蝶（*Papilio machaon*）的幼虫在它们所选择的植物上显示出了相当高的存活率。然而，在这两类偏好之间确实也存在一些不对称性（表2.6）。

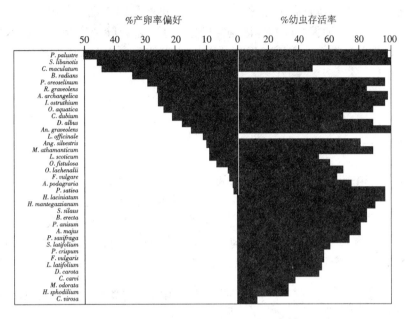

图2.6　雌性凤尾蝶产卵偏好的不同等级

许多植物可作为幼虫的食物，虽然雌虫可以在幅花芹上产卵，但幼虫却不能在此生长。幼虫在某些植物上可以有很高的存活率，但是此处却不是雌虫的产卵之地（维克隆德，1975）[127]

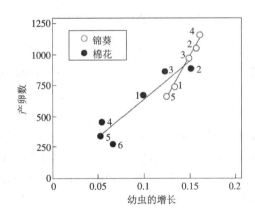

图2.7　卷蛾（*Crocidosema plebegana*）在锦葵（*Malva parviflora*）和棉花（*Gossypium hirsutum*）两种寄主植物上的产卵数量与幼虫增长速度（r_c）间的关系图

在非选择性实验中，所用的植物处在不同的生长阶段，从发芽（1）到停止生长（锦葵，5）或者结荚（棉花，6）。植物作为昆虫产卵的地点，它的可接受幼虫寄生的程度与其生长阶段有较强的关系，反映了它的营养是否能够满足幼虫的发育（哈密赖和祖克乐，1993）[47]

对此一些学者提出了几种解释[39,50,116]。其中一种被广泛接受：大多数产卵的优先选择和幼虫存活状况并不相符，这意味着产卵偏好与生理学的适应能力已在植食性雌虫间普遍化。

从自然选择的角度引申出的相关问题——是否成虫要选择最好的寄主给后代[60]，一个关于产卵地的选取和幼虫生存生长的研究确实显示了这两个参数呈正相关[105]。同一种蛱蝶(*Euphydryas editha*)种群的不同雌虫显示了对不同寄主植物的产卵偏好。后代在不同植物上的适应和母本的产卵偏好是有一定关联的。因此，幼虫可在母本选择的植物上生长得更好[105]。

不同的实例反映了虫体具有不完善的适应性，这可能是由于在等位基因上缺乏足够的基因变异所致，以致造成产卵偏好或其他限制性问题。也可能是除了幼虫生长发育外，仍有不为人知的其他因素控制着寄主植物的进化造成的[122]，也可能是实验本身的一些瑕疵造成的。

2.3 植物各部分的特化

2.3.1 取食植物地上部分的昆虫

昆虫可以取食植物的任何部位，除了对寄主植物有所选择之外，昆虫对摄食的植物部位仍有选择。通常情况下，昆虫在植物的各个部位不会同样的繁盛。许多昆虫是取食叶片的，如毛虫、甲虫和蚱蜢。其他的昆虫则有各自特定的需要。臭虫总是刺入植物表皮细胞获取细胞内容物，蚜虫主要吸食植物韧皮部筛管内流动的汁液，沫蝉和叶蝉总是吸食木质部内的汁液[117]。潜叶蛾在幼虫期吃叶片上下表皮间的组织(图2.8)。

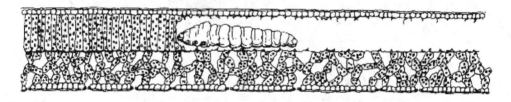

图2.8 甲虫幼虫在寄主植物中取食栅栏组织(palisade parenchyma)的横截面图

不同的昆虫食用叶片不同层面的组织，例如，当桦树叶被两种潜叶蛾侵袭时，一种是潜叶蜂(*Fenusa pumila*)，它全部吃叶肉，而另一种叶蜂(*Messa nana*)的幼虫则只吃栅栏组织(palisade parenchyma)[28,97]。更进一步的来说：潜叶蛾的不同种类偏好叶子的不同部位，一些洞穴接近叶子中脉，而另一些却在叶缘(图2.9)。

因此，叶片的不同部位味道不同，物理性质也不尽相同，不单对潜叶蛾存在影响，整个食叶昆虫都受其影响。有几种蛾类的幼虫(例如裳夜蛾属和舞毒蛾)可以区别它们所食用的植物叶子的基部、侧缘和端部，且并不偏好叶子的基部(图2.10)[42]。植物的茎可以庇护茎生蛀虫，主要包括鳞翅目、双翅目和鞘翅目幼虫(图2.11)。木本植物的树皮经常被甲虫侵食，如小蠹科及其他昆虫。

木质部中可能含有的一些物质能使鳞翅目、鞘翅目以及膜翅目的昆虫适应这种极端不均衡的食物。其他一些昆虫转化为仅取食植物的花、果实或者种子。有些昆虫会使植物的不同部位形成虫瘿[128]。以上的这些例子均可证明植物的所有部位均可成为昆虫的食物。

也许由于营养因素，昆虫为适应植物特定的组织而不断发生变化。不同部位、不同组织具有不同的营养价值。因此，当发现大多数小型昆虫都是专食性时就不奇怪了。植食性昆虫越小，它遇到最适植物组织的概率就越大。例如广食性蓓带夜蛾(*Mamestra configurata*)幼虫，当被放在油菜豆荚中时,存活率很小,死亡速率与喂食叶片的同种幼虫相比快30%[11]。而瘿蚊(*Dasineura barassicae*)的幼虫只取食油菜豆荚并且仅能以此存活[2]。当然,营养因素并不是取食专化性定位的唯一标准。同时必须考虑许多其他的生理和生态因素,这些会在以后的章节提及。

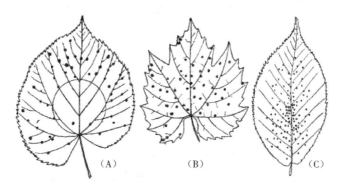

图 2.9 三种潜叶昆虫的分布

(A)50 头叶吉丁(*Brachys*)在椴树(*Tilia* sp.)叶上;(B)50 头斑潜蝇(*Antispila viticordifoliella*)在葡萄(*Vitis vinifera*)叶上;(C)100 头细蛾(*Lithocolletis ostryarella*)在角木(*Ostya* sp.)上(弗罗斯特,1942,经许可)[42]

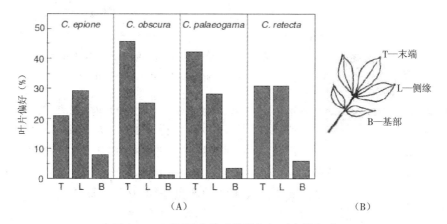

图 2.10 不同昆虫的叶片偏好与毛虫的复叶

(A)四种毛虫(*Catocala* spp.)取食山核桃(*Carya*)叶片基部(basal,B)的组织最少,取食末端和侧缘的较多;(B)毛虫(*C. ovate*)的复叶

2.3.2 取食植物地下部分的昆虫

近来有数据表明,生活在地下的动物的总生物量远比生活在地上的多。植物根系的大量物质是我们所不可见的,昆虫亦是如此。森林和草原地下每平方米含有的昆虫(含其他节肢动物)大约有 100 000 到 500 000 种。它们中的大部分取食植物的根部。昆虫和植物之间的关系十分紧密,这也能直接映射出地上昆虫与植物之间的关系。一些取食根部的昆虫生活在土壤中,例如,吃幼根的蛴螬或直接蛀入根部的昆虫(如洋葱蝇、胡萝卜蝇、卷心菜根蝇),反之,某些蝉和一些蚜虫则是刺穿根部取食液汁[130]。

植物根部被毁坏后就会导致植物根部吸收水、营养和矿质元素的能力降低,因而造成地上部分生长减缓,更有甚者会造成庄稼的减产[72]。取食根部的昆虫通过改变寄主植物的生理生化特性影响地上相对应的部分(反之亦然)[121]。

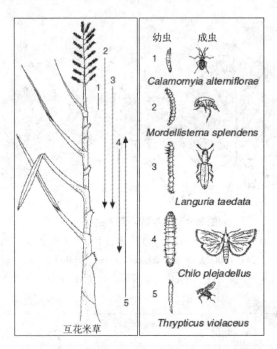

图 2.11 蛀干害虫与互花米草,在植物生长的不同阶段,不同的蛀干害虫
占据了茎干的不同位置(斯特朗等,1984,经许可)[113]

例如,通过毁坏水稻的根部,水稻水象鼻虫(*Lissorhoptrus oryzophilus*)显著地降低了取食水稻叶片的行军虫(*Spodoptera frugiperda*)的生长率。相反,行军虫(*S. frugiperda*)造成水稻大量落叶,同样也会对象鼻虫(*L. oryzophilus*)的生存产生一定的负面影响[118]。植食性昆虫啃食根部会间接地影响植物的防御体系。根部遭到线虫(*Agriotes lineatus*)侵害的棉花(*Gossypium herbaceum*),会使其自身产生的花外蜜比未受侵害时的产量增加 10 倍。而花蜜可以吸引诸如蚂蚁之类的捕食者,它们可以保护植物地上部分免遭其他昆虫的侵害[124]。

虽然不断有证据表明:取食根部的昆虫往往会对植株的健康造成巨大破坏,但是由于此类虫害发生在地下,因而具有不明显性,所以与地上侵害相比并没有引起足够的关注[10]。

就像发现地上植食性昆虫具有寄主植物专化性一样,地下的植食性昆虫也存在此特点。因此,寄主植物有两个维度:一是宿主植物的种类,二是宿主植物的部位。与其他动物相比,仅通过这两个特征使昆虫进化成了物种丰富的类群。显然,对寄主植物专化性根本机制的研究不再停留在研究限制昆虫取食植物部位这个层面上了,而是开始研究更为细化的问题了。

2.4 每种植物上昆虫的数目

据保守估计,植食性昆虫种类数量已超过维管植物的种类(图 2.1)。除了严格意义上的单食性昆虫之外,在一种植物上会出现多种昆虫。每株植物会庇护许多不同种的昆虫,这是在自然界易于观察到的。不同种的昆虫生活在同一植株上不一定会成为竞争对手。除了生存空间的不同,就像前面所提及的,不同种的昆虫也会因为生物气候的不同导致生活的时间不同。例如,刺荨麻(*Urtica dioica*)上有 8 种昆虫同时生存,而这 8 种昆虫的生活史周期模式大不相同,因而它们种群的形成随季候的不同而不同。所以,它们种群的峰值只会在有限的时期出现重叠的现象[27]。(图 2.12)

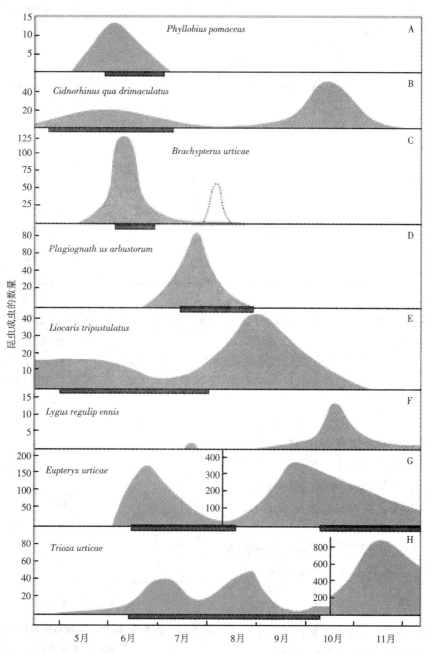

图 2.12　三种鞘翅目昆虫(A—C)、三种异翅亚目昆虫(D—F)以及两种同翅类昆虫(G,H)的生活周期

注:成虫数量来自每周样本统计,加粗线表示现阶段有成虫和卵(戴维斯,1983)[27]。

与其他种群相比,某些植物上寄生着巨大的昆虫群。例如,艾菊(*Tanacetum vulgare*)上就有143种昆虫(表2.5)。尽管不都是植食性昆虫,至少也有110种与刺荨麻有关。它们有31种是该植物的"专食者"[27]。有研究发现,有423种昆虫取食两类橡树。与之相比,红豆杉(*Taxus baccata*)上仅有6种昆虫[62]。当然,在所有植物上都会连续不断地有大量不同种类的昆虫来"拜访",但是,只有少数的"拜访者"可与植物建立一种永久的关系。例如,科根曾报道,现有超过400种植食性昆虫寄生在美国伊利诺角州的一块大豆田里,但是实际记录却显示该地区种群只有不到40种[65]。

表2.5 艾菊(*Tanacetum vulgare*)上的食草昆虫的取食策略(施米茨,1998)[98]

	n	%
物种数量	140	
单食性	19	14
寡食性	64	46
多食性	57	41
取食组织种类数量	89	62
吸食汁液种类数量	54	38
薄壁组织	28	20
韧皮部	23	16
木质部	3	2
外食性	92	64
内食性	51	36

当蕨类植物和被子植物进行比较时,不难发现,昆虫种类在不同植物门之间存在着显著差异。蕨类植物尽管在进化上早于被子植物,但是被子植物上的昆虫要比蕨类植物的昆虫多将近30倍[20]。可以想象,蕨类植物上的昆虫类群要少于被子植物上的。然而,这两大植物上的昆虫种类丰富度之所以存在如此巨大的差异,恐怕最根本的原因还要从生理和生态水平进行研究探索。

与特定植物相关的昆虫,其数量差异可以归因于诸多不同的因素,包括植物生活史、植物物种丰富度(图2.13)、共栖的进化适应度、植株大小与架构、防御机制的效果[12,62]。这些因素与植物之间的关系会在第10、11章讲解。

然而,大家应该可以从先前讨论的问题中明白:在一类植物上寄生的全部昆虫并不能在同一株植株上找到。正因为多种昆虫数量稀少,所以热带地区植食性昆虫出现的平均频率在每10株(甚至更多植株上)为一只。有数据表示在热带地区该数值可能更低[91]。

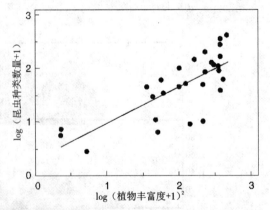

图2.13 在英国,植物丰富度与昆虫种类数量的关系
(肯尼迪和塞尔本南木,1984)[62]

2.5 植食性昆虫：它们是植物的分类学家？

寄主植物专化性要求昆虫必须能够寻找并识别寄主，即使该种植物生长在物种丰富的地区当中。法布尔(J. H. Fabre)曾在他的一本著作中提到有关昆虫行为方面的一些研究，包括产卵雌虫拥有一套"植物定位系统"，这能帮助它们准确地找到寄主植物[33]。后来，这种说法被略加修改并使用，来表示寡食性昆虫可以用某种方法或手段辨别寄主植物[100]。当用叶蝉(*Aphrophora alni*)作为实验对象进行测试时，把8种寄主植物摆放在其面前，它会像植物分类学家一样去辨别植物种类[84]。在早期进行植物分类时，某些专食性昆虫的"植物定位系统"在某些时候会帮助植物学家发现错误。例如，窗蛾幼虫(*Thyridia* sp.)取食玄参科(*Brunsfelsia* spp.)植物。当分类学家认识到全部已知窗盾蝶属(*Thyridia*)种类生活在茄科植物上，番茉莉属(*Brunsfelsia* spp.)的分类地位即被重新鉴定，最后被划分为茄科的一个属[55]。许多其他的实例也曾有报告：专食性昆虫为植物分类提供了重要的线索[114]。已经成功使用蚜虫和木虱解决植物系统分类学的相关问题或植物分类学家存在已久的有关近缘种的鉴定困惑(例如，科杨属(*Populus*)复合体)[30,56]。通过检查两种杨树(*Populus fremontii* 和 *P. angustifolia*)的杂交体和回交体，基因分析和利用植食性昆虫进行分类的结果一致性水平达到98%。这个结果表明，与利用某种单一的特征(如形态或化学特征)区分关系紧密的植物群系相比，昆虫生物鉴定也许是一种更严格、更精确的方法[1,37]。

由以上观察结果可得到一个结论：单食性昆虫和寡食性昆虫是十分出色的植物分类学家，它们依赖自身神奇的"植物定位系统"，无误地识别植物世界的分类关系。然而，我们现有的植物化学知识能在最大程度上解释昆虫识别植物的能力，因为种间关系经常可以等同于生物化学上的联系。昆虫不会像我们一样，把植物分成种、属、科，它们只是通过化学模式来寻找适合的植物。这种模式对寻找单一种的植物来说，比较狭小、严格、有限，对寻找隶属于属或科的植物来说，可能稍微广泛、多变些。通过这种解释，我们已经涉及了植物—昆虫关系的中心论题：植物的化学组成成分与昆虫相互作用的主要因素。在第4至7章将会对这部分内容进行更深入的讨论。

2.6 寄主植物多于被取食植物

寄主植物并不仅为昆虫提供食物，也为昆虫提供居所。肯尼迪(J. S. Kennedy)提出寄主植物会为昆虫提供生活条件，包括生物的和非生物的因素，而非仅仅提供食源。生活在同一植株上的昆虫会遭遇很多问题，包括来自竞争者、天敌、特殊的微环境、植物病原体造成的影响等。

例如，虎蛾(*Platyprepia virginalis*)是一种广(杂)食性昆虫。毒芹是其寄主植物之一，在其上捕捉到的幼虫中有83%被一种蝇寄生，而在同等条件下，从羽扇豆植株上捕捉的幼虫，寄生率只有50%[31]。假设这两种植物营养含量水平等同，如果该虫在毒芹植株上造成的损失不能用其他生理的或生态的优势来弥补的话，该虫也会做出相适应的反应而降低自身被寄生的危险，而发展出对毒芹的躲避反应，以保存种群数量。

昆虫更喜欢营养丰富而又不吸引天敌的寄主植物，也被说成是"无天敌空间"。与生活在营养含量高且昆虫自身易被寄生的植物相比,"无天敌空间"能够更好地保证昆虫的生存率[79,86,106]。对于一例食草昆虫的复合体(complexes)与它们寄主植物之间的关系分析表明,属于17个属的森林树木在对寄生性天敌的吸引水平上存在巨大的差异[69]。这些研究表明,在寄主植物上被寄生的危险是对昆虫生存的一种强烈影响,这种影响在植食性昆虫取食范围的进化过程中是一个潜在的强有力选择作用。

不同植物间在寄生水平上观察到的差异可能是由于植物形态的不同,例如,是否存在表层毛状体(trichomes)或茸毛。其他特征和这个特征一样,都会影响第三营养级成员(即昆虫的寄生者和捕食者)的取食。这个问题将在第3章和第10章进行讨论。

灯蛾(Grammia geneura)是一种广食性昆虫,被寄生后会取食次优植物来降低种群的死亡率。它的幼虫取食多种植物,但在被寄生后[103]仅取食某些特定的植物,虽然这些植物营养含量较低,但会提高它们存活的概率。

从上述事例中,我们可以得出这样的结论:对寄主植物的取食偏好不仅取决于营养质量,还取决于环境因素。

2.7 植物周围的微气候

植物为它的共生体提供独特的微气候(microclimate)。这种微气候在很大程度上区别于那种将植物作为整体进行研究的标准气象测量。植物表面有相对静止的空气层,在那里由于摩擦力的作用,空气流动相对较弱。由于受光合作用和蒸腾作用的影响,温度和相对湿度显著区别于周围环境。而且,即使气孔关闭[81],也会从角质层释放出相对高浓度单松烯类物质于边界层(Leaf boundary layer)。尽管叶片边缘界层的范围仅以毫米或者至多厘米计(图2.14),要根据风速和叶片的大小而定,然而它们却对栖息在这一区域内的所有昆虫至关重要。叶片边缘层的厚度不仅取决于叶片的大小和结构,也取决于片层和气流的流动。

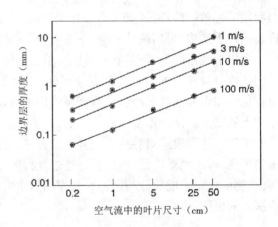

图2.14 叶片边界层厚度是受风速和叶片大小的影响(横纵坐标均取对数)(菲特和海,1987)[36]

叶片表面的温度取决于蒸腾速率、叶片的大小、形状、反射系数和距地面的高度。叶片的下表面通常比上表面温度低、更湿润。叶表的温度有时能高达10℃,而通常是高于或低于周围空气的温度(图2.15;表2.6),同样地,靠近叶子表面的相对湿度会远远超过周围空气中的相对湿度。因此,像与蚜虫和食叶昆虫的早龄幼虫类似的小型昆虫必然经历微气候环境,这种环境能显著影响它们生理[129]所必需的两大基本要素——温度和水的平衡。

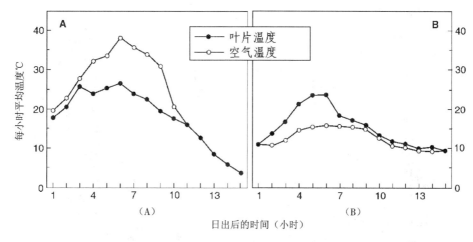

图 2.15　日出时间与每小时平均温度变的关系
（A）：在炎热多云的夏季测量的苹果树叶下表面和周围环境的温度；（B）：在凉爽晴朗的天气测量的苹果树叶下表面和周围环境的温度（法罗等，1979）[34]

表 2.6　来自温带、热带和荒漠地区的植物的叶片表面温度与周围空气温度的差值（Δt）（司陶特和巴克曼，1992）[112]

物种名称	生长地域	空气温度（℃）	Δt
欧洲女贞（*Ligustrum vulgare*）	荷兰	24.3	9.5
田旋花（*Convolvulus arvensis*）	荷兰	18.1	14.2
杜鹃花（*Rhododendron javanicum*）	印尼的爪哇岛（1500 m）	21.8	9.1
甘蔗（*Saccharum officinarum*）	印尼的爪哇岛低地	31.5	3.1
药西瓜（*Citrullus colocynthis*）	撒哈拉沙漠	50.0	-13.0

微气候也会在植物整体水平或自然植被（图 2.16）和农田作物中进行研究。在有植被覆盖下的土地表面的微气候条件与光秃土壤表面的微气候相比存在很大差异。牧场上的植被发育大约是光秃土壤上的 20～40 倍[43]，这应该能解释为植被具有降低地面辐射量的功能。此外，植被还造就了风速、温度和湿度上的梯度变化（图 2.17）。

因此，像生活在完全暴露于自然环境下的高耸草本植物花序上的禾缢管蚜（*Rhopalosiphumpadi*），其生存环境就完全不同于生活在三叶草叶片背面的苜蓿斑蚜（*Therioaphis trifolii*）生存的微气候，尽管这两种昆虫生活的区域可能仅仅相距几十厘米。

2.8　昆虫对自然生态系统和农业生态系统中的植物的危害程度

研究昆虫和植物关系的学者通常都会观察到矛盾的结果——自然界中的大多数植物虽然有大量的植食性昆虫寄生，但很少甚至没有表现出明显的病态。

植物完全落叶的现象只是偶然发生。据估测，昆虫约消耗植物年生物量的 10%[4,21,26]。当然，这个数据也是随着植物种类、时间和地域的不同而发生变化的。例如，热带干旱森林中的

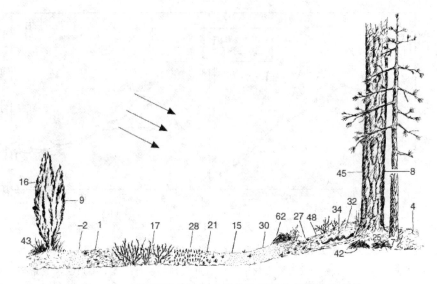

图2.16 微气候的差异与太阳的位置和植物的种类有关

图中所示为荷兰,1976年3月3日的中午,晴,垂直于一个森林朝南边缘的切面的表面温度。高度为一米时的空气温度是11.8 ℃。曾有记录,当柏树丛叶面温度高达62 ℃时,距离其只有几米远的柏树丛树荫下的地面却仍在冻结。图中箭头指的是光线的方向(司陶特和巴克曼,1992)[112]

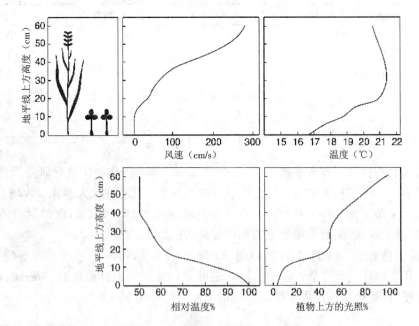

图2.17 草场植被的微气候变化(考克斯等,1973)[23]

植食性昆虫消耗(平均值为14%)要高于温带森林中的植食性昆虫消耗(平均值为7%)[19]。

由于昆虫吸食植物汁液(sap-feeding)造成的损失很难去计量,但据估测,可能达到净初级生产总量(NPP)的5%甚至更多。尽管吸食汁液的昆虫一般体形较食叶昆虫小,但是它们的消耗量要大于食叶昆虫(以平均每克体重的消耗量计算)。因此,一般说来,吸食汁液的昆虫可能会消耗与食叶昆虫[19]同样多的生物量。甚至有时它们所消耗的会超过食叶昆虫。例如,这种情况会发生在早期的落叶林中,在那里,吸食汁液的昆虫造成光合产物的损失要远远超过食叶昆虫造成的损失[101]。

测量植食性昆虫的密度通常很困难[19,101]。使用不同方法估计植食性昆虫造成的损失量是不同的。因此,研究不同植食性昆虫—植物关系的数据,在许多情况下很难进行比较。下面给出一些单株植物受害程度的例子,仅是为读者提供了参考,但是读者可以通过查询一些优秀的综述文章获取更多的信息[19,88,101]。

举例来说,当使用杀虫剂消除两种吸食汁液的缘蝽(coreid bugs)来保护两种桉树(*Eucalyptus*)时,与其他未被保护的树种相比,两种桉树在 12 个月的采样期间分别表现出 8.5% 和 39% 的高度优势[5]。然而澳大利亚桉树却长期遭受着严重的昆虫危害,叶片损失量占叶片生产量的 10% ~ 50%[71],而其他植物例如楝树(*Azadirachta indica*)、柏树(*Juniperus* spp.)和杜鹃花(*Rhododendron* spp.)则很少有死于昆虫危害的。即使亲缘关系相近的植物之间也显示出植食性昆虫造成损失量存在巨大的差异(图 2.18)。

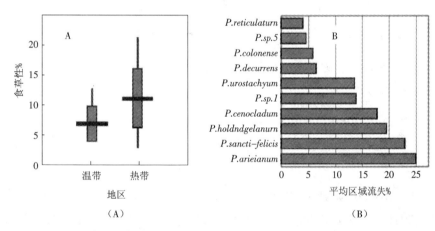

图 2.18　在不同气候区域中,森林和攀缘植物胡椒属的植食性动物比率

(A)在温带和热带阔叶林中每年被食的叶片面积,画线区域为平均值 ± 方差以及范围(科利和艾德,1991)[18];(B)每种昆虫均喂养至少 50 片新鲜叶片时,不同昆虫对叶片造成损失的平均值(马奎斯,1991)[75]

以年度衡量昆虫对森林树木造成的损失是巨大的。从已统计的数据得知,美国和加拿大的森林树木年损失量分别达到了 14% 和 22%[48]。更严重的危害也时有发生,例如在芬诺斯堪迪亚,隶属于欧普属和冬蛾属的蛾类造成了大范围的落叶现象和桦树林的死亡[115]。显然,森林害虫在世界范围内都有重要的危害,而考虑到现在限制杀虫剂使用的政策,由昆虫所引起的损失很可能仍将保持很高的水平。有趣的是,生长在都市或具观赏性的植物比自然丛林中的植物所遭受的昆虫侵害要少得多[83]。

有人提出质疑,10% 的损伤程度对植物而言究竟是可以忽略还是会造成严重危害,是否会严重影响植物的适应性。而一些迹象表明:低程度的虫害可能也会对植物有显著的影响。例如,据统计,多年生草本植物其净同化物(net assimilate)的 1% ~ 15% 用于繁殖,而对一年生草本植物来说,其净同化物的 15% ~ 30% 是用于繁殖的[49]。因此,初步可知,由于虫害所损失的 10% 的同化量会占据植物原本应分配给繁殖所需的那部分生物量[77]。所以,虫害最大程度上损失的能量近似等于植物用于繁殖所需的能量。这些大量的数据表明,虫害所造成的损失不可被忽略。当然,这10%的损失量对于昆虫而言并非全部用于繁殖,而是或多或少分配给了生命活动。

一项关于橡树的研究表明:即使中度的昆虫侵害也会显著地降低种子的产量。由于定期使用杀虫剂,实验用的橡树材料脱叶率一般低于5%,而对照组只浇水的树木却显示出其两倍的脱叶率。树木的生长,可以从年轮中发现规律。但是经过杀虫剂处理的橡树枝条所结的橡子数量,是未经杀虫剂处理的四倍以上[25]。然而,减产的橡树果量是否会对现有的橡树密度造成不良影响,目前仍未有定论。

刺吸式口器的昆虫,如果数量巨大也会对种子的产量造成不良影响(图2.19)。因此,有蚜虫侵染的(*Senecio sylvaticus*)植株会比无蚜虫滋生的植株减少50%的种子产量[32]。树木的生长也会由于吸食树液的昆虫(如介壳虫)的存在而显著减缓。当松树(*Pinus edulis*)免遭松蚧(*Matsucoccus acalyptus*)侵害时,比起长期受高密度(*M. acalyptus*)攻击的种群,树木的增长加快了25%~35%[119]。

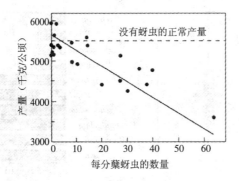

图2.19　小麦的产量与每蘖枝蚜虫(*Sitobion avenae*)峰值数的关系(维尔捷克,1979)[123]

另一个实验评估了昆虫侵害对种子产量(包括非自然落叶)的影响。在新热带地区,胡椒灌木丛经常遭受几类象鼻虫的严重侵袭。这些植物被取食后产生的种子数量更少,这种影响会延续到下一年,致使贮存分配的量减少(图2.20)。因此,昆虫数量往往对种子产量产生巨大的影响。

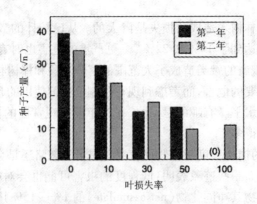

图2.20　通过修剪胡椒(*Piper arieianum*)灌木对其种子产量的影响

当30%或更大面积的叶片区域被剪除后,在之后的一年里,种子数目明显减少,随后的年份里也出现同样的情况(马奎斯,1984)[74]

统计昆虫造成的叶面损伤很可能低于实际损失量,因为许多小伤口可以带来比树叶完全脱落更大的影响,在一片破损的树叶中,虽然未被损害的组织可以进行光合作用,但其光合效率却被大大降低[132]。这是由于伤口造成的生理影响会被系统性地传递到植物的其他部位(见第4章),因此,很有可能破损部位的数目比破损的总面积更加重要。在12个植物种类中,所有被昆虫破坏的叶片被标记后,结果是:平均87%的树叶受到破坏,并且在一些植物中所有的叶片都受到不同程度的破坏(图2.21)[26]。这个数据与10%的破坏水平相差太大以至于不能被忽视。

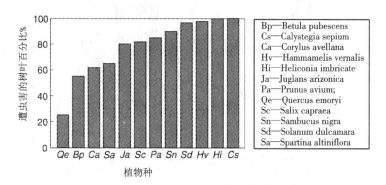

图2.21　该图显示出在12个双子叶植物物种中遭昆虫破坏的树叶所占的百分比

应注意许多给定物种的破坏程度评估是基于仅仅一个或两个植物样本,那么落叶实验的准确程度就取决于所选实验样本是否具有典型性(达曼,1993)[26]

然而另外一些研究发现:在一些植物群落中仅有少数植株被昆虫破坏。普赖期等人发现,尽管热带草原上有丰富的毛毛虫,但是每个物种的数量都很少[91]。平均来看,每10株植物中仅能发现一个鳞翅目幼虫(四种树木高1~2米)。显然,昆虫侵害的强度可能在不同的植物群落和植物物种之间有很大差异,然而目前的研究使我们无法进行全面的归纳和概括。

昆虫对农业生态系统造成的破坏往往大于对自然环境的破坏[88]。尽管使用了高强度杀虫剂,但是在美国,由于昆虫取食所导致的农作物损耗总计达到13%(图1.3),并且,该数值在全球范围内已达到15%或者更高。正如本章前面所讨论的那样,寄主专一性现象加剧了害虫数量的增加。在美国,农作物遭受约1 000种昆虫物种的侵袭。世界范围内,此数据增加至约9 000个物种,虽然被认为危害性较大的害虫数目少于5%[89],就目前昆虫的种类来看,这是一个相对较小的数字。然而,我们的农业依赖于世界植物群落中一个非常小的子集,只有4个主要的及26个次要的农作物种类所提供的营养含量占人类所需总营养量的95%(图2.22)[96],这些栽植的植物覆盖了大面积的土地,而且这种现状已持续了近千年。众所周知,它们为具高程度适应能力的昆虫提供了过量的食物。同样,昆虫中的有害昆虫种类也是占优势的专食性取食者:有害昆虫中,在温带有75%、在热带有80%的鳞翅目害虫是单食性或寡食性昆虫[3]。这些比率与前文提及的昆虫物种在天然植物中出现的数据(见2.1)惊人的一致。

讨论完丰富的植食性昆虫对植物净初级生产量NPP的损耗数据后,将这些数值与全球人类对NPP的使用作比较是很有意思的。人类占地球生物量的0.5%左右[58],但是却使用了陆生NPP总量的20%左右。

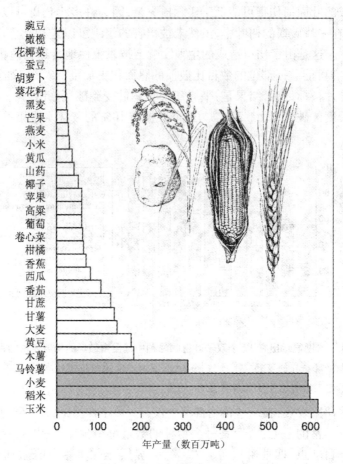

图 2.22　不同作物的年产量

地球上高等植物 250 000 个物种中,有 30 种可食用农作物,其产量占人类所需总营养量的 95%,超过 50% 的食物营养是从以下四种作物中获得:玉米、水稻、小麦和马铃薯(联合国粮食农业组织,2001)

2.9　植物对植食性昆虫所造成损伤的补偿

几乎没有植物能躲避植食性昆虫的损伤,但是植物能靠自身机制来减少植食性昆虫的有害影响。只要昆虫不侵袭它们的分生组织或顶端,植物的其他组织绝大部分具有非凡的再生能力。从形态学和进化的观点来看,植物和动物的存在有本质上的区别。结构上,植物是模块化(modular)的有机体,也就是说它们由重复的多细胞单位组成,每个多细胞单位有自己的分生组织。对于植物整体而言,这些单位都不是至关重要的,因为一个单位的损坏并不影响植物整体生命活动的正常进行。这种模块化性质在很大程度上减小了植食性昆虫带来的不利影响,同时也便于组织缺失后的恢复。这与单位化(unitary)的有机体相反,比如昆虫,当躯体的一部分受损坏后,它们会死去或至少严重残疾。植物被食用后出现令人惊讶的恢复现象,它们能激活分生组织(往往处于休眠状态)以及重新吸收原料(例如营养物质和光合产物)进行组织再生[51]。在适度或良好的资源条件下,植物能够局部地或者完全得到补偿恢复,有时甚至为了弥补昆虫破坏的损失,部分组织出现补偿过度的情况(补偿过度被定义为产出的生物量多于因为植食性动物食用而损失的生物量[6])。然而,想找到一个关于过度补偿现象的确凿实证比我们想象的要困难得多,因为这要求在自然条件下,受昆虫侵害的植物相比于未遭受侵害的植物要更为健康。由于许多因素,包括结构特征(分生组织的存在,不同部位维管的集合)、生殖方式、取食时机、有机物的存储以及可用的水、营养物质和光,这些有助于(部分地)增强植物弥补组织损耗的能力[29,126]。因此,就我们目前所掌握的知识水平来说,对生态或农业补偿反应的重要性下结论还为时过早(表2.7)。

表2.7 植物特点和决定植株对植食性昆虫补偿作用的环境因素[126]

不完全补偿	完全或过度补偿
植食性发生在季末	植食性发生在季初
匮乏的水位、营养和/或光照	充足的水位、营养和光照
激烈竞争	低频竞争
存在分生组织制约	不存在分生组织制约
生长缓慢	生长快速
不完整的植株模块	完整的植株模块
多年生木本植株	一年生或一年两生类植株

在室内生长条件下,对农作物的研究证实有过度补偿的现象存在,但在自然农田条件下结果如何仍持有争议。近期对81个受昆虫侵害后植物生长状况实例的综合分析指出,确切的补偿或过度补偿的出现占记录的35%。令人惊讶的是,就过度补偿的最佳条件(光、水和营养物)来说,双子叶植物和单子叶植物有差异。双子叶草本植物在资源丰富的条件下,被取食后长得更多,然而对于单子叶草本植物和木本植物来说,在资源匮乏的条件下也会有显著的恢复[52]。这种差异很可能是由于两类植物分生组织部位的不同而造成的,也就是说,生理差异造成的结果[52]。

对没有摧毁植物光合作用系统所进行的补偿反应更加难以衡量。然而却为光合作用现象中的变化提供了线索。通过测量受介壳虫(Coccus sp.)侵扰的冬青树(Ilex aquifolium)的光合作用效率,相比对照植株来说,冬青树的光合作用效率呈现出上升的趋势,从而也说明它们对被取食后损失营养物的补偿也在增加。这种影响扩大到受损叶片旁的完好叶片,其光合作用也被激发[95]。

由于我们对植食性昆虫影响方面的知识只源于个别极端事例,而忽视低密度植食性昆虫与其他植食性昆虫造成的诸多影响之间的关系[3]。因此对自然植株补偿反应重要性的评价受到了限制。

2.10 结论

本章的重点是植食性昆虫表现出普遍的对食物高度的专一性。这个概念将会成为本书以后章节中所有对昆虫和植物相互作用关系深入探索的核心。自然生态系统中由植食性昆虫对植物造成相对较少的破坏是由调查提出的中心问题之一,尽管它们有大量的种类和令人惊讶的生殖能力。与此同时,植食性昆虫的普遍存在已经被假定来代表在植物结构和功能进化方面主要的选择作用。这种想法被最近有关植物数量动态的模型研究所证实,研究表明,通过种子产量的减少说明植食性昆虫影响比以前所预期的更加强烈[73]。

显然,植物通常在受到昆虫侵袭后有很好的自我保护。保护机制的本质可以运用到发展栽植农作物的过程中,从而减少昆虫的破坏。

2.11 参考文献

1. Abrahamson, W. G., Melika, G., Scrafford, R., and Csoka, G. (1998). Gall-inducing insects provide insights into plant systematic relationships. *American Journal of Botany*, 85, 1159–65.
2. Åhman, I. (1985). Larval feeding period and growth of *Dasineura brassicae* (Diptera) on *Brassica* host plants. *Oikos*, 44, 191–4.
3. Barbosa, P. (1993). Lepidopteran foraging on plants in agroecosystems: constraints and consequences. In *Caterpillars. Ecological and evolutionary constraints on foraging* (ed. E. Stamp and T. M. Casey), pp. 523–66. Chapman & Hall, New York.
4. Barbosa, P. and Schultz, J. C. (1987). *Insect outbreaks*. Academic Press, San Diego.
5. Bashford, R. (1992). Observations on the life history of two sap-feeding coreid bugs and their impact on growth of plantation eucalypts. *Tasforests*, 4, 87–93.
6. Belsky, A. J. (1986). Does herbivory benefit plants? A review of the evidence. *American Naturalist*, 127, 870–92.
7. Bernays, E. A. (2001). Neural limitations in phytophagous insects: implications for diet breadth and evolution of host affiliation. *Annual Review of Entomology*, 46, 703–27.
8. Bernays, E. A. and Chapman, R. F. (1978). Plant chemistry and acridoid feeding behaviour. In *Biochemical aspects of plant-animal coevolution* (ed. J. Harborne), pp. 99–141. Academic Press, New York.
9. Bernays, E. A. and Graham, M. (1988). On the evolution of host specificity in phytophagous arthropods. *Ecology*, 69, 886–92.
10. Blossey, B. and Hunt-Joshi, T. R. (2003). Belowground herbivory by insects: Influence on plants and aboveground herbivores. *Annual Review of Entomology*, 48, 521–47.
11. Bracken, G. K. (1984). Within plant preferences of larvae of *Mamestra configurata* (Lepidoptera: Noctuidae) feeding on oilseed rape. *Canadian Entomologist*, 116, 45–9.
12. Brändle, M. and Brandl, R. (2001). Species richness of insect and mites on trees: expanding Southwood. *Journal of Animal Ecology*, 70, 491–504.
13. Byrne, D. N. and Bellows, T. S. (1991). Whitefly biology. *Annual Review of Entomology*, 36, 431–57.
14. Cates, R. G. (1980). Feeding patterns of monophagous, oligophagous, and polyphagous insect herbivores: the effect of resource abundance and plant chemistry. *Oecologia*, 46, 22–31.
15. Chapman, R. F. (1990). Food selection. In *Biology of grasshoppers* (ed. R. F. Chapman and A. Joern), pp. 39–72. Wiley, New York.

16. Chapman, R. F. and Sword, G. A. (1997). Polyphagy in the Acridomorpha. In *The bionomics of grasshoppers, katydids and their kin* (ed. S. K. Gangwere, M. C. Muralirangan, and M. Muralirangan), pp. 183 – 95. CAB International, Wallingford.
17. Cock, M. J. W. (1986). *Bemisia tabaci, a literature survey on cotton whitefly with an annotated bibliography.* C. A. B. Institute of Biological Control, Ascot.
18. Coley, P. D. and Aide, T. M. (1991). Comparison of herbivory and plant defenses in temperate and tropical broad-leaved forests. In *Plant-animal interactions. Evolutionary ecology in tropical and temperate regions* (ed. P. W. Price, T. M. Lewinsohn, G. W. Fernandes, and W. W. Benson), pp. 25 – 49. Wiley, New York.
19. Coley, P. D. and Barone, J. A. (1996). Herbivory and plant defenses in tropical forests. *Annual Review of Ecology and Systematics*, 27, 305 – 35.
20. Cooper-Driver, G. A. N. (1978). Insect-fern associations. *Entomologia Experimentalis et Applicata*, 24, 110 – 16.
21. Coupe, M. D. and Cahill, J. F. (2003). Effects of insects on primary production in temperate herbaceous communities: a meta-analysis. *Ecological Entomology*, 28, 511 – 21.
22. Courtney, S. P., Kibota, T. T., and Singleton, T. A. (1990). Ecology of mushroom-feeding Drosophilidae. *Advances in Ecological Research*, 20, 225 – 74.
23. Cox, C. B., Healy, I. N., and Moore, P. D. (1973). *Biogeography, an ecological and evolutionary approach.* Blackwell, Oxford.
24. Craig, T. P. and Ohgushi, T. (2002). Preference and performance are correlated in the spittlebug *Aphrophora pectoralis* on four species of willow. *Ecological Entomology*, 27, 529 – 40.
25. Crawley, M. J. (1985). Reduction of oak fecundity by low-density herbivore population. *Nature*, 314, 163 – 4.
26. Damman, H. (1993). Patterns of interaction among herbivore species. In *Caterpillars. Ecological and evolutionary constraints on foraging*, (ed. N. E. Stamp and T. M. Casey), pp. 131 – 69. Chapman & Hall, New York.
27. Davis, B. N. K. (1983). *Insects on nettles.* Cambridge University Press, Cambridge.
28. DeClerck, R. A. and Shorthouse, J. D. (1985). Tissue preference and damage by *Fenusa pusilla* and *Mesanana* (Hymenoptera: Tenthredinidae), leaf mining sawflies on white birch (*Betula papyrifera*). *Canadian Entomologist*, 117, 351 – 62.
29. Delaney, K. J. and Macedo, T. B. (2001). The impact of herbivory on plants: yield, fitness, and population dynamics. In *Biotic stress and yield loss*, (ed. R. K. D. Peterson and L. G. Higley), pp. 135 – 60. CRC Press, Boca Raton.
30. Eastop, V. (1979). Sternorrhyncha as angiosperm taxonomists. *Symbolae Botanicae Upsaliensis*, 22, 120 – 34.
31. English-Loeb, G. M., Brody, A. K., and Karban, R. (1993). Host-plant-mediated interactions between a generalist folivore and its tachinid parasitoid. *Journal of Animal Ecology*, 62, 465 – 71.
32. Ernst, W. H. O. (1987). Impact of the aphid *Aulocorthum solani* Kltb, on growth and repro-

duction of winter and summer annual life forms of *Senecio sylvaticus* L. *Acta Oecologica*, 8, 537 – 47.
33. Fabre, J. H. (1886). *Souvenirs entomologiques*, Vol. 3. Delagrave, Paris.
34. Ferro, D. N., Chapman, R. B., and Penman, D. R. (1979). Observations on insect microclimate and insect pest management. *Environmental Entomology*, 8, 1000 – 3.
35. Fiedler, K. (1996). Host-plant relationships of lycaenid butterflies: large-scale patterns, interactions with plant chemistry, and mutualism with ants. *Entomologia Experimentalis et Applicata*, 80, 259 – 67.
36. Fitter, A. H. and Hay, R. K. M. (1987). *Environmental physiology of plants* (2nd edn). Academic Press, London.
37. Floate, K. D. and Whitham, T. G. (1995). Insects as traits in plant systematics: their use in discriminating between hybrid cottonwoods. *Canadian Journal of Botany*, 73, 1 – 13.
38. Foss, L. K. and Rieske, L. K. (2003). Species-specific differences in oak foliage affect preference and performance of gypsy moth caterpillars. *Entomologia Experimentalis et Applicata*, 108, 87 – 93.
39. Foster, S. P. and Howard, A. J. (1999). Adult female and neonate larval plant preferences of the generalist herbivore, *Epiphyas postvittana*. *Entomologia Experimentalis et Applicata*, 92, 53 – 62.
40. Frost, S. W. (1942). *General entomology*. Mc-Graw Hill, New York.
41. Futuyma, D. J. (1976). Food plant specialization and environmental predictability in Lepidoptera. *American Naturalist*, 110, 285 – 92.
42. Gall, L. F. (1987). Leaflet position influences caterpillar feeding and development. *Oikos*, 49, 172 – 6.
43. Geiger, R. (1965). *The climate near the ground*. Harvard University Press, Cambridge, MA.
44. Görnitz, K. (1954). Frassauslösende Stoffe für polyphag an Holzgewächsen fressende Raupen. *Verhandlungen der Deutschen Gesellschaft für Angewandte Entomologie*, 13, 38 – 47.
45. Gratton, C. and Welter, S. C. (1998). Oviposition preference and larval performance of *Liriomyza helianthi* (Diptera: Agromyzidae) on normal and novel host plants. *Environmental Entomolology*, 27, 926 – 35.
46. Grevillius, A. Y. (1905). Zur Kenntnis der Biologie des Goldafters *Euproctis chrysorrhoea* L. *Botanisches Zentralblatt*, *Beiheft*, 18, 222 – 322.
47. Hamilton, J. G. and Zalucki, M. P. (1993). Interactions between a specialist herbivore, *Crocidosema plebejana*, and its host plants *Malva parviflora* and cotton, *Gossypium hirsutum*: oviposition preference. *Entomologia Experimentalis et Applicata*, 66, 207 – 12.
48. Harausz, E. and Pimentel, D. (2002). North American forest losses due to insects and plant pathogens. In *Encyclopedia of pest management* (ed. D. Pimentel), pp. 539 – 41. Dekker, New York.
49. Harper, J. L. (1977). *Population biology of plants*. Academic Press, London.
50. Harris, M. O., Sandanayaka, M., and Griffin, W. (2001). Oviposition preferences of the

Hessian fly and their consequences for the survival and reproductive potential of offspring. *Ecological Entomology*, 26, 473 – 86.

51. Haukioja, E. (1991). The influence of grazing on the evolution, morphology and physiology of plants as modular organisms. *Philosophical Transections of the Royal Society*, B333, 241 – 7.
52. Hawkes, C. V. and Sullivan, J. J. (2001). The impact of herbivory on plants in different resource conditions: a meta-analysis. *Ecology*, 82, 2045 – 58.
53. Hebert, P. D. N., Penton, E. H., Burns, J. M., Janzen, D. H., and Hallwachs, W. (2004). Ten species in one: DNA barcoding reveals cryptic species in the neotropical skipper butterfly *Astraptes fulgerator*. *Proceedings of the National Academy of Sciences of the USA*, 101, 14812 – 17.
54. Hendrix, S. D. (1980). An evolutionary and ecologicalper perspective of the insect fauna of ferns. *American Naturalist*, 115, 171 – 96.
55. Hering, M. (1926). *Biologie der Schmetterlinge*. Springer, Berlin.
56. Hille Ris Lambers, D. (1979). Aphids as botanists? *Symbolae Botanicae Upsaliensis*, 22, 114 – 19.
57. Howard, J. J., Raubenheimer, D., and Bernays, E. A. (1994). Population and individual polyphagy in the grasshopper *Taeniopoda eques* during natural foraging. *Entomologia Experimentalis et Applicata*, 71, 167 – 76.
58. Imhoff, M. L., Bounoua, L., Ricketts, T., Loucks, C., Harriss, R., and Lawrence, W. T. (2004). Global patterns in human consumption of net primary production. *Nature*, 429, 870 – 3.
59. Janz, N. (2002). Evolutionary ecology of oviposition strategies. In *Chemoecology of insect eggs and egg deposition* (ed. M. Hilker and T. Meiners), pp. 349 – 76. Blackwell, Berlin.
60. Karban, R. and Agrawal, A. A. (2002). Herbivore offense. *Annual Review of Ecology and Systematics*, 33, 641 – 64.
61. Kato, M. (2002). First record of herbivory on Lycopodiaceae (Lycopodiales) by a dipteran (Pallopteridae) leaf/stem-miner. *Canadian Entomologist*, 134, 699 – 701.
62. Kennedy, C. E. J. and Southwood, T. R. E. (1984). The number of species of insects associated with British trees, a reanalysis. *Journal of Animal Ecology*, 53, 455 – 78.
63. Kennedy, J. S. (1953). Host plant selection in Aphididae. *Transactions of the 9th International Congress of Entomology (Amsterdam)*, 2, 106 – 13.
64. Klausnitzer, B. (1983). Bemerkungen über die Ursachen und die Entstehung der Monophagie bei Insekten. *Verhandlungen SIEEC X*, Budapest, pp. 5 – 12.
65. Kogan, M. (1986). Plant defense strategies and hostplant resistance. In *Ecological theory and integrated pest management practice* (ed. M. Kogan), pp. 83 – 134. Wiley, New York.
66. Lawrey, J. D. (1987). Nutritional ecology of lichen/moss arthropods. *In Nutritional ecology of insects, mites, spiders and related invertebrates* (ed. F. Slansky and J. G. Rodriguez), pp. 209 – 33. John Wiley, New York.
67. Leather, S. R. and Awmack, C. S. (2002). Does variation in offspring size reflect strength of

preference performance index in herbivorous insects? *Oikos*, 96, 192 – 5.
68. Liebhold, A. M., Gottschalk, K. W., Muzika, R. M., Montgomery, M. E., Young, R., O'Day, K. et al. (1995). *Suitability of North American tree species to gypsy moth: a summary of field and laboratory tests*. General Technical Report NE-211. USDA Forest Service, Randor, PA.
69. Lill, J. T., Marquis, R. J., and Ricklefs, R. E. (2002). Host plants influence parasitism of forest caterpillars. *Nature*, 417, 170 – 3.
70. Lindström, J., Kaila, L., and Niemelä, P. (1994). Polyphagy and adult body size in geometrid moths. *Oecologia*, 78, 130 – 2.
71. Lowman, M. D. and Heatwole, H. (1992). Spatial and temporal variability in defoliation of Australian eucalypts. *Ecology*, 73, 129 – 42.
72. Maron, J. L. (2001). Intraspecific competition and subterranean herbivory: individual and interactive effects on bush lupine. *Oikos*, 92, 178 – 86.
73. Maron, J. L. and Gardner, S. N. (2000). Consumer pressure, seed versus safe-site limitations, and plant population dynamics. *Oecologia*, 124, 260 – 9.
74. Marquis, R. J. (1984). Leaf herbivores decrease fitness of a tropical plant. *Science*, 226, 537 – 9.
75. Marquis, R. J. (1991). Herbivore fauna of *Piper* (Piperaceae) in a Costa Rican wet forest, diversity, specificity, and impact. In *Plant-animal interactions. Evolutionary ecology in tropical and temperate regions* (ed. P. W. Price, T. M. Lewinsohn, G. W. Fernandes, and W. W. Benson), pp. 179 – 99. Wiley, New York.
76. Mattson, W. J., Lawrence, R. K., Haack, R. A., Herms, D. A., and Charles, P. J. (1988). Defensive strategies of woody plants against different insectfeeding guilds in relation to plant ecological strategies and intimacy of association with insects. In *Mechanisms of woody plant defenses against insects* (ed. W. J. Mattson, J. Levieux, and C. Bernard-Dagan), pp. 3 – 38. Springer, Berlin.
77. Mooney, H. A. (1972). The carbon balance of plants. *Annual Review of Ecology and Systematics*, 3, 315 – 46.
78. Moskowitz, D. P. and Westphal, C. (2002). Notes on the larval diet of the painted lichen moth *Hypoprepia fucosa* Hübner (Arctiidae, Lithosiinae). *Journal of the Lepidopterists' Society*, 56, 289 – 90.
79. Mulatu, B., Applebaum, S. W., and Coll, M. (2004). A recently acquired host plant provides an oligophagous insect herbivore with enemy-free space. *Oikos*, 107, 231 – 8.
80. Niemelä, P., Hanhimäki, S., and Mannila, R. (1981). The relationship of adult size in noctuid moths (Lepidoptera, Noctuidae) to breadth of diet and growth form of host plant. *Annales Entomologici Fennici*, 47, 17 – 20.
81. Niinemets, Ü, Loreto, F., and Reichstein, M. (2004). Physiological and physicochemical controls on foliar volatile organic compound emissions. *Trends in Plant Science*, 9, 180 – 6.
82. Novotny, V., Basset, Y., Miller, S. E., Weiblen, G. D., Bremer, B., Cizek, L., et al.

(2002). Low host specificity of herbivorous insects in a tropical forest. *Nature*, 416, 841 – 4.
83. Nuckols, M. S. and Connor, E. F. (1995). Do trees in urban or ornamental plantings receive more damage by insects than trees in natural forests? *Ecological Entomology*, 20, 253 – 60.
84. Nuorteva, P. (1952). Die Nahrungspflanzenwahl der Insekten im Lichte von Untersuchungen an Zikaden. *Annales Academiae Scientiarum Fennicae. Ser. A, IV Biologica*, 19, 1 – 90.
85. Odegaard, F. (2000). How many species of arthropods? Erwin's estimate revised. *Biological Journal of the Linnean Society*, 71, 583 – 97.
86. Ohsaki, N. and Sato, Y. (1994). Food plant choice of *Pieris* butterflies as a trade-off between parasitoid avoidance and quality of plants. *Ecology*, 75, 59 – 68.
87. Pashley, D. P. (1986). Host-associated genetic differentiation in fall armyworm (Lepidoptera, Noctuidae), a sibling species complex? *Annals of the Entomological Society of America*, 79, 898 – 904.
88. Peterson, R. K. D. and Higley, L. G. (eds) (2001). *Biotic stress and yield loss*. CRC Press, Boca Raton, FL.
89. Pimentel, D. (1991). Diversification of biological control strategies in agriculture. *Crop Protection*, 10, 243 – 53.
90. Potter, D. A. and Held, D. W. (2002). Biology and management of the Japanese beetle. *Annual Review of Entomology*, 47, 175 – 205.
91. Price, P. W., Diniz, I. R., Morais, H. C., and Marques, E. S. A. (1995). The abundance of insect herbivore species in the tropics, the high local richness of rare species. *Biotropica*, 27, 468 – 78.
92. Prins, A. H. and Verkaar, H. J. (1992). Defoliation, do physiological and morphological responses lead to (over) compensation? In *Pests and pathogens. Plant responses to foliar attack* (ed. P. G. Ayres), pp. 13 – 31. Bios Scientific Publishers, Oxford.
93. Robinson, A. G. (1985). *Macrosiphum (Sitobia) species on ferns (revised edn)*. Proceedings of the International Aphidology Symposium, Jablonna (1981), pp. 471 – 4. Ossolineum, Warsaw.
94. Ruehlmann, T. E., Matthews, R. W., and Matthews, J. R. (1988). Roles for structural and temporal shelter changing by fern-feeding lepidopteran larvae. *Oecologia*, 75, 228 – 32.
95. Retuerto, R., Fernandez-Lema, B., Roiloa, R., and Obeso, J. R. (2004). Increased photosynthetic performance in holly trees infested by scale insects. *Functional Ecology*, 18, 664 – 9.
96. Sattaur, O. (1989). The shrinking gene pool. *New Scientist*, 29 July 1989, 37 – 41.
97. Scheirs, J., Bruyn, L. de, and Verhagen, R. (2001). Nutritional benefits of the leaf-mining behaviour of two grass miners, a test of the selective feeding hypothesis. *Ecological Entomology*, 26, 509 – 16.
98. Schmitz, G. (1998). The phytophagous insect fauna of *Tanacetum vulgare* L. (Asteraceae) in Central Europe. *Beiträge zur Entomologie*, 48, 219 – 35.
99. Schoonhoven, L. M. (1977). Feeding behaviour in phytophagous insects, on the complexity of the stimulus situation. *Colloques Internationaux du Centre National de la Recherche Scientifique*,

265, 391 – 8.
100. Schoonhoven, L. M. (1991). Insects and host plants, 100 years of 'botanical instinct'. *Symposia Biologica Hungarica*, 39, 3 – 14.
101. Schowalter, T. D. (2000). *Insect ecology*, Chapter 12: Herbivory. Academic Press, San Diego.
102. Scriber, J. M. (1995). Overview of swallowtail butterflies, taxonomic and distributional latitude. In *Swallowtail butterflies. Their ecology and evolutionary biology* (ed. J. M. Scriber, Y. Tsubaki, and R. C. Lederhouse), pp. 3 – 8. Scientific Publications, Gainesville.
103. Singer, M. S. and Stireman, J. O. (2003). Does antiparasitoid defense explain host-plant selection by a polyphagous caterpillar? *Oikos*, 100, 554 – 62.
104. Singer, M. C., Ehrlich, P. R., and Gilbert, L. E. (1971). Butterfly feeding on lycopsid. *Science*, 172, 1341 – 2.
105. Singer, M. C., Ng, D., and Thomas, C. D. (1988). Heritability of oviposition preference and its relationship to offspring performance within a single insect population. *Evolution*, 42, 977 – 85.
106. Singer, M. S., Rodrigues, D., Stireman, J. O., and Carrière, Y. (2004). Roles of food quality and enemy-free space in host use by a generalist insect herbivore. *Ecology*, 85, 2747 – 53.
107. Slansky, F. and Scriber, J. M. (1985). Food consumption and utilization. In *Comprehensive insect physiology, biochemistry and pharmacology*, Vol. 4 (ed. G. A. Kerkut and L. I. Gilbert), pp. 87 – 163. Pergamon, Oxford.
108. Smith, R. M., Young, M. R., and Marquiss, M. (2001). Bryophyte use by an insect herbivore. Does the crane-fly *Tipula montana* select food to maximise growth? *Ecological Entomology*, 26, 83 – 90.
109. Spencer, K. A. (1990). *Host specialization in the world Agromyzidae (Diptera)*. Kluwer, Dordrecht.
110. Spencer, K. C., Hoffman, L. R., and Seigler, D. S. (1984). Host shift of *Ecpantheria defrorata* (Arctiidae) from an angiosperm to a liverwort. *Journal of the Lepidopterists' Society*, 38, 192 – 3.
111. Stork, N. E., Hammond, P. M., Russell, B. L., and Hadwen, W. L. (2001). The spatial distribution of beetles within the canopies of oak trees in Richmond Park, UK. *Ecological Entomology*, 26, 302 – 11.
112. Stoutjesdijk, P. and Barkman, J. J. (1992). *Microclimate, vegetation, and fauna*. Opulus, Knivsta.
113. Strong, D. R., Lawton, J. H., and Southwood, T. R. E. (1984). *Insects on plants. Community patterns and mechanisms*. Blackwell, Oxford.
114. Tempère, G. (1969). Un critère méconnu des systématiciens phanérogamistes, l'instinct des insects phytophages. *Botaniste*, 50, 473 – 82.
115. Tenow, O. (1972). The outbreaks of *Oporinia autumnata* Bkh. and *Operophtera* spp. (Lep., Geometridae) in the Scandinavian mountain chain and northern Finland 1862 – 1968. *Zoologiska Bidrag från Uppsala*, Suppl. 2, 1 – 107.

116. Thompson, J. N. (1994). *The coevolutionary process.* University of Chicago Press, Chicago.
117. Thompson, V. (1994). Spittlebug indicators of nitrogen-fixing plants. *Ecological Entomology*, 19, 391 – 8.
118. Tindall, K. V. and Stout, M. J. (2001). Plant-mediated interactions between the rice water weevil and fall armyworm in rice. *Entomologia Experimentalis et Applicata*, 101, 9 – 17.
119. Trotter, R. T., Cobb, N. S., and Whitham, T. G. (2002). Herbivory, plant resistance, and climate in the tree ring record: interactions distort climatic reconstructions. *Proceedings of the National Academy of Sciences of the USA*, 99, 10197 – 202.
120. Uvarov, B. (1977). *Grasshoppers and locusts*, Vol. 2. Centre for Overseas Pest Research, London.
121. Van der Putten, W. H., Vet, L. E. M., Harvey, J. A., and Wäckers, F. L. (2001). Linking above-and below ground multitrophic interactions of plants, herbivores, pathogens, and their antagonists. *Trends in Ecology and Evolution*, 16, 547 – 54.
122. Van Nouhuys, S., Singer, M. C., and Nieminen, M. (2003). Spatial and temporal patterns of caterpillar performance and the suitability of two host plant species. *Ecological Entomology*, 28, 193 – 202.
123. Vereijken, B. H. (1979). Feeding and multiplication of three cereal aphid species and their effect on yield of winter wheat. *Verslagen Landbouwkundige Onderzoekingen* 888, Centre for Agricultural Publishing and Documentation, Wageningen.
124. Wäckers, F. L. and Bezemer, T. M. (2003). Root herbivory induces an above-ground indirect defence. *Ecology Letters*, 6, 9 – 12.
125. Wheeler, A. G. (1975). Insect associates of *Ginkgo biloba. Entomological News*, 86, 37 – 44.
126. Whitham, T. G., Maschinski, J., Larson, K. C., and Paige, K. N. (1991). Plant responses to herbivory, the continuum from negative to positive and underlying physiological mechanisms. In *Plant-animal interactions. Evolutionary ecology in tropical and temperate regions* (ed. P. W. Price, T. M. Lewinsohn, G. W. Fernandes, and W. W. Benson), pp. 227 – 56. Wiley, New York.
127. Wiklund, C. (1975). The evolutionary relationship between adult oviposition preferences and larval host plant range in *Papilio machaon* L. *Oecologia*, 18, 186 – 97.
128. Williams, M. A. J. (1994). *Plant galls, organisms, interactions, populations.* Clarendon, Oxford.
129. Willmer, P. (1986). Microclimatic effects on insects at the plant surface. In *Insects and the plant surface* (ed. B. Juniper and T. R. E. Southwood), pp. 65 – 80. Arnold, London.
130. Wolfe, D. W. (2001). *Tales from the underground.* Perseus, Cambridge, MA.
131. Young, O. P. (1986). Host plants of the tarnished plant bug, *Lygus lineolaris* (Heteroptera, Miridae). *Annals of the Entomological Society of America*, 79, 747 – 62.
132. Zangerl, A. R., Hamilton, J. G., Miller, T. J., Crofts, A. R., Oxborough, K., Berenbaum, M. R., *et al.* (2002). Impact of folivory on photosynthesis is greater than the sum of its holes. *Proceedings of the National Academy of Sciences of the USA*, 99, 1088 – 91.

第3章 植物结构：坚固的抗虫害防护

3.1 昆虫的取食系统
3.2 叶面
　3.2.1 角质蜡层
　3.2.2 毛状体
3.3 叶片韧性
　3.3.1 上颚的磨损
　3.3.2 C_3植物与C_4植物
3.4 互利共生关系中所包含的结构
3.5 植物虫瘿
3.6 植物的结构
3.7 结论
3.8 参考文献

如果我们想要对昆虫和植物的关系有一个清晰的了解，那么就需要掌握有关植物的结构和对大多数昆虫有抵御作用的化学物质等方面的知识。本章主要讲述植物影响昆虫取食或产卵的形态学因素（物理因素），第4章则主讲影响昆虫行为和生理的植物化学物质。植物的物理特征可以对昆虫起到阻碍作用，例如叶片韧度、表皮蜡层、茸毛或植株构造，最后是基因表达调控的生化过程。因此，形态和化学防御因素在交替地起着防御的作用。

为了能够精确地评价植物的物理特征在昆虫的取食行为中起到的重要作用，我们首要讨论一下植食性昆虫取食系统的主要结构。

3.1 昆虫的取食系统

昆虫取食行为显著的特征是食物的选择、取食的方式和摄食率。在本节中我们主要讨论的是取食的方式，而摄食率和对食物的选择分别在第5章和第7章中进行详细的讨论。

昆虫主要有两种取食方式：咀嚼[17]和吸食[92]。咀嚼式口器昆虫，众所周知，其口器具颚，为最古老、最普遍的口器类型。一般有3对附肢，或多或少会彼此抑制（图3.1）。一对上颚用来切割和磨碎食物。它们具齿，用以切断和粉碎食物。上颚的下面是下颚。下颚具分节附肢，下颚须上具有化学感受器。下颚帮助控制和引导食物向嘴里输送。上唇形成了口前腔的顶面。其腹面称为内唇，往往具有味觉感受器。下唇形成口腔底面，它有一对承载机械感受器的下唇须。

在一些昆虫类群中，原始具颚的口器的某些附肢已经发生转换，形成具其他功能的口器。它们被称为吸食性口器（haustellate），用来刺破植物组织以便吸食液体食物。吸食性口器为多次起源，出现在半翅目（异翅亚目和同翅类）、缨翅目及鳞翅目成虫中[58]。半翅目昆虫的上、下颚形成口针,下颚须消失。下唇发展为一个特殊结构,形成一个前壁凹陷的唇槽,将上、下颚口针包藏其中。在植食性异翅亚目中,口器的末端有嗅觉感受器[86],但是蚜虫仅具机械感受器[91]。两下颚以双筒方式形成一个连锁,口针的背管称为食道,运输食物;腹面称为唾液道,运输唾液(图3.2)。口针可以穿透植物角质层和细胞壁,一旦进入植物细胞组织内,就会在不同方向寻找一个最佳的吸食位置。有关宿主植物以及蚜虫取食的进一步认识,详情见第7.8.8。

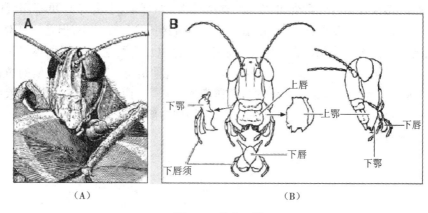

图 3.1　蝗虫口器

(A)蝗虫为具咀嚼式口器的典型昆虫,以三叶草为食(弗罗斯特,1959)[33];(B)蝗虫口器的正面和侧面图(罗斯,1948)

食物从基底进入食窦。在食窦内,肌肉的运动使食窦形成负压。在植物筛管内有时存在极高的流体静力压,可达 0.2~1 MPa(2~10 大气压)。这有利于吸取韧皮部的汁液。在蚜虫取食过程中,可观察到由于植物韧皮部膨胀压的作用,汁液能以极快的速度流动,例如在一种蚜虫(*tuberolargrus saligmus*)内可达 1~2 ul/h[56]。然而,许多同翅类昆虫可以在植物无膨胀压或者在木质部内存在负压时仍能取食人工饲料。为了产生足够的吸力来克服负压,取食木质部的昆虫,都具有肌肉发达的食窦泵[56]。

具有吸食性口器的昆虫吸食不同的植物汁液。大多数同翅类昆虫吸食韧皮部细胞液,而大多数异翅亚目和部分同翅类吸食薄壁组织和木质部汁液。蓟马的摄食器官和摄食方式是昆虫中独有的[41]。口器附肢进行整合后,形成一个口锥,通过短管基部的口针(一对下颚口针和左上颚口针)进行取食。蓟马以表皮细胞或薄壁组织的细胞液为食。

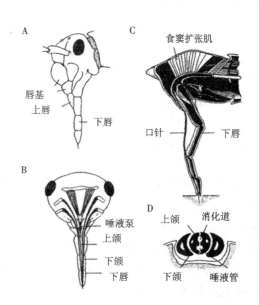

图 3.2　半翅目刺吸式口器

(A)侧面;(B)半翅目头部正观图;(C)蝽取食时头和口器纵切面;(D)口针横切面示意图(戴维斯,1988,经许可)[24]

有些植物可同时为咀嚼式口器昆虫和吸食性口器昆虫提供食物。例如,据记载,两种橡树为335种咀嚼式口器昆虫和88种吸食性口器昆虫提供食物[54]。口器结构的不同形成了不同的取食方式,这是保证取食专化性和昆虫多样性的重要前提。显然,不论哪种取食方式都存在各自的优缺点。吸食性口器的昆虫发展出更巧妙的取食策略,就是限制口器的大小;如吸食性口器的昆虫,其口器一般较咀嚼式昆虫的小。它们往往对寄主植物的机械损伤比较小,从而比同类的咀嚼式口器昆虫能开发更多的资源[68]。然而,吸食性口器的昆虫会造成嫩枝严重变形和发育迟缓,例如,蝽仅仅通过吸食就可以杀死整株植株(如耳草,小麦等),更不用说蚜虫和叶蝉等昆虫还可传播病毒和支原体[39,79]。由蚜虫造成的直接损伤往往较小,但是沫蝉对其寄主造成的影响会比食叶昆虫造成的伤害更大[62]。而咀嚼式昆虫也不可避免地摄取不可消化的植物组织结构以及有毒物质,吸食性口器的昆虫可以避免这种情况。例如,与其他的植物组织相比,韧皮部汁液的营养物质中存在化感物质的概率比较低[80]。而且,吸食韧皮部汁液的昆虫有一些额外的保护措施,在它们吸取汁液之前会向植物韧皮部内注射一些唾液分泌物,先对食物中的化感物质进行解毒[63]。

3.2 叶面

植物表面有着数量庞大且形式多样的微结构和由表皮产生的单细胞及多细胞产物。这些物质因为结构很小,通常不能用人眼辨别,但是对昆虫和其天敌来说是极其重要的。

3.2.1 角质蜡层

大多数维管束植物的角质层覆盖着一层薄薄的疏水成分。这些角质蜡层在防止自身脱水、抵御虫害和阻止植物病毒的入侵等方面发挥着极其重要的作用。蜡层的厚度不同,且数量只占植物干重的百分之一到百分之几。蜡层由化学和物理性质均不同的多个层面构成[49]。

蜡质晶体可呈现多种不同形态。叶片的表层为光滑的非晶体蜡质膜,与空气直接接触(图3.3)[7,48]。覆盖大量蜡质晶体的树叶使昆虫的取食变得更为困难,这样的例子有很多。例如,桉树(*Eucalyptus globulus*)嫩叶上的角质蜡层增加了叶子表面的光滑度,从而妨碍了两种植食性木虱在嫩叶上附着,这样会使昆虫由于饥饿导致存活率明显下降,而在"脱蜡"的嫩叶和"低蜡"的老叶上则可以生存[14]。角质蜡层也许不能一直抵御昆虫。许多实例表明,角质蜡层或具毛表层减少的农作物变种都表现出对害虫的敏感性降低的特性(表3.1)。许多因素似乎都可以解释这种现象。但迄今为止,我们对某些问题的理解仍然有限,例如,控制叶片光滑性状的基因以及提高防御力的机制[27]。

作为对植食性昆虫的间接影响因素,密集且易磨损的微蜡质晶体可能会削弱昆虫的附着和运动能力,也会影响捕食者控制植食性昆虫种群数量的效率。例如,在不具蜡质晶体和具蜡质晶体的两种卷心菜菜叶上,草蛉(*Chrysoperla plorabunda*)幼虫在降低小菜蛾(*Plutella xylostella*)种群数量的作用上,前者表现出来的作用更强(图3.4)[29]。同样地,与瓢虫捕食者一样,瓢虫(*Coccinellid plorabunda*)在豌豆植株的少蜡质叶片上比在正常蜡质叶片上捕食的豌豆蚜数量要多[16,96]。

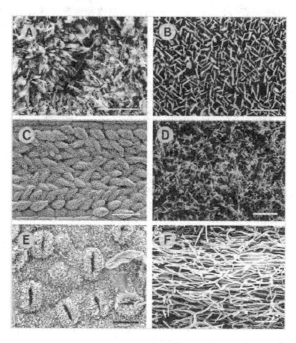

图 3.3 植物表面的扫描电子显微照片

(A)甘蓝(*Brassica oleracea*var. capitata,十字花科),比例为 10 μm;(B)羊茅(*Festuca arundinacea*,禾本科植物),图为近轴叶片表面顶端角质蜡层,蜡质晶体以平面形式排列,属于典型的长链伯醇结晶物,晶体以边缘相接,以优先取向 120 度排列,比例为 2 μm;(C)掌脉石楠(*Cyathodes colonsoi*,掌脉石楠科),图为远轴叶片表面一组短的蜡质晶体外壳,表皮毛状物在气孔复合体上成拱形,比例为 80 μm;(D)阿拉斯加云杉(*Picea sitchensis*,松科),图为近轴叶片表面,即角质蜡层管表面,其中有填充气室的抗蒸腾的蜡质塞,比例为 4 μm;(E)毛竹栎(*Quercus pubescens*,壳斗科)远轴叶表面,气孔复合体主要镶以伯醇蜡质晶体,而近轴叶表面则缺乏这层厚厚的角质蜡层,比例为 20 μm;(F)迷迭香(*Rosmarinus officinalis*,唇形科),远轴叶表面有许多分叉茸毛组成的密毛被,比例为 100 μm(杰弗里,爱丁堡大学,英国)

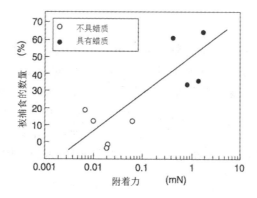

图 3.4 中华通草蛉(*Chrysoperla plorabunda*)对具不同角质蜡层的甘蓝(*Brassica oleracea*)的附着力与对小菜蛾幼虫(*Plutella xylostella*)种群数量的关系

附着力通过使用离心机在叶片表面水平方向产生的加速度测试而得(艾根布罗特等,1999)[29]

表 3.1 具光滑表型(glossy phenotypes)的农作物遭昆虫取食的敏感度

农作物	害虫	对光滑表型的影响	是否对植物有益
洋葱(Allium cepa)	烟蓟马(Thrips tabaci)	较低感染	是
油菜(Brassica napus)	菜缢管蚜(Lipaphis erysimi)	抵抗	是
白菜型油菜(B. campestris)	萝卜蚜(L. erysimi)	种群数量少	是
	萝卜蚜(L. erysimi)	敏感性	不是
	跳甲(Phyllotreta albionica)	更敏感	不是
	跳甲(P. cruciferae)	取食危害更大	不是
	甘蓝蚜(Brevicoryne brassicae)	数量少	是
	甘蓝地种蝇(Delia radicum)	较少的卵	是
甘蓝(B. oleracea)	菜粉虱(Aleyrodes brassicae)	危害较少	是
	烟粉虱(Bemisia tabaci)	种群数量少	是
	烟蓟马(Thrips tabaci)	危害较少	是
	桃蚜(Myzus persicae)	有时种群数量增加	不是
	小菜蛾(Plutlla xylostella)	危害少,卵少,幼虫成活率低	是
	菜粉蝶(Pieris rapae)	危害少,种群数量少	是
芸苔属(Brassica spp.)	菁淡足跳甲(Phyllotreta nemorum)	叶片减少	是
槠豆(Glycine max)	墨西哥豆瓢虫(Epilachna varivestis)	抵抗	是
大麦(Hordeum vulare)	四种蚜虫	种群联合加强	不是
	麦二叉蚜(Schizaphis graminum)	低偏好	是
高粱(Sorghum)	草地贪夜蛾(Spodoptera frugiperda)	低危害	是
	高粱芒蝇(Atherigona soccata)	抵抗	是
	玉米禾螟(Chilo partellus)	抵抗	是
小麦(Triticum aestivum)	麦长管蚜(Sitobion avenae)	种群数量减少	是

光滑具蜡质表皮的植物会使昆虫面临一系列严峻的问题[28],但是许多昆虫进化成不同的结构来解决这些问题。例如,许多金花虫科甲虫跗垫上的微刚毛能够分泌一种黏附物质,可以贴在光滑叶片上[36],多种叶蝉的跗垫形成吸盘[60],还有许多鳞翅目幼虫可以在植物表面吐丝作为"绳梯"或"踏脚点"。因此,昆虫进化出多种结构来克服由于叶片角质蜡层导致的无法停留在光滑叶片上的问题[27]。

3.2.2 毛状体

毛状体,或植物茸毛是由植物表皮细胞演化而来的单细胞或多细胞的附属结构,它们在形状、位置及功能上显示出巨大的差异性(图 3.5)[95]。英文 pubescence 就是指植物表面多毛的现象。茸毛有非腺体茸毛和腺体茸毛两种,非腺体茸毛可以提高植物遭受虫害的抵抗力。例如,茸毛可以起到阻碍初孵幼虫在叶表面移动的作用。密集的毛被可以阻止小的刺吸式昆虫的口器刺入植物表层[88]。茸毛也可以阻碍雌虫产卵[18],但对偏爱毛被的雌虫则有相反的作用[82]。虽然我们知道的许多实例都证实毛被是一种阻碍因素,但有时无毛被对某些昆虫也表现出阻碍作用(表 3.2)。

第3章 植物结构:坚固的抗虫害防护

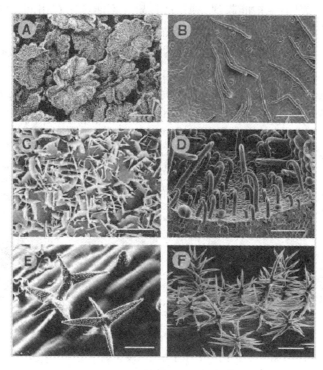

图 3.5 毛状体的扫描电子显微图

(A)杜鹃花(*Rhododendron callostrotum*,杜鹃花科),叶片远轴端的蜡质盾形茸毛,比例为 100 μm;(B)欧洲山毛榉(*Fagus sylvatica*,山毛榉科)的叶片在早夏刚完全展开,其上茸毛呈线状,表层细胞覆有一光滑蜡质膜,比例为 100 μm;(C)山毛榉(*Quercus pubescens*,山毛榉科)叶片表层蜡质晶体层的细微结构,比例为 4 μm;(D)菜豆(*Phaseolus vulgaris*,豆科)初生叶远轴端,其上的钩状茸毛,就像不能移动的植食性昆虫跗关节上的套钩,比例为 60 μm;(E)薰衣草(*Lavandula spicata*,唇形科)的叉状茸毛,表皮角质层的局部膨大产生了细胞瘤状表面,比例为 30 μm;(F)石蒜(*Anigoxanthus flavidus*,石蒜科)芽表面树杈状茸毛,由它们形成了密集的毛被,可阻挡太阳辐射,比例为 200 μm(杰弗里特译,爱丁堡大学,英国)

表 3.2 毛被在不同农作物中对不同目的昆虫起到的抵抗作用(诺里斯,科根,1980)[70]

	抵抗	敏感
小麦	C,Hy,D,Ho,D	A,D
水稻	L	
玉米	C	L
珍珠黍	L,L	
高粱	D	
甘蔗	Ho,Ho,L	
黄豆	C,	Ho,C,D,D,D,L
木豆	L	
苜蓿	Ho,C,Hy,Ho	
棉花	L,C,He,L,	Ho,He,He,L
豆类	Ho,Ho,Ho,Ho	
卷心菜	C	
全部植物	36 种昆虫	15 种节肢动物

注:A——蜱螨;C——鞘翅目;D——双翅目;He——异翅亚目;Ho——同翅类;Hy——膜翅目;L——鳞翅目;T——缨翅目。

然而,实验室研究表明:毛被能起到保护作用,却并不能证明在田间环境下对昆虫的抵抗力是由于其对天敌的后继影响。茸毛能够减慢捕食者和寄生者搜寻的速度,或者使天敌到不了植食性昆虫所在的位置。有实例表明,粉虱的寄生率与寄主植物茸毛的有无密切相关。丽蚜小蜂(*Encarsia formose*)是一种体形较小的寄生蜂,它在光滑的植物表面可以更有效地寻找到它的寄主——粉虱幼虫(图3.6)。在光滑叶片上,由于可以更快地移动,寄生蜂对粉虱幼虫的寄生率比在有茸毛叶片上高20%[94]。

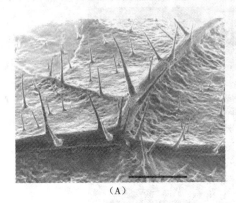 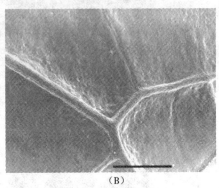

图3.6 两种黄瓜叶片表皮的扫描电子显微照片

(A)具毛的叶片;(B)光滑的叶片,具毛叶片茸毛长度是成年丽蚜小蜂(*Encarsia Formosa*)的尺寸,比例1 mm(由伦特恩提供,瓦赫宁恩大学,荷兰)

某些昆虫能够解决茸毛导致的问题,这在昆虫世界是非常普遍的。例如,角斑五蚜虫(*Myzocallis schreiberi*)有一对爪和一对灵活的爪间突,这能够帮助它在被有短茸毛的寄主植物棕树(*Quercus ilex*)叶片上生活[53]。

另一研究表明,茸毛可以作为植物防御的武器。通过观察,茸毛密度的变化不仅与植物生长因素有关,而且与遭虫害植株的危害程度有关。许多一年生植物的茸毛特征根据昆虫的取食而发生变化[23,93]。黑芥(*Brassica nigra*)幼株被菜青虫(*Pieris rapae*)或拟尺蠖(*Trichoplusia ni*)取食后,新展开的叶片上就会生长出更多茸毛。尽管这种反应依赖于植食性昆虫和叶片的位置,但茸毛密度有时会增加2倍多[93]。

具腺体茸毛可以通过分泌化感物质对多种植食性昆虫产生有害的影响,这种情况可看作一种化学保护,这部分内容将在第7章进行讨论。

3.3 叶片韧性

植物细胞壁由于大分子物质(如纤维素、木质素、木栓素、胼胝质)与厚壁纤维联合而变得坚硬。这可以使植物抵抗机械损伤,如具颚昆虫的啃咬或吸入式口器以及产卵管的刺入等。因此,叶片硬度是减少植食性昆虫损伤的有效措施。叶片强度是预测植食性昆虫种间取食变异的最好指标[22],这一观点被科利发现并支持。

3.3.1 上颚的磨损

取食坚硬的植物组织会使昆虫上颚磨损[17],即便是拥有高硬度上颚的某种昆虫也是如此。

坚硬的植物组织应该由于其表皮中含有锌元素或锰元素[85]。植物叶片的韧性和硬度是不同的。例如,牧草的韧性几乎是其他草本植物的 3 倍(表 3.3)[12]。此外,不同气候带植物种群的叶片平均韧性也是不同的。例如,热带植物叶片的韧性是温带植物的 3 倍。这些差异可能(部分)归因于热带地区植食性昆虫有较强的选择作用,在那里食叶昆虫可达 16.6%,而在温带为 7%[26]。

表 3.3 不同生长模式下的植物叶片坚硬度和韧性的对比
以双子叶草本植物叶片为标准 1(伯尼斯,1991)[12]

植物类型	被测试物种数	相对硬度
双子叶草本,全部叶片	166	1.0
木本,新叶	25	1.7
C_3 牧草,所有韧面	42	3.1
C_4 牧草,所有韧面	34	6.2
木本,全部伸展叶片	89	6.3
棕榈树,伸展复叶	8	9.8

不同的叶片韧性(专业术语和测量技术参考霍秋利[43]和桑松等[83])对昆虫取食和生长的影响比我们以往想象的要大得多(图 3.7)。例如,杂食性的甜菜夜蛾(*Spodoptera exigua*)取食西芹(*Apium gravenopodium*)的时间要比取食藜(*Chenopodium murale*)的时间多 3 倍,因为西芹的硬度是藜(*Chenopodium murale*)叶片硬度的 1.5 倍。对于各种栽培玉米变种来说,其叶片韧性与抵抗欧洲玉米螟的能力密切相关[10]。

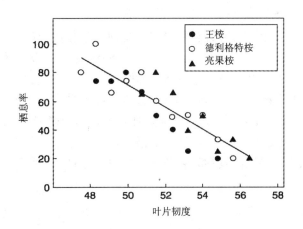

图 3.7 初孵甲虫(*Chrysophthatta biomaculate*)幼虫根据
三种桉树(*Eucalyptus*)叶片的韧度选择宿主

在韧度低于 46 g 的叶片上,幼虫的栖息率为 100%,而在高于 56.5 g 的叶片上没有幼虫栖息(豪利特等,2001)[45]

具咀嚼式口器的昆虫所食叶片的大小与昆虫的大小(龄期)和叶片的韧度有关[44]。因此,天蚕蛾幼虫的消化道能够消化相对较大的硬质叶片,而天蛾幼虫通常取食与其体型大小相当的软质草本植物叶片(图 3.8)。

通常昆虫对植物细胞壁的消化是有限的[44],我们猜想,低效率的消化吸收可能与昆虫大口取食叶片的习性有关。然而,事实并非如此,一种鳞翅目昆虫弄蝶(*Paratrytone melane*)的粪便中有76%～86%的非消化叶片。但是其可溶性碳水化合物和蛋白质的近似消化率仍可达78%和88%。因此推测大部分营养物质很可能是在细胞膜被消化后通过胞间连丝和壁孔流出后被吸收[4]。

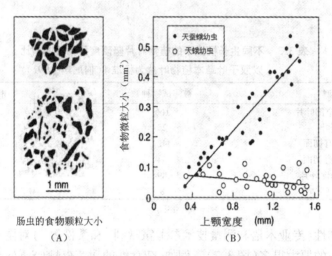

图3.8 上颚宽度与食物微粒大小的关系

(A)五龄的天蚕蛾幼虫(图顶)(*Rothschidia lebeau*)和天蛾(*Pachylia ficus*)幼虫(图底)(*Pachylia ficus*)肠中的食物颗粒图;(B)10只天蚕蛾幼虫和15只天蛾幼虫的上颚宽度、大小与其肠中食物微粒的大小之间的关系(不同龄期)。天蚕蛾取食厚而坚硬的成熟叶片,而天蛾取食薄而柔软的叶片,不分新叶片与老叶片(伯尼斯和詹曾,1988)[13]

在某种限制条件下,昆虫的头部形态可能与食物韧度是相适应的,例如,尽管身体大部分构造都相似,但取食硬质草的黏虫(*Pseudaletia unipunctata*)幼虫的头部和用于咀嚼的肌肉组织是食用人造柔软食物的幼虫的2倍[11]。取食水百合的水百合甲虫(*Galerucella nymphaeae*)的上颚会比同种取食酸模(*Rumex hydrolapathum*)的大得多,而酸模(*R. hydrolapathum*)的叶片组织比水百合柔软3倍。我们还并不清楚这种显著性差异是基于遗传还是对宿主植物的一种感应反应[77]。此外,种间各种各样的上颚形态,也许这就是不同昆虫取食其适合类型食物的原因。例如,蚱蜢在食用不同硬度的食物时会展现与之相符合的上颚形态。

图3.9 取食不同食物的两种蚱蜢的上颚

(A)西苯蝗(*Brachystola magna*)食用非禾本草本类植物,用于磨碎叶片的左颚和右颚上的小齿间是留有空隙的,产生了一个高效率的咀嚼机制;(B)大草原蝗虫(*Mermiria maculipennis*)左上颚的切齿几乎是融合的,组成了连续的切割边缘,而右上颚齿节组成的是斜切的边缘,产生了一个剪式切割的机制(艾斯利,1944)[47]

与食用较软的叶片组织相比,咀嚼坚硬的叶片组织需要更多的能量,并会严重磨损口器。因此,作为细胞壁的重要成分,纤维素可能会成为大范围防御植食性昆虫的一大要素。尤其是食用像禾本科(草和谷类)、莎草科(莎草)、棕榈科和木贼草科(木贼),这些含硅元素多(大于干重的15%,比其他无机物含量都要高)的植物更易过度磨损口器。非晶型硅($SiO_2 \cdot H_2O$)微粒储存在细胞壁和细胞腔中作为一种粗糙的研磨剂可使颚齿在咀嚼过程中完全磨损,最终导致昆虫饿死[31]。小麦[66]、大米[55]、甘蔗[52]和其他禾本科作物中硅含量的提高有利于多种植物害虫的防治(图3.10)。

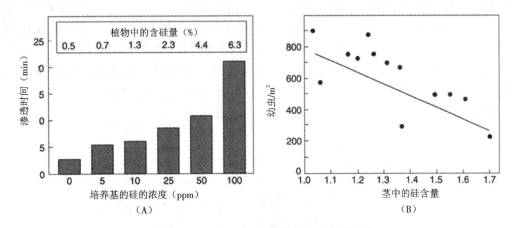

图3.10 硅的浓度对植食性昆虫存活率的影响

(A)生长在含硅量不同的营养液中的水稻植株被初孵三化螟(Scirpophaga incertulas)幼虫蛀入茎中的时间(可汗和拉马钱德朗,1989)[55];(B)13种意大利黑麦草(Lolium multiforum)对蛀干害虫瑞典麦秆蝇(Oscinella frit)幼虫的敏感度与它们残株中硅元素的含量的关系(摩尔,1984)[65]

物理防御并不需要植物中所有细胞和组织都具有相似的机械特性。像禾本科中的硅元素,它并不是等同分布的,而冬青(Ilex aquifolium)只在叶片的周边区域才有厚壁组织,这样可以使其变得坚硬,能防止大量的昆虫取食(图3.11)[61]。

图3.11 枯叶蛾取食冬青

冬青(Ilex aquifolium)的叶片边缘太坚硬以至不能被取食,图中的毛虫为枯叶蛾(Lasiocampa quercus),当多刺的边缘被剪掉后,毛虫开始取食叶片的其余部分

3.3.2 C_3植物与C_4植物

基于不同的光合作用路径,可将植物分为囊括多数温带种类的C_3植物和C_4植物。C_4植物

的光合能力要高于C_3植物。与C_3植物相比,C_4植物可以大大降低细胞间的CO_2浓度,因此能在一定气孔导度和少水量的情况下,达到较高的光合作用率。它们对水的需求量大约是C_3植物的一半,这便可以解释它们为何多生长于(亚)热带和干燥自生地带(图3.12)。C_4植物几乎都为稻科植物,大多喜欢干热的生长环境。包括重要的经济作物,如玉米、高粱、粟和甘蔗(但小麦、稻米和大麦是C_3植物)。C_4途径植物是由C_3途径植物进化而来,这很有可能是对中新世时期(距今2500百万年到500万)众多环境变化的一种适应[75]。固碳过程的不同造成了生理和形态不同。C_4植物的新陈代谢是随着被称为花环结构(Kranz anatomy)的一种特殊叶片构造完成的。这些植物的叶脉由大而厚的维管束鞘细胞包围。现已发现这些构造的改变可影响昆虫的取食,可食用的鞘细胞由许多昆虫不能消化的半纤维素加固。许多研究表明,植食性昆虫趋于躲避C_4植物。构造上的特征增加了它们的坚硬度(表3.3),这种物理上的限制可明显防止昆虫取食C_4植物和在其上排卵[84]。C_4植物因营养成分的不同,作为一种选择,所庇护的昆虫要比C_3植物少[5,6]。然而有来自生物力学方面的确凿证据解释,这归因于我们对其不同营养价值的认知较少。当然,这两种机制并不是互相排斥的。

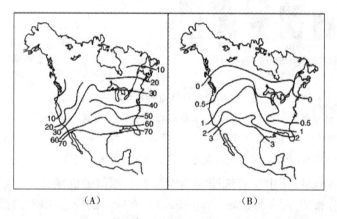

图3.12 等位线指示

(A)牧草类群比例;(B)使用C_4路径进行光合作用的北美双子叶植物类群比例(弗特和嗨,1984)

C_4植物总体不易受植食性昆虫的影响,却不能说此类植物缺乏侵食者。玉米、蔗糖、高粱和粟等重要的C_4经济作物也是害虫的寄主。至少在一定程度上,这是由特别的产量构成因素(yield components)和农业行为选择造成的。

3.4 互利共生关系中所包含的结构

虫菌穴(Domatia)和花外蜜腺(extrafloral nectaries)是植物为捕食性节肢动物提供的庇护所,它们对植食性昆虫的影响是间接性的。它们普遍存在于被子植物中,并在植物和昆虫相互作用关系中扮演重要角色(图3.13)。关于花外蜜腺的分布和功能将在第10章进行详细讨论。

叶片上的虫菌穴(希腊语称其为"小房间")是一种在木质植物中普遍存在的结构[73]。在420科植物中,90科以上的植物通常具有虫菌穴,以小毛丛、袋式小孔或凹陷的形式,分布在表皮下叶脉上。它也称为虱巢(Acarodomatia),虽然其直径只有1~2 mm,但其能给捕食或食真菌的螨类提供一个抵抗不利环境[37]的庇护所,或使它们免受其他捕食者的危害[30,71]。因此虫菌穴的占领者在减少植食性螨虫和真菌寄生物的数量等方面起着非常重要的作用[74]。实验表明,通常情况下去除地中海荚蒾(Viburnum tinus)叶片上的虫菌穴,捕食螨的数量会降低,特别是在相对湿度较低的条件下这种情况尤为明显。密度越低对植食性螨虫的捕食率越低[37]。还有另一个处理巧妙的实验,模仿虫菌穴的结构应用于棉花的叶片,实验表明与未经处理的植株相比,实验植株上有更多的肉食性节肢动物和更少的植食性螨虫。实验结果使得具虫菌穴的植株在收获量上有着30%的惊人增长率[1]。

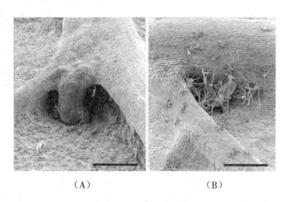

图3.13 两种植物叶片虫菌穴的扫描电子显微照片

(A)青楷槭(Acer tegmentosum);(B)具植绥螨(phytoseiidmite)的野茉莉(Styrax japonica),比例为0.5 mm

另外一些具虫菌穴的典型例子是空的茎干和刺。数百种热带植物都具有这种结构,可为蚂蚁提供庇护所和巢穴。由于蚂蚁是肉食性动物,它们可有效地帮助宿主植物抵抗植食性昆虫的危害[46]。此外,蚁穴的存在还可节省新陈代谢消耗的能量,原因为植株依靠蚂蚁便无须在化学防护方面耗费太多。

依据这些例子我们不难得出结论,虽然虫菌穴的建造是高消耗的[34],但它是植物和节肢动物之间基于形态而形成互惠共生现象的普遍存在形式。

3.5 植物虫瘿

植物发生某些强烈的变化是由虫瘿引起的[87,97]。由于虫瘿在形式和形成机制方面的不同,因此很难为虫瘿下一个明确的定义。由雷德芬[81]提出的定义看起来似乎简单易懂:"植物虫瘿是由植物细胞过度生长(扩大)和/或增殖的产物,且是由另一种生物导致的,虫瘿为该种生物提供营养物/食物和栖息处。"各种昆虫都可能会引起虫瘿,但结构最复杂、营养最丰富的虫瘿通常是由瘿蚊(双翅目:瘿蚊科)或瘿蜂(膜翅目:瘿蜂科)产生的。虽然真正的植物肿瘤一般是无组织结构和无固定形态的细胞复合体,然而昆虫引起的虫瘿却展示了独特的形态结构和生理机能。事实证明,昆虫的种类在决定虫瘿的外形中有着重要作用。近缘种昆虫在相同植物上产生的虫瘿在构造和功能方面也存在巨大的差异(图3.14)。例如,实蝇(Urophora quadrifasciata)在矢车菊(Centaurea jacea)上造了一个原始的瘦果虫瘿,但它的同属实蝇昆虫

(*U. jaceana*)却可产生一个复杂的多室虫瘿[15]。

虫瘿内的昆虫不但可以控制虫瘿结构的发展,还可以控制营养物质的移动和吸收。一项借助$^{14}CO_2$对蒲公英(*Taraxacum officinale*)进行光合作用吸收能量分配的研究表明,虫瘿可作为光合物质吸收和代谢的装置,这依赖于每株植物的虫瘿量,对于碳的固定最多可达到寄主植物碳固定量的70%[3]。同样地,由一种做瘿蚜虫在(*Sorbus commixta*)的叶片上做瘿,能引起韧皮部汁液中的氨基酸浓度发生变化。具虫瘿的叶片被切开后所产生的分泌物中的氨基酸含量是不具虫瘿叶片的5倍,这可能与植物叶蛋白的分解与昆虫自身的受益相关[57]。

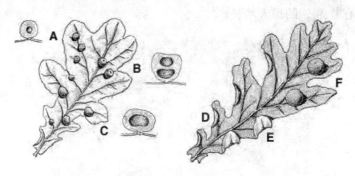

图3.14 橡树(欧洲栎)(*Quercus robur*)的树叶上有4种瘿蜂(cynipid)(A,B,C,F)和2种瘿蚊(cecidomyid)(D,E)的虫瘿

(A)瘿蜂(*Trichonaspis synaspis*);(B)瘿蜂(*Cynips disticha*);(C)瘿蜂(*C. divisa*);(D)瘿蚊(*Macrodiplossis volvens*);(E)瘿蜂(*M. dryobia*);(F)瘿蜂(*Neuropterus baccarum*)。三个种类的虫瘿横切片(A—C)显示出种特异性(鲁文,1957)

一项对于柳树的酚类化学物质的研究表明,瘿虫也干扰宿主防御系统的化合物的生物合成。叶峰(*Eupontania* spp.)可在不同种柳树上产生虫瘿,具虫瘿叶片所包含的酚类物质比正常叶片中的含量低。虫瘿能以某种方式在宿主的异常组织中改变酚类物质的质量和数量[72]。

不同的昆虫可以形成形态各异的虫瘿,这就引起了一个问题,即由虫瘿所产生的虫瘿特质鉴别以及在寄主植物上虫瘿发生的分子机制。虽然我们在这方面知之甚少,但由一些昆虫衍生的有丝分裂脂质能够在缺乏昆虫时刺激虫瘿形成,也许有利于解析这些错综复杂的昆虫—植物相互作用关系[42,89]。更多关于宿主植物对由昆虫诱导产生的反应会在第4章讨论。

所有的植物都难以防御做瘿昆虫吗?当然不是,因为许多植物以一种局部超敏性反应来应对入侵昆虫,并且经常能杀死它们。这是由一项8种木本植物受虫瘿危害致死率调查研究证实的。调查表明,12%～94%(平均59%)的入侵昆虫屈服于宿主植物的超敏性反应[32]。显然,植物对昆虫实施了强选择压力,虫瘿制造者和宿主植物之间不断进行着进化调整,这将促进宿主植物的特化。与不做瘿的鳞翅目相比,做瘿鳞翅类昆虫有相对较窄的宿主范围[64]。

以上信息可以得出的观点是:虫瘿的位置、大小、形状甚至生理机能是由昆虫而非宿主植物决定的。尽管虫瘿由宿主植物组织构成,但其发展主要由昆虫的基因控制,因此虫瘿可以被视为昆虫"扩展的表型"(道金斯提出[25])。

3.6 植物的结构

植物大小和结构均会影响以其为生的昆虫的数量。很明显,栎树比苔藓能庇护更多的昆虫。植物的结构(plant architecture)作为描述植物的大小和生长型的术语,包括在某特定时间内的树冠大小、茎、叶片、花型和面积、分叉角度与表面复杂性(纹理和软毛)等。在广义的解释中,该术语也包含一种植物随季节变化的发展和保留[59]。植物结构模型对于分析植物结构对昆虫的作用是非常有效的[35,38]。

在桂叶芫花(*Daphne laureola*)的不同个体植株间,由夜蛾幼虫引起的损伤程度存在巨大的差异,幼虫发生率似乎与叶轮的数量呈正相关,而与茎底部的平均直径呈负相关。夜蛾幼虫的不规则分布必定是由它们母体对植物叶片结构的区别对待造成的[2]。

其他的一些研究表明,一些特殊的植物结构可以通过影响昆虫天敌而影响植食性昆虫。例如,瓢虫幼虫在落芒草(*Oryzopsis hymenoides*)上捕获的蚜虫也比在大麦草(*Agropyron desertorum*)上捕获的蚜虫多 2.5 倍。这 2 种天然宿主除叶片形状外的整体结构几乎相似[19]。而其他捕食性昆虫,如草蛉幼虫,在落芒草(*Oryzopsis hymenoides*)上捕获的蚜虫也比在大麦草上捕获的昆虫多 2 倍。原因是蚜虫优先在相对隐蔽的场所取食,例如成熟叶片的叶—鞘结合处,而落芒草(*Oryzopsis hymenoides*)能提供的这种相对隐蔽的场所较多[20]。通常宿主植物的结构特点对寄生性昆虫搜寻寄主方式的也存在显著影响。研究发现,桔粉蚧对一种膜翅类寄主的攻击率与植物的几种特点呈负相关,例如大小、高度、叶片数量、叶表面积和分支数等[21]。

当比较天敌对植食性昆虫影响两种不同植物或同一植物的不同层面上后,会发现植物结构所起的作用与一些其他(未知)因素是很难分割的。植物结构对捕食性昆虫的捕食效率有明显的影响。豌豆蚜(*Acyrthosiphon pisum*)只侵染两个豌豆品系(*Pisum sativum*),且这两个品系只是等位基因的两个位点有所不同,这些等位基因系在叶片存在与否上是不同的,瓢虫以豌豆蚜(*Acyrthosiphon pisum*)为食。豌豆的结构变异对瓢虫的捕食效率有显著影响,继而影响豌豆蚜的种群数量[51]。这几个例子说明,宿主植物的结构可以影响其天敌,其可能是造成植食性昆虫种群变化的一个重大因素。

形态和结构上更为复杂的植物倾向拥有更高的昆虫种类丰富度。这是基于以下事实得出的:(1)大多数植食性昆虫对其宿主的取食部位具有局限性;(2)生态位的范围与植物的复杂性密切相关。所以,从单子叶植物到灌木、乔木,其大小和结构复杂度都在增加,因此其上相应的昆虫多样性也在增长[59,90]。芬兰的乔木和灌木上生存的巨鳞翅目昆虫比草本植物多 10 倍以上[69]。

一般认为,对比较小的同种个体而言,长得更为高大的植物能庇护数量更多的昆虫,帚石楠(*Calluna vulgaris*)就是一个例证,其上植食性昆虫的物种丰富度随株高呈现惊人地增长。英格兰和苏格兰的帚石楠(*Calluna*)株高在 5 cm 与 55 cm 之间变动。帚石楠(*Calluna*)的株高每增加 20 cm,鳞翅目幼虫的密度近乎成倍增加。也许因为较高的帚石楠(*Calluna*)可以使某些在矮植株上不能生存的物种生存[40]。这种现象也存在于一种生长于热带雨林中的锥头麻(*Pourouma bicolor*)。幼树(株高 < 4 m)与成熟树木(株高 17~30 m)的昆虫丰富度明显不同(图 3.15)[8]。当然,对于不同时期的同种植物,不仅只与个体大小有关,一些其他结构特征以及生理因素也随年龄的增长而变化,这就难以判断是何种因素导致了植食性昆虫的数量变化。

一项对仙人掌属(*Opuntia*)28 种植物的研究表明,五种结构特征会影响植食性昆虫的生态密度[67]。株高似乎是关键因素,因为它与昆虫的数量($r = 0.59$)显著相关。此外,植株叶状枝的数量是一种更为重要的植物特征($r = 0.73$)。如果共同考虑三种附加变量,植物的结构与

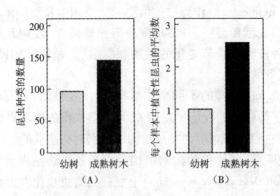

图 3.15 在幼树与成熟树木上的昆虫种类数和每个样本中植食性昆虫的平均数

(A)昆虫多样性的比较(植食性昆虫物种的数量);(B)锥头麻(*Pourouma bicolor*)中幼树和成熟树的昆虫丰富度(每个样本中植食性昆虫的平均数)数字取自 2 000 片叶片,面积均取 0.36 m²(巴西特等,2001)[8]

昆虫的生态密度之间的相关系数将达到一个更高的数值($r = 0.83$)。因此,植株大小似乎不能被当作唯一的重要变量(图 3.16)。

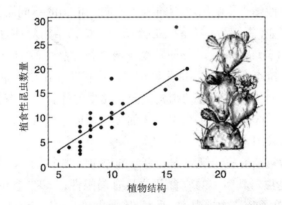

图 3.16 植食性昆虫数量与植物结构的关系图

样本是来自北美和南美的 28 种仙人掌(*Opuntia*),结构比例是以下(1) – (5)变量的总和:(1)成熟植株株高;(2)叶状枝的平均数量;(3)叶状枝的大小(cm²);(4)木质茎的发育程度;(5)叶状枝的复杂度,即叶状枝表面的质量、刺的有无以及密度(摩尔,1980)[67]

3.7 结论

尽管近十几年来,有越来越多的学者研究植物和植食性昆虫之间的相互作用,相对于物理特性而言,他们更关注于化学因素的作用。然而植物物理因素,例如韧性和纤维含量,相对于化学因素而言,可以更好地阻碍植食性昆虫的侵害[26,83]。这种观点不仅仅是对昆虫种群与它们宿主之间关系进行的判断,同时也是基于对寄生在特定植物上的植食性昆虫种群的生态密度的研究。因此,通过对一个桉树林中的植食性昆虫群落组成的分析得知,相比于叶子组分,植食性昆虫种群与叶子结构特征之间的关联更为密切[78]。

依靠经典的育种方法或基因工程来提高农作物的抗虫性,需要对植食性昆虫取食行为进行更精确详细地了解。为此,通过定量分析植物物理或化学方面的防御措施的重要性因素,我们将获取至关重要的信息[50]。对植物组织韧性这项能有效防御植食性昆虫的机制的关注,也是非常重要的。这是因为在培育改良的栽培变种过程中,这些特性往往是要被淘汰的,尤其是那些被昆虫吃掉叶子和果实的农作物种类[76]。

显然,植物形态特征不仅会直接影响植食性昆虫,还会间接影响植食性昆虫的天敌。尽管这项研究很有前景,但仍未引起我们足够的重视。

诺里斯和科根[70]、潘达和库西[76]已经考察了许多与植物抗虫性有关的形态学因素的实例。

3.8 参考文献

1. Agrawal, A. A. and Karban, R. (1997). Domatia mediate plant-arthropod mutualism. *Nature*, 387, 562–63.
2. Alonso, C. and Herrera, C. M. (1996). Variation in herbivory within and among plants of *Daphne laureola* (Thymelaeaceae): correlation with plant size and architecture. *Journal of Ecology*, 84, 495–502.
3. Bagatto, G., Paquette, L. C., and Shorthouse, J. D. (1996). Influence of galls of *Phanacis taraxaci* on carbon partitioning within common dandelion, *Taraxacum officinale*. *Entomologia Experimentalis et Applicata*, 79, 111–17.
4. Barbehenn, R. V. (1992). Digestion of uncrushed leaf tissues by leaf-snipping larval Lepidoptera. *Oecologia*, 89, 229–35.
5. Barbehenn, R. V. and Bernays, E. A. (1992). Relative nutritional quality of C_3 and C_4 grasses for a graminivorous lepidopteran, *Paratrytone melane* (Hesperidae). *Oecologia*, 92, 97–103.
6. Barbehenn, R. V., Karowe, D. N., and Chen, Z. (2004). Performance of a generalist grasshopper on a C_3 and a C_4 grass: compensation for the effects of elevated CO_2 on plant nutritional quality. *Oecologia*, 140, 96–103.
7. Barthlott, W., Neinhuis, C., Cutler, D., Ditsch, F., Muesel, I., Theisen, I., et al. (1998). Classification and terminology of plant epicuticular waxes. *Botanical Journal of the Linnean Society*, 126, 237–60.
8. Basset, Y. (2001). Communities of insect herbivores foraging on saplings versus mature trees of *Pourouma bicolor* (Cecropiaceae) in Panama. *Oecologia*, 129, 253–60.
9. Berdegue, M. and Trumble, J. T. (1996). Effects of plant chemical extracts and physical characteristics of *Apium graveolens* and *Chenopodium murale* on host choice by *Spodoptera exigua* larvae. *Entomologia Experimentalis et Applicata*, 78, 253–62.
10. Bergvinson, D. J., Arnason, J. T., Hamilton, R. I., Milton, J. A., and Jewel, D. C. (1994). Determining leaf toughness and its role in maize resistance to the European corn borer (Lepidoptera: Pyralidae). *Journal of Economic Entomology*, 87, 1743–8.
11. Bernays, E. A. (1986). Diet-induced head allometry among foliage chewing insects and its importance for graminivores. *Science*, 231, 495—7.
12. Bernays, E. A. (1991). Evolution of insect morphology in relation to plants. *Philosophical*

Transactions of the Royal Society, London, B333, 257—64.
13. Bernays, E. A. and Janzen, D. H. (1988). Saturniid and sphingid caterpillars: two ways to eat leaves. *Ecology*, 69, 1153—60.
14. Brennan, E. B. and Weinbaum, S. A. (2001). Stylet penetration and survival of three psyllid species on adult leaves and 'waxy' and 'de-waxed' juvenile leaves of *Eucalyptus globulus*. *Entomologia Experimentalis et Applicata*, 100, 355—66.
15. Burkhardt, B. and Zwölfer, H. (2002). Macroevolutionary trade-offs in the tephritid genus *Urophora*: benefits and costs of an improved plant gall. *Evolutionary Ecology Research*, 4, 61—77.
16. Chang, G. C., Rutledge, C. E., Biggam, R. C., and Eigenbrode, S. D. (2004). Arthropod diversity in peas with normal or reduced waxy bloom. *Journal of Insect Science*, 4(18), 1—11.
17. Chapman, R. F. (1995). Mechanics of food handling by chewing insects. In *Regulatory mechanisms in insect feeding*, (ed. R. F. Chapman and G. de Boer), pp. 3—31. Chapman & Hall, New York.
18. Chiang, H. -S. and Norris, D. M. (1983). Morphological and physiological parameters of soy bean resistance to Agromyzid bean flies. *Environmental Entomology*, 12, 260—5.
19. Clark, T. L. and Messina, F. J. (1998a). Plant architecture and the foraging success of ladybird beetles attacking the Russian wheat aphid. *Entomologia Experimentalis et Applicata*, 86, 153—61.
20. Clark, T. L. and Messina, F. J. (1998b). Foraging behavior of lacewing larvae (Neuroptera: Chrysopidae) on plants with divergent architectures. *Journal of Insect Behavior*, 11, 303—17.
21. Cloyd, R. A. and Sadof C. S. (2000). Effects of plant architecture on the attack rate of *Leptomastix dactylopii* (Hymenoptera: Encyrtidae), a parasitoid of the citrus mealybug (Homoptera: Pseudococcidae). *Environmental Entomology*, 29, 535—41.
22. Coley, P. D. (1983). Herbivory and defensive characteristics of tree species in a lowland tropical forest. *Ecological Monographs*, 53, 209—33.
23. Dalin, P. and Björkman, C. (2003). Adult beetle grazing induces willow trichome defence against subsequent larval feeding. *Oecologia*, 134, 112—18, 554.
24. Davies, R. G. (1988). *Outlines of entomology* (7th edn). Chapman & Hall, London.
25. Dawkins, R. (1982). *The extended phenotype*. Oxford University Press, Oxford.
26. Dyer, L. A. and Coley, P. D. (2002). Tritrophic interactions in tropical versus temperate communities. In *Multitrophic level interactios* (ed. T. Tscharntke and B. A. Hawkins), pp. 67—88. Cambridge University Press, Cambridge.
27. Eigenbrode, S. D. (2004). The effects of plant epicuticular waxy blooms on attachment and effectiveness of predatory insects. *Arthropod Structure and Development*, 33, 91—102.
28. Eigenbrode, S. D. and Espelie, K. E. (1995). Effects of plant epicuticular lipids on insect herbivores. *Annual Review of Entomology*, 40, 171—94.

29. Eigenbrode, S. D., Kabalo, N. N., and Stoner, K. A. (1999). Predation, behavior, and attachment by *Chrysoperla plarabunda* larvae on *Brassica oleracea* with different surface wax-blooms. *Entomologia Experimentalis et Applicata*, 90, 225—35.
30. English-Loeb, G., Norton, A. P., and Walker, M. A. (2002). Behavioral and populations consequences of acarodomatia in grapes on phytoseiid mites (Mesostigmata) and implications for plant breeding. *Entomologia Experimentalis et Applicata*, 104, 307—19.
31. Epstein, E. (1999). Silicon. *Annual Review of Plant Physiology and Plant Molecular Biology*, 50, 641—64.
32. Fernandes, G. W. and Negreiros, D. (2001). The occurrence and effectiveness of hypersensitive reaction against galling herbivores across host taxa. *Ecological Entomology*, 26, 46—55.
33. Frost, S. W. (1959). *Insect life and insect natural history* (2nd edn). Dover Publications, New York.
34. Gartner, B. L. (2001). Multitasking and tradeoffs in stems, and the costly dominion of domatia. *New Phytologist*, 151, 311—13.
35. Godin, C., Costes, E., and Sinoquet, H. (1999). A method for describing plant architecture which integrates topology and geometry. *Annals of Botany*, 84, 343—57.
36. Gorb, E. V. and Gorb, S. N. (2002). Attachment ability of the beetle *Chrysolina fastuosa* on various plant surfaces. *Entomologia Experimentalis et Applicata*, 105, 13—28.
37. Grostal, P. and O'Dowd, D. J. (1994). Plants, mites and mutulism: leaf domatia and the abundance and reproduction of mites on *Viburnum tinus* (Caprifoliaceae). *Oecologia*, 97, 308—15.
38. Hanan, J., Prusinkiewicz, P., Zalucki, M., and Skirvin, D. (2002). Simulation of insect movement with respect to plant architecture and morphogenesis. *Computers and Electronics in Agriculture*, 35, 255—69.
39. Harris, K. F., Smith, O. P., and Duffus, J. E. (ed) (2001). *Virus-insect-plant interactions*. Academic Press, San Diego.
40. Haysom, K. A. and Coulson, J. C. (1999). The Lepidoptera fauna associated with *Calluna vulgaris*: effects of plant architecture on abundance and diversity. *Ecological Entomology*, 23, 377—85.
41. Heming, B. S. (1993). Structure, function, ontogeny, and evolution of feeding in thrips (Thysanoptera). In *Functional morphology of insect feeding* (ed. C. W. Schaefer and R. A. B. Leschen), pp. 5—41. Entomological Society of America, Lanham.
42. Hilker, M., Rohfritsch, O., and Meiners, T. (2002). The plant's response towards insect egg deposition. In *Chemoecology of insect eggs and egg deposition* (ed. M. Hilker and T. Meiners), pp. 205—33. Blackwell, Berlin.
43. Hochuli, D. F. (1996). The ecology of plant/insect interactions: implications of digestive strategy for feeding by phytophagous insects. *Oikos*, 75, 133—41.
44. Hochuli, D. F. and Roberts, F. M. (1996). Approximate digestibility of fibre for a graminivorous caterpillar. *Entomologia Experimentalis et Applicata*, 81, 15—20.

45. Howlett, B. G., Clarke, A. R., and Madden, J. L. (2001). The influence of leaf age on the oviposition preference of *Chrysophtharta bimaculata* (Olivier) and the establishment of neonates. *Agricultural and Forest Entomology*, 3, 121—27.

46. Huxley, C. R. and Cutler, D. F. (ed) (1991). *Ant-plant interactions*. Oxford University Press, Oxford.

47. Isely, F. B. (1944). Correlation between mandibular morphology and food specificity in grasshoppers. *Annals of the Entomological Society of America*, 37, 47—67.

48. Jeffree, C. E. (1986). The cuticle, epicuticular waxes and trichomes of plants, with reference to their structure, functions and evolution. In *Insect and the plant surface*, (ed. B. E. Juniper and T. R. E. Southwood), pp. 23—64. E. Arnold, London.

49. Jetter, R., Schäffer, S., and Riederer, M. (2000). Leaf cuticular waxes are arranged in chemically and mechanically distinct layers: evidence from *Prunus laurocerasus* L. *Plant and Environment*, 23, 619—28.

50. Kanno, H. and Harris, M. O. (2001). Leaf physical and chemical features influence selection of plant genotypes by Hessian fly. *Journal of Chemical Ecology*, 26, 2335—54.

51. Kareiva, P. and Sahakian, R. (1990). Tritrophic efects of a simple architectural mutation in pea plants. *Nature*, 345, 433—4.

52. Keeping, M. G. and Meyer, J. H. (2002). Calcium silicate enhances resistance of sugarcane to the African stalk borer *Eldana saccharina* Walker (Lepidoptera: Pyralidae). *Agricultural and Forest Entomology*, 4, 265—74.

53. Kennedy, C. E. J. (1986). Attachment may be a basis for specialization in oak aphids. *Ecological Entomology*, 11, 291—300.

54. Kennedy, C. E. J. and Southwood, T. R. E. (1984). The number of species of insects associated with British trees: a re-analysis. *Journal of Animal Ecology*, 53, 455—78.

55. Khan, Z. R. and Ramachandran, R. (1989). Studies on the biology of the yellow stem borer, *Scirpophaga incertulas*. *ICIPE Annual Report*, 1989, 24—5.

56. Kingsolver, J. G. and Daniel, T. L. (1995). Mechanics of food handling by fluid-feeding insects. In: *Regulatory mechanisms in insect feeding* (ed. R. F. Chapman and G. de Boer), pp. 32—73. Chapman & Hall, New York.

57. Koyama, Y., Yao, I., and Akimoto, S. I. (2004). Aphid galls accumulate high concentrations of amino acids: a support for the nutrition hypothesis for gall formation. *Entomologia Experimentalis et Applicata*, 113, 35—44.

58. Labandeira, C. C. (1997). Insect mouthparts: ascertaining the paleobiology of insect feeding strategies. *Annual Review of Ecology and Systematics*, 28, 153—93.

59. Lawton, J. H. (1983). Plant architecture and the diversity of phytophagous insects. *Annual Review of Entomology*, 28, 23—29.

60. Lee, Y. L., Kogan, M., and Larsen, J. R. (1986). Attachment of the potato leafhopper to soybean plant surfaces as affected by morphology of the pretarsus. *Entomologia Experimentalis et Applicata*, 42, 101—8.

61. Merz, E. (1959). Pflanzen und Raupen. Über einige Prinzipien der Futterwahl bei Großschmetterlingsraupen. *Biologisches Zentralblatt*, 78, 152—88.
62. Meyer, G. A. (1993). A comparison of the impacts of leaf-and sap-feeding insects on growth and allocation of goldedrod. *Ecology*, 74, 1101—16.
63. Miles, P. W. (1999). Aphid saliva. *Biological Reviews*, 74, 41—85.
64. Miller, W. E. (2004). Host breadth and voltinism in gall-inducing Lepidoptera. *Journal of the Lepidopterists' Society*, 58, 44—7.
65. Moore, D. (1984). The role of silica in protecting Italian ryegrass (*Lolium multiflorum*) from attack by dipterous stem-boring larvae (*Oscinella frit* and other related species). *Annals of Applied Biology*, 104, 161—6.
66. Morae, J. C., Goussain, M. M., Basagli, M. A. B., Carvalho, G. A., Ecole, C. C., and Sampaio, M. V. (2004). Silicon influence on the tritrophic interaction: wheat plants, the greenbug *Schizaphis graminum* (Rondani)(Hemiptera: Aphididae), and its natural enemies, *Chrysoperla externa* (Hagen)(Neuroptera: Chrysopidae) and *Aphidius colemani* Viereck (Hymenoptera: Aphidiidae). *Neotropical Entomology*, 33, 619—24.
67. Moran, V. C. (1980). Interactions between phytophagous insects and their *Opuntia* hosts. *Ecological Entomology*, 5, 153—64.
68. Mullin, C. A. (1986). Adaptive divergence of chewing and sucking arthropods to plant allelochemicals. In: *Molecular aspects of insect-plant associations* (ed. L. B. Brattsten and S. Ahmad), pp. 175-209. Plenum Press, New York.
69. Niemelä, P., Tahvanainen, J., Sorjonen, J., Hokkanen, T., and Neuvonen, S. (1982). The influence of host plant growth form and phenology on the life strategies of Finnish macrolepidopterous larvae. *Oikos*, 39, 164-70.
70. Norris, D. M. and Kogan, M. (1980). Biochemical and morphological bases of resistance. In *Breeding plants resistant to insects* (ed. F. G. Maxwell and P. R. Jennings), pp. 23-61. Wiley, New York.
71. Norton, A. P., English-Loeb, G., and Belden, E. (2001). Host plant manipulation of natural enemies: leaf domatia protect beneficial mites from predators. *Oecologia*, 126, 535-42.
72. Nyman, T. and Julkunen-Tiitto, R. (2000). Manipulation of the phenolic chemistry of willows by gallinducing sawflies. *Proceedings of the National Academy of Sciences of the USA*, 97, 13184-7.
73. O'Dowd, D. J. and Pemberton, R. W. (1998). Leaf domatia and foliar mite abundance in broadleaf deciduous forest of north Asia. *American Journal of Botany*, 85, 70-8.
74. O'Dowd, D. J. and Willson, M. F. (1991). Associations between mites and leaf domatia. *Trends in Evolotion and Ecology*, 6, 179-82.
75. Pagani, M., Freeman, K. H., and Arthur, M. A., (1999). Late Miocene atmospheric CO_2 concentrations and the expansion of C_4 grasses. *Science*, 285, 876-9.
76. Panda, N. and Khush, G. S., (1995). *Host plant resistance to insects*. CAB International, London.

77. Pappers, S. M., Van Dommelen, H., Van der Velen, G., and Ouborg, N. J. (2001). Difernces in morphology and reproductive traits of *Galerucella nymphaeae* from four host plant species. *Entomologia Experimentalis et Applicata*, 99, 183 – 91.
78. Peeters, P. J. (2002). Correlations between leaf stuctural traits and the densities of herbivorous insect guilds. *Biological Journal of the Linnean Society*, 77, 43 – 65.
79. Plumb, R. T. (ed.) (2002). Plant virus vector interactions. *Advances in Botanical Research*, 36, 1 – 221.
80. Raven, J. A. (1993). Phytophages of xylem and phloem. *Advances in Ecological Research*, 13, 135 – 234.
81. Redfern, M. (1997). Plant galls: an intimate association between animals and plants. *Antenna*, 21, 55 – 63.
82. Robinson, S. J., Thurston, R., and Jones, G. A. (1980). Antixenosis of smooth leaf cotton *Gossypium* spp. to the ovipositional response of the tobacco budworm *Heliothis virescens*. *Crop Science*, 20, 646 – 49.
83. Sanson, G., Read, J., Aranwela, N., Clissold, F., and Peeters, P. (2001). Measurement of leaf biomechanical properties in studies of herbivory: opportunities, problems and procedures. *Austral Ecology*, 26, 535 – 46.
84. Scheirs, J., de Bruyn, L., and Verhagen, R. (2001). A test of the $C_3 – C_4$ hypothesis with two grass miners. *Ecology*, 82, 410 – 21.
85. Schofield, R. M. S. Nesson, M. H., and Richardson, K. A. (2002). Tooth hardness increases with zinccontent in mandibles of young adult leaf-cutter ants. *Naturwissenschaften*, 89, 579 – 83.
86. Schoonhoven, L. M., and Henstra, S. (1972). Morphology of some rostrum receptors in *Dysderus* spp. *Netherlands Journal of Zoology*, 22, 343 – 46.
87. Shorthouse, J. D. and Rohfritsch, O. (eds) (1992). *Biology of insect-induced galls*. Oxford University Press, New York.
88. Southwood, T. R. E. (1986). Plant surfaces and insects—an overview. In *Insects and the plant surface* (ed. B. Juniper and T. R. E. Southwood), pp. 1 – 22. E. Arnold, London.
89. Stone, G. N. and Schönrogge, K. (2003). The adaptive significance of insect gall morphology. *Trends in Ecology and Evolution*, 18, 512 – 22.
90. Strong, D. R., Lawton, J. H., and Southwood, T. R. E. (1984). *Insects on plants. Community patterns and mechanisms*. Blackwell, Oxford.
91. Tjallingii, W. F. (1978). Mechanoreceptors of the aphid labium. *Entomologia Experimentalis et Applicata*, 24, 731 – 7.
92. Tjallingii, W. F. (1995). Regulation of phloem sap feeding by aphids. In *Regulatory mechanisms in insect feeding* (ed. R. F. Chapman and G. de Boer), pp. 190 – 209. Chapman & Hall, New York.
93. Traw, M. B. and Dawson, T. E. (2002). Differential induction of trichomes by three herbivores of black mustard. *Oecologia*, 131, 526 – 32.

94. Van Lenteren, J. C., Hua, L. Z, Kamerman, J. W., and Xu, R. (1995). The parasite-host relationship between *Encarsia formosa* (Hym., Aphelinidae). and *Trialeurodes vaporariorum* (Hom., Aleyrodidae). XXVI. Leaf hairs reduce the capacity of *Encarsia* to control greenhouse whitefly on cucumber. *Journal of Applied Entomology*, 119, 553 – 9.
95. Werker, E. (2000). Trichome diversity and development. *Advances in Botanical Research*, 31, 1 – 36.
96. White, C. and Eigenbrode, S. D. (2000). Leaf surface waxbloom in *Pisum sativum* influences predation and intra-guild interactions involving two predator species. *Oecologia*, 124, 252 – 59.
97. Williams, M. A. J. (ed.) (1994). *Plant galls: organisms, interactions, populations*. Clarendon, Oxford.

第4章 植物化学:无穷无尽的变化

4.1 植物生物化学
　4.1.1 植物的初生代谢物质
　4.1.2 植物的次生代谢物质
4.2 生物碱
4.3 萜类化合物和类固醇
4.4 酚类化合物(酚醛树脂)
4.5 芥子油苷
4.6 生氰类化合物
4.7 叶表面的化学物质
4.8 植物的挥发物质
4.9 植物次生代谢物浓度
4.10 生产成本
4.11 区室作用
4.12 时间异质性
　4.12.1 季节影响
　4.12.2 昼夜影响
　4.12.3 年间变化
4.13 地理位置和肥料的影响
　4.13.1 日晒和遮阴
　4.13.2 土壤因素
4.14 诱导抗性
　4.14.1 直接诱导
　4.14.2 间接诱导
　4.14.3 植食诱导变化引起的变异
　4.14.4 食草性昆虫引发的基因变异和代谢变化
　4.14.5 系统影响
　4.14.6 长期反应
　4.14.7 信号传导
　4.14.8 食草动物诱导和病原体诱导的相互作用
　4.14.9 植物间的相互作用
4.15 基因型差异
　4.15.1 植物个体间的化学差异
　4.15.2 植物个体内部的化学变化
　4.15.3 植物性别对昆虫的敏感性的影响

4.16 结论
4.17 文献
4.18 参考文献

从表面上看，植物既不能抵抗也不能逃避，通常还具有较长的生命周期和较低的重组率，当受到植食动物的捕食时，它们似乎处于劣势。尤以昆虫为例，由于具有体型小、生命周期相对较短以及繁殖能力较强等特点，它们往往能够很快地适应周围不断变化的环境。此外，昆虫的飞翔能力为它们四处分散和入侵潜在食物提供了保障，哪怕是在距离昆虫出生地和幼虫期居所很远的地方。尽管当植食动物侵袭时，植物似乎表现得十分脆弱，然而地表植物群仍逐渐形成了一个绿色、多元的区域。毫无疑问，植物拥有一个以物理、化学和生长特征相结合为基础的高效防御系统。术语抵抗resistance（在昆虫—植物相互作用的背景下）是用来描述一种植物避免或降低植食性昆虫所带来的伤害的能力。但抵抗并不等同于防御（defence），因为防御隐含着一些由进化带来的新性状，进而表明植物种群中逐渐形成或维持着的抵抗特性是由植食性昆虫或者其他天敌的选择作用造成的。抵抗这个术语具有经验性（第13章），在植物性状防御功能未经证实的情况下更偏向于使用这个词[151,213]。

植物的化学特性吸引了昆虫—植物相互作用关系领域众多学者的高度兴趣，他们就这一课题发表了大量的著作。目前已得到公认的是，植物能产生种类繁多的次生代谢产物。在目前已知的天然化合物中，有80%以上都属植物源[119]。由于除了物理防御，植物产生的化学物质在控制昆虫行为方面起主要作用，因此本章主要介绍植物次生代谢物质（secondary plant substances）的性质和动态。植物化学的基础知识是全面理解昆虫—植物相互作用关系的必要前提。

昆虫学家们意识到了昆虫在形态学和行为学上存在着巨大的差异，但他们认为不同的植物对于植食性昆虫来说具有同质的来源，并且认为植物内部或植物个体间的某一特定部位，如叶子，它们的化学成分是相似的。这种理解是错误的。在本章中，我们将看到宿主植物在时间和空间上有着高度差异性。

植物化学成分和结构的差异对于阻止植食性昆虫取食宿主植物至关重要，这个观点也被越来越多的证据证实。昆虫通常具有高度的专化性并且只取食特定的食物，它们以直接或间接途径控制多种成分来应对降低的适合度[65]。

4.1 植物生物化学

植物，同其他生物一样，通过一定量的生物化学反应来维持它们的基本代谢，即参与形成和分解一定种类的化学物质。这些化学物质包括核酸、蛋白质，以及它们的前体物质、碳水化合物、羧酸等[38]。在初生代谢基础上，植物通过次生代谢途径制造多种次生代谢物质。植物主要通过3条生物发生路径产生次生代谢物质，每条途径都会产生一种或几种主要的代谢物质，再经过转移酶的作用形成许多衍生物[125]。迄今为止，次生代谢物质产生的路径仍未被研究透彻。通常，它们的形成过程都十分复杂，以紫杉醇（taxol）的合成为例，这是一种存在于紫杉树叶和树皮中，对昆虫取食具有很强阻碍抑制性的物质[61]。它的合成包含了20种酶的转化过程[283]。

虽然"初生"和"次生"这两个词可能会使人认为这两种代谢系统间存在显著差异,其实不然。山梨糖醇,在蔷薇科植物中很少存在,但在山楂(*Crataegus monogyna*)、苹果(*Malus domestica*)以及其他物种中作为主要的可溶性碳水化合物存在。山梨糖醇在叶片中的含量占干重的11%以上,起主要能量载体的作用[99,159],这样看来,似乎很难将这种化合物与"次生"这个词联系起来。另外,初生代谢过程和次生代谢过程是极其错综复杂地联系在一起的,因此植物初生和次生物质没有明确的界定,只是为了方便而已。事实上,这两种系统协同发挥作用[23]。

4.1.1 植物的初生代谢物质

光合作用是绿色植物获取太阳能并将其以糖的形式储存起来的过程,是植物体内最基本的化学能量来源。这些能量一部分用于将氮转化成氨基酸,后者是构成蛋白质的主要成分。同时糖类也是构成细胞壁结构的物质之一。绝大部分的植物生物量构成植物初生代谢物质,其中一些是大量存在的,例如木质纤维素,它是地球上最丰富的有机高分子。纤维素和半纤维素(都是复杂的多糖)还有木质素(酚醛化聚合物)合计(根据干重)大约占落叶木的90%,占草的66%,占落叶的50%[2]。植物初生代谢物的重要组分,即参与植物基本生理过程的蛋白质、糖类、脂类——形成植食性动物的必需营养物。因此,植物初生代谢化合物的定性和定量变异都将会对昆虫的偏好和行为产生重大影响[22]。我们将在第5章讨论这个问题。

4.1.2 植物的次生代谢物质

植物次生代谢物质可以被定义为:"在高等植物中不常见,仅在特定植物类群中存在,或者在某些特定植物类群中存在得更为集中,但在初级代谢过程中不发挥(明显)作用的植物化合物。"尽管人类很早就认识到不同植物种间存在化学成分的差异,但直到20世纪,人们才逐渐对这些化学成分功能的差异有所认识。弗伦克尔在一篇重要的论文中强调:作为防御系统,植物次生代谢物质在对抗昆虫及其他天敌方面起着重要的作用[95]。尽管已有许多有力的证据证实这个假说,但是仍有很多人提出质疑,他们认为许多植物次生物质似乎在植物内部有其他(额外的)作用。他们提出,这些化合物的防御作用只是在特定环境压力的选择作用下,一种由基因控制的抵抗因子的多效性影响。因此,种内个体和种间个体(抑他作用)间的竞争、营养匮乏(例如生物碱以氮形式存储)、干旱和紫外线照射等环境因子刺激生化反应,进而产生植物次生代谢物质。详见第11章。

根据其生态作用,植物次生代谢物质可称为"种间交感物质",此术语由惠特克提出。种间交感物质被定义为:某一物种个体所产生的,对另一物种的生长、健康、行为乃至生物种群产生影响的非营养化学物质[290]。与"次生"这个形容词表现的意思正相反,植物次生代谢物质在植物内扮演着重要的生态角色。与相对单一的初生代谢状况相比,植物能产生数量惊人的次生代谢物(图4.1)。即使单个物种也可以制造出大量复杂的化学物质。以长春花(*Catharanthus roseus*)为例,它含有100种以上不同的类单萜吲哚生物碱[38],而葡萄(*Vitis vinifera*)的浆果富含200种以上与葡萄糖结合的苷类衍生物[237]。由于一个物种中含有大量次生代谢物和次生代谢物生成过程中的多步酶促反应。我们猜想,一个物种中代谢物的含量超过了控制生物合成的基因数目[235]。

据估计,植物界能合成成百上千种植物次生代谢物质。现在已知的已超过10万种,而且每天都有文献报道新物质的发现[29,235]。毫无疑问,植食性昆虫所处的化学世界本身就极其复杂。

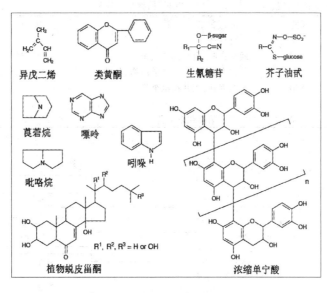

图 4.1 植物次生代谢物质主要成分的一般分子结构

为植物次生代谢物建立一个合理的分类很难,例如根据分子结构进行分类。次生代谢物一般由现存的前体物质衍化而来,通常是乙酰辅酶A、氨基酸或莽草酸,因此根据它们的生物合成途径来进行分类(图 4.2)似乎适用于大多数情况[38,176,223]。

从基础化学物质开始,次生化合物的合成通常就像一个"多维网络"——不同的分支途径在某些方面是相互联系的。可将其简单分为:(1)含氮化合物;(2)萜类化合物;(3)酚类化合物;(4)乙炔化合物(表 4.1)。

表 4.1 昆虫—植物相互作用关系中发挥主要作用的植物次生化合物(哈伯纳,1993)

种类	已知化合物数目	在维管植物中的分布	生理活性
含氮化合物			
生物碱	16 000	广泛分布在被子植物的根部、叶子和果实	多数有毒,苦涩
胺类	100	广泛存在于被子植物花中	多数具趋避作用
氨基酸(非蛋白)	400	普遍存在于豆科植物的种子中	多数有毒
氰苷类	60	在果实和叶子中零散存在	有毒(HCN)
芥子油式	120	十字花科和其他 10 个科中	辛辣且苦(似异硫氰酸酯)
萜类	30 000		
单萜	1 000	广泛存在于植物精油中	很好的气味
倍半萜烯	6 500	在被子植物中,特别是在菊科、在精油和树脂中	一些有苦味,有毒
双萜	30 000	广泛存在,尤其是在橡胶和树脂	多数有毒
皂苷	600	存在于 70 多个科中,尤其是龙胆科、茄科、玄参科	毒性(溶血)
柠檬苦素类似物	300	主要是在芸香科、楝科中	苦味
葫芦素	50	主要存在葫芦科	苦味且有毒
强心苷	150	在 12 个被子植物科中,特别是在夹竹桃科和萝藦科	有毒和苦涩

续表

种类	已知化合物数目	在维管植物中的分布	生理活性
类胡萝卜素	650	普遍存在在叶子、花和果实	色素
其他酚类	1500	广泛存在	
简单酚类	200	通常在叶,也在其他组织	抗菌
黄酮类(含单宁)	8000	通常在被子植物、裸子植物、蕨类	
醌类	800	普遍存在,尤其在鼠李科	色素
聚乙酸酯			
聚乙炔	750	主要在菊科和伞形科	部分有毒

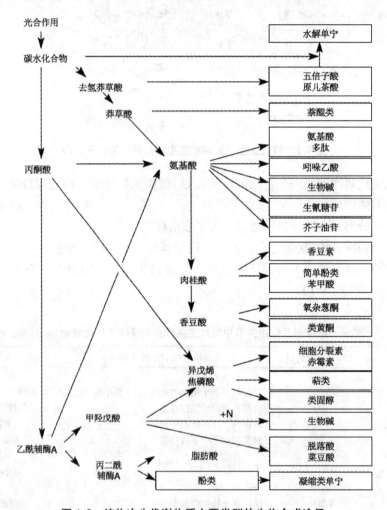

图 4.2　植物次生代谢物质主要类群的生物合成途径

4.2　生物碱

生物碱是在一些生命体中有限分布的环状含氮化合物。它包含了大量通常在结构上不相关的化学物质。生物碱通常以前体分子区分。大多数来源于常见的氨基酸,例如赖氨酸、酪氨酸、色氨酸、组氨酸和鸟氨酸[89]。如烟碱,是由鸟氨酸和烟酸结合产生的。其中最具代表性的苄基异喹啉类生物碱是罂粟碱、小檗碱,还有吗啡。大部分箭毒碱也属于这一类,包括筒箭毒碱。很多生物碱尤其是茄科特有的,都属于托烷生物碱一类。如存在于颠茄中的阿托品和东莨菪碱。可卡因以及其他存在于古柯植物(Erythroxylon coca)中的相关生物碱都是同一类型,却不存在于茄科。所谓的吲哚生物碱是因其具吲哚环。大家熟知的两种化合物,士的宁和奎宁,味苦且对许多昆虫都具有强大威慑力,也属于生物碱类。吡咯生物碱是酯类生物碱,在千里光属植物中,研究人员对它们的生物合成研究最为广泛。千里光宁碱是一种有毒的大环酯。哌啶类生物碱中赖氨酸的衍生物,常被称为羽扇豆碱,因为它们通常存在于羽扇豆属植物中。多羟基生物碱近来被认为是仿造糖化合物,从而干扰糖苷酶。它们是各种昆虫的摄食阻碍剂[92]。一些其他的生物碱来自烟酸、嘌呤、邻氨基苯甲酸、聚丙烯酸酯和萜烯,也包括嘌呤生物碱,例如咖啡因[13]。

将近20%的被子植物都能产生生物碱。而在裸子植物(如松柏科植物)或隐花植物(如蕨类)中生物碱则很少见。对多数昆虫来说,当生物碱的药液浓度超过0.1% w/w时就会引起虫体的摄食抑制和/或毒杀作用。

4.3 萜类化合物和类固醇

萜类化合物是最大的次生代谢物质类群(目前大概有30 000种),呈现出了超乎想象的结构多样性,这类化合物通过两种途径实现生物合成:一种途径是甲羟戊酸在细胞液中合成倍半萜烯、三萜烯、固醇类和复萜类;另一种途径是最近才发现的,即右旋脱氧木酮糖合成异戊二烯、单萜、双萜和类胡萝卜素。大部分萜类化合物被认为是以异戊二烯为单位组成的(图4.1),异戊二烯以不同的方式和类型环化,它们在色饱和度和官能团上各有差异。许多植物都能释放出大量的异戊二烯,尤其是在高温条件下。异戊二烯的含量可作为萜类化合物的分类依据(表4.2)。

表4.2 萜类化合物的主要类别

萜类物质(分子式)	植物来源	主要类型
半萜类(C_5H_8)	精油	郁金香甙
单萜类($C_{10}H_{16}$)	精油	环烯醚萜
倍半萜类化合物($C_{15}H_{24}$)	精油、树脂	倍半萜烯内酯
双萜类($C_{20}H_{32}$)	树脂、苦味物质提取物	二萜、赤霉素
三萜系化合物($C_{30}H_{48}$)	树脂、橡胶、软木、表皮素	甾醇、强心甙(cardenolides)、植物性蜕皮甾类、葫芦素、皂甙
四萜($C_{40}H_{64}$)	色素	胡萝卜素、叶黄素
多萜[$(C_5H_8)_n$]	乳液	古塔胶(滴状物)橡胶、树胶

此表是利用现有知识对已知次生代谢物质所进行的粗略估计,然而随着新物质不断被发现,此表的数据需不断更新。

大部分单萜类化合物是挥发性化合物,是精油的主要组成成分。它们广泛存在于各种被子植物和裸子植物,可以用它们制成具有独特气味的纸巾。单萜类化合物可以是非环状的(即开环的),如醇;也可以是单环的,如柠檬烯;或者是二环的,如松萜。为了阻止自毒作用,植物内部需要专门的存储物质以便隔离单萜类化合物。

萜类中最大的一类包括倍半萜，它们通常也存在于精油中。补身烷(drimane-type)中较常见的是全草含水蓼二醛(polygodial)和活乐木醛类(warburganal)，对多种昆虫的摄食行为有显著的抑制作用。倍半萜烯内酯有五个内酯环，如灰青斑鸠菊内酯A。它们经常被发现存在于菊科植物的毛腺里或乳腺管。棉酚是从棉花(棉属)和与其相关的锦葵科中发现的一类常见的酚醛树脂倍半萜烯二聚物。单萜烯和倍半萜烯这种烃类的气味对人来说较不敏感，却对昆虫起到非常重要的诱导作用。

双萜类化合物包括松柏类的树脂酸(例如松香酸)和二萜，比如从印度的一种树中提取的大青素和从筋骨草(*Ajuga remota*)叶片中提取的筋骨草素二萜。二萜是许多昆虫摄食行为的有效抑制剂。

三萜类化合物分布广、种类多，存在于树脂、角质层和木栓层中。它们包含柠檬苦素类似物(其中包括强效抑制昆虫摄食的印楝素)、总岩茨烯和葫芦素类(如葫芦素)。对人类来说后者的混合物尝起来有强烈的苦味，它们可以阻止植食昆虫的摄食。另一方面，一些专生在葫芦上的昆虫也可以利用葫芦素作为识别宿主的有力线索。

皂苷包含一个多环的糖苷配基并在三萜类化合物(C_{30})或类固醇(C_{27})的结构之间附加一个糖基。三萜类化合物和类固醇的代表为七叶皂苷和薯蓣皂苷，它们分别存在于七叶树(七叶树属 *Aesculus*)和山药(薯蓣属的所有种)中。现已在80个科的植物中检测皂苷，并它们大部分可以干扰昆虫的生长和发育。昆虫无法自身合成大量的甾族化合物，因此必须在食物中获得胆固醇或谷甾醇来合成甾类激素，例如蜕皮激素。5%~6%的植物种能产生蜕皮激素和一种被称作植物蜕皮甾醇的类似衍生物(如图4.1)。特别是一些蕨类植物和裸子植物，当中的蜕皮激素或其衍生物浓度可能比昆虫中的浓度多5个数量级。蕨类植物的根状茎(欧亚多足蕨)普遍都含有蜕皮激素，例如，苍白秤钩风(*Diploclisia glaucescens*，防己科 Menispermaceae)的干燥茎含有1%以上的β-蜕皮激素(β-ecdysone，即重要的昆虫蜕化类固醇 ecdysteroid)和3.2%左右的植物蜕皮类固醇。由于蜕皮激素在植物中真正的生理学作用是未知的，它被人们视为植物抵抗植食性动物的基本防御机制。实验证明，到目前为止这个设想是缺乏根据的。

一些化合物属于萜类化合物，在结构上却增加或失去若干个碳原子，例如在高等植物中作为荷尔蒙制剂的赤霉素，在种子油中充当抗氧化剂的生育酚(如维生素E)和大麻的有效成分(如大麻二酚)。

萜类化合物常作为复杂混合物存在于植物。例如，胡萝卜叶中的精油可能包含30多种不同的萜类化合物，从挪威云杉(欧洲云杉)茎内的树脂中已经提取出了23种萜类化合物。

4.4 酚类化合物(酚醛树脂)

酚类化合物普遍存在于植物[120,285]。它们含有一个或多个羟基的芳香环，以及其他组成成分。这类化合物的母体物质是酚——一种简单的芳香剂，但是大部分都包含一个以上的羟基(多酚类)。酚类化合物可直接按照基本骨架上的碳原子数量进行种类划分(表4.3)。

第4章 植物化学：无穷无尽的变化

表4.3 植物中酚醛树脂的主要类别（哈本，1994）[120]

C 原子的数目		种类	例子
C_6	6	简单酚类	
苯醌		邻苯二酚、氢醌	
$C_6 - C_1$	7	酚酸	2,6 - 二甲氧基苯醌
$C_6 - C_2$	8	苯乙酮	
苯乙酸氨基酸		溴代水杨酸	
		羟基酸	咖啡酸，阿魏酸
		苯丙烯类	肉豆蔻醚，丁香酚
$C_6 - C_3$	9	香豆素	伞形花内酯，秦皮乙素
		异香豆素类（异）	岩白菜素
		色酮	番樱桃素
$C_6 - C_4$	10	萘醌类	胡桃，白花丹素
$C_6 - C_1 - C_6$	13	氧杂蒽酮	杧果苷
$C_6 - C_2 - C_6$	14	对称二苯代乙烯	
蒽醌类		半月苔酸（Lunularic）酸	
		大黄素	
$C_6 - C_3 - C_6$	15	黄酮类	槲皮素，锦葵花甙（malvin）
		异黄酮类	染料木黄酮
$(C_6 - C_3)_2$	18	木脂素类	足叶草毒素
$(C_6 - C_3 - C_6)_2$	30	黄酮	穗花双黄酮
$(C_6 - C_3)_n$	9n	木质素	
$(C_6)_n$	6n	邻苯二酚黑色素	
$(C_6 - C_3 - C_6)_n$	15n	缩合单宁（Flavolans）	

相对简单的酚醛树脂包括水杨酸（如香草酸）、羟肉桂酸（如咖啡酸）和香豆素。香豆素属于伞形花内酯类，普遍存在于伞形科、东莨菪、茄属和一些其他属。

迄今为止，种类和数量最多的植物酚醛树脂是类黄酮，它普遍存在于高等植物。因此，几乎每个植食性动物都能在摄食时摄取到这种次生代谢物。一种植物通常有几种这类化合物，而且几乎每种植物有特有的类黄酮结构。类黄酮都有一个基础结构 $C_6 - C_3 - C_6$（图4.1，如山奈酚）。类黄酮的中心通常会连接一个糖基，进而形成一个水溶性苷。大部分类黄酮贮存在植物细胞的液泡中。类黄酮可再细分成黄酮（如毛地黄黄酮）、黄烷酮类（如柚苷配基）、黄酮醇（如山奈酚）、花青素和查耳酮类。许多黄酮、黄烷酮类以及黄酮醇可吸收可见光，因此它们能使植物的花和其他部分呈现嫩黄色或乳白色。这一类别中的许多无色化合物作为摄食抑制剂（如儿茶酸）或昆虫毒素（如天然的杀虫剂鱼藤酮）有着重要的意义。

例如，曾有记载表明，菜豆素（Phaseolin）是有效的摄食抑制剂。在对食根类甲虫新西兰金龟子（Costelytra zealandica）的幼虫进行实验时，这种化合物的 FD50（在喂养时将浓度降到控制值的50%）降低到百万分之 0.03[168]。另一方面，有几种类黄酮已被证实用于单食性或寡食性昆虫识别寄主植物和刺激摄食（表7.4）等研究领域[240,273]。

花青素含有大量天然的红色和蓝色色素，一般存在于植物的花、果实和叶片[113]。它们含有以葡萄糖为最常见糖配基的苷类。花色素苷是花青色素葡萄糖内酯。

丹宁酸为多酚化合物（分子量为 500～20 000 道尔顿），几乎存在于所有种类的维管植物，并且存在浓度较高（表4.4）。它们通常以可溶性化合物的形式存在于树液的活细胞。丹宁酸的酚羟基几乎与所有可溶性蛋白质结合起来，产生不易溶解的高分子聚合物。在这种情况下复合作用生成的酶活性明显降低。此外,与丹宁酸结合的蛋白质不能被消化道的酶分解。因此,普遍认为丹宁酸能降低植物组织的营养价值(第 5 章)。丹宁酸还能与核酸以及多糖交叉连接在一起,从而阻碍它们的生理功能。

表 4.4 植物中一些次级代谢化合物的浓度

化合物	类别	植物种类	浓度(% 干重)	参考文献
长春新碱(Vincristine)	生物碱	长春花(Catharanthus roseus)(叶片)	0.000 2	255
黑芥子硫苷酸(Sinigrin)	芥子甙	甘蓝(Brassica oleracea)(叶片)	0.03~0.3	53
洋地黄毒苷(Digitoxin)	强心甾	毛地黄(Digitalis purpurea)(叶片)	0.06	188
香柑内酯(Bergaptan)(及其他)	香豆素	防风草(Pastinaca sativa)(叶片)	0.1	297
马兜铃酸(Aristolochic acid)	生物碱	马兜铃(Aristolochia philippinensis)(叶片)	0.1	185
秋水仙碱(Colchicine)	生物碱	百合(Merendera montana)(叶片)	0.1	109
芸苔葡糖硫苷(Glucobrassicin)	芥子甙	甘蓝(Brassica oleracea)(叶片)	0.3~3	53
金丝桃素(Hypericin)	醌	毛金丝桃(Hypericum hirsutum)(花序)	0.3	214
金丝桃素(Hypericin)	醌	贯叶连翘(Hypericum perforatum)(叶片)	1.4	243
桃叶珊瑚甙(Aucubin)	环烯醚萜苷	长叶车前(Plantago lanceolata)(叶片)	0.4	180
番茄苷(Tomatine)	糖苷生物碱	番茄(Lycopersicum esculentum)(叶片)	0.5~5.1	242
喹唑啉生物碱(Quinolizidine alkaloid)	生物碱	灌木羽扁豆(Lupinus arboreus)(叶片)	0.8~2	5
喹唑啉生物碱(Quinolizidine alkaloid)	生物碱	灌木羽扁豆(Lupinus arboreus)(种子)	2~14	5
吡啶生物碱(Pyrrolizidine alkaloid)	生物碱	千里光(Senecio jacobaea)(叶片)	1.6~5.5	178
氰(Cyanogen)	氰	Ryparosa sp. nov.(种子)	1	287
尼古丁(Nicotine)	生物碱	烟草(Nicotiana tabacum)(整个植株)	1~6	244
苦杏仁甙(Amygdalin)	氰苷	樱桃巴旦杏(Prunus amygdalus)(种子)	3~5	98
环烯萜类(Ligustaloside)	多元酚	女贞(Ligustrum vulgare)(叶片)	4.1	222
小檗碱(Berberine)	生物碱	欧洲小檗(Berberis vulgaris)(树皮)	5	232
梓甙(Catalposide)	环烯醚萜苷	美国木豆树(Catalpa bignonioides)(叶片)	5.3	33
丹宁酸(Tannins)	多元酚	英国栎(Quercus robur)(叶片)	0.6~6	91
丹宁酸(Tannins)	多元酚	Englerina woodfordioides(叶片)	15	284
丹宁酸(Tannins)	多元酚	糖槭(Acer saccharum)(叶片)	30	208
丹宁酸(Tannins)	多元酚	茶树(Thea sinensis)(叶片)	<30	279
L-二羟基苯丙氨酸 L-Dopa	氨基酸	血藤属(Mucuna)(种子)	5~10	20
松醇(Pinitol)	糖醇	冰叶日中花(Mesembryanthemum)日中花(Crystallinum)(叶片)	10	209
特里杨甙(Tremulacin)(及其他)	酚苷	三角叶杨(Populus trichocarpa)(叶片)	23	21 259
树脂(Resin)	酚醛苷	(Mimulus aurantiacus)(叶片)	>30	122
树脂(Resin)	酚醛苷	(Larrea cuneifolia)(嫩叶)	44	218

丹宁酸通常被划分为两类:易水解丹宁酸和不易水解丹宁酸(或缩合丹宁酸)。易水解丹宁酸在加热条件下遇稀酸水解。它们仅存在于被子植物。相反,缩合丹宁酸则广泛存在于植物界(表 4.5)。最常见的水解丹宁酸是由没食子酸(gallic acid)和六羟二酚酸(hexahydroxydiphenic acid)以及糖一起形成的酯。缩合丹宁酸是类黄酮聚合物(图 4.1),由碳键连接,不易被水解。在植物组织成熟期间,如水果成熟时,丹宁酸通常会进一步聚合而变得难以溶解,最终降低果实的涩味。

表 4.5　水解丹宁酸和缩合丹宁酸在植物中的分布(斯维因,1979)[256]

分类	含有丹宁酸植物种类的比重	
	水解的	缩合的
松叶蕨纲(原始的蕨类植物)	0	0
石松纲(石松)	0	0
楔叶蕨纲(楔叶类)	0	28
蕨纲(蕨类植物)	0	92
裸子植物	0	74
被子植物	13	54
单子叶亚纲	0	29
双子叶植物	18	62

4.5　芥子油苷

芥子油苷是一个体积虽小但具有明确定义的化合物类群[90,191]。它们的基本结构已在图 4.1 中给出。所有芥子油甙或芥末油糖苷都包括硫黄和氮原子。它们可以是无环的,比如黑芥子硫苷酸钾(sinigrin)(或异硫氰酸盐 allyl isothiocyanate),也可以是芳香族的,比如白芥子硫苷(sinalbin)。芥子酶(myrosinase)促进了芥子油甙的水解,在这一过程中会根据酸碱度和其他环境差异产生异硫氰酸盐(或芥子油)、腈类或其他混合物。在植物组织破裂时能迅速水解,但在正常代谢时,发生的水解速度很慢。芥子油甙主要存在于十字花科植物,但这并不是绝对的。自维夏菲尔特用黑芥子硫苷酸钾和甘蓝白粉蝶进行实验后,这类物质吸引了许多研究昆虫—植物关系的学者的兴趣[191]。对于一般的昆虫和那些不以十字花科植物为食的昆虫来说,芥子油甙味道难吃且有毒[190,215]。对于许多专以十字花科植物为食的昆虫来说,芥子油苷是很好的饲料和产卵兴奋剂(图 4.3)[50]。

4.6　生氰类化合物

几乎所有植物都能合成生氰苷类(cyanogenic glycosides)化合物(基本分子式见图 4.1),但在大多数物种中,由于新陈代谢的作用,它们并不会积蓄在植物体内。仅有约 11% 的植物中含有数量可观的生氰类物质。例如桉树(Eucalyptus cladocalyx)叶片可能分配 15% 的氮元素用于合成一种基本的生氰苷类——洋李甙(prunasin)[40]。

由于碳原子羟基化的偏光力,生氰类化合物具有光学活性。由于发生在李属(Prunus spp.),因而被命名洋李甙(prunasin),是在接骨木属(Sambucus spp.)的接骨木甙中出现的典型立体异构物。植物细胞的液泡通常作为一个储藏库(如在多数蔷薇科植物中)。当植物组织被破坏时,生氰类化合物会发生酶促水解并形成含有剧毒的氰化氢(hydrogen cyanide,HCN)[108]。例如稠李(Prunus padus)的植物叶片被压碎时,其散发的苦杏仁味是很容易辨别的。除了作为还原态氮的贮藏所之外,在对抗植食性动物和病原体的侵袭时,生氰类所起的防护作用是很吸引人进行研究的[138]。这一事实证实 HCN 是不同种类昆虫的摄食抑制剂,这对植物可能起到保护的作用。

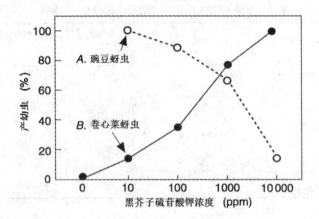

图 4.3 使用黑芥子硫苷酸钾(一种硫代葡萄糖苷)经系统处理过的蚕豆(*Vicia faba*)叶片上豌豆蚜虫(*Acyrthosi phonpisum*)和卷心菜蚜虫(*Brevicoryne brassicae*)的总量

实验对照物为未经处理的萝卜叶上卷心菜蚜虫的数量以及未经处理的蚕豆叶片上豌豆蚜虫的数量。黑芥子硫苷酸钾抑制不以十字花科植物为食的物种进行繁殖,相反却刺激专以十字花科植物为食的昆虫物种进行繁衍,即使这种化合物产生于非宿主植物时也不例外(纳特和斯泰尔,1972)[199]。

4.7 叶表面的化学物质

当昆虫落在叶面或碰到叶面时,它们和植物的第一次触碰便发生了。除了在第 3 章中讨论的物理特征外,植物表面的化学特征也会影响昆虫随后的行为(第 7 章)。因此,叶表面的化学物质值得引起我们的高度重视。

叶表面的蜡状物或树脂形成了植物的第一道防线。不同种类的植物,其角质层蜡质的结构和化学组成也各不相同。微观形态[186]也存在很大的差异,从无定形的薄膜状到蜡状管、条状、盘状等。化学组分包括长链碳氢化合物、脂类、醇类和脂肪酸等[88]。从小麦叶片中提取的蜡状物,可能会含有多达 50 种不同的成分[46],而覆盖在王紫萁(*Osmunda regalis*)叶状体上的蜡状物可能包含 139 种成分[141]。蜡状物成分变化非常大,不仅存在于同类物种,例如 8 种罂粟科(Papaveraceae)的蜡状物成分各有差别[140],而且同一物种的不同基因型可能也不相同[46]。

通常,蜡质涂层包括植物的初级和次级代谢物质,尽管通常含量较小。当完整的植株暂时浸泡在水或有机溶剂中时,很多化合物通常会被冲洗掉[249]。冲洗叶片表面的洗涤液中含有糖[67,173]、一些氨基酸[67,247]和植物次生代谢物质,比如根皮苷(phloridzin)(在苹果叶中的二氢查尔酮(dihydrochalcon)[161]),芸苔葡糖硫苷(glucobrassicin)(卷心菜的芸苔葡糖硫苷[272]),香豆素(furanocoumarins)[250]和生物碱(alkaloids)[140](表 7.3)。洗涤技术可以释放角质层内的蜡状物(即嵌入聚合的角质层内的蜡状物)和角质层蜡状物(即覆盖在角质层表面的薄膜)。以上两种蜡状物的化学成分有很大不同[143]。目前一种新技术的出现可以分离角质层蜡状物并进行化学成分分析[142]。用这种方法可能会更为准确地分析出昆虫与植物叶面发生触碰时,最先接触到的是哪些化学物质。

植物同其他生命有机体一样,存在着一定的动力系统(dynamic systems),因此叶片表面的化学物质随季节的变化而变化也是不足为奇的。例如胡萝卜(*Daucus carota*),其叶表面的黄酮类物质的浓度在从生长阶段向繁殖阶段转变期间达到最大值,这表明控制植物生长的因素能够影响化合物在表皮中的运输。

与叶表面化学物质的接触通常会阻碍昆虫对植物更进一步接触。例如,在一株完整植物上,飞蝗仅根据触觉进行的一次试咬,将有可能会抑制接下来的摄食活动。然而在叶片上的蜡状物被去掉时,昆虫会在决定拒绝非宿主植物之前进行一次或多次试咬(图4.4)。另一方面,许多叶表面的化学物质会帮助昆虫识别它们特定的宿主植物。一些甲虫类昆虫对食物的摄取会受它们各自宿主植物的蜡状物成分的刺激[3]。

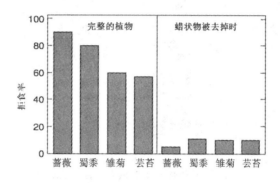

图4.4 飞蝗五龄幼虫对四种植物试尝后的拒食率(伍德黑德和查普曼,1986)[296]

腺状的(或分泌的,图4.5)毛状体附着在表皮上,主要在唇形科、茄科、菊科和牻牛儿科中存在。它们含有高度专化的分泌细胞,可以合成和累积大量不同的萜油和其他必需的油类。一些腺状毛状体从产生开始便能进行永久性分泌。还有些腺状毛状体有敏感的触觉机制,一旦受到破坏,它们便会释放有黏性的分泌物并困住且杀死小型昆虫(图4.6)[84,229]。通常较大的昆虫也会被分泌物阻止,或者使它们的摄食受阻碍,由此,种群的发展受到了限制[154]。

有关植物表面化学物质的研究在瞻博和索思伍德[146],查普曼和伯尼斯[48],斯莱特和罗西尼[249]以及德瑞杰[66]的综述中得到充分的论述。

4.8 植物的挥发物质

所有植物都能释放大量化学性质相异的挥发性碳氢化合物[82]。许多植物次级代谢物质和初级新陈代谢的中间产物会被释放到空气中,形成较高的蒸汽压,并会对其他有机物产生影响。许多分子量在100~200Da的萜类化合物、芳香类、酚类、醇类、醛类等均容易挥发,且当植物组织受损时会完全释放。完整的植株也会通过打开的气孔、叶表皮和腺体渗透释放这样的挥发性化合物,但是这样的释放效率很低。以往对植物挥发物质的鉴定主要通过粉碎或浸制的植物材料的抽取。在过去的30年中,"顶空提取"技术的发展实现了从未毁损(或毁损)的植物中获得挥发物质的可能。这项技术真正实现了植物释放到周围空气中的挥发物成分的分析,通过与气相色谱法相结合,这一技术比组织提取法更能提供关于自然排放混合物的可靠信息。例如,棉花植株的顶部空气中有54种化学物,但在棉花芽精油中存在的58种混合物中只含其中6种[129]。

植物挥发物可以分成普通挥发物和特殊挥发物。最普通的"绿叶挥发物"[280]是受损叶片释放的有割草气味(cut grass)的气体[158],大部分是六碳结构的饱和或不饱和的醇类和醛类,具

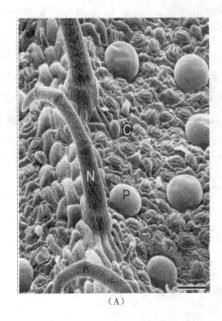 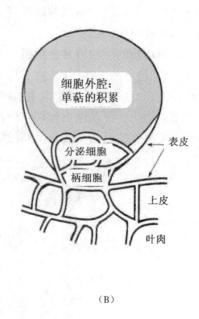

图4.5 叶片表面分泌单萜类化合物的腺毛

（A）薄荷（*Mentha piperita*）远轴叶表面的扫描电镜照片，P，盾状毛状体；C，头状的毛状体；N，非腺状的毛状体，比例尺为50um；（B）薄荷叶表面的腺状毛状体的横截面图解，展现的是由分泌细胞产生的单松烯类物质未被排到细胞外之前（赫尔申宗和克罗托，1991）

图4.6 腿上粘黏着胶块的桃蚜虫（*Myzus persicae*）被固定在叶片上

当野生土豆（*Solanum berthaultii*）叶片上的毛状体破裂时，会分泌一种透明液体，一旦接触空气，黏度便会增加，粘黏小型昆虫（斯密德，瓦格宁根大学，荷兰）

不同结构的同分异构体。许多这类物质的衍生物也被纳入这类名下（例如醋酸盐）（图4.7）。在叶脂质的氧化作用下，它们被大量生产[126]。不饱和的醇类和醛类，亚麻酸前体物质的质量常占叶子干重的1%以上。对于特定的植物来说，其释放的气味化合物的数量是特定的。

有些昆虫能够感知这种物种特异性，并用它们来区分寄主植物和非寄主植物。例如，科罗拉多马铃薯甲虫对马铃薯枝叶释放的混合挥发物反应强烈，但当自然混合物中的一种成分浓度提高，打破原有的平衡时，甲虫对混合挥发物的反应也会消失[280]。

大部分植物均能散发出具有其分类特征的挥发物，但到目前为止，对于这类挥发物的研究

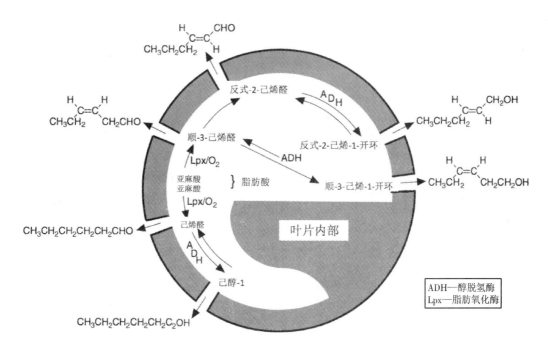

图 4.7　植物绿叶挥发物源自叶片内部的脂肪酸并且通过
气孔渗透到周围的空气中(维瑟和艾维，1978)

仅有少数的案例。许多昆虫能够对大部分植物挥发物做出反应，利用它们可以作为一种信号，从而在空中发现或者避开某些植物(第6章)。

在植物周围的空气中存在的挥发性物质可能有上百种，然而混合挥发物可能会由一种或几种主要成分组成(图4.8)。例如，玉米叶片周围的空气中至少含有24种挥发物质，但是主要部分(75%)只有7种重要物质。

如图4.8所示，对化合物的主要类别进行分析可以得出一些结论。在所有植物中，主要挥发物是绿叶挥发物(醛或酯)或者萜类化合物。当其被破坏时，这些主要成分的比例有时会上升(如大豆、茄子)，有时会略微上升(如卷心菜、拟南芥)，有时会大幅下降(如红豆、胡椒)。在所有植物中，由于螨虫或者毛虫破坏而引发释放出的挥发物质，并没有在无损伤的植物中被发现，即便发现了含量也是微乎其微的。另一个有趣的现象是，相同科的植物却有着明显不同的挥发性物质(如红豆和黄豆)。在茄子中检测到大量的 N–肟(N-oximes)，而在拟南芥中检测到两种大量的环庚烯。显而易见的是，在受到损伤时，植物的挥发曲线常常发生显著的变化(图4.8、图4.9)[262,282]。另外，由机械损伤引发的花香类挥发性物质与植食动物损伤引发的挥发性物质有着极大差别[261,263]。

从这些例子中可以看出，不同植物的挥发物之间往往存在很大的定性和定量差异，这种不同也表现在同种植物的不同栽培品系上。例如，在3种菊花栽培品系释放的43种化合物中，只有14种是这3个品种所共有的[251]。本章主要围绕植物营养体的挥发性物质展开讨论，但会顺带提及花的挥发性物质，其常由100多种化合物组成。

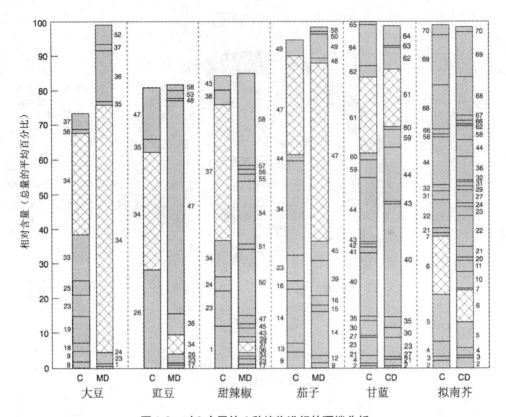

图4.8 对3个属的6种植物进行的顶端分析

醛:1.（E）-2-已烯醛;2.已烯醛;3.庚醛;4.辛烷;5.壬醛;6.癸醛;7.十一（烷）醛;8.（E）-4-氧-2-已烯醛;

含氮化合物:9. 2 - methylbutanenitrile;10. 5 -（甲硫基）丁基氰;11. 6,7 - dithiooctanenitrile;12. 2 - methylpropanal , O - methyloxime;13.（syn正）或（反）- 2 - methylbutanal,肟;14. 2 - methylbutanal , O - methyloxime;15. 3 - methylbutanal , O - methyloxime;16.苯乙醛 , O - methyloxime（暂定的）酮类;17. 2 - 丁酮;18. 4 - 甲基 - 3 - 五亚乙基六胺 - 2 - 一;19. 4 - 羟基 - 4 - 甲基 - 2 - 戊酮;20. 1 - 五亚乙基六胺 - 3 - 一;21. 3 - 戊酮;22. 4 - 甲基 - 3 - 庚酮;

醇类:23.（Z）- 2 - 己醛 - 1 - 开环;24. 1 - 己醇;25. 2,4 - 戊二醇,2 - 甲基;26. 1 - octen - 3 - 开环;27. 1 - 五亚乙基六胺 - 3 - 开环;28. 2 - 五亚乙基六胺 - 1 - 开环;29. 1 - 戊醇;30. 2 - 乙烷基 - 1 - 己醇;31. 1 - 壬醇;32. 1 - 十二烷醇

羧酸:33.己酸;

酯类:34.（Z）- 3 - hexen - 1 - 开环,醋酸盐;35.已基醋酸盐;36.甲基水杨酸盐;37.（Z）- 3 - hexen - 1 - 开环,butanoate;38.已基 butanoate;39.（Z）- 3 - hexen - 1 - 开环,2 - methylbutanoate;40.（Z）- 3 - hexeyl 酸;

硫化物:41. dimethyldisulphide;42. dimethyltrisulphide;

萜类化合物:43.沉香醇;44.柠檬烯;45(Z) - β - 罗勒烯;46. α - 松萜;47.（3Z）- 4,8 - 二甲基 - 1,3,7 - nonatriene;48.（3E）- 4,8 - - 二甲基 - 1,3,7 - nonatriene;49. α - bergamotene;50.（E）- β - 榄香烯;51.（Z）- β - 榄香烯;52.（E,E）- α - 金合欢烯;53.（E）- β - 金合欢烯;54.吉马烷A;55. α - 芹子烯;56. β - 芹子烯;57.（3E,7Z）或（3Z,7Z）- 4,8 - trimrthyl - 1,3,7 - 十三烷;58.（3E,7E）- 4,8,12 - trimrthyl - 1,3,7,11 - 十三烷;59. α - 侧柏烯;60. β - 松萜;61.桧烯;62.月桂烯;63. β - 水芹烯;64. 1,8 - 桉树脑;65.（E）- 桧烯水合物;66.长叶烯;67. β - 紫罗酮;

其他:68. 6 - [（Z）- 1 - 丁烯基] - 1,4 - 环庚烷;69. 6 - 丁基 - 1,4 - 环庚烯;70.十四醇。

干净叶片(对照物,C)的顶端构成与被叶螨(二斑叶螨)(MD)或毛毛虫(菜青虫)(CD)损坏的叶片进行比较。超过挥发混合物总量0.5%的化合物在大豆(*Glycine max*)、豇豆(*Vigna unguiculata*,豆科,Fabaceae)、甜辣椒(*Capsicum annuum*)、茄子(*Solanum melalonga*)、甘蓝(*Brassica oleracea*)和拟南芥(*Arabidopsis thaliana*)中均有所体现。在未受损的对照植物的挥发性混合物的主要成分都被用交叉图像表示,以强调未受损植物与受植食动物损害的植物之间的差别(旺当等,2004[271];布莱克模等,1994[26];万·波克等,2001[276])

昆虫对植物造成的破坏,可以大大地促进植物挥发物的释放。由植食性动物引发的植物挥发物比未被破坏的完整植株释放的挥发物多出近2.5倍[282]。值得注意的是,萜类化合物通常在损坏植株释放的混合物中占主要地位(图4.8)。两个无环亚甲基萜烯,E-4,8-二甲基-1,3,7-壬三烯(18)和4,8,12-三甲基-1,3(E),7(E),11-十三碳四烯值得引起注意,因为它们常被发现于虫害植物的顶端[30]。这些化合物的数量随着植食性昆虫物种的不同而发生变化。所以,在受六点黄蜘蛛(Panonychus suomi)侵扰的苹果叶顶端存在49%的4,8-二甲基-1,3(E),7-壬三烯,而在受到另一种蜘蛛(Tetranychus urticae)侵扰时,以上化合物的含量只有9%。更有意思的是,这些差异足以吸引不同种类的蜘蛛。显然这些捕食者能对这些气味做出反应。

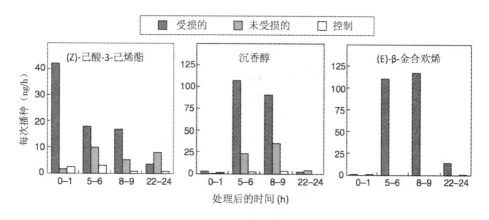

图4.9　在不同时期玉米幼苗受不同损伤后释放的挥发物数量

玉米幼苗经受人为损伤后又受到毛毛虫咀嚼(目的是模仿食草行为)处理,得出的不同时期释放出的叶片挥发性物质中三种主要成分的数量尽管有所延迟,但是受损植株上未受损叶片也系统地释放出一些挥发性物质,以未受损害的植株释放的挥发性物质作为对应参照物,注意由损害引起挥发的气味混合物的构成随时间变化而变化

有些化合物始终保留在植物体内,必要时候还可以进行循环再造,与之相反,被释放到空气中的化学物质将会导致永久性的能量损耗。植物在合成化合物的过程中有目的地或不可避免地释放能量,这与生成的化合物的质量和种类有直接关系。现有的有限数据表明,在某些情况下,这种生产成本是不容忽视的。例如,在伯乐树(Bretschneidera sinensis)生产过程中产生的异硫氰酸酯的挥发增长量可达到每天0.7%(表示为总净重的增长比例)[32]。

木本植物和蕨类植物会排放大量碳氢化合物,特别是异戊二烯和单萜烯类。在光照下,异戊二烯的释放会损失最新通过光合作用合成的碳能量,直接损失估计在0.5%和2%之间,这个数值在高温条件下将会更高[123]。

4.9　植物次生代谢物浓度

在植物中发现的次生代谢化合物的相对含量不仅明显不同(表4.4),而且它们极不均匀地分布在各种植物的各个部分。后者与植食性昆虫的关联非常密切,因为它们往往以特定的细胞或组织为食(参考第2章)。从植物角度来看,这似乎是合乎逻辑的,当次生代谢物质发挥保护作用时[95,96]，其防御性化学物质会被分配到被昆虫损害程度最大的地方[270]。不同的植物会有不同水平的保护剂，例如，植物种子受到的损害对植物的影响程度远大于旧叶受到破坏的影响程度。按照这个概念，野生欧洲防风草(*Pastinaca sativa*)的果实避难的防护能力是其叶子的4倍，是其根部的800倍[297]。同样，毛金丝桃(*Hypericum hirsutum*)的花中含有的金丝桃素是其叶片的6~11倍，同时幼叶及其他正在生长的植株部分比成熟的组织更能受到来自次生代谢化合物的保护[214]。因此，红花琉璃草(*Cynoglossum officinale*)最嫩的叶子中含有的吡咯生物碱是其老叶的53倍(图4.10)[268]。

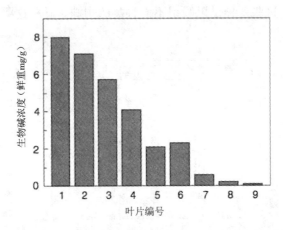

图4.10　红花琉璃草(*Cynoglossum officinale*)每片叶子中生物碱的浓度
(1号显示的是幼叶部位)(万·戴姆等,1994)[268]

即使是同一棵植物的同一枝干，其保护水平也不同。例如，由多酚含量不同导致一些寄生毛虫在林冠中进行的长途觅食旅行只为取食庇荫部分的叶片。很明显，很多森林昆虫像鳞翅目、叶蜂幼虫都很小心地选择它们的食物[100,234]。

甚至在同一片叶子不同的位置上保护性化学物质的浓度也有所不同，以白杨叶片为例，保护性化学物质的含量从叶子根部到顶端呈递增趋势(图4.11)。因此，寄生瘿蚋不会随意寄生在整个叶子上，而是集中在酚类物质浓度最低的叶子根部[298]。同样，烟草植物叶子中的尼古丁含量通常是从基部到叶的顶端部分呈2~3倍的增长。烟草天蛾(*Manduca sexta*)喜欢在尼古丁含量低的叶片部位上寄生生成虫瘿；与之相反，番茄天蛾(*M. quinquemaculata*)则喜欢尼古丁含量高的叶片部位[156]。

在植物器官的不同成分中明显体现出次生代谢化合物含量的不同。可以说明这一点的例子是光亮可乐果(*Cola nitida*)。豆荚外皮的咖啡因含量是微量的，其占种子外皮含量的0.44%(干重)，却占种子含量的1.85%[204]。不同的叶片部分，还有不同的叶片组织，通常在具抵抗性的化学物质上表现出很大的差异。因为当新生的幼虫开始摄食时，通常先接触的是表皮细胞，这对于一种植物来说，将抵抗性化学物质集中在表皮似乎是一个好策略。在禾本科植物中，生氰糖苷集中在表皮细胞中代表着表皮组织中的可溶性碳水化合物含量的90%[211]。美洲冬青(*Ilex opaca*)叶肉的栅栏组织中含有38%(干重)的皂苷，但在叶片的其余部分中，皂苷的平均含量只有1.3%。尽管栅栏组织内含有大量的皂苷，它们在植食性动物的体内充当着蛋白酶抑制剂的作用，但植潜蝇(*Phytomyza ilicicola*)的幼虫很专一地寄居在这一特定组织内，这一组织中蛋白质浓度是剩余叶片组织蛋白质浓度的10倍[157]。相反，一些寄居在橡树上的潜

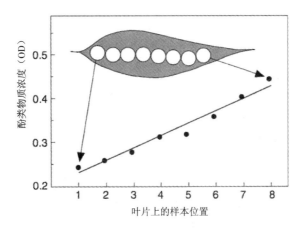

图 4.11 叶片上的位置与酚类物质浓度的关系

在窄叶杨(*Populus angustifolia*)一张叶片上的 8 个子样品中酚类物质的浓度(以光密度 OD 表示),瘿蚜虫更喜欢将虫瘿建在叶片根部,因为那个部位的酚类物质浓度最低(惠瑟姆,宋克,1982)[298]

叶蝇幼虫更喜欢取食叶片的海绵组织而避开叶片的栅栏层,因为后者的丹宁酸含量很高[91]。

在许多情况下,植物次生代谢物质会在一些组织内积累,但是其产生过程是在其他组织中完成的。一些生物碱和其他化合物,例如在烟草中的尼古丁,是在根部合成的,并且通过木质部运输到叶子[15]。生物碱通常在新生植物组织中含量较高。

当然,植物个体间的次生代谢化合物的数量也有所差别。这种差异是相当明显的,不仅体现在数量上,而且体现在质量上。对在哥斯达黎加地区几公顷的金合欢(*Acacia farnesiana*)叶片进行研究发现,个体间的生氰糖苷含量的差别可以达到 20 倍,而黄酮类化合物在种类和质量方面保持稳定[238]。这种变化可能具有重要的生态意义,尽管直到最近生物学家才意识到植食动物对同属物种不同的植物个体会有不同的反应[150]。

考虑到一株植物所含的精油可包含大量成分,生物合成途径的多样性大概要追溯到非常古老的生命形式[45,223],植物实现这一特性是很难的。显然,植物次级化合物的多样性取决于多种植物普遍含有的化合物,而不仅是一个或几个特征物质。例如,新的茶树嫩芽包含 24 种以上的酚类复合物[230],而生产精油和树脂的植物叶片上,通常有 30~40 种萜类化合物并且每种成分的浓度不低于 1%[169]。

从地理区域层面研究植物的多样性时,一些研究表明,次生化合物的数量和多样性随纬度变化存在一个明显的梯度变化。约 16% 的温带植物物种含有生物碱,而在热带植物中这一比例则超过 35%。热带树木成熟叶片的丹宁酸浓度要高出三倍,尽管简单的酚类物质似乎不随着纬度变化。对新老叶片中次生代谢化合物的浓度进行比较,会呈现显著的差别。热带树木的新叶比老叶含有更多的酚类、萜烯类和生物碱类物质,然而以温带植物为例,新叶中浓缩的丹宁酸含量仅是老叶的一半[86]。

以上均说明,植物除了具有结构性和时间性的维度特征外,还具有另一个维度特征——化学构成。

4.10 生产成本

植物次生代谢物质的产生需要物质和能量。特别是在植物中已含数量相当可观的次生化合物时,对它们的合成和贮藏所需成本的测量十分困难。生理学家能够使用生化术语解释化学抵抗的消耗,例如,产生一定数量的次生代谢物所需的能量。表4.6显示的是通过这种方法得出的各组化合物,以表明次生化合物的生产成本比大多数初级代谢产物的生产成本要高一些。

萜类化合物的生产成本很高,因为在生产过程中涉及化学物质的大量消耗和酶的大量转换。一些单萜烯化合物合成需要9步,而环烯醚萜苷的形成则需要23个步骤[103]。当然,合成这些具有抗性的化学物质的总成本不仅取决于合成的成本(相对较小),还取决于植物中现有化学物质的实际数量、它们的周转率、运输和储存成本,以及避免自体中毒的消耗。在"处理"过程中的成本消耗也是值得重视的。表4.6显示的数值与自然条件下在植物粘猴花(*Diplacus aurantiacus*)中测得的数据相吻合。在该物种叶表层中,酚醛树脂的数量和增长率之间呈负相关。每生产出1克树脂,地上部分生物量的干重便会减少2.1克。因此,树脂生产比植株生长要消耗更多的成本[117]。

生态学家可以根据防御化学物质所需的能量确定生物维持自身生长能量的消耗量,他们重点关注由于植食性动物取食植株特定器官或组织而释放的次生代谢化合物消耗的能量对植物生长发育的影响。这种方法被应用于研究号角树(*Cecropia peltata*)幼树的生长,这种树常生长于新热带地区。研究发现植株生长率与叶片中次生代谢化合物(丹宁酸)的浓度成反比(图4.12)。

表4.6 各种初级和次级植物代谢化合成分的消耗
(以每克化合物中葡萄糖的克数为单位计算)

	消耗
初级化合物	
碳水化合物	1.07
有机酸	0.73
脂类	3.10
核苷酸	1.59
氨基酸	2.09
次生化合物	
萜类化合物	3.18
酚醛树脂	2.11
生物碱	3.24
其他含氮次生化合物	2.27

此表给出了每组化合物中的平均值,更多详细的信息,请参见赫尔申宗(1994b)[104],上表就以其为基础制作而成。

研究表明,丹宁酸对植食动物具有阻抗力。在标准条件下生长18个月且丹宁酸含量不同的植株,被放置于一个大森林里,经受十天自然发生的植食动物的侵害,结果表明丹宁酸含量低的植物受到损害的程度明显高于丹宁酸含量高的植物。通常产生丹宁酸的成本是相当大的。在伞树科(Cecropia)植物中,实现丹宁酸含量从1%到5%的增长比实现叶片数量同比增长要多消耗30%的能量[55]。同样地,在白桦树(Betula pendula)中证实了树的高度和叶片的黄酮含量之间呈负相关的关系,这表明抗植食动物物质的产生是以植物的增长为代价的。

我们知道绝大多数情况下植物次生代谢物不仅具有单一功能。例如,号角树(Cecropia peltata)叶中的丹宁酸含量增加会减少病虫害,这可能仅是有利影响之一。这可以被合理解释为投入和产出要实现平衡,多成分构成的投入也会导致多成分的产出。

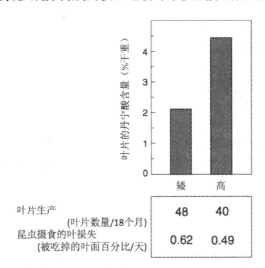

图4.12　丹宁酸浓度不同时,号角树(Cecropia peltata)幼苗的叶片产量和昆虫摄食损害程度的比较(科利,1986)[55]

一些研究结果表明,植物次生代谢物质的制造成本较高[104,281],尽管也有其他研究证实这些化学阻抗剂的消耗作用很小甚至没有[253,269]。这种矛盾的结论可能缘于不同的衡量标准。例如,在植物的生物量增长被用来作为参数时,大麦(Hordeum vulgare)生产出抵抗绿色小虫的物质的消耗就大,但在植株顶端的叶原基数量作为参数时,结论恰恰相反[44]。很显然,想要充分理解植食动物阻抗剂的消耗机制,就需要以更多的实验数据和理论发展为前提[241]。

值得注意的是,植物的生长通常受限于有效氮而非能量(光合作用、葡萄糖)。在这种情况下,含氮化合物的产生,例如生物碱,是以植物的生长和繁殖(图4.13)为代价的。这便可以解释为什么叶片中生物碱的含量比其他化合物(如酚醛树脂)要低。化学物质的防护程度是决定这种物质产量的生态因素。换句话说,通常情况下生物碱的毒性比酚醛树脂强,而自然生物碱的浓度相对较低。

最近开展的33项以植食动物阻抗剂的消耗成本为对象的研究显示,在大多数情形下,需要消耗很高的成本[253]。然而,我们应该意识到,对生产化学阻抗剂的投入不能与其他投入完全分离。在对植物其他功能的投入中,一些投资可以增强化学防御系统的功效,而有些投入则可能会减弱其功效[128]。

赫尔申宗[104]和西姆斯[241]对这个生产消耗的问题已经进行了较为详细地讨论。

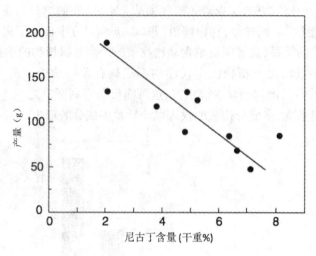

图4.13 8种本地或原生烟草(*Nicotiana tabacum*)以及2种移植的烟草栽培种的叶片数量和尼古丁含量之间的关系($r = -0.86$)(范登堡和玛卿格,1970)[277]

4.11 区室作用

一种植物能以化学阻抗剂为抵制策略来阻止植食性昆虫或病原体侵入,然而随之产生的问题是,任何能够有效抵抗各种侵害有机体的化学物质都具有毒性,极易引发自毒。这个问题可以通过以下两种方式解决:

(1)植物不会积累高毒性的化合物,却会储存毒性较低的前体物质,这些前体物质仅在必要的时候才会转化为有毒物质,例如受到植食性昆虫侵害时。

(2)有毒化学物质被储存在细胞隔间(即细胞壁和液泡),远离新陈代谢过程。

这两种机制普遍存在,彼此之间往往存在一定的联系。液泡中有毒化合物的浓度非常高。例如,很多白屈菜(*Chelidonium majus*)的细胞液泡中,黄连素生物碱的含量超过0.25M。利用生理机制将化合物从被合成的场所转移至专门的储存场所,需要特定的膜载体使化合物在储存器官中不断积累并防止泄漏到其他场所。一些执行这类传送工作的机制已经被一些研究阐释[294]。显而易见,这些过程需要消耗新陈代谢产生的能量。

与糖类的结合使许多化合物的毒性减弱,溶解度增加,因此大量的化合物能被储存在细胞液泡中。只有在叶片受损时,这些葡萄糖苷结合在一起后,在特殊降解酶的作用下产生毒素。高粱(*Sorghum bicolor*)的幼枝有超过30%(干重)的蜀黍苷,它是一种氰苷。大多数以较高含量被储存在表皮细胞的液泡中。叶肉细胞的叶绿体中含有β-葡萄糖苷酶,这些细胞的细胞液中有腈裂解酶。当叶片受损时,蜀黍苷与这两种酶结合后导致蜀黍苷快速降解并生成氢化氰(图4.14)[108]。众所周知,这种物质的毒性一般对多数生命体都有影响。同样,当白花草木樨(*Melilotus alba*)的叶片受到损害时,通过水解作用形成香豆素(图4.14)。在白花草木樨中,自毒作用被两层膜所阻碍:液泡(液泡膜)和原生质膜,这两层膜的作用就像一些挡板隔在酶作用物和酶之间。芥子油甙分布于细胞的液泡并存在于所有植物器官,然而,芥子酶则主要分布于分散的黑芥子硫苷酸细胞,这类细胞似乎不含芥子油甙。完好无缺的芥子油甙是没有毒性的,但是,一旦植物组织受到损伤,例如被昆虫咀嚼后,它们便与相邻细胞中的芥子酶结合。然后,芥子油甙被转换为芥末油,对许多种类的昆虫来说,这是一种强力驱虫剂和潜在毒剂(图4.14)[183,291]。

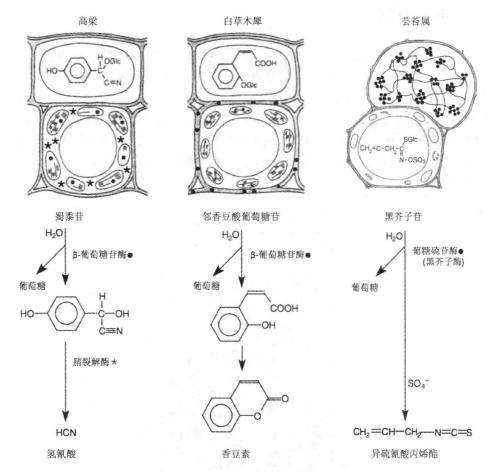

图4.14　高粱、白草木樨和山葵三种植物中有毒植物化合物的前体物质及其降解酶的区室作用

　　这三种植物中都有葡萄糖苷酶(●),但是它的分布集中于不同组织的细胞区室中,另一种酶——腈裂解酶(＊)出现在高粱的叶肉细胞中(玛蒂绘制,1984)[183]

另一种形式的区室作用的例证为:小分子量的萜类和其他具易挥发性的油类在腺毛和异形细胞中的累积,树脂在树脂管道内的累积,以及橡胶乳汁在乳汁细胞的累积[85]。分别属于40个科的20000余种被子植物,都可以产生乳胶,这是一种包含不同种类的橡胶、树脂、精油等物质的悬浮液或乳浊液[171]。像树脂管道这些特殊结构的存在完全是为了实现毒素的存储和排泄。

虽然许多植物次生代谢物质被储存在液泡中,但它们并不是具有惰性的最终产物,其并没有完全脱离植物的新陈代谢过程。相反,细胞液泡形成一个动态的环境,代谢产物通过细胞液可以重新进入细胞质。

强烈的昼夜变化对次生代谢化合物含量的影响也证实了代谢活动要受到内生因素(如发育阶段)和环境因素(如季节、气候、光照量)的共同控制。然而,近来的结果表明许多情况下生物碱和萜烯类的周转率比之前想象的要低得多[103,125,233]。在白杨幼苗中,酚苷和丹宁酸/酚却经历了显著的快速循环[160],然而在柳树幼苗中的水杨酸盐则没有发生类似的变化[227],由此可以证明人们对植物次生化合物的认知仍不够全面。

4.12 时间异质性

植物的化学特性并不是恒定不变的,也可能会出现广泛的时间异质性(图表4.15)[165]。成熟的植物在许多方面与幼年植物和衰老的植物组织不同,在整个生命阶段显示出定性和定量的变化。许多方面的变化与季节相关联,这在一年生植物的老化过程中表现得尤为明显,但是对于多年生植物来说,即使它们可以把生殖期与季节同步,并且表现出形态和生理方面的适应性,如叶片脱落和养分储存,其化学构成也在不断地改变。一些物理因素例如叶子的韧性[170]、表面粗糙度[186]、水分含量[91],随着年龄的增长表现出相应的变化,但在初级代谢产物以及次生代谢产物方面的变化往往更为显著。

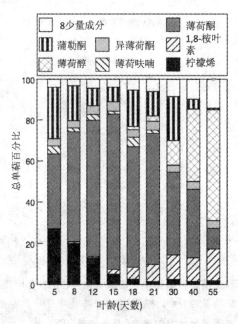

图4.15 薄荷叶片生长过程中单萜成分的变化(赫尔申宗等,2000)[106]

4.12.1 季节影响

季节变化引起的叶片中营养成分和化感物质含量的变化对于以该植物为食的昆虫来说是极为重要的。在一个经典的研究中菲尼发现,生活在橡树叶上的多种昆虫集中在早春时期取食,因为此时叶片的营养价值是最高的,并且叶片的营养价值会随着叶片成熟而不断下降[91]。在夏季,水和蛋白质含量会减少,同时丹宁酸会不断累积(图4.16)。最近的一项研究证实了菲尼的结论,研究发现在季节的初期和末期,动物种群的密度和多样性与含有高浓度单宁酸的叶片密度呈负相关[94]。

春天,在白杨(*Populus trichocarpa*)嫩叶中,酚醛树脂的含量会快速增加,浓度会达到相当高的水平(图4.17),这无疑会对具有攻击性的昆虫产生生理影响。相反,约翰·史密斯的结论却表明,在冬季,植物碱基数量比夏季高20倍以上[248]。

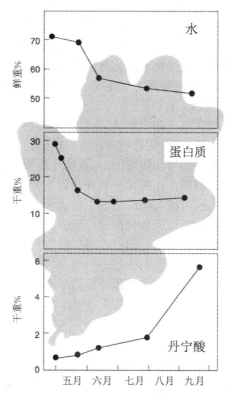

图4.16 橡木(*Quercus robur*)的阳生叶上的水、蛋白质和丹宁酸含量表现出的季节变化(非尼重绘,1970)[91]

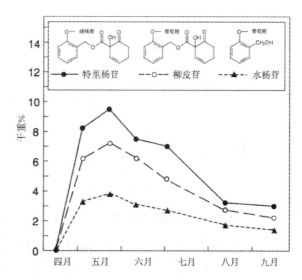

图4.17 白杨(*Populus trichocarpa*)叶上三种酚类浓度随季节发生的变化

同样,许多植物嫩叶中的萜类化合物的含量远远高于成熟叶片中的含量。其他组织器官中也有类似情况,例如枝干和树根,在幼嫩的组织器官中萜类化合物的含量通常是成熟组织官的2~10倍[105]。通常植物次生代谢物质在开花期和种子生产期会有进一步的激增。然而，许多取食嫩叶的昆虫比取食同一植株上成熟或衰老叶片的昆虫发育得更好，从而可以达到更高的繁殖水平，因为这些嫩叶具有更高的营养价值[139]，然而也有相反例子的存在。因此，菜青虫(*Pieris rapae*)幼虫更喜欢取食嫩叶，而拟尺蠖(*Trichoplusia ni*)则更喜欢成熟的叶片[35]。具刺吸式口器的昆虫，其寄主的养分含量变化可能比具咀嚼式口器的昆虫的寄主的养分含量变化更大[231]，最为突出的例子就是柳树，在春季它们的韧皮部汁液含有0.4%的自由氨基酸，而在夏季，这个值将下降到0.05%左右。

然而，许多被认为是毒素的次生代谢物质，其含量会随着叶龄增加而降低，而被指定为降低植食性动物消化能力或起阻碍因素的化合物却通常表现出相反的情况。这些物质如丹宁酸和树脂类，在许多情况下会随着叶龄增加而增加，并可能导致植物的叶片不适合植食性昆虫进行取食。在这些植物中，早期起阻碍作用的化合物往往以含氮化合物为主要物质，如生物碱、生氰化合物、非蛋白质氨基酸。这可能与当时土壤中氮含量的增加有关，在生长初期，不仅要保证植物生长对含氮化合物的需求，而且要保证储存足够的氮用于后期生长。与此同时，在正在生长的幼嫩组织中，碳的含量非常有限。但是在随后的生长过程中，碳可能提供大于植物生长所需的能量从而允许植物产生阻抗物质，例如丹宁酸。"资源有效性假设"[56]（见第11章）已经成为较吸引人的解释，其记录了含氮和含碳阻抗物质随季节变化而发生的改变，但是其有效性仍存在争议[201]。

4.12.2 昼夜影响

从另外一个时间标度看，许多植物次级代谢物质的水平似乎每天都在发生波动，由于植物的光合作用和代谢活动的变化，植食性昆虫的饮食习惯在夜间与白天存在显著差异。在已有对木薯中的氰化合物的报道中，一个昼夜周期中次生化合物的数量变化可能高达35%[205]。白羽扇豆(*Lupinus albus*)（图4.18）中叮类生物碱含量的日变化较大。据报道，挥发物的释放量也有日变化[174]。

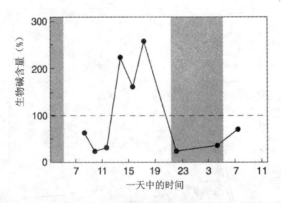

图4.18 白羽扇豆(*Lupinus albus*)叶片中的羽扇烷宁和叮类生物碱的昼夜平均含量百分比

显然，即使在一天内，植物的化学成分也不是恒定不变的，这种显著变化不仅表现在绝对的种类数量上，而且表现在不同化合物组成的比率上[220]。这种昼夜波动，可能会为一些古老药用植物的收割总结出一些基本规律。狄奥弗拉斯图(Theophrastos)报道，那个时期（公元前四世纪）的采集药物规定："有些有根植物应该在晚上采集，有些在白天，有些则在阳光照射之前。"[220]例如罂粟(*Papaver somniferum*)胶囊，上午9点时乳胶中的吗啡含量是晚上9点的4倍。

以上内容可以解释为什么许多昆虫是夜行性取食的,为什么许多昆虫在不同时间段取食不同的植物组织[153]。一些熟知的毛毛虫是典型例子,它们的高龄幼虫会从日食性转变为夜食性[59]。然而,目前我们还不清楚是否还有食物质量或一些其他生态因素影响这种觅食策略。对通常在夜间取食的植食性舞毒蛾幼虫(*Lymantria dispar*)而言,食物质量确实起到一定的作用。它们放弃了白天取食营养含量少的寄主植物而形成夜行性取食的规律。当这种昆虫用人工饲料喂养时,取食行为通常发生在夜晚,除非食物中含有2%的丹宁酸,取食行为才会发生在白天。摄食规律的变化可能是为了适应由脱叶诱导(defoliation-induced)引起的食物质量的变化,就像种群爆发时摄食规律也会发生变化一样[167]。

4.12.3 年间变化

已有研究证实,对一株多年生植物而言,如气候变化、养分供应等外部因素,均会在年间该植物影响化感物质的化学成分在数量和质量上发生变化。通过三个草种的平均酚类物质含量随年份变化的例子(图4.19)显示,酚类物质的数量在年间发生明显变化。在须芒草(*Andropogon scoparius*)中,发现其最高浓度是最低浓度的2.5倍[192]。

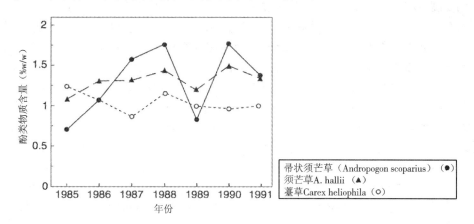

图 4.19　三个草种的平均酚类物质含量随年份的变化(摩尔和纳恩,1993)[192]

4.13　地理位置和肥料的影响

地理位置包括直接暴露于阳光和土壤中的物理化学特性等,可能对植物化学组分有极大的影响。

4.13.1 日晒和遮阴

阳光对绿色植物来说是一个必备的基本条件,植物组织将太阳能固定为有机物。毫无疑问,光强度通常能够影响植物初级或者次级新陈代谢。当光强度降低时,光合作用强度与碳水化合物的产量就会随之下降。

当一植株被完全遮阴或部分遮阴时,如果矿质营养吸收不受影响,那么C/N平衡就会净减。这会导致基于C的代谢产物的含量降低,如酚类物质(图4.20)[83,193]。通常,阴凉也会影响植物的形态特征,如叶的韧度和叶脉间隔。阴面的叶子通常较大,但比日照下的叶片薄。

我们所熟知的许多实例都已证实,在户外生长的植物和在遮阴条件下生长的植物存在显

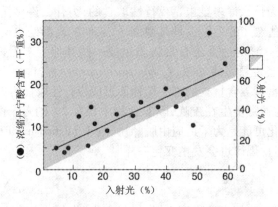

图4.20 到达羽叶金合欢(*Acaciapennata*)叶片的可利用光与叶片中的浓缩丹宁酸含量之间的关系(摩尔等,1988)[193]

著差异。例如,生长在阴处的欧洲蕨(*Pteridium aquilinum*)所包含的生氰化合物比生长在开阔地的植物要高50%,而生长在日照地区的植株往往含有高浓度的黄酮类化合物和丹宁酸。这些差异均对植食性昆虫以及植食性哺乳动物产生显著影响[58]。即使是同一植物的不同部位,由于暴露在阳光下的程度不同,其化学成分也会存在显著差异[254]。如前面提到的(2.1节),用于选择测试的舞毒蛾幼虫,同一棵桤木树首选南侧的叶子多于北侧的叶子(图2.5)。同样,另一个臭名昭著的杂食性昆虫——日本甲虫(*Popillia japonica*),相比遮阴条件下的食物更喜取食日照下的叶片。这所以得知,即使是杂食性动物也会表现出对食物选择的不同倾向[226]。同样,杂食性的天幕毛虫(*Malacosoma disstria*)的幼虫倾向于取食糖槭(*Acer saccharum*)阳面的叶片,因为阳面的叶片中含有更高水平的可溶性糖[208]。也有一些相反的例子,杂食性尺蠖(*Epirrita autumnata*)的幼虫更喜其寄主高原桦阴面的树枝,其取食遮阴叶片比取食非遮阴叶片生长得更好,这是因为遮阴叶片中含有更高含量的蛋白质、自由氨基酸和水分。同时阴面叶片有更低水平的总酚含量且叶片韧性较低[133]。这些例子均说明,避光能够改变叶片的化学性质,并因此影响植食性昆虫的行为。

具刺吸式口器的昆虫也会随寄主暴露于阳光的程度不同而发生改变。桉树在阳面受到木虱(*Cardiaspina densitexta*)的危害明显比阴面要严重,因为在背光条件下,若虫吸食次优营养的植物汁液,这将导致巨大的死亡率[288]。

在野荠菜(*Cardamine cordifolia*)中发现了一个与上述情况相反的例子。相对而言,植食性昆虫取食背光叶片较多。在阳光下生长的植物叶片含有更高的糖分,而蛋白质的含量却相对较低。果蝇(*Scaptomyza nigrita*)幼虫在阴面叶片上造成的损害是阳面叶片的两倍。芥子油苷在阳面叶片和阴面叶片间的含量并没有区别,因此得到的结论为,在阴面取食的昆虫对增高的蛋白质含量做出了积极的反应[57]。

高强度光照通常可以刺激次生代谢水平。结果导致整株植物或者部分植物组织在光照条件下比背光条件下包含更多的次级代谢产物。有一个典型的例子是欧洲赤松(*Pinus sylvestris*)中二萜类化合物的浓度和二萜树脂的总含量,与背光的叶片相比可能会增加100%[115]。也有一些其他研究证实了入射光和次生化合物的生产之间存在正相关关系[284]。

4.13.2 土壤因素

影响植物生长的另外一个重要的环境因素是土壤的性质,包括土壤的矿物质状况。许多关于昆虫生长和丰富度的研究也涉及对其寄主植物的土壤化学成分的研究。在农业方面,肥料一般用于提高植物生长速度,促进植物健康生长和提高产量。施肥主要影响植物生理功能,也可以诱导植物形态和行为的改变。生理上的反应通过养分结构的变化来表现,例如蛋白质水平。次级代谢也会受到影响,这会导致次级代谢物含量的增加或降低[102]。一份引用了147篇文献的分析报告提出了具有高度说服力的证据,即高氮肥能够降低以碳为基础的植物次级代谢物质的浓度[164]。应值得注意的是,当次级代谢物质的增长率比生物量增长率低时,其浓度也会降低[163]。一些情况下,昆虫主要对其寄主植物营养物质的化学组成改变做出回应,而在另外一些情况下,化感物质的变化也会占主导地位。

肥料不仅影响植物的营养状况,还影响叶表面的化学性质和外观。因此,当昆虫寻找寄主植物产卵时,它们对已施肥和未施肥的同种植物可能会呈现出不同的反应。在菜粉蝶(*Pieris rapae*)中观察到一个对肥料存在敏感性的现象。待产卵的雌性似乎能够区分24小时内施肥和未施肥的寄主植物[198]。这就证实了不仅植物可以迅速对肥料做出反应,昆虫也可以察觉到寄主植物之间微妙的不同。

许多昆虫由于植物营养状况的改善而获益。然而,随着对昆虫—植物相互作用关系研究的深入,植物的肥料对昆虫的数量有消极影响的例子也有许多相关报告,例如植物可以通过缩短对昆虫攻击的易感时间段来提高自身活力。然而这一领域的结论存在太多的例外,并非缺乏科学的证据,而是由于昆虫那不可预测的灵活性,它们巨大的物种多样性以及昆虫和植物之间过于微妙的关系。

氮是最能限制作物产量的重要元素,对田间作物施用氮肥,更多的消耗量以及更高的使用率,似乎常可以刺激昆虫的数量(图 4.21)[196]。斯克赖伯曾报告,在约115项研究中发现,对作物破坏的加重与害虫寄主植物的氮含量增长有关[236]。在木本观赏植物中同样证明了施肥会造成昆虫营养质量的提高或会诱导植物刺激代谢物质的积累,这会增加昆虫的攻击性[134]。奇怪的是,对于一些森林树木存在相反的结论,这些区别依然有待于解释。

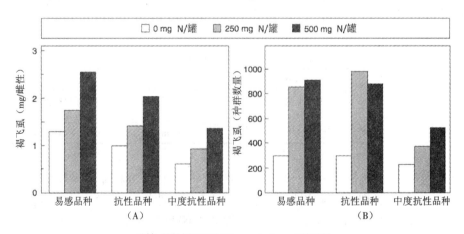

纵轴:雌性褐飞虱(*Nilaparvata lugens*)的重量

图 4.21 褐飞虱重量与种群数量的变化

(A)雌性褐飞虱(*Nilaparvata lugens*)的重量变化;(B)根据不同的氮肥率,对实验笼中种植的三个选定的水稻品种分别施以不同浓度的氮肥,褐飞虱种群数量的变化(海因里希斯和梅德拉诺,1985)[132]

目前,还没有关于磷、钾或其他有机肥料对昆虫数量影响的结论。可以认为:在很多情况下,肥料对植食性昆虫有深远的影响,尽管我们现有的关于土壤—植物—昆虫相互联系的知识还很匮乏[60]。

当作物分别生长于田间和温室中时,环境对次级代谢的影响会呈现巨大的差异。温室生长的卷心菜的葡萄糖含量仅为田间卷心菜葡萄糖含量的10%[52]。然而,在番茄植株中发现了相反的结果。在这种情况下,在温室种植的植物中生物碱的含量比在野外生长的植物高出2~4倍[19]。关于引起这些显著差异的因素,目前缺乏更详细的信息,而目前的例子再次表明可以将这些因素运用到植物生理学上,包括在次级代谢的资源分配方面。这也提醒我们不能以温室实验为背景推断关于昆虫—植物与自然环境的联系。

4.14 诱导抗性

一种植物会被许多可运动的物种伤害,因为植物不会逃离。然而,它所要面对的攻击者种类是不可预测的,因此它不需要预先消耗能量去防卫。而创伤可以诱导植物的抗性,植物可以对不同的昆虫种类及不同的创伤做出不同的反应。下面将要对此进行具体解释。

在19世纪80年代生化学家[114]和生态学家[127]发现了植物诱导抗性对植食性动物的反应现象,随后也被植物生理学家证实,分子生物学家和昆虫学家也对此极度着迷。因此,大量的相关文献相继出现(例如凯邦堡和鲍尔温[151],阿格拉沃尔等[6],迪克和希尔克[74])。现在的观点普遍认为:植物的化学构成不仅受非生物因素影响,在一定程度上也受到生物因素的影响,例如植食性动物。这对大量次级代谢物质也同样适用,所有的次级代谢物的生物合成很可能被植食性昆虫诱导,如关于酚醛树脂类、烯类、生物碱类、芥子甙类、烃类和生氰糖甙诱导的例子[151]。另外,酶类和其他蛋白质也可能被诱导。这种诱导的变化不仅会影响植食性昆虫(直接影响植食性昆虫的诱导)而且也会影响它们的天敌,例如食肉动物和拟寄生物(间接地影响植食性昆虫的诱导)。因此,学者们发现了两种典型的植物诱导形式(表4.7)。

表4.7 植物对不同植食性动物的抗性

	基本抗性	诱导抗性
直接抗性	对植食性昆虫产生负面影响	诱导取食;植食性昆虫产生负面影响
间接抗性	对植食性昆虫的天敌的行为产生积极影响	诱导取食;对植食性昆虫的天敌的行为产生积极影响

直接性诱导[151]和间接性诱导[71]已经被许多来自不同种群的植物证实,因此成为植物防御的基本形式。这两种类型的诱导抗性有效地防御了不同食性的昆虫,例如咀嚼、刺入细胞和韧皮部的昆虫(如毛虫和蚜虫),以及内部取食者(如潜叶蝇和天牛)。

4.14.1 直接诱导

在植物基本化学抗性中的所有机制可以通过诱导而被激活。例如,通过马铃薯甲虫诱导被破坏的马铃薯来合成高水平的蛋白质抑制剂[114]。蛋白质抑制剂干扰这种甲虫或其他植食性昆虫的食物消化能力。当转入植物诱导过程中起关键作用的基因为反义基因时,烟草天蛾(*Manduca sexta*)的重量明显提高(图4.22)。

第4章 植物化学:无穷无尽的变化

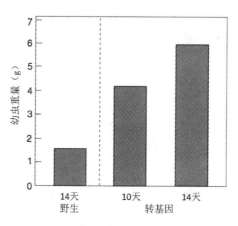

图 4.22　烟草天蛾 (*Manduca sexta*) 幼虫重量的增长

这种机制也可以有完全不同的过程。一些转基因马铃薯对马铃薯甲虫的卵高度敏感。围绕着卵块的植物组织死亡,卵会变得干燥以致穿过树叶掉落在地上,最终被该区域土壤中的捕食者消灭掉[14]。

显而易见,直接性诱导在植物界是普遍存在的,在34个科的100多种植物中已被记录[151]。

4.14.2　间接诱导

植物也可利用化学物质吸引植食性昆虫的天敌。遭到棉叶螨(*Tetranychus urticae*)取食破坏后,利马豆植物会散发出挥发性混合物,这种混合物在成分上不同于未破坏或遭机械性破坏的利马豆所释放的混合物,含有许多萜类和水杨酸甲酯,这些物质吸引棉叶螨的捕食者智利捕植螨(*Phytoseiulus persimilis*)消灭棉叶螨(图4.23)[271]。

在13科20种植物中都发现了其挥发物质对天敌有引诱作用[71]。由植食性昆虫诱导产生的物质不仅可以吸引捕食者还可以吸引寄生性天敌[264]。

植物上的虫卵沉积物也可诱导植物挥发物的产生。榆叶甲虫(*Xanthogaleruca luteola*)在榆树叶上产卵,挥发物会吸引其卵寄生天敌啮小蜂(*Oomyzus galleruca*)[187]。半翅目植食性昆虫稻绿蝽(*Nezara viridula*)产卵会诱导豆科植物释放挥发物,吸引其卵寄生性天敌沟卵蜂(*Trissolcus basalis*)[51]。有趣的是,由产卵诱导的挥发物与直接被取食诱导的挥发物是很相似的[136]。最近有关于毛虫足迹诱导植物化学物质变化的报告[34],但昆虫足迹是否会诱导挥发物的释放仍在研究之中。

机械性损伤不能像昆虫取食或产卵一样有诱导抗性。显然,诱导抗性是由植食性昆虫诱导的。对天敌的吸引经常发生在几天之内,挥发物的诱导不仅在受害部位产生,也会通过系统传送信号在不被植食性昆虫接触的其他树叶上产生[75,136]。

除了寄生性天敌和捕食者的影响之外,昆虫诱导产生的植物挥发物可能也影响植食性昆虫。一些植食性昆虫被挥发物吸引[31,149],而一些植食性昆虫被排斥[64,68]。因此,虫害诱导植物挥发物对不同昆虫的影响可能会存在差异。

4.14.3　植食诱导变化引起的变异

植食性昆虫对不同种植物及不同基因型的植物诱导产生的化合物会存在差异。因此,在

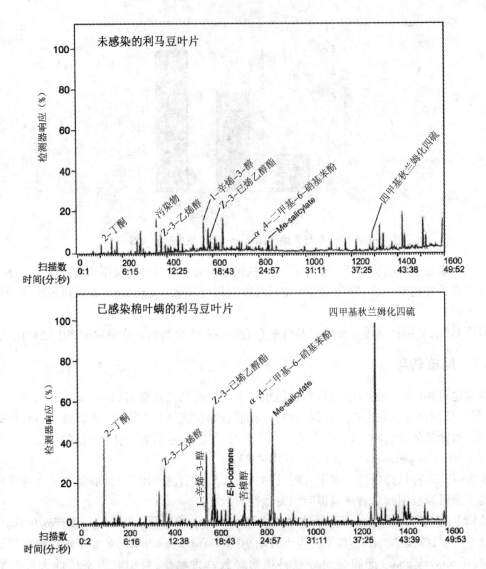

图 4.23　利马豆叶片和感染了棉叶螨(*Tetranychus urticae*)的利马豆叶片
释放的挥发性物质的气相色谱——质谱分析图

不同基因型的玉米中,关于挥发性物质排放量的巨大差异已被记录(图 4.24)[111]。此外,一些非生物因素也有显著影响,如光照、湿度、有效养分等。例如,玉米植株对模拟毛毛虫取食而释放的挥发物数量与接受的光照量呈正相关,并且在土壤低湿度和高有效养分条件下的释放量会更高[110]。

一种植物对植食性昆虫的反应与其对机械损伤的反应不同,甚至对于不同物种的植食性昆虫的反应都会不同[75]。例如,在秋天,玉米植株被大量甜菜夜蛾(*Spodoptera exigua*)幼虫危害,玉米植株散发出的几种萜类化合物可以吸引寄生蜂(*Cotesia marginiventris*)。相反的,受到机械损伤的植物并不能散发出这些挥发性物质。然而,将毛虫置于植物受机械损伤的部位时,植物做出的回应与其对毛虫的直接损伤做出的回应相同[263]。

有许多植食性昆虫天敌的例子,它们能区别不同植食性昆虫诱导产生的挥发性物质[70]。

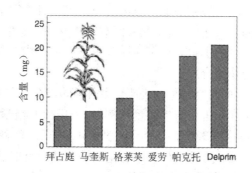

图 4.24　不同的玉米品种经机械损伤后,放置灰翅夜蛾(*Spodoptera littoralis*)幼虫,
9 个小时后搜集到的挥发性物质的量(苟英谷那等,2001)[111]

例如,蚜虫的寄生性天敌(*Aphidius ervi*)可以受蚕豆被蚕豆蚜(*Acyrthosiphon pisum*)侵染后释放的挥发物质吸引。而对被甜菜蚜(*Aphis fabae*)侵染的植株所释放的物质没有反应。显然,植物应对不同的植食性昆虫释放不同的挥发物质,这涉及三级营养关系的层面。最近,捕食性螨虫——智利捕植螨(*Phytoseiulus persimilis*)的例子充分说明了天敌可以区分其捕食对象和非捕食对象引发的植物挥发物。当豆科植物感染棉叶螨(*Tetranychus urticae*)后会释放大量的水杨酸甲酯和萜类化合物。而其感染甜菜夜蛾(*Spodoptera exigua*)后却只能释放微量的物质。捕食者可以区分受棉叶螨(*Tetranychus urticae*)和甜菜夜蛾(*Spodoptera exigua*)感染后的豆科植物所挥发的物质。然而在豆科植物感染甜菜夜蛾后释放的物质中添加水杨酸甲酯和萜类化合物,可以起到对捕食者的引诱作用[63]。这个结果显示了不同的植食性昆虫可以引发植物释放不同的混合物,而混合物中的某些特定成分可被捕食性昆虫用于区分特定猎物。

4.14.4　食草性昆虫引发的基因变异和代谢变化

食草性昆虫可能也会导致植物化学成分发生巨大的变化。例如,利马豆植株被红蜘蛛损坏后会释放出约 30 种挥发性化合物,特别是萜类化合物,这些物质并不由未被损坏的植物散发[77]。植食性昆虫也会导致植株发生更多的生理改变,不仅局限于挥发性物质。基因组的分析表明,植物为应对植食性昆虫,在基因水平上也发生了改变[216]。受烟草天蛾(*Manduca sexta*)刺激的烟草植物(*Nicotiana attenuata*)应答转录包含超过 500 种不同的信使 RNA[135]。这些结果显示,在受植食性昆虫的刺激下,植物的生理机能会随之发生强烈的变化。在光合作用中,转录被强烈的抑制,而对于应激、损伤、病原体的反应和那些与碳和氮转移防御的反应都被强烈地调节[135]。在拟南芥(*Arabidopsis thaliana*)被研究的 7 200 个基因中,有 114 个基因的表达会随菜粉蝶(*Pieris rapae*)幼虫的破坏而发生改变[217]。在受病原体感染后,与那些未被感染的拟南芥相比,2 375 个基因的表达会发生改变。当然,并不是所有改变的基因都会影响植物化学物质的产生,但是许多物质确实直接或间接地对其进行影响。

植食性昆虫会引发植物释放挥发物质,参与这一生物合成过程的基因是快速作用的。不同的玉米栽培品系释放萜类物质的差异是由萜烯合成酶基因中的一个等位基因变异引起的[162]。目前已在豆科植物的百脉根(*Lotus japonicus*)中鉴定出由齿虱类诱导的(E)-b-罗勒烯合成酶基因[12]。此外,通过结合代谢组学和转录组学的方法,在黄瓜植株中也发现了由齿虱类诱导的、参与萜类挥发物生物合成的基因[189]。

目前,人们已经利用拟南芥(*Arabidopsis thaliana*)的基因组信息克隆了许多萜烯合酶基因。它们对不同压力有不同的表达模式,由此人们对于挥发性物质生物合成基因的关闭与否,所受的环境控制因素有了更多的了解。许多萜烯合酶基因似乎参与花挥发物的合成,但也有一些参与叶挥发物或根化合物的合成[49]。

在植物—昆虫系统中,不同研究结果均表明食草性昆虫导致了植物发生普遍的化学变化。现代技术也可以对这种化学变化进行直接测量。例如,利用代谢组学的方法能分析不同基因型拟南芥(*Arabidopsis thaliana*)中 326 个叶片的化学成分差异[93]。这种技术的进一步发展将为植物在植食作用引发的化学变化方面提供一个更全面的评估。目前昆虫学家们已经开展了广泛的实验,需要确定这些变化与每个生物的相关性。

显然,分子技术的出现提供了新的重要的分析方法,例如,将突变和转基因技术应用到一个单基因中,可以发现对增强抗性的单因素的贡献[79,155,275]。然而,抗性诱导很少是由单一性状的变化引起的:这是由草食性昆虫介导的大规模转录改变的结果。对于这些由植食性昆虫介导的,重要调控要素的基因改变,继而引发复杂成分的改变,对假定有防御特性的功能进行分析,可能是最好的方法[221]。这些调控要素可能包括能够控制十八碳酸(octadecanoid)路径表达的基因。

4.14.5 系统影响

除了局部影响之外,局部虫害通常也会对系统造成影响。这种间隔效应与直接诱导抗性和间接诱导抗性有关。对番茄单一叶片的取食损伤会导致同一植株的其他叶片产生蛋白酶抑制剂[228]。同样,鳞翅目幼虫取食玉米叶将导致同一植株的不同叶子释放出吸引其拟寄生蜂的挥发物[262]。在某些情况下,系统影响甚至是局部影响都是与生俱来的。烟草植物中的尼古丁在根部合成。鳞翅目幼虫损坏烟草叶片,导致植物根部诱导合成尼古丁,并且随后将尼古丁运输到已损坏和未损坏的叶片中[16]。这种系统影响的明确机制目前仍然没有揭开。有证据表明化学诱导因子[228]对液压[179]和电信号[292]起重要作用,系统性信号可能是基于这三种模式的结合。

系统信号的传输依赖于叶片相连的脉管"管道"[206]。在对棉花植物的研究中已阐明,杨树的每 5 个叶片共享一个脉管管道。通过导管系统(图 4.25)直接与受损叶子相连,叶片经抗性诱导后发生的改变是很大的[145],而且在系统反应中呈上升趋势。位于植株下层的叶片是原叶(source leaves),而植株高层的叶片为库叶(sink leaves),植物的能量主要流向库叶。这也反应在系统性抗性中。如天幕毛虫(*Malacosoma disstria*)在白杨树上取食,调节基因在植株高层的叶片中表达,而不在下层叶片中表达[11]。

总之,植食性昆虫在不同的叶片系统中可能导致不同的改变,因此,取食导致在空间上表达不同的表型。这意味着植食性昆虫会在局部取食造成植物损伤后面临资源发生空间变化的局面。

4.14.6 长期反应

以上都是关于植物迅速应答反应的讨论,这种应答反应所需的时间相对较短。研究发现也存在一些诱导延迟反应的例子,多出现在同一季节的晚期、下一季节或是木本植物后生的叶片中[202]。例如,植食性昆虫在早期的取食过程中会诱导热带植物巴豆(*Croton pseudoniveus*)和巴豆(*Bursera instabilis*)释放酚醛树脂和浓缩丹宁酸,而这些物质直到季末才会对植食性昆虫

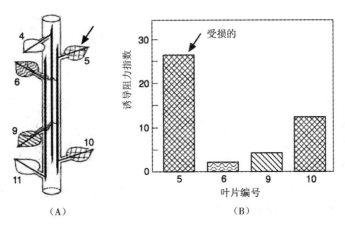

图4.25　通过机械损伤东方杨树(*Populus deltoides*)的一个叶片(箭头指示),诱导抗性的强度反应了不同叶片之间的脉管联系

(A)叶片5和10有共同的脉管联系,而叶片6和9很少直接与叶片5联系;(B)诱导模型反应了叶子之间的脉管联系(琼斯等,1993)[145]

造成影响[29]。白桦树也是一个很好的例证,在芬兰森林中白桦树繁盛且分布广泛。豪基奥亚和他的同事们记录了在叶片上有鳞翅目幼虫存在的时候,叶片中酚醛树脂的浓度增加。这种变化对食叶动物而言,营养价值降低了。在这种情况下,化学物质的改变显示出两个不同的时间尺度:在几小时到几天内,酚醛树脂含量的升高为短期效应,持续几个月到几年为长期效应[200]。白桦树受到秋白尺蛾(*Epirrita autumnata*)幼虫危害而导致落叶,在两年(或更多年)后,秋白尺蛾幼虫的增长和繁殖都会受到抑制(图4.26)。即使在几年之后,由于叶片脱落而造成的营养质量的降低都会对植食性昆虫造成负面影响。

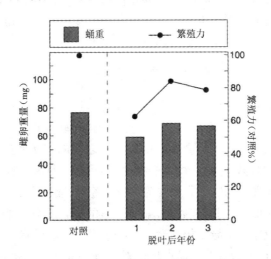

图4.26　白桦树(*Betula pubescens*)自开始脱叶后,对秋白尺蛾(*Epirrita autumnata*)繁殖力及雌蛹的重量的影响

秋白尺蛾幼虫以白桦树自开始脱叶2~4年后的叶片喂养,呈现出生长和繁殖力降低的趋势。(内乌沃宁和哈乌基奥,1991)[200]

植物对植食性昆虫的抵抗存在时间延迟,并且会产生周期性波动。因此,红杉小卷蛾(*Zeiraphera diniana*)的周期性爆发至少可以部分解释为其寄主欧洲落叶松(*Larix decidua*)在开始脱叶后的4年至5年中松针的化学成分和形态结构发生了变化[21]。

4.14.7 信号传导

由于植食性昆虫会诱导植物化学物质发生变化,我们目前已对这种变化的信号转导过程有了更深入的了解。一些研究结果显示,昆虫的分泌物会更强烈地诱导植物化学反应。植物可以区分由枝剪和昆虫口器造成的损伤[124],对于昆虫取食危害和机械损害的区别,可以解释为是否存在昆虫衍生的诱导因子。将昆虫反刍过的物质放置于受机械损伤的植物表面,可以模拟昆虫取食危害[112,116]。一些反刍物质中的诱导因子已经被鉴定出来,包含了粘多糖、一些脂肪酸—氨基酸偶联物[116],β-葡糖苷酶[184]和葡萄糖氧化酶[195]。

植物的诱导反应主要包含三个信号转导路径:(1)十八碳酸(octadecanoid)路径,包括茉莉酸;(2)莽草酸路径,主要包括水杨酸;(3)乙烯路径[76]。其中十八碳酸(octadecanoid)路径是参与昆虫诱导抗性的主要信号转导路径,其他两个也参与诱导过程。例如,青豆植物由于叶螨的损伤而释放挥发性物质,这是由茉莉酸和乙烯协同调节的[137]。除此之外,水杨酸甲酯也被认为是一种诱导挥发物。而由叶螨诱导产生的大部分挥发物都含有茉莉酸,却不包含水杨酸甲酯[77]。水杨酸甲酯是否为水杨酸对抗茉莉酸的作用代谢产物,这个问题至今仍是一个未解之谜。在番茄中,茉莉酸和乙烯与蛋白酶抑制剂协同作用[203],但在野生烟草释放的尼古丁中,乙烯对茉莉酸产生的效果起拮抗作用[147]。

在昆虫—植物相互作用关系中,不同路径的组合可能会被激活。感染了菜青虫(*Pieris rapae*)的拟南芥(*Arabidopsis*)会激活十八碳酸(octadecanoid)路径和水杨酸路径,用以吸引盘绒茧蜂(*Cotesia rubecula*)。这已经在拟南芥(*Arabidopsis*)的转基因植株中得到证实[275]。失去任意一种信号传导路径的拟南芥转基因植株,在经菜青虫取食后对盘绒茧蜂的吸引力都会降低。任意一种信号传导路径都不能完全控制拟南芥挥发物的诱导。

4.14.8 食草动物诱导和病原体诱导的相互作用

植物暴露于各种危险当中,不仅有食草昆虫,还有病原微生物。植物病原体,如细菌和真菌,也会诱导植物释放挥发物质,也由上述三个主要的信号转导途径介导[210]。因此,植物由病原体和草食昆虫引发的诱导过程可能会通过信号通路之间的交叉效应相互作用。例如,在番茄植株中,由病原体引发的诱导作用是由水杨酸途径介导的,而这将对由虫害诱导的茉莉酸途径产生负面影响[258]。已发现病原体诱导通常为水杨酸途径,而昆虫诱导主要为茉莉酸途径,这在多种植物中被证实。然而,近期研究表明,如此严格的途径区分可能不存在,也存在一些病原体诱导茉莉酸途径以及昆虫诱导水杨酸途径的例子[274]。当植物使用不同的途径对病原体或昆虫的攻击做出应对时,可能为食草性昆虫提供对自身诱导的产物进行破坏的机会。我们发现玉米夜蛾(*Helicoverpa zea*)幼虫诱导大豆中产生水杨酸途径而非茉莉酸途径。水杨酸途径不会引发大豆对玉米夜蛾的抗性,玉米夜蛾操纵了大豆的信号传递方式,从而抑制了茉莉酸途径的诱导抗性[195]。这个例子表明植物对多种攻击的防御是相互作用的,这是一个令人兴奋的话题,需要更深入的研究。

4.14.9 植物间的相互作用

在20世纪80年代,人们第一次发现诱导化学反应不仅发生在植物损伤部位,同样也会发生在邻近的植物中[17,219]。在过去的十年里已经累积了充足的证据来支持这个观点[72,78]。在实验室和田间实验的证据都可以证明,在同种和异种植物之间,存在直接诱导和间接诱导两种诱导抗性。例如,对赤杨进行研究时发现,在受损植物和其临近的同种未受损植物中,均产生诱导抗性。在实验室内对这种现象的潜在机制进行了研究[260]。将未受虫害的利马豆叶片置于同种被叶螨取食后的叶片释放的挥发物中,会导致其抗性基因的表达并且能引诱叶螨的天敌——捕食性螨类[10,37]。

关于异种植物间的相互作用,有一个美国蒿属植物的例子,对蒿属植物的损害会诱导其邻近的野生烟草(*Nicotiana attenuata*)植物产生抗性。这种抗性可能是对蒿属植物释放的挥发性物质茉莉酸甲酯做出的生理反应[152,212]。

目前,大多数研究都关注气生植物(aerial plant)挥发物在植物间的相互作用中所起的作用。最近两项研究显示,食草昆虫取食植物地下部分,可以导致植物地下部分化学物质的改变,继而诱导邻近植物的化学物质发生改变。豆类植物的根感染了蚜虫或叶螨后会诱导产生抗性物质,引发邻近未经破坏的豆类植物根部释放对天敌有引诱作用的物质。显然,这种植物能够预先采取防御措施来预防草食昆虫的危害[47,73]。

关于邻近植物的诱导效应的相关数据,暗示了植物对食草性昆虫的防御是有空间性的。对临近植物的影响与受损植物距离呈梯度关系。

4.15 基因型差异

4.15.1 植物个体间的化学差异

不同植物释放的次生化合物在数量和成分上都不同,这主要是受基因控制的,尽管环境因素会也产生一定影响。次生化合物中超过50%的成分都是由基因决定的,基因的影响超过了表型变异性[24,165]。可以通过选择实验证明由基因控制次生化合物的浓度。以油菜(*Brassica rapa*)为例,只通过3代的选择实验就筛选出比亲代含量高60%或低40%的硫代葡萄糖苷品系。在可选择的情况下,菜青虫(*Pieris rapae*)幼虫和粉纹夜蛾(*Trichoplusia ni*)对低硫代葡萄糖苷含量的植物叶片取食更多[252]。

不仅次级代谢水平受基因的控制,植物受损后诱导产生的化合物也受基因的控制,目前对于此类的研究相对较少。在红花琉璃草(*Cynoglossum officinale*)中发现的可诱导的吡咯里西啶生物碱,遗传因素占了35%[267]。植物的次级化合物发生持续性的变化,通常是基因控制的(通常包含几个或多个基因)。在自然种群中基因型的差异可能是巨大的。例如,柳树不同个体叶子中的水杨苷的浓度,从0.05%到5%(干重)以上,变化了100倍,而标准偏差应为平均变异水平的二分之一。

由于植物次级代谢物的浓度一般受到严格的基因控制,可经选择改变其产出量。因此,人工栽培品种与野生品种的次级化合会有很大的不同。含有较低的化感物质的品系会对昆虫攻击的敏感度发生改变。棉子酚(一种酚类倍半萜烯色素)对许多昆虫都能产生抗性,相对于棉子酚含量多的棉花(*Gossypium hirsutum*)品系,昆虫更喜棉子酚含量较少的棉花(图4.27)。含有三萜系化合物葫芦素的黄瓜会带点苦味,这种葫芦素是由单个基因控制的。人工繁殖育种都会选择非苦味品种以满足人类的口味。葫芦素是强效的威慑物并且对很多物种(包括人类)有毒,但低葫芦素品种很容易受到叶螨(*Tetranychus urticae*)和一些其他昆虫的侵害。

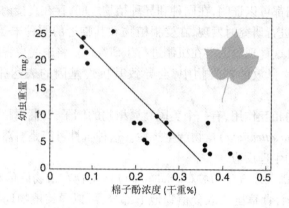

图4.27 一龄烟草夜蛾幼虫五天内体重变化与20个不同棉子酚浓度的棉花栽培品种的关系(埃丁等,1983)[130]

植物育种专家通过对植物天然的变异品种进行选择来开发特定的理想品种。由于各品种间基因的差异,它们对虫害的敏感程度也不同(图4.28)。选择快速生长发育和更高产量的品系,往往会致使植物的抵抗力降低。结果是食用品种变得比野生品种更易受到攻击。如今,植物育种专家们更倾向于选择保留天然化学抗性的食用种,而那些经常被食用的组织、果实、种子除外。这种类型的选择目前已经实现,例如马铃薯块茎[144]。抗性育种的话题将会在第13章进行更详细讨论。

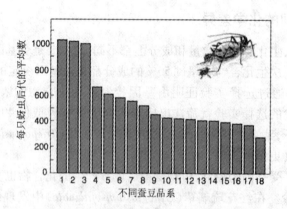

图4.28 蚜虫(*Aphisfabae*)于14天内在18个不同的蚕豆(*Vicia faba*)品系上产生的后代的平均数(戴维斯,1922)[62]

4.15.2 植物个体内部的化学变化

尽管对大部分动物(包括昆虫)而言,个体的概念通常不会出现混淆。但是对植物而言,这个概念却没有明显的界线。将植物个体定义为模块化有机体(Modular organisms),使得对个体的定义更为复杂。一棵树或许可以被定义为次级生物体单位(模块)的群体,每一部分都具有各自的生长和死亡时间。也可以将树描述成一个与茎和根相关的模块单位[286]。然而关于世代周期的概念就变得更不明确了,一些植物(一般意义上)的世代周期很长,和昆虫相比它们的基因重组率非常低。可存活200年或更久的阔叶树,因为它们的寿命很长,可以被看成是活化石。在自然状态下进行无性繁殖的植物,其基因类似且都能活很久。一些进行无性繁殖的蕨类植物(*Pteridium aquilinum*)预计可以活1000年或更久。基于分子的证据发现,源于罗马的英国大叶榆(*Ulmus procera*)可以存活2000年。19世纪70年代,榆树病在荷兰迅速蔓延的原因也许可以用基因的同质性解释,这是由榆树甲虫(*Scolytus*)携带的真菌引起的传染病[107]。起源于更新世的草本被子植物秋麒麟(*Solidago missouriensis*)或木本植物山杨树(*Populus tremuloides*)的分布范围都很广[289]。为克服长寿命无性繁殖遗传一些刚性的缺点,可以用体细胞突变进行补偿,这种突变可以通过自然有性及无性繁殖进行继承。体细胞突变的积累可能会导致一个无性繁殖个体发展成一个具有遗传多样性的个体[87]。

由于细胞突变,在长寿植物的生长发育过程中,它们可能被嵌入其他分支或同种植物的不同基因来进行基因重组。以下为一个关于基因嵌入的例子,白杨树的不同枝条对瘿蚜侵染的敏感程度不尽相同。虫瘿的分布并非是随机的,而是反映了这种镶嵌模型寄主抗性的表现(图4.29)。同一株植物相同光度(magnitude)的枝条对球蚜敏感度存在高度的变化[289]。另外一个关于"镶嵌"抗性的例子如图4.30所示。

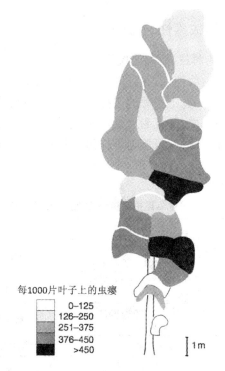

图4.29 约53 000个瘿蚜(*Pemphigus betae*)(蚜科)虫瘿,在20.1米高的杨树(*Populus angustifolia*)的20个枝条上的分布情况,每个枝条的大小代表其总叶面积(惠瑟姆,1983)[289]

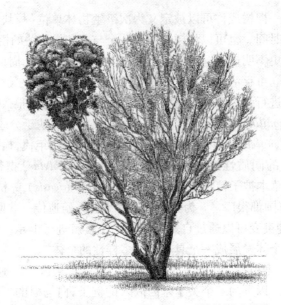

图 4.30 桉树（*Eucalyptus meliodora*）的镶嵌式抗性

在食根虫（*Anoplognatus montanus*）爆发期内，树上的叶子能完全掉光。因为挥发物成分的不同，一些枝条（有时甚至是整棵树）能够免受这种灾难。这种具有抗性的部分可能含有新出现的体细胞突变分生细胞，具有抗性的枝条会产生携带抗性基因的种子（爱德华，1990）[87]

4.15.3 植物性别对昆虫的敏感性的影响

所有开花植物中约有 6% 为雌雄异株，雌雄异株即一个植株上仅有雄花或仅有雌花，昆虫学家们注意到一些昆虫能辨别植物性别，并且昆虫更倾向在雄性植株上产卵和捕食（表 4.8）。食草昆虫在植物两性间施加潜在的、不均等的选择压力，结果可能影响植物繁育系统的演化。由于雌雄异株植物往往为大型植株，并且为风媒传粉，这种生殖策略最需要的是降低近亲繁殖的风险。食草昆虫对雄性植株的偏好，可能与雄株氮含量较高，叶子的韧性较低，与雌性相比防御能力较低有关[8]。

表 4.8 不同食草昆虫对植物性别进行选择的研究

植物	昆虫种名和目名	选择性	参考文献
小酸模（*Rumex acetosella*）蓼科	烟草金针虫（*Conoderus vespertinus*）鞘翅目	雌株面临更大的伤害	175
褪色柳（*Salix discolor*）杨柳科	角叶蜂（*Phyllocolpa leavitti*）膜翅目	雌株的存活率更高	97
乳香木（*Pistacia atlantica*）漆树科	蚜虫（*Slavum wertheimae*）同翅类	雄株的状况更好	295
云梅（*Rubus chamaemorus*）	四种昆虫	雄性植株的叶片受到的伤害更大	7
灰毛柳（*Salix cinerea*）杨柳科	多种昆虫	雄性植株受到的伤害更大	9
长叶麻黄（*Ephedra trifurca*）麻黄科	瘿蚊（*Lasioptera ephedrae*）双翅目	雄性植株上有更多的虫瘿	27

植物	昆虫种名和目名	选择性	续表 参考文献
柳属（*Salix*） 杨柳科	5种叶蜂 膜翅目	雄株上发现更多的昆虫	28
香根菊（*Baccharis halimifolia*） 菊科	叶甲（*Trirhabda bacharidis*） 鞘翅目	昆虫喜欢在雄株上取食	166

4.16 结论

在相对一致的初级代谢基础上，植物又产生了复杂而多变的次级代谢化合物。目前对这些化合物的多重作用还知之甚少，但毫无疑问，它们为植物提供了对抗严酷环境的保护作用，防止病原微生物的入侵以及植食昆虫的侵害。在图4.13中已进行了总结。

在自然界中探索一般规律时，一般而言，科学家们往往必须忽视生物现象中的一些细微变化。然而，在这一章中，我们一直把重点放在植物物种（或个体）水平的化学成分变化上。植物化学成分的变化可能受昆虫、光、土壤中的营养物质或大气等其他条件的影响。这些由基因型和环境因素所造成的时间和空间变化，不能被视为正常或标准偏差，而是代表了植物为了增大生存机会所采取的基本生存策略特征。

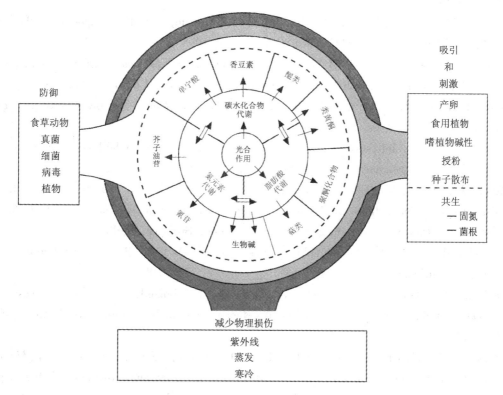

图4.31　植物的次级物质主要来源于植物初级代谢物

注：它们表现出的许多功能，涉及许多生物和非生物环境因素（哈特曼重绘于1996年）[125]。

4.17 文献

目前关于植物的次级代谢产物有大量的文献。维克瑞[279]和汉森[118]提供了详细的介绍。弗罗内、延森[98]和史密斯[246]报道了植物的化学分类。这本书由卢臣泰、詹曾[225]和贝伦鲍姆[224]提供了很多与昆虫有关的植物化合物信息。著名的默克索引(*Merck Index*)[188]包括 10 000 种化合物的结构式和信息。其他最近的数据由哈本、巴克斯特[121]和白金汉姆[39]提供。

4.18 参考文献

1. Abe, M. and Matsuda, K. (2000). Feeding responses of four phytophagous lady beetle species (Coleoptera: Coccinellidae) to cucurbitacins and alkaloids. *Applied Entomology and Zoology*, 35, 257–64.

2. Abe, T. and Higashi, M. (1991). Cellulose centered perspective on terrestrial community structure. *Oikos*, 60, 127–33.

3. Adati, T. and Matsuda, K. (1993). Feeding stimulants for various leaf beetles (Coleoptera: Chrysomelidae) in the leaf surface wax of their host plants. *Applied Entomology and Zoology*, 28, 319–24.

4. Adel, M.M., Sehnal, F., and Jurzysta, M. (2000). Effects of alfalfa saponins on the moth *Spodoptera littorali*. *Journal of Chemical Ecology*, 26, 1065–78.

5. Adler, L.S. and Kittelson, P.M. (2004). Variation in *Lupinus arboreus* alkaloid profiles and relationships with multiple herbivores. *Biochemical Systematics and Ecology*, 32, 371–90.

6. Agrawal, A.A., Tuzun, S., and Bent, E. (1999). *Induced plant defenses against pathogens and herbivores*. APS Press, St Paul, MI.

7. Ågren, J. (1987). Intersexual difference in phenology and damage by herbivores and pathogens in dioecious *Rubus chamaemorus* L. *Oecologia*, 72, 161–9.

8. Ågren, J., Danell, K., Elmqvist, T., Ericson, L., and Hjältén, J. (1999). Sexual dimorphism and biotic interactions. In *Gender and sexual dimorphism in flowering plants* (ed. M.A. Geber, T.E. Dawson, and L.F. Delph), pp. 217–46. Springer, Berlin.

9. Alliende, M.C. (1989). Demographic studies of a dioicous tree. II. The distribution of leaf predation within and between trees. *Journal of Ecology*, 77, 1048–58.

10. Arimura, G., Ozawa, R., Shimoda, T., Nishioka, T., Boland, W., and Takabayashi, J. (2000). Herbivory-induced volatiles elicit defence genes in lima bean leaves. *Nature*, 406, 512–15.

11. Arimura, G., Huber, D.P.W., and Bohlman, J. (2004a). Forest tent caterpillars (*Malacosoma disstria*) induce local and systemic diurnal emissions of terpenoid volatiles in hybrid poplar (*Populus trichocarpa x deltoides*): cDNA cloning, functional characterization, and patterns of gene expression of germacence D synthase, PtdTPS1. *Plant Journal*, 37, 603–16.

12. Arimura, G., Ozawa, R., Kugimiya, S., Takabayashi, J., and Bohlmann, J. (2004b).

Herbivore-induced defense reponse in a model legume. Two-spotted spider mites induce emission of $(E)-\beta$-ocimene and transcript accumulation of $(E)-\beta$-ocimene synthase in *Lotus japonicus*. *Plant Physiology*, 135, 1976–83.

13. Ashihara, H. and Crozier, A. (2001). Caffeine: a well known but little mentioned compound in plant science. *Trends in Plant Science*, 6, 407–13.
14. Balbyshev, N. F. and Lorenzen, J. H. (1997). Hypersensitivity and egg drop, a novel mechanism of host-plant resistance to Colorado potato beetle (Coleoptera: Chrysomelidae). *Journal of Economic Entomology*, 90, 652–7.
15. Baldwin, I. T. (1994). Chemicial changes rapidly induced by folivory. In *Insect-plant interactions*, Vol. 5 (ed. E. A. Bernays), pp. 1–23. CRC Press, Boca Raton.
16. Baldwin, I. T. (1999) Inducible nicotine production in native *Nicotiana* as an example of adaptive phenotypic plasticity. *Joournal of Chemical Ecology*, 25, 3–30.
17. Baldwin, I. T. and Schultz, J. C. (1983). Rapid changes in tree leaf chemistry induced by damage: evidence for communication between plants. *Science*, 221, 277–9.
18. Balkema-Boomstra, A. G., Zijlstra, S., Verstappen, F. W. A., Inggamer, H., Mercke, P. E., Jongsma, M. A., *et al.* (2003). Role of cucurbitacin c in resistance to spider mite (*Tetranychus urticae*) in cucumber (*Cucumis sativus* L.). *Journal of Chemical Ecology*, 29, 225–35.
19. Barbour, J. D. and Kennedy, G. G. (1991). Role of steroidal glycoalkaloid α-tomatine in host-plant resistance of tomato to Colorado potato beetle. *Journal of Chemical Ecology*, 17, 989–1005.
20. Bell, E. A. and Janzen, D. H. (1971). Medical and ecological considerations on L-dopa and 5-HTP in seeds. *Nature*, 229, 136–7.
21. Benz, G. (1974). Negative Rückkoppelung durch Raum und Nahrungskonkurrenz sowie zyklische Veränderungen der Nahrungsgrundlage als Regel prinzip in der Populationsdynamik des Grauen Lärchenwicklers, *Zeiraphera diniana* (Guenee) (Lep., Tortricidae). *Zeitschrift für angewandte Entomologie*, 76, 196–228.
22. Berenbaum, M. R. (1995). Turnabout is fair play: secondary roles for primary compounds. *Journal of Chemical Ecology*, 21, 925–40.
23. Berebbaum, M. and Seigler, D. (1992). Biochemicals: engineering problems for natural selection. In *Insect chemical ecology* (ed. B. D. Roitberg and M. B. Isman), pp. 89–121. Chapman & Hall, New York.
24. Berenbaum, M. and Zangerl, A. R. (1992). Genetics of secondary metabolism and herbivore resistance in plants. In *Herbivores. Their interactions with secondary plant metabolites*, Vol. II (2nd edn) (ed. G. A. Rosenthal and M. R. Berebaum), pp. 415–38. Academic Press, San Diego.
25. Bingaman, B. R. and Hart, E. R. (1993). Clonal and leaf age variation in *Populus* phenolic glycosides: implications for host selection by *Chrysomela scripta* (Coleoptera: Chrysomelidae). *Environmental Entomology*, 22, 397–403.

26. Blaakmeer, A., Geervliet, J. B. F., van Loon, J. J. A., Posthumus, M. A., van Beek, T. A., and de Groot, Æ. (1994). Comparative heaspace analysis of cabbage plants damaged by two species of *Pieris* caterpillars: consequences for in-flight host location by *Cotesia* parasitoids. *Entomologia Experimentalis et Applicata*, 73, 175 – 82.
27. Boecklen, W. J. and Hoffman, M. T. (1993). Sex-biased herbivory: the importance of sex-by-environment interactions. *Oecologia*, 96, 49 – 55.
28. Boecklen, W. J., Price, P. W., and Mopper, S. (1990). Sex and drugs and herbivores: sex-biased herbivory in arroyo willow (*Salix lasiolepis*). *Ecology*, 71, 581 – 8.
29. Boege, K. (2004). Induced responses in three tropical dry forest plant species – direct and indirect effects on herbivory. *Oikos*, 107, 541 – 8.
30. Boland, W., Feng, Z., Donath, J., and Gäbler, A. (1992). Are acyclic C11 and C16 homoterpenes plant volatiles indicating herbivory? *Naturwisschenschaften*, 79, 368 – 71.
31. Bolter, C. J., Dicke, M., Van Loon, J. J. A., Visser, J. H., and Posthumus, M. A. (1997). Attraction of Colorado potato beetle to herbivore-damaged plants during herbivory and after its termination. *Journal of Chemical Ecology*, 23, 1003 – 23.
32. Booufford, D. E., Kjaer, A., Madsen, J. O., and Skrydstrup, T. (1989). Glucosinolates in Breschneideraceae. *Biochemical and Systematic Ecology*, 17, 375 – 9.
33. Bowers, M. D., (2003). Hostplant suitability and defensive chemistry of the catalpa sphinx, *Ceratomia catalpae*. *Journal of Chemical Ecology*, 29, 2359 – 67.
34. Bown, A. W., Hall, D. E., and MacGregor, K. B. (2002). Insect footsteps on leaves stimulate the accumulation of 4 – aminobutyrate and can be visualized through increased chlorophyll fluorescence and superoxide production. *Plant Physiology*, 129, 1430 – 4.
35. Broadway, R. M. and Colvin, A. A. (1992). Influence of cabbage proteinase inhibitors *in situ* on the growth of larval *Trichoplusia ni and Pieris rapae*. *Journal of Chemical Ecology*, 18, 1009 – 24.
36. Brooks, J. S. and Feeny, P. (2004). Seasonal variation in *Daucus carota* leaf-surface and leaf-tissue chemical profiles. *Biochemical Systematics and Ecology*, 32, 769 – 82.
37. Bruin, J., Sabelis, M. W., and Dicke, M. (1995). Do plants tap SOS signals from their infested neighbours? *Trends in Ecology and Evolution*, 10, 167 – 70.
38. Buchanan, B, B., Gruissen, W., and Jones, R. L. (2000). *Biochemistry and molecular biology of plants*. American Society of Plant Physiologists, Rockville, MD.
39. Buckingham, J. (1993). *Dictionary of natural products*, Vols. 1 – 8. Chapman & Hall, London.
40. Burns, A. E., Gleadow, R. M., and Woodrow, I. E. (2002). Light alters the allocation of nitrogen to cyanogenic glycosides in *Eucalyptus cladocalyx*. *Oecologia*, 133, 288 – 94.
41. Buttery, R. G. and Ling, L. C. (1984). Corn leaf volatiles: identification using Tenax trapping for possible insect attractants. *Journal of Agricultural and Food Chemistry*, 32, 1104 – 6.
42. Caballero, C., Castañera, P., Ortego, F., Fontana G., Pierro, P., Savona, G., *et al.* (2001). Effects of ajugarins and related neoclerodance diterpenoids on feeding behavious of

Leptinotarsa decemlineata and *Spodoptera exigua* larvae. *Phytochemistry*, 58, 249 – 56.
43. Callaway, R. M. (2002). The detection of neighbors by plants. *Trends in Ecology and Evolution*, 17, 104 – 5.
44. Catro, A. M., Rumi, C. P., and Arriaga, H. O. (1998). Influence of greenbug on root growth of resistant and susceptible barley genotypes. *Environmental and Experimental Botany*, 28, 61 – 72.
45. Cavalier-Smith, T. (1992). Origins of secondary metabolism. In *Secondary metabolites: their function and evolution* (ed. D. J. Chadwick and J. Whelan), pp. 64 – 87. John Wiley, Chichester.
46. Cervantes, D. E., Eigenbrode, S. D., Ding, H. J., and Bosque-Pérez, N. A. (2002). Oviposition responses by Hessian fly, *Mayetiola destructor*, to wheats varying in surface waxes. *Journal of Chemical Ecology*, 28, 193 – 210.
47. Chamberlain, K., Guerrieri, E., Pennacchio, F., Pettersson, J., Pickett, J. A., Poppy, G. M., *et al.* (2001). Can aphid-induced plant signals be transmitted aerially and through the rhizosphere? *Biochemical Systematics and Ecology*, 29, 1063 – 74.
48. Chapman, R. F. and Bernays, E. A. (1989). Insect behavior at the leaf surface and learning as aspects of host plant selection. *Experientia*, 45, 215 – 22.
49. Chen, F., Tholl, D., D'Auria, J. C., Farooq, A., Pichersky, E., and Gershenzon, J. (2003). Biosynthesis and emission of terpenoid volatiles from *Arabidopsis* flowers. *Plant Cell*, 15, 481 – 94.
50. Chew, F. S. and Renwick, J. J. A. (1995). Host-plant choice in *Pieris* butterflies. In *Chemical ecology of insects*, Vol. 2. (ed. R. T. Cardé and W. J. Bell), pp. 214 – 38. Chapman & Hall, New York.
51. Colazza, S., Fucarino, A., Peri, E., Salerno, G., Conti, E., and Bin, F. (2004). Insect oviposition induces volatile emission in herbaceous plants that attracts egg parasitids. *Journal of Experimental Biology*, 207, 47 – 53.
52. Cole, R. A. (1994). Locating a resistance mechanism to the cabbage aphid in two wild Brassicas. *Entomologia Experimentalis et Applicata*, 71, 23 – 31.
53. Cole, R. A. (1997). The relative importance of glucosinolates and amino acids to the development of two aphid pests *Brevicoryne brassicae* and *Myzus persicae* on wild and cultivated brassica species. *Entomologia Experimentalis et Applicata*, 85, 121 – 33.
54. Coleman, J. S. and Jones, C. G. (1991). A phytocentric perspective of phytochemical induction by herbivores. In *Phytochemical induction by herbivores* (ed. D. W. Tallamy and M. J. Raupp), pp. 3 – 45. John Wiley, New York.
55. Coley, P. D., (1986). Costs and benefits of defense by tannins in a neotropical tree. *Oecologia*, 70, 238 – 41.
56. Coley, P. D., Bryant, J. P., and Chapin, T. (1985). Resource availability and plant antiherbivore defense. *Science*, 230, 895 – 9.
57. Collinge, S. K. and Louda, S. M. (1988). Herbivory by leaf miners in response to experimen-

tal shading of a native crucifer. *Oecologia*, 75, 559 – 66.
58. Cooper-Driver, G., Finch, S., and Swain, T. (1977). Seasonal variation in secondary plant compounds in relation to palatability of *Pteridium aquilinum*. *Biochemical and Systematic Evolution*, 5, 177 – 83.
59. Costa, J. T., Gotzek, D. A., and Janzen, D. H. (2003). Late-instar shift in foraging strategy and trail pheromone use by caterpillars of the neotropical moth *Arsenura armida* (Cramer) (Saturniidae: Arsenurinae). *Journal of the Lepidopterists' Society*, 57, 220 – 9.
60. Dale, D. (1988). Plant-mediated effects of soil mineral stresses on insects. In *Plant stress-insect interactions* (ed. E. A. Heinrichs), pp. 35 – 110. John Wiley, NewYork.
61. Daniewski, W. M., Gumulka, M., Anczewski, W., Masnyk, M., Bloszyk, E., and Gupta, K. K. (1998). Why the yew tree (*Taxus baccata*) is not attacked by insects. *Phytochemistry*, 49, 1279 – 82.
62. Davidson, J. (1922). Biological studies of *Aphis rumicis* Linn. Reproduction on varieties of *Vicia faba*. *Annals of Applied Biology*, 9, 135 – 42.
63. De Boer, J. G., Posthumus, M. A., and Dicke, M. (2004). Identification of volatiles that are used in discrimination between plants infested with prey or non-prey herbivores by a predatory mite. *Journal of Chemical Ecology*, 30, 2215 – 30.
64. De Moraes, C. M., Mescher, M. C., and Tumlinson, J. H. (2001). Caterpillar-induced nocturnal plant volatiles repel nonspecific females. *Nature*, 410, 577 – 80.
65. Denno, R. F. and McClure, M. S. (1983). *Variable plants and herbivores in natural and managed systems*. Academic Press, San Diego.
66. Derridj, S. (1996). Nutrients on the leaf surface. In *Aerial plant surface microbiology* (ed. C. E. Morris, P. C. Nicot, and C. Nguyen-The), pp. 25 – 42. Plenum Press, New York.
67. Derridj, S., Wu, B. R., Stammitti, L., Garrec, J. P., and Derrien, A. (1996). Chemicals on the leaf surface, information about the plant available to insects. *Entomologia Experimentalis et Applicata*, 80, 197 – 201.
68. Dicke, M. (1986). Volatile spider-mite pheromone and host-plant kairomone, involved in spaced-out gregariousness in the spider mite *Tetranychus urticae*. *Physiological Entomology*, 11, 251 – 62.
69. Dicke, M. (1994). Local and systemic production of volatile herbivore-induced terpenoids: their role in plant-carnivore mutualism. *Journal of Plant Physiology*, 143, 465 – 72.
70. Dicke, M. (1999a). Are herbivore-induced plant volatiles reliable indicators of herbivore identity to foraging carnivorous arthropods? *Entomologia Experimentalis et Applicata*, 92, 131 – 42.
71. Dicke, M. (1999b). Evolution of induced indirect defence of plants. In *The ecology and evolution of inducible defenses* (ed. R. Tollrian and C. D. Harvell), pp. 62 – 88. Princeton University Press, Princeton, NJ.
72. Dicke, M. and Bruin, J. (2001). Chemical information transfer between plants: back to the future. *Biochemical Systematics and Ecology*, 29, 981 – 94.

73. Dicke, M. and Dijkman, H. (2001). Within-plant circulation of systemic elicitor of induced defence and release from roots of elicitor that affects neighbouring plants. *Biochemical Systematics and Ecology*, 29, 1075–87.
74. Dicke, M. and Hilker, M. (2003). Induced plant defences: from molecular biology to evolutionary ecology. *Basic and Applied Ecology*, 4, 3–14.
75. Dicke, M. and Van Loon, J. J. A. (2000). Multitrophic effects of herbivore-induced plant volatiles in an evolutionary context. *Entomologia Experimentalis et Applicata*, 97, 237–49.
76. Dicke, M. and Van Poecke, R. M. P. (2002). Signalling in plant-insect interactions: signal transduction in direct and indirect plant defence. In *Plant signal transduction* (ed. D. Scheel and C. Wasternack), pp. 289–316. Oxford University Press, Oxford.
77. Dicke, M., Gols, R., Ludeking, D., and Posthumus, M. A. (1999). Jasmonic acid and herbivory differentially induce carnivore-attracting plant volatiles in lima bean plants. *Journal of Chemical Ecology*, 25, 1907–22.
78. Dicke, M., Agrawal, A. A., and Bruin, J. (2003). Plant stalk, but are they deaf? *Trends in Plant Science*, 8, 403–5.
79. Dicke, M., Van Loon, J. J. A., and De Jong, P. W. (2004). Ecogenomics benefits community ecology. *Science*, 305, 618–19.
80. Dinan, L. (2001). Phytoecdysteroids: biological aspects. *Phytochemistry*, 57, 325–39.
81. Du, Y., Poppy, G. M., Powell, W., Pickett, J. A., Wadhams, L. J., and Woodcock, C. M. (1998). Identification of semiochemicals released during aphid feeding that attract parasitoid *Aphidius ervi*. *Journal of Chemical Ecology*, 24, 1355–68.
82. Dudareva, N., Pichersky, E., and Gershenzon, J. (2004). Biochemistry of plant volatiles. *Plant Physiology*, 135, 1893–902.
83. Dudt, J. F. and Shure, D. J. (1994). The influence of light and nutrients on foliar phenolics and insect herbivory. *Ecology*, 75, 86–98.
84. Dussourd, D. E. (1995). Entrapment of aphids and whiteflies in lettuce latex. *Annals of the Entomological Society of America*, 88, 163–72.
85. Dussourd, D. E. and Hoyle, A. M. (2000). Poisoned plusiines: toxicity of milkweed latex and cardenolides to some generalist caterpillars. *Chemoecology*, 10, 11–16.
86. Dyer, L. A. and Coley, P. D. (2002). Tritrophic interactions in tropical versus temperate communities. In *Multitrophic level interactions* (ed. T. Tscharntke and B. A. Hawkins), pp. 67–88. Cambridge University Press, Cambridge.
87. Edwards, P. B., Wanjura, W. J., Brown, W. V., and Dearn, J. M. (1990). Mosaic resistance in plants. *Nature*, 347, 434.
88. Eigenbrode, S. D. and Espelie, K. E. (1995). Effects of plant epicuticular lipids on insect herbivores. *Annual Review of Entomology*, 40, 171–94.
89. Facchini, P. J. (2001). Alkaloid biosynthesis in plants: biochemistry, cell biology, molecular regulation, and metabolic engineering applications. *Annual Review of Plant Physiology*, 52, 29–66.

90. Fahey, J. W., Zalcmann, A. T., and Talalay, P. (2001). The chemical diversity and distribution of glucosinolatesand isothiocyanates among plants. *Phytochemistry*, 56, 5 – 51.
91. Feeny, P. (1970). Seasonal changes in oak leaf tannins and nutrients as a cause of spring feeding by winter moth caterpillars. *Ecology*, 51, 565 – 81.
92. Fellows, L. E., Kite, G. C., Nash, R. J., Simmonds, M. S. J., and Schofield, A. M. (1989). Castanospermine, swainsonine and related polyhydroxy alkaloids: structure, distribution and biological activity. *Recent Advances in Phytochemistry*, 23, 395 – 427.
93. Fiehn, O. (2002). Metabolomics—the link between genotypes and phenotypes. *Plant Molecular Biology*, 48, 155 – 71.
94. Forkner, R. E., Marquis, R. J., and Lill, J. T. (2004). Feeny revisited: condensed tannins as anti-herbivore defences in leaf-chewing herbivore communities of *Quercus*. *Ecological Entomology*, 29, 174 – 87.
95. Fraenkel, G. S. (1959). The raison d'être of secondary plant substances. *Science*, 129, 1466 – 70.
96. Fraenkel, G. S. (1969). Evaluation of our thoughts on secondary plant substances. *Entomologia Experimentalis et Applicata*, 12, 473 – 86.
97. Fritz, R. S, Crabb, B. A., and Hochwender, C. G. (2003). Preference and performance of a gall-inducing sawfly: plant vigor, sex, gall traits and phenology. *Oikos*, 102, 601 – 13.
98. Frohne, D. and Jensen, U. (1973). *Systematik des Pflanzenreichs*. G. Fischer, Stuttgart.
99. Fung, S. Y. and Herrebout, W. M. (1988). Sorbitol and dulcitol in celastraceous and rosaceous plants, hosts of *Yponomeuta* spp. *Biochemical and Systematic Ecology*, 16, 191 – 4.
100. Gall, L. F. (1987). Leaflet position influences caterpillar feeding and development. *Oikos*, 49, 172 – 6.
101. Geiger, D. R. and Servaites, J. C. (1994). Diurnal regulation of photosynthetic carbon metabolism in C_3 plants. *Annual Review of Plant Physiology and Plant Molecular Biology*, 45, 235 – 56.
102. Gershenzon, J. (1984). Changes in the levels of plant secondary metabolites and water and nutrient stress. *Recent Advances in Phytochemistry*, 18, 273 – 320.
103. Gershenzon, J. (1994a). Metabolic costs of terpenoid accumulation in higher plants. *Journal of Chemical Ecology*, 20, 1281 – 354.
104. Gershenzon, J. (1994b). The cost of plant chemical defense against herbivory: a biochemical perspective. In *Insect-plant interactions*, Vol. 5 (ed. E. A. Bernays), pp. 105 – 73. CRC Press, Boca Raton.
105. Gershenzon, J. and Croteau, R. (1991). Terpenoids. In *Herbivores. Their interactions with secondary plant metabolites*, Vol. 1 (2nd edn) (ed. G. A. Rosenthal and M. R. Berenbaum), pp. 165 – 219. Academic Press, San Diego.
106. Gershenzon, J., McConkey, M. E., and Croteau, R. B. (2000). Regulation of monoterpene accumulation in leaves of peppermint. *Plant Physiology*, 122, 205 – 13.
107. Gil, L., Fuentes-Utrilla, P., Soto, Á., Cervera, M. T., and Collada, C. (2004). English

elm is a 2000-year-old Roman clone. *Nature*, 431, 1053.

108. Gleadow, R. M. and Woodrow, I. E. (2002). Constraints on effectiveness of cyanogenic glycosides in herbivore defense. *Journal of Chemical Ecology*, 28, 1301 – 13.

109. Gómez, D., Azorín, J., Bastida, J., Viladomat, F., and Codina, C. (2003). Seasonal and spatial variations of alkaloids in *Merendera montana* in relation to chemical defense and phenology. *Journal of Chemical Ecology*, 29, 1117 – 26.

110. Gouinguené, S. P. and Turlings, T. C. J. (2002). The effects of abiotic factors on induced volatile emissions in corn plants. *Plant Physiology*, 129, 1296 – 307.

111. Gouinguené, S., Degen, T., and Turlings, T. C. J. (2001). Variability in herbivore-induced odour emissions among maize cultivars and their wild ancestors (teosinte). *Chemoecology*, 11, 9 – 16.

112. Gouinguené, S., Alborn, H., and Turlings, T. C. J. (2003). Induction of volatile emissions in maize by different larval instars of *Spodoptera littoralis*. *Journal of Chemical Ecology*, 29, 145 – 62.

113. Gould, K. S. and Lee, D. W. (ed) (2002). Anthocyanins in leaves. *Advances in Botanical Research*, 37, 1 – 212.

114. Green, T. R. and Ryan, C. A. (1972). Wound induced proteinase inhibitors in plant leaves. *Science*, 175, 776 – 7.

115. Gref, R. and Tenow, O. (1987). Resin acid variation in sun and shade needles of Scots pine (*Pinus sylvestris*). *Canadian Journal of Forest Research*, 17, 346 – 9.

116. Halitschke, R., Schittko, U., Pohnert, G., Boland, W., and Baldwin, I. T. (2001). Molecular interactions between the specialist herbivore *Manduca sexta* (Lepidoptera, Sphingidae) and its natural host *Nicotiana attenuata*. III. Fatty acid-amino acid conjugates in herbivore oral secretions are necessary and sufficient for herbivore-specific plant responses. *Plant Physiology*, 125, 711 – 17.

117. Han, K. and Lincoln, D. E. (1994). The evolution of carbon allocation to plant secondary metabolites: a genetic analysis of cost in *Diplacus aurantiacus*. *Evolution*, 48, 1550 – 63.

118. Hanson, J. R. (2003). *Natural products: the secondary metabolites*. Royal Society of Chemistry, Cambridge, UK.

119. Harborne, J. B. (1993). *Introduction to ecological biochemistry* (4th edn). Academic Press, London.

120. Harborne, J. B. (1994). Phenolics. In *Natural products. Their chemistry and biological significance* (ed. J. Mann, R. S. Davidson, J. B. Hobbs, D. V. Banthorpe, and J. B. Harborne), pp. 362 – 88. Longman, Harlow.

121. Harborne, J. B. and Baxter, H. (1993). *Phytochemical dictionary. A handbook of bioactive compounds from plants*. Taylor & Francis, London.

122. Hare, J. D. (2002). Seasonal variation in the leaf resin components of *Mimulus aurantiacus*. *Biochemical Systematics and Ecology*, 30, 709 – 20.

123. Hartley, P. C., Monson, R. K., and Lerdau, M. T. (1999). Ecological and evolutionary as-

pects of isoprene emission from plants. *Oecologia*, 118, 109 – 23.

124. Hartley, S. E. and Lawton, J. H. (1991). Biochemical aspects and significance of the rapidly induced accumulation of phenolics in birch foliage. In *phytochemical induction by herbivores* (ed. D. W. Tallamy and M. J. Raupp), pp. 105 – 32. John Wiley, New York.

125. Hartmann T. (1996). Diversity and variability of plant secondary metabolism: a mechanistic view. *Entomologia Experimentalist et Applicata*, 80, 177 – 88.

126. Hatanaka, A. (1993). The biogeneration of green odour by green leaves. *Phytochemistry*, 34, 1201 – 18.

127. Haukioja, E. and Hakala, T. (1975). Herbivore cycles and periodic outbreaks. Formulation of a general hypothesis. *Report of the Kevo Subarctic Research Station*, 12, 1 – 9.

128. Haukioja, E, Ossipov, V., and Lempa, K. (2002). Interactive effects of leaf maturation and phenolics on consumption and growth of a geometrid moth. *Entomologia Experimntalis et Applicata*, 104, 125 – 36.

129. Hedin P. A., Thompson, A. C., and Gueldner, R. C. (1975). Survey of the air space volatiles of the cotton plant. *Phytochemistry*, 14, 2008 – 90.

130. Hedin P. A., Jenkins, J. N., Collum D. H., White, W. H., and Parrott, W. L. (1983). Multiple factors in cotton contributing to resistance to the tobacco budworm, *Heliothis virescens* F. In *Plant resistance to insects* (ed. P. A. Hedin), pp. 347 – 65. American Chemical Society Symposium 208, Washington, DC.

131. Hegnauer, R. (1962 – 2001). *Chemotaxonomie der Pflanzen*, Vols 1 – 11. Birkhauser, Basel.

132. Heinrichs, E. A. and Medrano, F. G. (1985). Influence of N fertilizer on the population development of brown planthopper(BPH). *International Rice Research Newsletter*, 10(6), 20 – 1.

133. Henriksson, J., Haukioja, E., Ossipov, V., Ossipova, S., Sillanpää, S., Kapari, L., et al. (2003). Effects of host shading on consumption and growth of the geometrid *Epirrita autumnata*: interactive roles of water, primary and secondary compounds. *Oikos*, 103, 3 – 16.

134. Herms, D. A. (2002). Effects of fertilization on insect resistance of woody ornamental plants: reassessing an entrenched paradigm. *Environmental Entomology*, 31, 923 – 33.

135. Hermsmeier, D., Schittko, U., and Baldwin, I. T. (2001). Molecular interactions between the specialist herbivore *Manduca sexta* (Lepidoptera, Sphingidae) and its natural host *Nicotiana attenuate*. I. Large-scale changes in the accumulation of growth and defense-related plant mRNAs. *Plant Physiology*, 125, 683 – 700.

136. Hilker, M. and Meiners, T. (2002). Induction of plant responses to oviposition and feeding by herbivorous arthropods: a comparison. *Entomologia Experimentalis et Applicata*, 104, 181 – 92.

137. Horiuchi, J., Arimura, G., Ozawa, R., Shimoda, T., Takabayashi, J., and Nishioka, T. (2001). Exogenous ACC enhances volatiles production mediated by jasmonic acid in lima bean leaves. *FEBS Lettrts*, 509, 332 – 6.

138. Hruska, A. J. (1988). Cyanogenic glucosides as defense compounds: a review of the evi-

dence. *Journal of Chemical Ecology*, 14, 2213 – 17.

139. Ikonen, A. (2002). Preference of six leaf beetle species among qualitatively different leaf age classes of three Salicaceous host species. *Chemoecology*, 12, 23 – 8.

140. Jetter, R. and Riederer, M. (1996). Cuticular waxes from the leaves and fruit capsules of eight Papaveraceae species. *Canandian Journal of Botany*, 74, 419 – 30.

141. Jetter, R. and Riederer, M. (2000). Composition of cuticular waxes on *Osmunda regalis* fronds. *Journal of Chemical Ecology*, 26, 399 – 412.

142. Jetter, R. and Schäffer, S. (2001). Chemical composition of the *Prunus laurocerasus* leaf surface. Dynamic changes of the epicuticular wax film during leaf development. *Plant Physiology*, 126, 1725 – 35.

143. Jetter, R., Schäffer, S., and Riederer, M. (2000). Leaf cuticular waxes are arranged in chemically and mechanically distinct layers: evidence from *Prunus laurocerasus. Plant, Cell and Environment*, 23, 619 – 28.

144. Johns, T. and Alonso, J. G. (1990). Glycoalkaloid change during domestication of the potato, *Solanum section Petota. Euphytica*, 50S, 203 – 10.

145. Johns, C. G., Hopper, R. F., Coleman, J. S., and Krischik, V. A. (1993). Control of systemically induced herbivore resistance by plant vascular architecture. *Oecologia*, 93, 452 – 6.

146. Juniper, B. E. and Southwood T. R. E. (1986). *Insects and the plant surface*. Edward Arnold, London.

147. Kahl, J., Siemens, D. H., Aerts, R. J., Gäbler, R., Kühnemann, F., Preston, C. A., *et al.* (2000). Herbivoreinduced ethylene suppresses a direct defense but not a putative indirect defense against an adapted herbivore. *Planta*, 210, 336 – 42.

148. Kainulainen, P., Nissinen, A., Piirainen, A., Tiilikkala, K., and Holopainen, J. K. (2002). Essential oil composition in leaves of carrot varieties and preference of specialist and generalist sucking insect herbivores. *Agriculture and Forest Entomology*, 4, 211 – 16.

149. Kalberer, N. M., Turling, T. C. J., and Rahier, M. (2001). Attraction of a leaf beetle (*Oreina cacaliae*) to damaged host plants. *Journal of Chemical Ecology*, 27, 647 – 61.

150. Karban, R. (1992). Plant variation: its effect on populations of herbivorous insects. In *Plant resistance to herbivores and pathogens* (ed. R. S. Fritz and E. L. Simms), pp. 195 – 215. University of Chicago Press, Chicago.

151. Karban, R. and Baldwin, I. T. (1997). *Induced responses to herbivory*. Chicago University Press, Chicago.

152. Karban, R., Baldwin I. T., Baxter, K. J., Laue, G., and Felton, G. W. (2000). Communication between plants: induced resistance in wild tobacco plants following clipping of neighboring sagebrush. *Oecologia*, 125, 66 – 71.

153. Karowe, D. (1989). Facultative monophagy as a consequence of prior feeding experience: behavioural and physiological specialization in *Colias philoice* larvae. *Oecologia*, 78, 106 – 11.

154. Kellogg, D. W., Taylor., T. N., and Krings, M. (2002). Effectiveness in defense against phytophagous arthropods of the cassabanana (*Sicana odorifera*) glandular trichomes. *Entomol-

igia *Experimentalis et Applicata*, 103, 187-9.

155. Kessler, A., Halitschke, R., and Baldwin, I. T. (2004). Silencing the jasmonate cascade: induced plant defenses and insect populations. *Science*, 305, 665-8.

156. Kester, K. M., Peterson, S. C., Hanson, F., Jackson, D. M., and Severson, R. F. (2002). The roles of nicotine and natural enemies in determining larval feeding site distributions of *Manduca sexta* L. and *Manduca quinquemaculata* (Haworth) on tobacco. *Chemoecology*, 12, 1-10.

157. Kimmerer, T. W. and Potter, D. A. (1987). Nutritional quality of specific leaf tissues and selective feeding by a specialist leaf miner. *Oecologia*, 71, 548-51.

158. Kirstine, W., Galbally, I., Ye, Y. R., and Hooper, M. (1998). Emissions of volatile organic compounds (primarily oxygenated species) from pasture. *Journal of Geophysical Research-Atmospheres*, 103, 10605-19.

159. Klages. K., Donnison, H., Wünsche, J., and Boldingh, J. (2001). Diurnal changes in non-structural carbohydrates in leaves, phloem exudate and fruit in 'Braeburn' apple. *Australian Journal of Plant Physiology*, 28, 131-9.

160. Kleiner, K. W., Raffa, K. F., and Dickson, R. E. (1999). Partitionng of ^{14}C-labeled photosynthate to allelochemicals and primary metabolites in source and sink leaves of aspen: evidence for secondary metabolite turnover. *Oecologia*, 119, 408-18.

161. Klingauf, F. (1971). Die Wirkung des Glucosids Phlorizin auf das Wirtswahlverhalten von *Rhopalosiphum insertum* (Walk.) und *Aphis pomi* de Geer (Homoptera: Aphididae). *Zeitschrift für Angewandte Entomologie*, 68, 41-55.

162. Kollner, T. G., Schnee, C., Gershenzon, J., and Degenhardt, J. (2004). The variability of sesquiterpenes cultivars emitted from two *Zea mays* cultivars is controlled by allelic variation of two terpene synthase genes encoding stereoselective multiple product enzymes. *Plant Cell*, 16, 1115-31.

163. Koricheva, J. (1999). Interpreting phenotypic variation in plant allelochemistry: problems with the use of concentrations. *Oecologia*, 119, 467-73.

164. Koricheva, J., Larsson, S., Haukioja, E., and Keinanen, M. (1998). Regulation of woody plant secondary metabolism by resource availability: hypothesis testing by means of meta-analysis. *Oikos*, 83, 212-26.

165. Krischik, V. A. and Denno, R. F. (1983). Individual, populational, and geographic patterns in plant defense. In *Variable plants and herbivores in natural and managed systems* (ed. R. F. Denno and M. S. McClure), pp. 463-512. Academic Press, San Diego.

166. Krischik, V. A. and Denno, R. F. (1990). Patterns of growth, reproduction, defense, and herbivory in the dioecious shrub *Baccharis halimifolia* (Compositae). *Oecologia*, 83, 182-90.

167. Lance, D. R., Elkinton, J. S., and Schwalbe, C. P. (1986). Feeding rhythms of gypsy moth larvae; effect of food quality during outbreaks. *Ecology*, 67, 1650-4.

168. Lane, G. A., Biggs, D. R, Sutherland, O. W. R., Williams, E. M., Maindonald, J. M.,

and Donnell, D. J. (1985). Isoflavonoid feeding deterrents for *Costelytra zealandica*: structure-activity relationships. *Journal of Chemical Ecology*, 11, 1713 – 35.

169. Langenheim, J. H. (1994). Higher plant terpenoids: a phytocentric overview of their ecological roles. *Journal of Chemical Ecology*, 20, 1223 – 80.

170. Larsson, S. and Ohmart, C. P. (1988). Leaf age and larval performance of the leaf beetle, *Paropsis atomari*. *Evolutionary Entomology*, 13, 19 – 24.

171. Lewinsohn, T. M. (1991). The geographical distribution of plant latex. *Chemoecology*, 2, 64 – 8.

172. Lichtenthaler, H. K. (1999). The 1-deoxy – D – xylulose – 5 – phosphate pathway of isoprenoid biosynthesis in plants. *Annual Review of Plant Physiology and Plant Molecular Biology*, 50, 47 – 65.

173. Lombarkia, N. and Derridj, S. (2002). Incidence of apple fruit and leaf surface metabolites on *Cydia pomonella* oviposition. *Entomologia Experimentalis et Applicata*, 104, 79 – 86.

174. Loughrin, J. H., Manukian, A., Heath, R. R., Turlings, T. C. J., and Tumlinson, J. H. (1994). Diurnal cycle of emission of induced volatile terpenoids herbivore-injured cotton plants. *Proceedings of the National Academy of Sciences of the USA*, 91, 11836 – 40.

175. Lovett Doust, J. and Lovett Doust, L. (1985). Sex ratios, clonal growth and herbivory in *Rumex acetosella*. In *Studies on plant demography: a festschrift for John L. Harper* (ed. J. White), pp. 327 – 41. Academic Press, London.

176. Luckner, M. (1990). *Secondary metabolism in microorganisms, plants, and animals* (3rd edn). Springer, Berlin.

177. Deleted.

178. Macel, M., Vrieling, K., and Klinkhamer, P. G. L. (2004). Variation in pyrrolizidine alkaloid patterns of *Senecio jacobaea*. *Phytochemistry*, 65, 865 – 73.

179. Malone, M., Palumbo, L., Boari, F., Monteleone, M., and Jones, H. G. (1994). The relationship between wound-induced proteinase inhibitors and hydraulic signals in tomato seedlings. *Plant Cell and Environment*, 17, 81 – 7.

180. Marak, H. B., Biere, A., and van Damme, J. M. M. (2000). Direct and correlated responses to selection on iridoid glycosides in *Plantago lanceolata* L. *Journal of Evolutionary Biology*, 13, 985 – 96.

181. Marion-Poll, F. and Descoins, C. (2002). Taste detection of phytoecdysteroids in larvae of *Bombyxmori*, *Spodoptera littoralis* and *Ostrinia nubilalis*. *Journal of Insect Physiology*, 48, 467 – 76.

182. Martin, D., Tholl, D., Gershenzon, J., and Bohlmann, J. (2002). Methyl jasmonate induces traumatic resinducts, terpenoid resin biosynthesis, and terpenoid accumulation in developing xylem of Norway spruce stems. *Plant Physiology*, 129, 1003 – 18.

183. Matile, P. (1984). Das toxische Kompartiment der Pflanzenzell. *Naturwissenschaften*, 71, 18 – 24.

184. Mattiacci, L., Dicke, M., and Posthumus, M. A. (1995). β-Glucosidase: an elicitor of her-

bivore-inducible plant odors that attract host-searching parasitic wasps. *Proceedings of the National Academy of Science of the USA*, 92, 2036 – 46.

185. Mebs, D. and Schneider, M. (2002). Aristolochic acid content of South-East Asian troidine swallowtails(Lepidoptera: Papilionidae) and of *Aristolochia* plant species (Aristolochiaceae). *Chemoecology*, 12, 11 – 13.

186. Mechaber, W. L., Marshall, DB., Mechaber, R. A., Jobe, R. T., and Chew, F. S. (1996). Mapping leaf surface landscapes. *Proceedings of the National Academy of Science of the USA*, 93, 4600 – 3.

187. Meiners, T. and Hilker, M. (2000). Induction of plant synomones by oviposition of a phytophagous insect. *Journal of Chemical Ecology*, 26, 221 – 32.

188. Merck Index (13th edn). (2001). Merck and Co., Rahway, NJ.

189. Mercke, P., Kappers, I. F., Verstappen, F. W. A., Vorst, O., Dicke, M., and Bouwmeester, H. J. (2004). Combined transcript and metabolite analysis reveals genes involved in spider mite induced volatile formation in cucumber plants. *Plant Physiology*, 135, 2012 – 24.

190. Mitchell, B. K., Justus, K. A., and Asaoka, K. (1996). Deterrency and the variable caterpillar: *Trichoplusiani* and sinigrin. *Entomologia Experimentalis et Applicata*, 80, 27 – 31.

191. Mithen, R. F. (2001). Glucosinolates and their degradation products. *Advances in Botanical Research*, 35, 214 – 62.

192. Mole, S. and Joern, A. (1993). Foliar phenolics of Nebraska sandhills prairie graminoids: betweenyears, seasonal, and interspecific variation. *Journal of Chemical Ecology*, 19, 1861 – 74.

193. Mole, S., Ross, J. A. M., and Waterman, P. G. (1988). Light-induced variation in phenolic levels of foliage of rain-forest plants. I. Chemical changes. *Journal of Chemical Ecology*, 14, 1 – 21.

194. Mordue (Luntz), A. J. and Blackwell, A. (1993). Azadirachtin: an update. *Journal of Insect Physiology*, 39, 903 – 24.

195. Musser, R. O., Hum-Musser, S. M., Eichenseer, H., Peiffer, M., Ervin, G., Murphy, J. B., et al. (2002) Caterpillar saliva beats plant defences—a new weapon emerges in the evolutionary arms race between plants and herbivores. *Nature*, 416, 599 – 600.

196. Muthukrishnan, J. and Selvan, S. (1993). Fertilization affects leaf consumption and utilization by *Porthesia scintillans* Walker (Lepidoptera: Lymantridae). *Annals of the Entomological Society of America*, 86, 173 – 8.

197. Mutikainen, P., Walls, M., Ovaska, J., Keinänen, M., Julkunen-Tiitto, R., and Vapaavuori, E. (2002). Costs of herbivore resistance in clonal saplings of *Betula pendula*. *Oecologia*, 133, 364 – 71.

198. Myers, J. H. (1985). Effect of physiological condition of the host plant on the ovipositional choice of the cabbage white butterfly *Pieris rapae*. *Journal of Animal Ecology*, 54: 193 – 204.

199. Nault, L. R. and Styer, W. E. (1972). Effects of sinigrin on host selection by aphids. *Ento-

mologia Experimentalis et Applicata, 15, 423 – 37.

200. Neuvonen, S. and Haukioja, E. (1991). The effects of inducible resistance in host foliage on birch-feeding herbivores. In *Phytochemical induction by herbivores* (ed. D. W. Tallamy and M. J. Raupp), pp. 277 – 91. John Wiley, New York.

201. Nitao, J. J., Zangerl, A. R., and Berenbaum, M. R. (2002). CNB: requiescat in pace? *Oikos*, 98, 540 – 6.

202. Nykanen, H. and Koricheva, J. (2004). Damageinduced changes in woody plants and their effects on insect herbivore performance: a meta-analysis. *Oikos*, 104, 247 – 68.

203. O'Donnell, P. J., Calvert, C., Atzorn, R., Wasternack, C., Leyser, H. M. O., and Bowles, D. J. (1996). Ethylene as a signal mediating the wound response of tomato plants. *Science*, 274, 1914 – 17.

204. Ogutuga, D. B. A. (1975). Chemical composition and potential commercial use of kola nut, *Cola nitida*, Vent. Schott and Endlicher. *Ghana Journal of Agricultural Science*, 8, 121 – 5.

205. Okolie, P. N. and Obasi, B. N. (1993). Diurnal variationof cyanogenic glucosides, thiocyanate and rhodanese in cassava. *Phytochemistry*, 33, 775 – 8.

206. Orians, C. M., Pomerleau, J., and Ricco, R. (2000). Vascular architecture generates fine scale variation in systemic induction of proteinase inhibitors in tomato. *Journal of Chemical Ecology*, 26, 471 – 85.

207. Orozco-Cardenas, M., McGurl, B., and Ryan, C. A. (1993). Expression of an antisense prosystemin gene in tomato plants reduces resistance toward *Manduca sexta* larvae. *Proceedings of the National Academy of Sciences of the USA*, 90, 8273 – 6.

208. Panzuto, M., Lorenzetti, F., Mauffette, Y., and Albert, P. J. (2001). Perception of aspen and sun/shade sugar maple leaf soluble extracts by larvae of *Malacosoma disstria*. *Journal of Chemical Ecology*, 27, 1963 – 78.

209. Paul, M. J. and Cockburn, W. (1989). Pinitol, a compatible solute in *Mesembryanthemum crystallinum* L. *Journal of Experimental Botany*, 40, 1093 – 8.

210. Pieterse, C. M. J. and Van Loon, L. C. (1999). Salicylic acid-independent plant defence pathways. *Trends in Plant Science*, 4, 52 – 8.

211. Pourmohseni, H., Ibenthal, W. D., Machinek, R., Remberg, G., and Vray, V. (1993). Cyanoglucosidases in the epidermis of *Hordeum vulgare*. *Phytochemistry*, 33, 295 – 7.

212. Preston, C. A., Laue, G., and Baldwin, I. T. (2001). Methyl jasmonate is blowing in the wind, but can it act as a plant-plant airborne signal? *Biochemical Systematics and Ecology*, 29, 1007 – 23.

213. Rausher, M. D. (1992). Natural selection and the evolution of plant-insect interactions. In *Insect chemical ecology* (ed. B. D. Roitberg and M. B. Isman), pp. 20 – 88. Chapman & Hall, New York.

214. Rees, C. J. C. (1969). Chemoreceptor specificity associated with choice of feeding site by the beetle, *Chrysolina brunsvicensis* on its foodplant, *Hypericum hirsutum*. *Entomologia Experimentalis et Applicata*, 12: 565 – 83.

215. Renwick, J. A. A. (2002). The chemical world of crucivores: lures, treats and traps. *Entomologia Experimentalis et Applicata*, 104, 35-42.
216. Reymond, P., Weber, H., Damond, M., and Farmer, E. E. (2000). Differential gene expression in response to mechanical wounding and insect feeding in *Arabidopsis*. *Plant Cell*, 12, 707-19.
217. Reymond, P., Bodenhausen, N., Van Poecke, R. M. P., Krishnamurthy, V., Dicke, M., and Farmer, E. E. (2004). A conserved transcript pattern in response to a specialist and a generalist herbivore. *Plant Cell*, 16, 3132-47.
218. Rhoades, D. F. (1977). The antiherbivore chemistry of *Larrea*. In *Creosote bush: biology and chemistry* of Larrea *in new world deserts* (ed. T. J. Mabry, J. H. Hunziker, and D. R. DiFeo), pp. 135-77. Dowden, Hutchinson & Ross, Stroutberg.
219. Rhoades, D. F. (1983). Responses of alder and willow to attack by tent caterpillars and webworms: evidence for pheromonal sensitivity of willows. In *Plant resistance to insects* (ed. P. A. Hedin), pp. 55-68. American Chemical Society Symposium 208, Washington, DC.
220. Robinson, T. (1974). Metabolism and function of alkaloids in plants. *Science*, 184, 430-5.
221. Roda, A. L. and Baldwin, I. T. (2003). Molecular technology reveals how the induced direct defenses of plants work. *Basic and Applied Ecology*, 4, 15-26.
222. Romani, A., Pinelli, P., Mulinacci, N., Vincieri, F. F., Gravano, E., and Tattini, M. (2000). HPLC analysis of flavonoids and secoiridoids in leaves of *Ligustrum vulgare* L. (Oleaceae). *Journal of Agricultural and Food Chemistry*, 48, 4091-6.
223. Romeo, J. T., Ibrahim, R., Varin, L., and De Luca, V. (ed.). (2000). Evolution of metabolic pathways. *Recent Advances in Phytochemistry* 34, 1-480.
224. Rosenthal, G. A. and Berenbaum, M. R. (ed.) (1991-1992). *Herbivores. Their interactions with secondary plant metabolites*, 2 Vols (2nd edn). Academic Press, San Diego.
225. Rosenthal, G. A. and Janzen, D. H. (ed.) (1979). *Herbivores. Their interaction with secondary plant metabolites*. Academic Press, San Diego.
226. Rowe, W. J. and Potter, D. A. (2000). Shading effects on susceptibility of *Rosa* spp. to defoliation by *Popillia japonica* (Coleoptera: Scarabaeidae). *Environmental Entomology*, 29, 502-8.
227. Ruuhola, T. M. and Julkunen-Tiitto, M. R. K. (2000). Salicylates of intact *Salix myrsinifolia* plantlets do not undergo rapid metabolic turnover. *Plant Physiology*, 122, 895-905.
228. Ryan, C. A. (2000). The systemin signaling pathway: differential activation of plant defensive genes. *Biochimica et Biophysica Acta—Protein Structure and Molecular Enzymology*, 1477, 112-21.
229. Ryan, J. D., Gregory, P., and Tingey, T. M. (1982). Phenolic oxidase activities in glandular trichomes of *Solanum berthaultii*. *Phytochemistry*, 21, 1885-7.
230. Sanderson, G. W. (1972). The chemistry of tea and tea manufacturing. *Recent Advances in Phytochemistry*, 5, 247-316.

231. Sandström, J. (2000). Nutritional quality of phloem sap in relation to host plant-alternation in the bird cherry-oat aphid. *Chemoecology*, 10, 17–24.
232. Schmeller, T. and Wink, M. (1998). Utilization of alkaloids in modern medicine. In *Alkaloids. Biochemistry, ecology, and medical applications* (ed. M. F. Roberts and M. Wink), pp. 435–59. Plenum, New York.
233. Schmelz, E. A., Grebenok, R. J., Ohnmeiss, T. E., and Bowers, W. S. (2000). Phytoecdysteroid turnover in spinach: long-term stability supports a plant defense hypothesis. *Journal of Chemical Ecology*, 26, 2883–96.
234. Schultz, J. C. (1983). Habitat selection and foraging tactics of caterpillars in heterogeneous trees. In *Variable plants and herbivores in natural and managed systems* (ed. R. F. Denno and M. S. McClure), pp. 61–90. Academic Press, San Diego.
235. Schwab, W. (2003). Metabolome diversity: too few genes, too many metabolites? *Phytochemistry*, 62, 837–49.
236. Scriber, J. M. (1984). Nitrogen nutrition of plants and insect invasion. In *Nitrogen in crop production* (ed. R. D. Hack), pp. 175–228. American Society of Agronomy, Madison, WI.
237. Sefton, M. A., Francis, I. L., and Williams, P. J. (1994). Free and bound volatile secondary metabolites of *Vitis vinifera* grape cv. Sauvignon blanc. *Journal of Food Science*, 59, 142–7.
238. Seigler, D. S. and Conn, E. E. (1979). Cited by D. H. Janzen in Ref. 225.
239. Sharkey, T. D. and Yeh, S. (2001). Isoprene emission from plants. *Annual Review of Plant Physiology and Plant Molecular Biology*, 52, 407–36.
240. Simmonds, M. S. J. (2001). Importance of flavonoids in insect-plant interactions: feeding and oviposition. *Phytochemsitry*, 56, 245–52.
241. Simms, E. L. (1992). Costs of plant resistance to herbivory. In *Plant resistance to herbivores and pathogens* (ed. R. S. Fritz and E. L. Simms), pp. 392–425. University of Chicago Press, Chicago.
242. Sinden, S. L., Schalk, J. M., and Osman, S. F. (1978). Effects of daylength and maturity of tomato plants on tomatine content and resistance to the Colorado potato beetle. *Journal of the American Society of Horticultural Science*, 103, 596–600.
243. Sirvent, T. M., Krasnoff, S. B., and Gibson, D. M. (2003). Induction of hypericins and hyperforins in *Hypericum perforatum* in response to damage by herbivores. *Journal of Chemical Ecology*, 29, 2667–91.
244. Sisson, V. A. and Saunders, J. A. (1982). Alkaloid composition of the USDA tobacco (*Nicotiana tabacum* L.) introduction collection. *Tobacco Science*, 26, 117–20.
245. Smiley, J. T., Horn, J. M., and Rank, N. E. (1985). Ecological effects of salicin at three trophic levels: new problems from old adaptations. *Science*, 229, 649–51.
246. Smith, P. M. (1976). *The chemotaxonomy of plants*. Edward Arnold, London.
247. Soldaat, L. L., Boutin, J. P., and Derridj, S. (1996). Species-specific composition of free amino acids from the leaf surface of four *Senecio* species. *Journal of Chemical Ecology*, 22,

1 – 12.

248. Southwell, I. A. and Bourke, C. A. (2001). Seasonal variation in hypericin content of *Hypericum perforatum* L. (St John's wort). *Phytochemistry*, 56,437 – 41.
249. Städler, E. and Roessingh, P. (1991). Perception of surface chemicals by feeding and ovipositing insects. *Symposia Biologica Hungarica*, 39, 71 – 86.
250. Stanjek, V., Herhaus, C., Ritgen, U., Boland, W., and Städler, E. (1997). Changes in the leaf surface chemistry of *Apium graveolens* (Apiaceae) stimulated by jasmonic acid and perceived by a specialist insect. *Helvetica Chimica Acta*, 80, 1408 – 20.
251. Storer, J. R., Elmore, J. S., and van Emden, H. F. (1993). Airborne volatiles from the foliage of three cultivars of autumn flowering chrysanthemums. *Phytochemistry*,34, 1489 – 92.
252. Stowe, K. A. (1998). Realized defense of artificially selected lines of *Brassica rapa*: effects of quantitative variation in foliar glucosinolate concentration. *Environmental Entomology*, 27, 1166 – 74.
253. Strauss, S. Y., Rudgers, J. A., Lau, J. A., and Irwin, R. E. (2002). Direct and ecological costs of resistance to herbivory. *Trends in Ecology and Evolution*, 17,278 – 85.
254. Suomela, J., Kaitaniemi, P., and Nilson, A. (1995). Systematic within-tree variation in mountain birch leaf quality for a geometrid, *Epirrita autumnata*. *Ecological Entomology*, 20, 283 – 92.
255. Svoboda, G. H. (1961). Alkaloids of *Vinca rosea* (*Catharanthus roseus*). IX. Extraction and characterization of leurosidine and leurocristine. *Lloydia*, 24,173 – 8.
256. Swain, T. (1979). Tannins and lignins. In *Herbivores. Their interaction with secondary plant metabolites* (ed. G. A. Rosenthal and D. H. Janzen), pp. 657 – 82. Academic Press, New York.
257. Tallamy, D. W., Stull, J., Ehresman, N. P., Gorski, P. M., and Mason, C. E. (1997). Cucurbitacins as feeding and oviposition deterrents to insects. *Environmental Entomology*, 26, 678 – 83.
258. Thaler, J. S., Fidantsef, A. L., and Bostock, R. M. (2002). Antagonism between jasmonate and salicylate-mediated induced plant resistance: effects of concentration and timing of elicitation on defense related proteins, herbivore, and pathogen performance in tomato. *Journal of Chemical Ecology*, 28,1131 – 59.
259. Thieme, H. and Benecke, R. (1971). Die Phenylglykoside der Salicaceen. Untersuchungen über die Glykosidakkumulation in einigen mitteleuropäischen *Populus* – Arten. *Die Pharmazie*, 26, 227 – 31.
260. Tscharntke, T., Thiessen, S., Dolch, R., and Boland, W. (2001). Herbivory, induced resistance, and interplant signal transfer in *Alnus glutinosa*. *Biochemical Systematics and Ecology*, 29, 1025 – 47.
261. Turlings, T. C. J. (1994). The active role of plants in the foraging successes of entomophagous insects. *Norwegian Journal of Agricultural Sciences*, 16,211 – 19.
262. Turlings, T. C. J. and Tumlinson, J. H. (1992). Systemic release of chemical signals by her-

bivore-injured corn. *Proceedings of the National Academy of Sciences of the USA*, 89, 8399 – 402.

263. Turlings, T. C. J., Tumlinson, J. H., and Lewis, W. J. (1990). Exploitation of herbivore-induced plant odors by host-seeking parasitic wasps. *Science*, 250, 1251 – 3.
264. Turlings, T. C. J., Loughrin, J. H., McCall, P. J., Rose, U. S. R., Lewis, W. J., and Tumlinson, J. H. (1995). How caterpillar-damaged plants protect themselves by attracting parasitic wasps. *Proceedings of the National Academy of Sciences of the USA*, 92, 4169 – 74.
265. Turlings, T. C. J., Alborn, H. T., Loughrin, J. H., and Tumlinson, J. H. (2000). Volicitin, an elicitor of maize volatiles in oral secretion of *Spodoptera exigua*: isolation and bioactivity. *Journal of Chemical Ecology*, 26, 189 – 202.
266. Van Beek, T. A. and de Groot, AE. (1986). Terpenoid antifeedants, part I. An overview of terpenoid antifeedants of natural origin. *Recueil des Travaux Chimiques des Pays Bas*, 105, 513 – 27.
267. Van Dam, N. M. and Vrieling, K. (1994). Genetic variation in constitutive and inducible pyrrolizidine alkaloid levels in *Cynoglossum officinale* L. *Oecologia*, 99, 374 – 8.
268. Van Dam, N. M., Verpoorte, R., and van der Meijden, E. (1994). Extreme differences in pyrrolizidine alkaloid levels between leaves of *Cynoglossum officinale*. *Phytochemistry*, 27, 1013 – 16.
269. Van Dam, N. M., Witte, L., Theuring, C., and Hartmann, T. (1995). Distribution, biosynthesis, and turnover of pyrrolizidine alkaloids in *Cynoglossum officinale* L. *Phytochemistry*, 39, 287 – 92.
270. Van Dam, N. M., De Jong, T. J., Iwasa, Y., and Kubo, T. (1996). Optimal distribution of defenses: are plants smart investors? *Functional Ecology*, 10, 128 – 36.
271. Van den Boom, C. E. M., Van Beek, T. A., Posthumus, M. A., De Groot, AE., and Dicke, M. (2004). Qualitative and quantitative variation among volatile profiles induced by *Tetranychus urticae* feeding on plants from various families. *Journal of Chemical Ecology*, 30, 69 – 89.
272. Van Loon, J. J. A., Blaakmeer, A., Griepink, F. C., van Beek, T. A., Schoonhoven, L. M., and de Groot, AE. (1992). Leaf surface compound from *Brassica oleracea* (Cruciferae) induces oviposition by *Pieris brassicae* (Lepidoptera: Pieridae). *Chemoecology*, 3, 39 – 44.
273. Van Loon, J. J. A., Wang, C.-Z., Nielsen, J. K., Gols, R., and Qiu, Y.-T. (2002). Flavonoids from cabbage are feeding stimulants for diamondback moth larvae additional to glucosinolates: chemoreception and behaviour. *Entomologia Experimentalis et Applicata*, 104, 27 – 34.
274. Van Oosten, V. R., De Vos, M., Van Pelt, J. A., Van Loon, L. C., Van Poecke, R. M. P., Dicke, M., *et al.* (2004). Signal signature of *Arabidopsis* induced upon pathogen and insect attack. In *Biology of plant-microbe interactions*, Vol. 4 (ed. I. Tikhonovich, B. Lugtenberg, and N. Provorov), pp. 199 – 202. International Society for Molecular Plant-Microbe Interactions, St Paul, MN.

275. Van Poecke, R. M. P. and Dicke, M. (2002). Induced parasitoid attraction by *Arabidopsis thaliana*: involvement of the octadecanoid and the salicylic acid pathway. *Journal of Experimental Botany*, 53,1793 – 9.
276. Van Poecke, R. M. P., Posthumus, M. A., and Dicke, M. (2001). Herbivore-induced volatile production by *Arabidopsis thaliana* leads to attraction of the parasitoid *Cotesia rubecula*: chemical, behavioral, and gene-expression analysis. *Journal of Chemical Ecology*, 27, 1911 – 28.
277. Vandenberg, P. and Matzinger, D. F. (1970). Genetic diversity and heterosis in *Nicotiana*. III. Crosses among tobacco introductions and flue-cured varieties. *Crop Science*, 10, 437 – 40.
278. Verschaffelt, E. (1910). The cause determining the selection of food in some herbivorous insects. *Proceedings Koninklijke Nederlandse Akademie van Wetenschappen*, 13, 536 – 42. (Reprinted in *Proceedings Koninklijke Nederlandse Akademie van Wetenschappen*, 100, 362 – 68, 1997)
279. Vickery, M. L. and Vickery, B. (1981). *Secondary plant metabolism*. Macmillan, London.
280. Visser, J. H. and Avé, D. A. (1978). General green leaf volatiles in the olfactory orientation of the Colorado potato beetle, *Leptinotarsa decemlineata*. *Entomologia Experimentalis et Applicata*, 24, 738 – 49.
281. Vrieling, K. and Van Wijk, A. M. (1994). Cost assessment of the production of pyrrolizidine alkaloids in ragwort (*Senecio jacobaea* L.). *Oecologia*, 97, 541 – 6.
282. Vuorinen, T., Reddy, G. V. P., Nerg, A. M., and Holopainen, J. K. (2004). Monoterpene and herbivore-induced emissions from cabbage plants grown at elevated atmospheric CO_2 concentration. *Atmospheric Environment*, 38, 675 – 82.
283. Walker, K., Long, R., and Croteau, R. (2002). The final acylation step in taxol biosynthesis: cloning of the taxoid C13 – side-chain N-benzoyltransferase from *Taxus*. *Proceedings of the National Academy of Science of the USA*, 99, 9166 – 71.
284. Waterman, P. G. and Mole, S. (1989). Extrinsic factors influencing production of secondary metabolites in plants. In *Insect-plant interactions*, Vol. 5 (ed. E. A. Bernays), pp. 107 – 34. CRC Press, Boca Raton.
285. Waterman, P. G. and Mole, S. (1994). *Analysis of phenolic plant metabolites*. Blackwell, Oxford.
286. Watson, M. A. (1986). Integrated physiological units in plants. *Trends in Ecology and Evolution*, 1,119 – 23.
287. Webber, B. L. and Woodrow, I. E. (2004). Cassowary frugivory, seed defleshing and fruit fly infestation influence the transition from seed to seedling in the rare Australian rainforest tree, *Ryparosa* sp. nov. 1 (Achariaceae). *Functional Plant Biology*, 31,505 – 16.
288. White, T. C. R. (1970). The nymphal stage of *Cardiaspina densitexta* (Homoptera: Psyllidae) on leaves of *Eucalyptus fasciculosa*. *Australian Journal of Zoology*, 18, 273 – 3.
289. Whitham, T. G. (1983). Host manipulation of parasites: within-plant variation as a defense against rapidly evolving pests. In *Variable plants and herbivores in natural and managed systems*

(ed. R. F. Denno and M. S. McClure), pp. 15-41. Academic Press, San Diego.

290. Whittaker, R. H. (1970). The biochemical ecology of higher plants. In *Chemical ecology* (ed. E. Sondheimer and J. B. Simeone), pp 43-70. Academic Press, New York.

291. Wittstock, U. and Halkier, B. A. (2002). Glucosinolate research in the *Arabidopsis* era. *Trends in Plant science*, 7, 263-70.

292. Wildon, D. C., Thain, J. F., Minchin, P. E. H., Gubb, I. R., Reilly, A. J., Skipper, Y. D., *et al.* (1992). Electrical signalling and systemic proteinase inhibitor induction in the wounded plant. *Nature*, 360, 62-5.

293. Wink. M. (1998). Chemical ecology of alkaloids. In *Alkaloids. Biochemistry, ecology and medicinal applicationsc* (ed. M. F. Roberts and M. Wink), pp. 265-326. Plenum, New York.

294. Wink, M. and Roberts, M. F., (1998). Compartmentation of alkaloid synthesis, transport, and storage. In *Alkaloids. Biochemistry, ecology and medical applications* (ed. M. F. Roberts, and M. Wink), pp. 239-62. Plenum, New York.

295. Wool D. and Bogen R. (1999). Ecology of the gall-forming aphid, *Slavum wertheimae*, on *Pistacia atlantica*: population dynamics and differential herbivory. *Israel Journal of Zoology*, 45, 247-60.

296. Woodhead, S. and Chapman, R. F. (1986). Insect behaviour and the chemistry of plant surface waxes. In *Insects and the plant surface*. (ed. B. E. Juniper and T. R. E. Southwood), pp. 123-35. Edward Arnold, London.

297. Zangerl, A. R. and Bazzaz, F. A. (1992). Theory and pattern in plant defense allocation. In *plant resistance to herbivores and pathogens* (ed. R. S Fritz, and E. L. Simms) pp. 363-91. University of Chicago Press, Chicago.

298. Zucker, W. V. (1982). How aphids choose leaves: the role of phenolics in host selection by a galling aphid. *Ecology*, 63, 972-81.

第5章 植物并非昆虫最理想的食物来源

5.1 植物是次优的食物
　　5.1.1 氮
　　5.1.2 水
5.2 人工饲料
5.3 消耗与利用
　　5.3.1 被取食的食物数量
　　5.3.2 利用
　　5.3.3 次优食物和补偿性摄食行为
　　5.3.4 化感物质及食物的利用率
　　5.3.5 植物化感物质的解毒作用
5.4 共生体
　　5.4.1 食物利用与补充
　　5.4.2 植物化感物质的解毒作用
5.5 微生物影响寄主植物特性
　　5.5.1 植物致病菌
　　5.5.2 内生真菌
5.6 寄主植物影响植食性昆虫对病原体和杀虫剂的敏感性
5.7 植物食物质量与环境因素的关系
　　5.7.1 干旱
　　5.7.2 空气污染
5.8 结论
5.9 参考文献

　　这章的主题——植物作为植食昆虫的食物——是昆虫与植物关系的核心。本章主要解答以下两个问题:第一,植物为动物提供了什么营养？第二,昆虫的生长繁殖需要什么营养物质？解答这两个问题需要考虑以下因素:(1)如前文所述,不同植物的化学组成在时间和空间上有很大差别;(2)不同种类的昆虫在不同生长阶段和不同环境条件下,对营养的需求差别很大。除了这些情况之外,在植物所供和昆虫所需之间,植物似乎只能给昆虫提供最低限度的食物。

　　昆虫和所有动物一样,需要将食物转变为体内物质和能量,我们可以将昆虫的化学构成同植物的进行比较。图5.1为昆虫和植物体内主要元素的浓度。

　　在昆虫体内发现了4/7的主要元素和3/4的微量元素,其平均浓度远远高于在植物中的平均浓度。氮含量应当被特别关注,因为相对于其他生物体的组成而言,相当一部分的植物氮不能被昆虫所利用,尽管也有例外存在[85]。动物的氮含量约占它们干重的8%~14%,而植物通常只含2%~4%(图5.2)。

第 5 章 植物并非昆虫最理想的食物来源

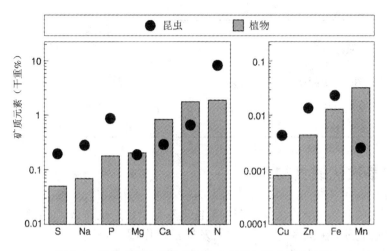

图 5.1 植物组织和昆虫组织中各元素平均含量的比较

种间含量的差异在植物中尤为明显,环境因素和植物(组织)年龄因素引起种间和种内的变化(氮元素含量见图 5.2),垂直刻度为对数(艾伦等,1974)[4]

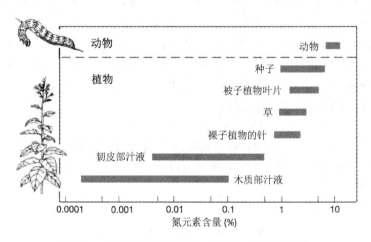

图 5.2 动植物不同部分的氮含量(%干重)的比较

注:木质部和韧皮部汁液浓度被表示为氮重/体积(马特森,1980)[112]。

同样的,动物的热值(caloric value)也超过了植物(陆生植物平均 18.9 J/mg;昆虫 22.8 J/mg)。这些比例表明植食性昆虫把食物转化为身体组织时必须富集氮等一些元素。因为相对于植物而言,昆虫生长缓慢却耗能巨大,植物的营养价值主要由氮元素的含量决定,因而它的热值没有那么重要(表 5.1)。

表 5.1 食草动物与一些哺乳动物饮食中,可获得的碳水化合物及脂肪(表示为葡萄糖的克数)与蛋白质的最适比(伯奈斯,1982)[25]

动物	蛋白质与葡萄糖的比例
蚕	1∶3
蚕(人工饲料)	1∶1.5

续表

动物	蛋白质与葡萄糖的比例
蝗虫（人工饲料）	1∶1
菜粉蝶幼虫	1∶1
牛犊（幼年期）	1∶4
牛	1∶7
水牛	1∶10
山羊	1∶15

5.1 植物是次优的食物

植食性昆虫的食物中含有大量营养价值低且不可消化的化合物,如纤维素和木质素,以及各种各样的化感物质(在很多情况下,它们发挥毒性作用,妨碍了消化作用或者阻碍摄食行为)。而更糟的是,植物所供营养与昆虫所需营养有极大的差别。在质量方面,昆虫所需的营养成分与其他动物大体上是相同的。然而区别于其他动物,昆虫不能合成固醇。因此,它们必须从食物中摄取一些固醇和其他几种必需的营养素(氨基酸、碳水化合物、脂类、脂肪酸、多种维生素、微量元素)[19]。不同种类的昆虫对营养的需求,无论是在数量还是质量方面,都相当特殊。昆虫最佳的生长、生存和繁殖能力都需要食物中蛋白质与碳水化合物的特定比率,这一比率在不同的物种间以及不同的生长阶段会有很大差异。在人工喂养条件下,食物中蛋白质与碳水化合物比例为 79∶21 时,杂食的玉米夜蛾(*Helicoverpa zea*)长势最好。相反地,东亚飞蝗(*Locusta migratoria*)幼虫所需要比例却为 50∶50[155]。形成这一明显对比的原因在于生物特征[177]。玉米棉铃幼虫长得很快,所以就需要蛋白质丰富的食物,而蝗虫具有生长慢但较为活跃的特性,因此就需要摄取大量的能量供肌肉消耗。

由于不完善的营养比例和体内含有有毒化感物质,植物成为次优的食物。人工饲料在促进昆虫生长方面优于天然植物。糖蛾(*Agrotis ipsilon*)幼虫经人工喂养后的体重是食用玉米叶片后体重的 12 倍(图 5.3)[136]。

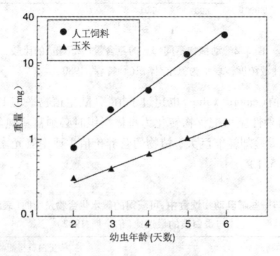

图 5.3 在敏感玉米植株上或人工饲养孵化后,球菜夜蛾(*Agrotis ipsilon*)幼虫 6 天内的生长情况
注：y 轴(幼虫体重)为对数值(里斯和菲尔德,1986)[136]。

通过对其他昆虫的观察显示,人工喂养条件下这些昆虫表现出更快的生长速度、更重的虫蛹重量以及更好的繁殖能力,以此证明表面上最易受损的植物实际对于昆虫的侵袭具有强大的防御力[136],同时从营养的角度来看,植物是昆虫不完善的养料来源。考虑到这个结论,贝伦鲍姆提出了一个有趣的假设,这个假设是以下列论点为基础的:养分贫瘠有可能是宿主植物产生抗性的决定因素[22]。这种情况下,植食动物对植物的选择性影响,可能是驱动植物初生代谢和结构多样性的因素,从而使植物不适合成为植食动物的食物来源。不幸的是,目前还无法证明此假说。相反,越来越多的证据表明,植物体内代谢的主要特征早于陆生植物的出现。然而,考虑到长期以来昆虫对植物世界的多样化影响,昆虫很有可能极大程度地影响了古代植物进化的基本特征,比如生化过程的新陈代谢。

5.1.1 氮

氮元素对普通昆虫的生长和繁殖极为重要。尽管大气层中氮含量很多,并且氮是地球上生命体所必要的元素,但是大气中的氮不能被直接利用,必须结合其他化合物才能被使用。蛋白质是昆虫的基本结构成分,是软组织和表皮的组成成分。表皮蛋白质的含量占表皮干重的50%。大部分的植物组织中包含碳水化合物,它同纤维素和半纤维素一起作为细胞壁的主要成分,此外还有木质素、角质、二氧化硅、细胞壁蛋白质。而构成植物蛋白的氨基酸的比例不同于昆虫的摄食需求(图5.4)。因为大量的芳香族化合物与表皮蛋白结合在一起。而相对于植物蛋白的氨基酸组成,昆虫需要更多的芳香族氨基酸,如苯丙氨酸和色氨酸等[25,40]。

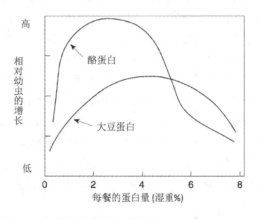

图5.4 食物中两种蛋白质含量与杂食性甜菜夜蛾(*Spodoptera exigua*)幼虫的增长的关系

甜菜夜蛾幼虫在不同蛋白质方案下的成长情况,与食物中含2.4%湿重的酪蛋白的人工饲料作为参照。甜菜夜蛾要实现最大化成长需要更高水平的大豆蛋白而非酪蛋白。植物蛋白(大豆蛋白)比起动物源蛋白能为昆虫生长提供更充足的养分(达菲等,1986)[59]

植物中的氮含量(与蛋白质含量相关)在不同的物种、器官、季节与其他环境等方面有很大的不同。一般来说,非草本植物叶片的可溶性蛋白质含量多于落叶植物和禾草类,而在常青树中的含量最低。通过对72篇文献进行分析可以得出,在仲夏时期8个不同科的植物叶片的含氮量存在明显差异。杜鹃花科和桃金娘科的木本种类的蛋白质含量较低,而草本豆类植物的含量最高[28]。这种差异可以决定某种特定植物对特定植食性昆虫的适合性(不适合性)。因此,寄生在温室植株上的两种杂食性牧草虫,在植株叶片的氮含量保持中等或偏高水平时开始侵害植物,而含氮量偏低的叶片则被证明不适合昆虫的生长和发育[145]。

各种昆虫的生长效率与植物的氮含量密切相关,氮含量与蛋白质含量直接相关。随着食物中的氮含量增加,昆虫将植物成分转化为自身组织的效率就更加高效(图5.5)。

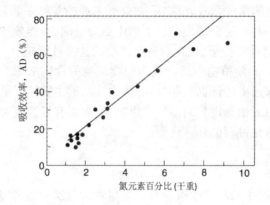

图5.5 植物中的氮含量(干重)与植食性昆虫(*Leptoterna dolabrata*)

因此,以氮含量为1%的植物为食的昆虫的摄食量是以氮含量为6%的昆虫摄食量的三倍以上[112]。然而,尽管在植食动物与植物的相互关系中植物的氮含量起到至关重要的作用,通常它却并不能完全衡量植物的营养价值。高的氮含量可能是一些不能代谢的含氮化合物造成的,如生物碱等,在某些情况下,导致消化功能降低的丹宁酸含量也会影响氮含量(见5.3.4节)。在一般情况下,以韧皮部汁液为食的昆虫比咀嚼类昆虫更有优势,因为韧皮汁液中所有的含氮化合物几乎都可以被吸收利用[56]。

相比韧皮部,木质部更不适合成为食物,因为木质部汁液中的氮浓度(少于0.1%)通常为韧皮部的1/10,而比叶片组织中的氮含量低两个数量级(图5.2),因此以木质部为食的叶蝉必须吸取大量的树液来满足它们对氮和碳水化合物的要求。其日平均摄食率是体重增长率的300倍至1 000倍[41]。尽管食物流经消化道的速率极快,提取养分的效率却能达到99%。对于叶蝉的测试结果表明,一个昆虫个体每天在摄食过程中要消耗3.9 ml水、57 mmol有机碳、21 mmol有机氮。碳的日摄取量的14%用来形成昆虫自身的碳元素。对于氮而言,高达29%的日摄取量转化为自身的氮元素。从植物的角度来看,水和养分流失会引起很大的问题。例如,在为期3周的实验中,16只叶蝉从大豆植株(*Glycine max*)中吸取的氮占植物体内氮总量的48%[7]。

虽然许多研究证明昆虫的行为与宿主植物氮浓度呈正相关,但是值得注意的是以上结果均来源于幼虫阶段的实验。一个以鳞翅目幼虫和成虫的生长状态为对象的研究发现,幼虫生长速度越快,虫蛹的成活率越低,并且成虫个体越小。在这种情况下,在氮含量丰富的植株上昆虫的整体状态与幼虫阶段更高的生长率并没有关联[66]。

在一些情况下,氮的确是一项衡量食物质量的重要指标,氮甚至是植物生长的限制因素[180]。对植株施用氮肥会对植食行为产生积极的影响(图4.21),但同时也存在消极的影响。斯夸伯的一篇文献报道列举出至少有115项研究表明增加植物氮含量有助于昆虫的生长[147]。而与此相反,至少有44项研究表明植物氮含量的增加会抑制昆虫的摄食行为。氮含量变化在昆虫身上体现出矛盾性的反应,这一现象可以用一些机制来解释。或许,昆虫在生理上能适应宿主植物正常(或者略高于正常)水平的氮含量。当冷杉的氮含量受育秧营养液影响时,由于营养液中氮浓度会发生变化,针叶中含氮量(干重)会在0.7%~5%之间浮动。云杉色卷蛾

(*Choristoneura occidentalis*)在宿主植物氮含量为 2.5% 时表现出最佳水平,这一浓度在自然界中较为常见[39]。用人工饲料喂养两种夜蛾,在不同蛋白质剂量的喂养方案试验中,两种夜蛾的生长也达到了最佳曲线(见图 5.4)[59]。

据推测,"旺盛的摄食者(flush feeders)"就是那些可以适应食物中高水平氮含量的昆虫,它们会积极响应氮含量的增加并将其运输到正在生长的组织器官,而"衰老的摄食者(senescence feeders)"则对自身器官中氮的流失消极对待[180]。此外,氮肥可能会改变多种植物生理和形态,并影响植物次生代谢,最终导致抗性物质产量的增加。对植物病原体和环境因素(如微气候和杂草生长)的敏感性,也有可能引起抗性物质产量变化。形态学的变化包括叶片表面积的增大及叶片厚度、节间长度、叶脉韧性的变化等,都可能消极地作用于植食昆虫。因此,氮引起的变化自然而然地改变了植物作为植食动物宿主及其天敌的重要性[67]。

尽管氮的作用至关重要却不能把氮同植物其他化学成分分开评估。这是从以下发现得出的:植物次生代谢物质,例如丹宁酸、酚醛树脂等,对昆虫生长的影响可能主要依赖于昆虫摄食中碳水化合物和蛋白质的比率[78,155]。

5.1.2 水

水是生命之源。昆虫同其他动物一样,都离不开水。尽管在传统意义上,水并不是一种营养物质,但是获得充足的水对大多数植食动物来说是一个重要的营养屏障[147,151]。叶片中的水含量为 45% 到 95%(鲜重)。鳞翅目幼虫食物中水的含量,是评估食物营养价值的重要指标。叶片中充足的水含量对于昆虫有重要的意义,有一项研究结果显示,在豆科植物和苜蓿科植物中,含水量高的植物上寄生的幼虫明显生长得更好。通过改变人工饲料中的水含量,可以观察到昆虫的生长速率与食物中的水含量呈正相关[147]。当叶片中的水含量下降时,其营养价值就下降了。把从植株切离且并不通过叶柄补充水分的叶片作为食物提供给 16 种毛虫,即使叶片并没有任何缺水迹象,这些毛虫的相对生长率却降低了 40%。较以草本(非禾本)植物叶片为食的昆虫相比,在以木本植物叶片为食的昆虫身上,这样的结果体现得更为显著,这可能是由于木本植物转化效率较低,在自然条件下木本植物叶片的氮、水含量要比草本植物中的低[82,146]。蝗虫对于宿主植物水分降低的耐受性较强。在 41 种植物中,水分压力使得其中的 12 种成为沙漠蝗虫非常美味的食物,而仅有 5 种是不适合食用的[29]。

叶片中的水含量和氮含量(特别是蛋白质和氨基酸)经常共同变化,在嫩叶中的含量要比在成熟的和开始衰老的叶片中含量要高。草本植物水分含量为 80%~90%,氮含量为 5%~6%,其营养价值往往比同等条件下的木本植物要高,木本植物通常包含 60% 的水和 2% 的氮(见 5.4.2 节)[161]。植物类群之间差异性的叠加是随季节性水(和氮)含量的变化而变化的,在夏季,植物的营养价值是逐渐减少的(图 5.6)。

虽然植食昆虫倾向取食含水量高的食物,但还是可以观察到许多类群倾向于喝露水或其他来源的自由水。毛虫可以感知 2 厘米范围内的水,并且可以随时饮用遇到的水滴[51,116]。相对而言,毛虫容易蒸发水分[187],而通常蚱蜢的日平均水分摄入量仅为食物的 60%。水对蝗虫的重要性可以从它们的水合状态影响食物选择中体现。在用较干的食物喂养一段时间后,蝗虫会比在正常情况下吃得多来补偿它们的缺水量[138]。显然,植食动物拥有通过选择性摄食来保持水分平衡的控制系统。

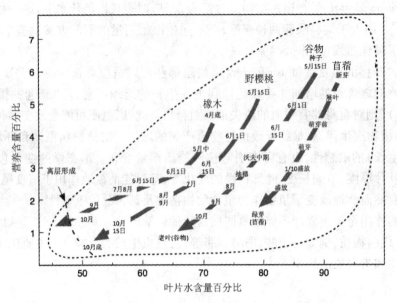

图5.6 草本植物,落叶植物叶片中氮、水的含量随季节而减少,
因此植物的营养价值与春天时比较会有所降低

5.2 人工饲料

对特定植物化合物或某种植物成分的营养价值的行为反应进行研究时,使用已知化学成分进行人工喂养已经被证实是一种不可缺少的手段。植物成分很难标准化,因为不同的植物和各个植物不同部位也许会随着季节、成长阶段等发生很大的变化。人工饲料考虑到对营养因素的精确控制。从20世纪50年代开始,饲料被开发为多个类别。如果植物的化学成分是已知的,草食昆虫的营养需求也是明确的,那么饲料的配置也会出现比预计的更加困难的情况。困难主要是由于人工饲养难以复制植物的两种特征。第一,尽管植物的含水量可以高达总重量的90%,但仍旧为植食动物提供了一个干燥的基体。食物本质上是在微型胶囊内包装的液体(细胞),因此赋予其干燥的外环境。第二,这些微型胶囊的物理和化学结构具有防止微生物侵袭其具防止微生物侵袭其具高营养内容物的细胞的功能。要仿造植物牢固的表面,人工饲料需要通过整合琼脂、纤维素,或者增加惰性营养物质来增强液体食物密度。这一过程也产生了膳食纤维,用以帮助食物顺利通过消化道。为了抑制细菌和霉菌腐烂,食物必须通过高温消毒或者加入抗生素。然而,也许通过影响摄食者的内脏微生物和解毒酶类,或者以其他方式来影响这些摄食者,因此这些化合物通常是受到排斥的。因此,发现有效的剂量且同时不伤害昆虫摄食者,已经成为饲料开发中的重要内容。由于不同的物种对营养要求往往略有不同,因此即使在饮食结构中发生微小的变化也可以对昆虫生长和繁殖产生剧烈的影响。此外,即使是杂食、不挑剔的昆虫,原料的种类和存储条件以及饲料准备中的微小变化都会严重影响饲料质量,这在舞毒蛾幼虫(*Lymantria dispar*)实验中已被详细描述[96]。

目前,一些有关人工饲料的优秀书籍往往具有一种"菜谱"的风格,给人一步一步地指导以便为特殊类群的昆虫准备食谱[49,159]。此外,一些面向特殊类群的"即食饲料"已经出现在市面上。一些人工饲料可适合大量不同类群的昆虫。它们通常包含蛋白质(酪蛋白或小麦胚芽)、糖类、脂类、甾醇、矿物质和维生素。琼脂和纤维素起到胶凝和膨胀作用,添加微生物抑制剂主要由于抑制微生物生长。此外,据植物专家分析,增加宿主植物材料是成功制成饲料的部分因素,但是在用途上,宿主植物作为特定饲养刺激物还是营养因子仍然是未知的[68]。

当然,人工饲料与天然食物在许多方面是不同的。多代以饲料为食,甚至仅有一代以饲料为食的昆虫可能会表现出行为上或生理上的变化。用人工饲料连续喂养欧洲玉米螟(*Ostrinia nubilalis*),一直到153个世代后,会发现它侵害易感玉米植株的能力开始减弱。然而,经过人工饲料喂养七个世代后,用玉米再喂养一个世代,它就会维持其对玉米的侵染力[72]。而其他昆虫,例如棉铃虫(*Helicoverpa zea*)经过人工饲料饲养275代后,在其宿主植株上的活力和食物利用特征仍然保持不变[48]。但是昆虫在短时期内,也有可能会发生一些变化。生长在一个相对舒适的人工喂养环境下的昆虫,与那些以坚硬植物叶片为食的昆虫相比,因为缺少运动,所以头部的肌肉系统和尺寸均产生了变化。这种变异也能在以不同韧性的宿主植物为食的毛虫身上发现[26]。以人工饲料为食的烟草天蛾(*Manduca sexta*)幼虫,与以植物组织为食的同种个体相比,行动更加缓慢,更不善于在垂直的植株上控制自己。因此,对昆虫饲料质量的控制是昆虫饲养程序的关键环节[8,98]。

5.3 消耗和利用

5.3.1 被取食的食物数量

昆虫在迅速生长时期能消耗很多食物。它们的食道在非繁殖阶段占据体腔大部分体积,并且适于处理大量食物。昆虫的消化速率很高,以叶片为食的昆虫只需几小时便能消化完全[167]。以沙漠蝗虫(*Schistocerca gregaria*)为例,在不断进食时只需1.5小时就能消化完全。蚜虫类的消化时间可以短至1个小时,有一种以木质部为食的叶蝉需要的时间可能稍微长些[53]。毛虫每天消耗的植物组织可能是它们体重的6倍,而成年蝗虫的进食量与其体重相同。沫蝉日摄取木质部汁液的质量是它们体重的100倍到1 000倍[164]。如果在没有食物的情况下,消化速会迅速下降。

由于幼虫具有阶梯式生长模式,因此确定其摄食和消化效率变得较为容易[164]。幼虫重量呈指数级增长,在每个龄期通常呈现双倍增长。生长曲线与食物摄取量增长指数通常保持一致(图5.7)。

因此,成熟幼虫的体重往往是其孵化时期的几千倍,例如,家蚕(*Bombyx mori*)老熟幼虫体重是刚出生时的10 000倍。鳞翅目幼虫消耗全部食物的94%~98%仅发生在生长发育的最后两个阶段[6]。

5.3.2 利用

1. 对宿主植物的利用

在昆虫与植物的研究中的一个主要问题是,食草昆虫对宿主植物的营养物质的利用效率如何?要回答这个问题,我们首先需要知道,所有被摄取食物中的哪一部分是真正用于虫体的生长发育和繁殖的。利用效率更高的食物可以被看作一个植物(或者植物一部分)具较高的

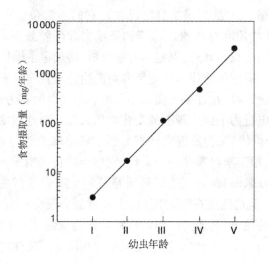

图 5.7　家蚕(*Bombyx mori*)幼虫在不同龄期的食物摄取量

营养指标。宿主植物的利用效率的差异也许可以用宿主植物的特异性来解释,在这一领域已积累了大量的文献,这一领域学科也被称作"营养生态学"[163]。昆虫营养领域的一个里程碑式的论文是由沃尔鲍尔在1968年撰写的,他总结早期的文献并提出规范的计量方法和参数[176]。这篇文章自1968年发表以来已被多次引用[161,163,164]。在昆虫营养生态学领域,1992年到2003年期间出现的论文数量比1982年到1991年增加了3倍[174]。目前对营养利用效率的定量研究结论普遍认为,主要的食草昆虫如木本植物和草本植物的摄食者们,它们的营养利用效率不同,而且正如上文所述,水和氮都是营养质量的决定因素。然而,如果确定到种,评估一种昆虫是否能够更有效利用一种植物远超于另一种植物,或者一种次生代谢产物是否会影响植物的利用效率,准确的测量结果往往比预计难得多[174]。

2. 利用率(utilization)和性能(performance)的关系及参数

沃尔鲍尔确定了3个利用率参数,现在通常叫作营养指数:

(1)近似消化率(Approximate digestibility,简称为 AD,也被称为吸收效率)。

(2)摄取食物的转化率(Efficiency of conversion of ingested food,ECI,也被称作增长效率)。

(3)消化食物的转化率(Efficiency of conversion of digested food,ECD,也被称作代谢效率)。

重量分析法(gravimetric method)是目前常用的量化食物的摄入量和利用率的方法,包括在实验开始和结束时称重食物、昆虫身体和粪便的重量。主要基于一个预算方程式 $C = G + R + FU$[130,142],其中 C 为消耗的食物量,G 为昆虫增加的生物量(即躯体和生殖生长,以及对躯体生长无价值的分泌排泄产物,如蜕皮、丝状物、消化酶等),R 为呼吸作用(呼出的二氧化碳量),FU 是粪便的量(尿中的废物和其他代谢产物与未消化的食物粪便一起排出)。近年来,作为一种选择方案,一种利用率双坐标平面图(bicoordinate utilization plots)开始被更多地应用[74,155]。这种平面图能表现摄食量和上文提到的营养预算方程式中各种组分之间的关系,包括摄入总量和特定的营养类别的关系,也能避免使用营养指数(nutritional indices)时会遇到的一些相关问题。

通常使用干物质单位(drymatter units)作为对物质的估算单位,因为食物、粪便和昆虫身体里的水分以水蒸气的形式蒸发掉了,而这一过程的流失量是难以量化的。因呼吸流失的干物质重量只在极少情况下可以被直接量化[173]。因此,估算的准确性因不易被检验而容易引起争议,特别是在生态能量学领域[184]。一个引起误差的重要原因,是错误地判断了食物干物质的含量(不论是植物性食物还是人工饲料)[10,141,172]。由于植物的呼吸作用(不同的物种和组织会使植物呼吸速率有差异),为了测量有效的营养指数,应该考虑植物呼吸作用造成的损失。然而为了弥补这种损失,提供过量的食物会使错误更加严重,尽管某种程度的过剩可以确保一种自然的实验状态[168]。例如,向摄食的昆虫提供的食物只有一半(以叶片为例)被消耗,那么估算这种食物的干重消耗的比例会出现轻微的误差。例如,估算结果是 14.5%,而实际数值为14%;ECD 的值被估算为 40%,而实际上是 50%。由于植物叶片及其他器官的干物质含量存在空间异质性,这些误差的出现更具有随机性而非系统的本质特性,因此会妨碍测量出可靠的差异数值。此外,如果不能合理考虑叶片呼吸率(respiration rates)的因素,这些错误可能会更加严重,就会推导出生理上不可能出现的新陈代谢速率,最终得出错误结论(见 5.3.2 节)[172,174]。不幸的是,在极少数的情况下,人们才会考虑到植物组织的呼吸效率,并采取适当的控制手段[173]。关于具体的测量、技术方法和误差分析等内容都超出了本书所涉及的范围,可以参考其他相关著作[98,172]。根据 C、G 以及 FU 值计算 AD、ECI 和 ECD 值的方法如图 5.8 中所示[2]。量化食物利用率的替代方法,是以标记物、成分估算、放射性示踪剂以及气体分析为基础的[98,166,172]。这些技术基于精密的化学或物理分析设备,因此至今只在少数的研究中应用。双标水法(doubly labelled water method)提供了在自然觅食条件下(即现实的生理生态条件下),研究植物释放的二氧化碳量的可能性[91]。这项技术已通过验证,并成功地应用于大黄蜂的觅食研究[186]。真实的测量植物为昆虫提供食物的营养质量,应在自然光照强度下进行,使光合作用正常发生。而这在目前的有关重量分析研究中均未提及[174]。在绝大多数情况下,已被证明昆虫可以通过取食行为影响(增加或者减少)植物光合速率[179]。机械损伤不能模拟任何情况下的昆虫的取食作用,这主要取决于昆虫的种类[129]。在一片叶子上,因摄食行为导致的光合速率下降的叶片面积,是摄食行为消耗叶片面积的 6 倍[190]。忽略光合作用在植物昆虫营养关系中的作用,会对重量分析结果产生严重的限制。从生态学的角度,我们对植物—昆虫间能量流动的动力模式了解甚少。想在这一领域取得重大进展必须依靠几种技术相结合。红外气体检测器(Infrared gas analysis,IRGA)可以与重量分析法结合,将内部的随机因素和系统的误差降到最低。IRGA 测量应结合双标记水法,以长期获得植物光合作用的代谢速率[174]。

性能(Performance),即食草动物能够实现的最大生长和繁殖的能力,优先成为一个速率参数。最常用的参数是相对生长率(relative growth rate,RGR),表示单位时间内单位体重(毫克/干物质)的增长量(毫克/干物质)。相对生长率是相对消耗率(relative consumption rate,RCR,毫克/毫克/天)和营养指数的乘积:$RGR = RCR \times AD \times ECD = RCR \times ECI$。

这个等式说明,以一定食物为来源,通过提高食物的摄食率,或者食物的利用率可以得到一个更高的 RGR。当消耗或者增长率和营养指数(定量分析)联系起来的时候,通常会出现相反的关系[81,147,164]。然而这种关系并没有体现因果之间的差别,即生长率低导致新陈代谢效率降低还是新陈代谢效率低导致生长率降低,限制植食昆虫生长率的因素被推断是营养成分而非能量(见 5.1.1)[143,162,175,183]。因此,对限制性营养(limiting nutrient,通常指的是氮和水)的次优利用,会降低植食昆虫生长率,增加其维持性消耗(maintenance costs),并且产生较低的新陈代谢效率。摄取限制性营养的昆虫会增加补偿性摄食,这在许多植食昆虫中都得到验证,这一内容将在下文详细论述[156]。当摄入营养量发生变化时,利用效率也会随之发生定量变化。但如果是其他因素(有毒的化感物质、其他的营养物、温度等)导致利用效率发生改变,就是定性的变化。定量变化和定性变化可以相互影响。使用利用率双坐标平面图可以区别定量变化、定性变化以及二者的相互影响(见5.3.4节)。

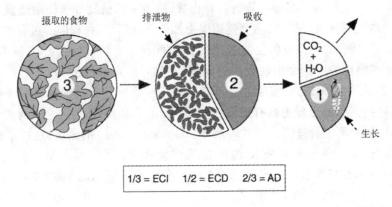

图5.8 摄取的食物(3)、部分转化为排泄物排出,部分被吸收(2)、被吸收的物质在随后用于生长(1)和呼吸(二氧化碳和水)之间的分配

3. 不同摄食群(feeding guilds)对植物性食物的利用率

当分析性能(performance)和利用率(utilizationvalues)的时候,会发现不同摄食群之间存在巨大的差异,如以草本或木本植物为食的同具咀嚼口器的类群,或具咀嚼口器和刺吸口器的不同类群(表5.2)。

表5.2 具咀嚼式口器或刺吸式口器的植食性昆虫的性能平均值(圆括号内为范围)以及营养利用指数(斯兰斯基和斯夸伯,1985)[164]

	AD(%)	ECD(%)	RCR(mg/mg/天)	RGR(mg/mg/天)	N
具咀嚼口器(鳞翅目)					
草本植物	53	41	2.0(0.27~6.0)	0.37(0.03~1.5)	26
稻科植物	43	45	2.0(0.07~4.8)	0.29(0.06~0.62)	6
木本植物	39	37	1.5(0.31~5.0)	0.17(0.03~0.51)	82
具刺吸式口器(同翅类)					
草本植物	60	65	1.0(0.90~1.6)	0.39(0.11~0.67)	3

注:AD——近似消化率;ECD——消化食物的转化效率;RCR——相对消耗率;RGR——相对生长率;N——被调查的昆虫数量。

取食木本植物的昆虫由于较低的RCR和AD值,因而体现出较低的RGR值。具刺吸式口器昆虫的RCR值平均为咀嚼式口器昆虫的1/2,而RGR却达到最高值,可以归因于AD和ECD的较高值。这些营养指数之间的差异可以解释为,植食性昆虫对这些植物不同组织的营养利用效果之间的差别。因此,一般而言,刺吸类昆虫生长速度快于咀嚼类昆虫。这种差异有可能会产生重要的生态影响。

4. 发育期间对食物利用的变化

昆虫发育过程中,消化性能的特征数值会有所改变。近似消化率(AD)的数值从低龄期向高龄期阶段逐渐减少[6,164],因为随着幼虫生长,其肠道体积会逐渐增加,因此摄食率也随之增长。而较短的保留时间和较大的食物团会使酶促降解反应速率降低,并且营养物通过肠壁被吸收的效率也随之降低。这对以特定植物或植物特定部位为食的昆虫对食物的利用能力产生很重要的影响。因此可以解释为什么在同种个体之间幼虫比成虫对食物更为挑剔。例如,烟芽叶蛾(Helicoverpa virescens)一龄幼虫表现出体重增加速率的降低,这是因为在食物中存在着浓缩的丹宁酸,其浓度大约为制约生长五龄幼虫的丹宁酸浓度的 1/10[121]。

研究一种植物对于特定昆虫的适合性有着非常重要的价值。一般而言,低龄期幼虫比高龄期幼虫能更高效地利用食物,而低龄期幼虫体内解毒酶的含量低于高龄期幼虫[2,31]。由于食物中营养成分及次生代谢化合物改变可能对低龄幼虫有巨大的影响,所以尽管难度很大,对营养需求和化感物质的影响的研究应从低龄幼虫开始[164]。

5. 生长消耗:代谢效率的决定因素

对宿主植物的利用率所需要考虑的问题是,利用效率的差异是如何产生的。根据代谢负荷(metabolic load)或生理效率假说(physiological efficiency hypothesis),对于能量处理(energetic processing)消耗(区别于维持消耗 maintenance costs)的增加是导致增长率降低的直接原因,说明在一系列代谢过程中产生的能量和促进植物生长的合成代谢所需的能量要互相权衡。虽然这个观点被反复提出,但是仍缺乏实验依据[135,158,164]。

在代谢负荷假说中,多亚基单加氧酶(polysubstrate mono-oxygenases,PSMOs)系统对食物中化感物质(见 5.3.4)所做出的应答反应被认为是重要的需能过程。尽管 PSMOs 酶发挥着良好的解毒作用,但因量太少而无法计算消耗量[122]。因此测定这种应答反应的消耗(重量分析)实验并不能提供有力证据。

对代谢进行直接的、长期的测量后却几乎得不到任何有效数据,如呼吸测量法。并且食物质量对代谢过程的影响目前仍无定论[174]。由于实验存在测量难度[171],目前仍没有有效的研究能够将长期呼吸速率的测量与干重增长的测量相结合,最终真正实现对重量分析预算(gravimetric budget)的检验[172]。东亚飞蝗(Locusta migratory)为渐变态的物种,较毛虫类如欧洲粉蝶(Pieris brassicae)幼虫(表 5.3)的生长速率低,并且每单位生长量需要更多的能量。两个物种的生长消耗值,即为散失到环境中的热量(H)和增长量(G)之间的比率(H/G;表 5.3)相差仅 1.5,尽管两个物种的成长速率、个体大小和生活习性都存在相当大的差异。

表 5.3 完全变态和渐变态昆虫在末龄幼虫阶段的生长消耗、呼吸运动数据(连续或重复)和生长重量数据(范隆,1991,1993)[172,173]

	食物	持续时间(小时)	G(mg)	RGR(mg/mg/天)	H(J)	H/G(J/mg)
欧洲粉蝶(Pieris brassicae)	卷心菜	90	88	0.640	1027	11.7
东亚飞蝗(Locusta migratoria)	小麦	240	258	0.124	4536	17.6

注:对昆虫的最优宿主植物进行摄食行为研究。G——生长(毫克干物质/毫克/天);RGR——相对生长率;H——热产量,是通过测量呼吸运动的数据计算。

在蚱蜢(Melanoplus sanguinipes)和一星黏虫(Pseudaletia unipunvta)之间也进行了类似的比较实验,在相同的条件下摄食相同的宿主植物(小麦),会得到相似的结果:一星黏虫总需氧量是身体质量的 2 倍,是蚱蜢的 2 倍[26]。而蚱蜢的生长需要更高的能量消耗,主要原因之一为个体发育期的延长,这意味着增加总耗能中用于维持能量(maintenance energy)的比例。对于直翅目而言，维持耗能的重要部分用于它们的角质层，消耗的能量是同等大小毛虫的 10 倍。从生理学角度考虑，根据重量测量方法，植食昆虫的 ECD 不可能有很大范围的变动，特别是在增长率几乎或者完全没有受到影响的情况下[174]。事实上，即使对于不同纲的变温动物来说，尽管个体大小相差很大，经测量得出的生长消耗却是非常相近的（7～9 J/毫克干重）[128]。

昆虫与植物营养生态学主要关注昆虫幼虫，因为幼虫长期寄居在宿主植物上而且消耗很少的时间和能量移动。在摄食时，代谢速率暂时性上升（毛虫的代谢速率为原来的 1.5～5 倍）[3,102,110]，蝗虫若虫的代谢速率为原来的 3～4 倍[70]。毛虫呼吸速率的上升可以归因于摄食行为引发的肌肉运动或者摄食后消化活动的增加[3,110]。而对于蝗虫若虫而言，由于摄食活动产生热量的影响似乎与消化吸收的能量消耗没有联系，而是归因于一套神经刺激[70]。在整个发育期间，代谢速率的短期增加能在多大程度上促进总体代谢消耗，对此目前仍未进行研究。

捕食类和寄生类的肉食昆虫比植食昆虫更具营养优势，因为它们的食物组成可完全满足它们成长和发育的需求。动物组织比植物组织的近似消化力（approximate digestibility，AD）要高。食叶昆虫的 AD 值大概为 40%～50%（见表 5.2），而食肉类的 AD 值一般能达到 80% 左右[164]。尽管如此，如果植食昆虫的食物不受限制，它们会增长得更快。根据表 5.2 中列举的一些组数据来看，植食昆虫的相对生长率（relative growth rate，RGR）尽管变化幅度较大，但通常很高。植食昆虫的相对成长率在 0.03 到 0.4 之间浮动，而肉食昆虫的相对成长率为 0.01 到 0.03 之间，二者形成了鲜明的对比。高速的生长与低能耗的取食，意味着植食昆虫用于自身组织和产卵所消化的食物的量是捕食性昆虫的两倍（图 5.9）。

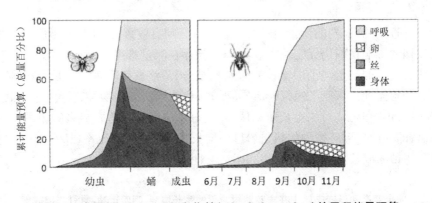

图 5.9 家蚕（*Bombyx mori*）和蜘蛛（*Oligolephus tridens*）的累积能量预算

能量消耗为一只雌性昆虫的一生中总吸收能量的百分比，注意维持代谢和生长储备方面能量分配的差异，蜘蛛需要更多的呼吸能量与捕食活动有关

5.3.3 次优食物和补偿性摄食行为

1. 补偿摄食的范围和机制

如上所述，植食昆虫的食物中营养较为贫乏。并且，植物化学成分会随时间和空间发生改变。为了克服这些困难，植食动物通过增加食物的摄入和改变它们的食物来弥补在营养上的不足，以达到最大生长率。它们会利用以下几种机制[92]。

第一，它们会离开之前作为宿主但是营养匮乏的植物，然后开始寻找其他可代替的食物。这种行为本质上源于营养反馈，以此为根据，昆虫们可以重新选择宿主植物，同时也交替地向不同的宿主植物取食[156]。这种现象就像是实验设计的"自助餐厅"一样,这一现象被称作"食物的自主选择"(见8.6.2)[177]。

第二,如上文提到的,植食昆虫在相同植物上能增加食物摄取率。这种补偿性行为可以抵消昆虫生长率的下降和用于维持消耗(maintenance costs)的增加(可能为一种功能反应),假设在更高的取食率下,用于额外的耗能较小。

已在大量研究中发现昆虫具有补偿性取食行为(昆虫通过调整取食率来接近或实现最大的增长率),而这种现象可能广泛存在于食草类昆虫中[156]。当食物的质量未达标准并且蛋白质成为其生长的限制性营养因素时,食物的消耗率会增加到原来的2.5~3倍[164,165]。以黑脉金斑蝶(Danaus plexippus)幼虫为例,当它们的寄主植物——乳草属植物生长在低氮肥的环境下时,它们叶子的氮含量仅为2.5%(干重),而生长在高氮肥环境下的植物的氮含量为6.4%。以低氮叶片为食的金斑蝶幼虫,单位时间内吃掉的叶片数量几乎是以高氮含量叶片为食的幼虫的两倍(鲜重)[104]。

第三,就理论而言,昆虫通过保持消耗速率恒定而使利用率达到最高。由于速率与效率之间的相互关系仍有待验证[172],所以目前仍没有确定的数据支持。

在试验中发现,使用人工饲料(特定成分)稀释后,消耗速率会增加(辛普森和辛普森综述)[156]。食物消耗速率的变化存在很大的跨度,这样的生理机制是相当复杂的[105,156]。最近,出现了一些模型,将各种机械感觉及化学感应(外围及内部的)整合到一起,但是这方面的讨论已经超出了本书的范围。

这些反馈的速度是超乎想象的。以东亚飞蝗(Locusta migratoria)为例,将一种混合8种氨基酸的注射试剂注入其血腔中,能显著地延迟下一餐的时间,这说明其体内存在一些未知的光化学传感物质,能监控血淋巴的构成,并且在几分钟之内对取食行为做出反馈[1]。当食物的营养不丰富时,在这种昆虫体内会发生补偿性自选取食[157]。

2. 补偿性取食的限制因素

尽管一些实验证实了补偿性取食能够缓解食物营养不足的影响,但是在生理上和生态上仍存在明显的制约因素。首先,在摄入速率和吸收效率间可能存在一个权衡。摄入速率的增加导致食物在肠内的停留时间减少,这将导致吸收效率相应降低[156]。其次,例如蝗虫摄入的蛋白质和碳水化合物是分别调控的[153],如果其中一种营养成分不足,会通过增加摄入的形式进行弥补,然而可能会导致过剩,并因此减少对其他营养物质的利用,从而部分抵消了补偿作用的效果[191]。最后,补偿性取食可能会导致昆虫中毒,因为增加取食就意味着增加了化感物质的吸收,而昆虫的解毒系统无法迅速应对进入机体的化感物质增长速度[165]。如果食草动物拥有检测化感物质的外部化感器(第7章),或通过"厌恶学习"来避免这些食物源(第8章),它们就可以预防中毒。

在生态学方面,首先假定取食时间较短的食草昆虫暴露于天敌威胁下的时间也最短[162,170]。而事实上,在野外环境下,通过对毛虫被捕食风险的研究可证明:取食的昆虫与休息的昆虫在选择压力上存在巨大差异。进食时,被捕食的风险比非进食期间高出100倍[27]。另一些研究提供了间接的证据,通过减少农作物中的氮含量以增加菜粉蝶(Pieris rapae)幼虫的摄食率[109]。假设这些幼虫的补偿性摄食增加了氮的含量[162],却会导致昆虫暴露时间的增加,进而增加被捕食的可能性。

5.3.4 化感物质及食物的利用率

化感物质对植食性昆虫的生理机能的消极影响表现在3个方面：
(1) 它们可以通过抑制摄食行为来减少食物的摄入。
(2) 食物被摄取后，它们可以降低食物的利用率。
(3) 它们可以通过干预重要的代谢过程来使昆虫中毒。

通常情况下，化感物质是通过结合上述三个机制而作用的。往往对摄食后的作用方式很难区分，它们或在内脏中，或经吸收后在身体的其他部位发生作用[32,74,160]。在这里，我们主要讨论化感物质对食物利用的影响，因为化感物质是降低食草动物生长和发育的主要因素之一，化感物质经常干扰植物的内在营养价值而使昆虫难以适应，最终使昆虫无法继续以这种植物为食。化感物会影响昆虫对食物的利用，尽管在昆虫选取植物作为寄主时，这种影响程度会变得更低。这或许可以解释为什么一些取食人工饲料的昆虫比取食自然宿主植物的昆虫生长得更好。

当人工饲料补充化感物质后，其食物的利用指数与对照值相比时，补充物的负（或者正）效应是可以被量化的。在对杂食性昆虫的研究中，添加5种植物次生性化合物，即使食物的摄入量没有减少，却都能发现昆虫生长抑制。不同的化合物对各种利用参数会有不同的影响，这表明它们会干扰消化或者吸收过程的不同方面[18]。鉴于这些结论的重要意义，需要进行严格的重复研究[141]。利用率平面图(Utilization plots)同样可以在这类的研究中提供有用的帮助（图 5.10），当东亚飞蝗（*Locusta migratoria*）接触到含不同浓度的丹宁酸食物时，这种化感物质的效果取决于蛋白质与碳水化合物的比率（P：C）。当这个比率低的时候，食物摄入率低，反之（蛋白质过量）氮元素利用率降低。这可以从利用率平面图的回归线中推导出结论，由于食物中不同的 P：C 而出现图形中的不同斜率（代表氮利用率）（图 5.10）[155]。也许在不久的将来，缺乏特殊化感物质的转基因植物，可为分析这些化合物的有害影响提供材料。

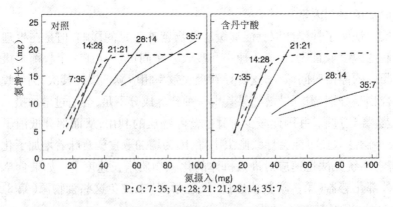

图 5.10 以不同的蛋白质与碳水化合物的比率为食的五龄期东亚飞蝗（*Locusta migratoria*）氮元素的摄取和转化的关系，及添加丹宁酸(3.3%、6.7% 或 10%)后的效果

根据96只昆虫的数据得出的回归线。两图的虚线均表明预期的氮元素的摄取和转化的关系。向具有低的 P:C 比率的食物中加入丹宁酸后，虽然摄入量和生长（斜率越高，氮转化效率越高）之间的关系保持不变，却由于丹宁酸造成了动物取食量减少，这表明了化感物质的一个量化效果。然而，当食物含高的 P:C 比率时，在消耗相同的蛋白质的前提下（蛋白质含量相当于 6.25 倍氮含量），蛋白质转化的增长率下降（斜率低），可以认为是化感物质的定性影响

第5章 植物并非昆虫最理想的食物来源

生物化学方法可为植物次生物质的作用方式提供更详细的解释。棉子酚,一种在棉花里存在的倍半萜类物质,可以抑制许多昆虫的进食和生长,灰翅夜蛾(Spodoptera litoralis)幼虫,在低棉子酚含量的棉花叶片上的生长速度远远高于在高棉子酚株上的生长速度。经人工饲料饲养证明,棉子酚的确阻碍昆虫生长。当用混合有棉子酚的食物喂养幼虫时,在一天内,肠道中的蛋白酶和淀粉酶活性开始下降(表5.4)。在胃肠道内棉子酚与蛋白质的结合已经被研究证实,它可以与食物中摄入的蛋白质,或者消化酶本身相结合。这两种情况都会抑制蛋白质的消化[115]。

表5.4 灰翅夜蛾(Spodoptera litoralis)幼虫(90~100毫克)连续两天取食含不同数量棉子酚醋酸盐的人工饲料后蛋白酶的活性

棉子酚醋酸盐浓度(%)	两天后幼虫平均体重(mg)	蛋白酶活性(%)
0(对照)	546	100
0.25	491	89
0.50	392	55

在许多植物的防御系统中发现了另一种蛋白酶抑制剂。这些化合物与昆虫肠壁上消化水解蛋白酶形成稳定复合物,从而减少了摄入蛋白质中的氨基酸释放。然而,昆虫可以通过改变中肠蛋白酶的结构使其对外来的蛋白酶抑制剂不敏感,昆虫可以以这种方式应对这种负面影响。在缺乏蛋白酶抑制剂的植物中也发现存在相同的作用方式,例如棉铃虫(Helicoverpa armigera)幼虫,它是一种杂食性昆虫,拥有大量编码蛋白酶的基因,这种基因能够产生消化不同特性蛋白质的酶。在应对特定植物食物中的蛋白酶抑制剂时,有些基因可能会上调,可以应对特定蛋白酶抑制剂的影响[34]。

已有一些实验对植物次级化合物和营养成分之间的相互关系进行了研究,在实验中,植物营养成分发生变化,同时特定的化感物质的数量也会有变化。芸香苷是一种广泛分布的黄酮类物质,不仅随植物中蛋白质的数量而变化,也与蛋白质的种类有关(图5.11)[59]。

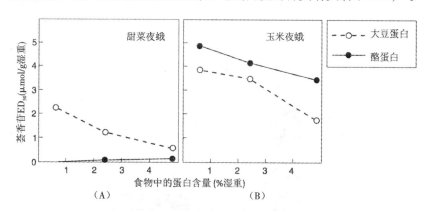

图5.11 植物中不同的蛋白质方案与芸香苷对两种夜蛾幼虫生长的抑制的关系

各种方案的生长抑制,表现为生长速率降低50%所需的芸香苷剂量(ED50),以生长于2.4%的酪蛋白的昆虫为对照,实验结果表明,以酪蛋白为食的玉米夜蛾(Helicoverpa zea)生长受到抑制的芸香苷的浓度比以大豆蛋白为食的低,而在甜菜夜蛾(Spodoptera exigua)中蛋白质的作用是相反的(达菲等,1986)[59]

菲尼有一篇经典论文,是关于松白条尺蠖蛾幼虫在年轻和成熟橡树上生长率的差异[62](图10.4),由此在很长时间内,引发了关于丹宁酸作为消化还原剂所起的作用的讨论。菲尼认为丹宁酸或者与叶片蛋白质形成复合物,或者与肠道中的消化酶形成复合物,从而降低消化效率,继而减缓生长[62]。虽然蛋白质与丹宁酸形成的复合物在一定程度上可以起到部分减缓生长的作用,然而另一种机制更被人们熟知,例如抑制取食、诱导中肠病变及药理毒性[14,30,63]。丹宁酸的抗营养作用的生化机制非常复杂,至今尚未完全阐明。

也有一些昆虫能适应丹宁酸含量丰富的食物,甚至还能通过刺激消化或其他因素,从食物中的丹宁酸受益(图2.4)[93]。对树蝗虫(*Anacridium melanorhodon*)的研究显示,当丹宁酸被添加到食物中时,会增加干物质消化率(AD)、生长效率(ECD)和由此产生的15%的增长速率[30]。能适应丹宁酸的昆虫具有几种避免丹宁酸危害的机制,包括PH=9的碱性肠道和围食膜对丹宁酸的吸收[30,62]。还有一些昆虫的肠道内拥有聚合机制[80],或者能够浓缩和排出食物中的多酚类物质[99]。

在第4章中,曾提到过植物一般会包含多种抗性物质。这些化学物质一旦进入昆虫体内[20],它们可能会以多种方式协同作用。从合成杀虫剂的实验中得知,昆虫可以很容易地对特定化学物质产生抗性。但发展对两种或更多不同模式的抗性是非常困难的。因此,如果在一种寄主植物中能够适应高水平的毒素,往往意味着在其他寄主植物中对其他化合物的耐受性较低[63]。由于昆虫具有的这些反应特性,植物无法只产生单一的次级化合物或是一类单一的化学物质。欧洲防风草可以通过至少7种不同的生物合成路径产生次生化合物。一类化合物的毒性很大程度上受其他化合物存在与否的影响[20]。增效烯是一种含官能团的木酚素,具有关键解毒酶的抑制剂(细胞色素,P450)。在伞形科植物中常常与光毒性的呋喃香豆素共同出现。

增效烯能够使花椒毒素对谷实夜蛾(*Helicoverpa zea*)的毒性增强5倍。因此,1毫克增效烯可以使植物少生产77毫克的花椒毒素(图5.12),可以实现产物(能量)在消耗上显著的节省,这是一种特有的适应性策略[24]。毋庸置疑,在植物的化学防护方面,化感物质间的协同作用已成为一个极为重要的因素。目前关于植物化合物协同作用的文献还非常有限,主要是由于对这种相互作用的检测和分析还存在着很大困难。目前可用的一些统计方法都较为复杂,对这一重要领域的化学生态学现象仍需进一步研究[60,123]。

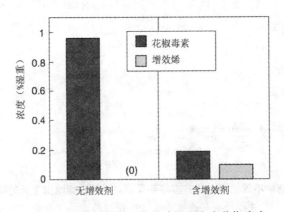

图5.12 增效剂的有无会影响植物生产花椒毒素

花椒毒素为一种在伞形科植物中发现的呋喃香豆素,当将花椒毒素与增效烯(伞形植物有机杀虫剂增效剂)或只将花椒毒素添加到人工饲料中时,导致谷实夜蛾(*Helicoverpa zea*)一龄幼虫死亡率达到50%时的浓度(湿重)

5.3.5 植物化感物质的解毒作用

布腊特斯顿在其论文中指出[35],在取食过程中,植食性昆虫都需面对大量有害化学物质,因此它们冒着每一餐都可能中毒的风险。然而,这些有毒物质可以被接纳,因为食草动物已进化出多种生理机制来避免其有害影响。它们可以迅速排泄掉不需要的化合物,或者进行酶促反应将其降解,或者在化感物质达到药物活性之前将其中和。作为最后的手段,它们演化出了靶位点不敏感策略,换言之就是由于靶位点(或临近位点)结构的改变,使有毒物质不能成功绑定目标[21]。由于食草性昆虫消耗多于其体重的食物(毛虫可以一天吃掉相当于它们体重5倍的食物),因而它们的解毒系统也必须高效。而事实上,昆虫似乎比高等生物更能忍受,如氢化氰(HCN,对所有高等生物均有毒性,在消化过程中从植物中释放出来)和生物碱(见4.6)。有趣的是,杂食性昆虫对于氢化氰的敏感程度是哺乳动物的百分之一左右(表5.5)。同样,生物碱对于非专食性的昆虫的毒性似乎比哺乳动物低一到两个数量级(图5.13)。

表5.5 氰化氢对于杂食性昆虫与哺乳动物的毒性的比较(伯奈斯,1982)[25]

动物	口服 LD50(mg/kg)
东亚飞蝗(*Locusta migratoria*,直翅目)	500
臭腹腺蝗(*Zonocerus variegatus*,直翅目)	1 000
海灰赤夜蛾(*Spodoptera littoralis*,鳞翅目)	800
南方灰翅夜蛾(*Spodoptera eridania*,鳞翅目)	1 500
哺乳动物(普通)	0.5~3.5

注:LD50,达到50%动物死亡时的剂量。

专食性昆虫也能应对它们食物中高浓度的化感物质[25]。然而,目前的争议主要在于体重的比较。当对毒性的比较主要基于代谢活性或身体表面特征而不考虑体重时,哺乳动物和昆虫之间的差异变得不那么引人注目,甚至可以忽略不计。

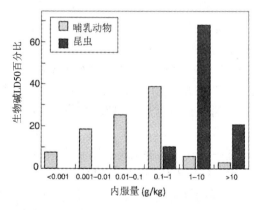

图5.13 生物碱对非专食性的哺乳类食草动物与非专食性昆虫的毒性比较,表现为在不同剂量范围的生物碱的LD50百分比(伯奈斯,1982)[25]

1. 生理性适应:迅速排泄

防止中毒的一种最有效的方法,就是让有毒物质无法到达目标位置。也可以防止潜在的毒素穿过肠壁。在生理条件下，许多植物次生化合物都为带电极性的大分子，或亲水化合物（例如糖类），往往难以穿透生物膜。再加上食草昆虫快速运动的肠道，意味着许多毒性物质几乎没有机会进入体腔[185]。

烟草天蛾（Manduca sexta）幼虫以含有尼古丁的烟草为食。对于其他动物而言，尼古丁是传统的杀虫剂及致命毒药。而与许多生物碱相反，尼古丁是亲脂性的，因此在大多数昆虫中很容易通过肠道上皮。这是由于它的毒性与乙酰胆碱（动物的中枢神经系统中关键的神经递质）在功能上有相似之处。通过模仿乙酰胆碱分子，尼古丁微妙地干扰了中枢神经系统的基本功能。烟草天蛾幼虫已经进化出一系列的防御机制来对抗尼古丁这个非常强的毒素。毒性累积到中毒量之前，它们会迅速地排泄尼古丁和其他摄入的生物碱。在一项实验中，天蛾幼虫被喂养含有已知剂量的尼古丁，93%的 0.5mg 剂量的尼古丁在两个小时之内被排出，然而在家蝇体内 90% 以上的剂量在昆虫体内保持长达 18 小时[149]。少量的尼古丁可以进入烟草天蛾幼虫的血淋巴，但不能穿过神经鞘，并可通过马氏管排出。尽管此时物种已演化出了生理屏障，而尼古丁已经到达中枢神经系统的神经细胞中，这些细胞似乎能包容这种化合物，为靶位点不敏感策略提供了范例。

这是一个深入研究的范例，昆虫具抵御化感物质的多重防护系统，而化感物质对于所有不适应的动物都是有剧毒的。快速排泄并非是一个专化的生理特性。以十个不同科的植物喂养杂食性的绿细纹蝶（Callophrys rubi）幼虫，能将其在恒定寄主植物染料木（Genista tinctoria）中取食的所有生物碱排出，当以非寄主植物，例如羽扇豆（Lupinus polyphyllus）为食时仍可以排出[65]。

2. 解毒酶

在很大程度上，多数食草昆虫依赖酶促降解来中和已吸收的植物化感物质。最为广泛研究的酶，是能有效代谢各种毒素的细胞色素 P450 单氧酶，也称作聚乙烯基质单氧酶（polysubstrate monooxygenases, PSMOs）或混合功能氧化酶（mixed-function oxydases, MFOs）。P450 酶类几乎存在于昆虫的所有组织中，在对寄主植物的化学物质代谢中起重要作用，将寄主植物的化学物质先转化为极性更强的化合物，然后通过二级酶进一步代谢。P450 的名字源于它的最大吸收光谱大约为 450 nm。每种昆虫的基因组携带约 100 个 P450 基因。由此说明了 P450 酶结构上的多样性以及在代谢途径中的多样化功能[64]。

在各食草动物中，细胞色素 P450 单氧酶的活性有很大不同。在一项对 58 种毛虫的研究中发现，这种酶的活性似乎与被取食的植物类型有关。以富有单萜类的植物（例如桃金娘科、芸香科、茄科的某些种类）为食的物种，比以其他植物如豆科、车前草科、禾本科为食的物种，其酶具有相对高的活性[139]。

除 P450 系统外，同时还有一些其他的酶系统也用于化感物质解毒。除通过氧化代谢外，还能通过水解分裂和接合作用进行毒素代谢（图 5.14）。

有些化感物质（被称作"化剂"）在和昆虫的消化系统相互作用过程中可能产生有毒的氧分。例如，呋喃香豆素在光化学作用下，同时产生过氧化物、单态氧以及水解的丹宁酸，水解丹宁酸经氧化之后可以产生活性氧。为了抵消氧化剂对身体组织和细胞外液的毒性作用，食草昆虫演化出一系列抗氧化酶，这些酶的数量随食物的种类和先前暴露于"化剂"的组织而变化。因为成熟的禾本科植物通常包含的化感物质数量少且种类有限，因而食草蚱蜢比杂食性蚱蜢肠道中的抗氧化酶含量更低。对两种不同取食习惯的蚱蜢进行了酶浓度（中肠组织和肠

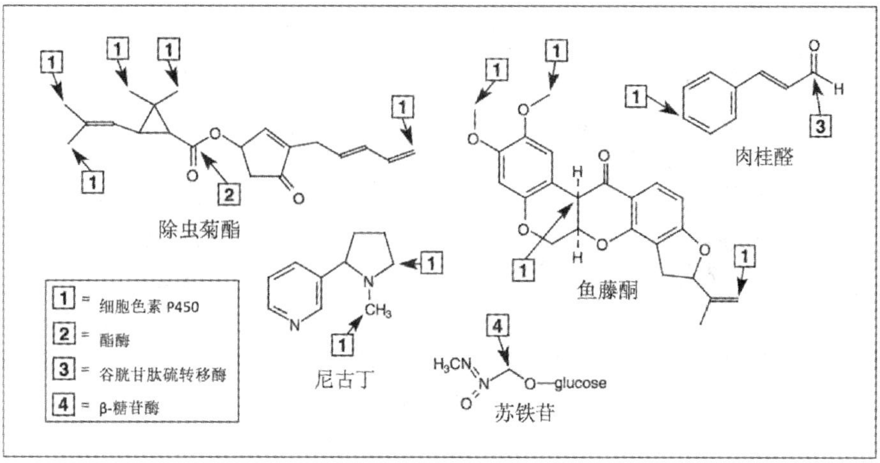

图 5.14 酶对某些植物化感物质进行攻击的位置（布拉特森，1988）[36]

道液体中)的比较，揭示了杂食性物种酶的含量更高[12,13]。

毒性化合物的分解产物可以在中间代谢途径中被回收，或被转化为易排泄的物质。虽然酶促降解反应的产物通常无毒，但有时降解产物的毒性比前体还大（表 5.15），随后再经其他酶代谢为无毒的物质。

当昆虫遭遇一种新的毒素时，解毒酶系（如 P450）的数量在几分钟内迅速增加。这一现象被称为"诱导"，取决于被基因激活的酶蛋白的合成过程。当取食人工饲料后，杂色地老虎（*Peridroma saucia*）幼虫显示出对 P450 的活性较低。喂食薄荷叶片后，它们的 P450 活性比原来的 45 倍以上，由于薄荷叶片中含有高浓度的单萜烯类物质。昆虫取食带有蒎烯或薄荷醇的人工饲料后，中肠中细胞色素 P450 的含量会显著地增加[188]。当以 11 个科的寄主植物喂养杂食性斜纹夜蛾（*Spodoptera litura*）的幼虫时，P450 活性水平会随植物的不同呈现出 20 倍范围内的变化[139]。此外，由 P450 系统控制的反应在不同的植物中各不相同，进一步表明了 P450 同工酶的存在（表 5.16）[64]。

欧洲防风草结网虫（*Depressaria pastinacella*）必须与许多类型的呋喃香豆素斗争，这种毒素大量存在于它们所有的伞形科寄主植物中。同时，昆虫的 P450 活性增强，对食物中存在的呋喃香豆素做出反应。但是，并没有检测出解毒酶对一种特定的香豆素或一种特定的混合香豆素进行特别的调整[45]。然而，以伞形科植物为食的黑凤蝶（*Papilio polyxenes*）幼虫具有可调控的 P450 基因，这种调节对特定的宿主化感物质有特殊的反应，表明了调节的专一性[23]。

植物化感物质能引起食草动物中 P450 抗性基因的激活，这是一个表型可塑性的例子，并且具有重要的生存价值。昆虫的防御策略具备更精准的细节，可以使 P450 酶对植物信号化合物做出反应，如茉莉酸和水杨酸。当植物受到昆虫（或病原菌）攻击和刺激后，这些化合物会诱导防御性化学物质的含量剧烈上升，这部分内容在第 4 章中也已提到过（见 4.14）。昆虫通过关注植物的信号分子来调节其解毒系统，在生理上已经准备好应对食物中化感物质增加的可能性[106]。

特定的昆虫对于杀虫剂的敏感度会随其寄主植物的种类不同表现出巨大区别，这可以用解毒酶的可诱导性进行解释（图 5.16），也可以解释为什么植食昆虫的天敌通常比它们的寄主

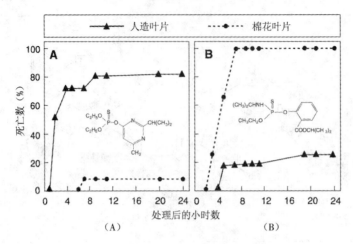

图 5.15 饲料对两种杀虫剂毒性的影响

新蜕皮的六龄草地夜蛾(*Spodoptera frugiperda*)幼虫提前两天喂以人工饲料或棉花叶片,然后用杀虫剂进行处理,(A)每只幼虫 2 微克二嗪磷;(B)每只幼虫 15 微克异柳磷,以棉花为食的幼虫,能将二嗪磷快速分解成毒性小的化合物,因而低降死亡率;以棉花为食的幼虫,因快速分解异柳磷而增加了死亡率,因为产生的代谢产物比前体化合物毒性更强

更容易受到杀虫剂处理的影响:它们通常在食物中很少摄取或根本不摄取毒素。在进化过程中,寄生者和捕食者并没有像植食性昆虫一样接触过多的植物次生物质。

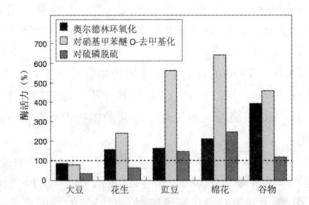

图 5.16 寄主植物对草地夜蛾(*Spodoptera frugiperda*)幼虫中肠氧化酶活性的影响

将三种不同的 P450 酶(由五种植物诱导)的活性与以人工饲料喂养的酶的活性进行比较(100%)。不同的 P450 活性数据表明了大量的细胞色素 P450 酶的存在

一旦诱导化学物质不再存在,酶的活性就开始下降到诱导前水平。由于这种灵活的诱导机制,食草动物能表现出高度可变的酶活性,这种酶活性的变化取决于被消耗的食物及在餐后多长时间进行检测。研究表明,维持高水平的解毒酶含量需要消耗一定的成本。但是,目前还没有证据表明,需要消耗重要的能量或营养成本(见 5.3.2)[160]。因此针对诱导的耗能情况,目前仍需深入研究。

昆虫的排毒机制只有不断进化,才可能打破植物的化学防御系统(自然中最多样的化学宝库)。在最近几篇综述中,对这些机制进行了广泛的讨论[37,64,107]。

5.4 共生体

地球上的生命不可避免地与细菌共存。食草昆虫也与看不见的微生物世界发生密切关系。当昆虫接触到植物时,它就接触到细菌及其代谢的产物。对于许多植食昆虫而言,它们利用植物食物的能力取决于共生微生物的存在。目前,证明微生物可对食叶昆虫消化和吸收起到一定作用的证据是非常稀少的[9],但可以确定的是,如果没有微生物,以植物的汁液为食的昆虫将很难存活。细菌、酵母和其他单细胞真菌或原生动物协助降解植物食物,并合成一些植物不能提供的或数量不足的必需营养(如固醇、一些维生素、20种氨基酸中的10种)[17 44 131]。共生体的第三个作用是对植物的次生物质进行协助解毒。

5.4.1 食物利用与补充

位于消化道中的细胞外共生体,其中一些在肠道内腔中自由运动,另一些位于中肠或后肠的盲囊中,许多的鞘翅目、半翅目、直翅目昆虫都是如此,鳞翅目昆虫则缺少盲囊(图5.17)。约10%的昆虫可以容纳细胞内共生体。

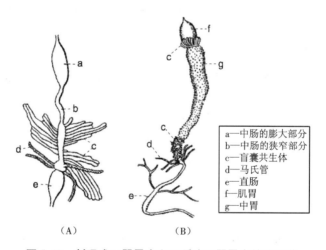

图 5.17 刺吸式口器昆虫和咀嚼式口器昆虫的消化道

(A)以种子为食的长蝽(*Aphanus* sp,长蝽科)的肠道;(B)以柳兰为食的叶甲(*Adoxus obscurus*)的肠道

a—中肠的膨大部分
b—中肠的狭窄部分
c—盲囊共生体
d—马氏管
e—直肠
f—肌胃
g—中胃

它们在肠壁细胞内部发生,并不断释放进内腔。这种情况也会发生在一些以木头为食的天牛科幼虫中,然而,通常它们被限制在特别的细胞(含菌细胞)中,这些含菌细胞通常散布在各种组织中,或者聚集到一个器官结构中。在所有的植食性动物中,对内共生体的研究最多的是半翅类和鞘翅目。如吸食韧皮部汁液的蚜虫类,无论食物营养是否不良,微生物都能帮助其克服营养阻碍[54]。

绿色桃蚜(*Myzus persicae*)肠道中的微生物能产生所有必需氨基酸(除了寄主植物提供的四种氨基酸)[117]。对于多种昆虫的研究表明,共生体可以为昆虫补充天然植物无法提供的或供给不足的营养[44]。最近的研究表明,共生体可以对单个氨基酸的生产或释放进行调控,适应其寄主的营养需求。而韧皮汁液的氨基酸组成随植物种类、物候、环境条件变化而变化,甚至同一植物不同筛管成分之间也是不同的。共生体针对不同的氨基酸产生的反应不同,黑草属

(Buchnera)的例子表明了必需氨基酸生产的灵活性,也许不是绝对必要的,但对蚜虫来说也是明显有利的[57]。

5.4.2 植物化感物质的解毒作用

可以在富含有毒化合物的寄主植物上生存的昆虫,必须具备内在的抗毒机制来防止中毒。基于最近的一些研究,共生体可以将寄主的化感物质转化为毒性较小的化合物[58]。例如,寄生于烟草甲虫(*Lasioderma serricorne*)的一种酵母菌,可以使多种有毒物质(包括一些黄酮类和丹宁酸)分解。依据这种生化机制,这种酵母菌共生体可能对其寄主成功寄生于多种植物有所帮助[150]。一个基于不同目昆虫的例子,同样说明了共生体的解毒作用为经常发生。根皮苷(phloridzin)为苹果实蝇(*Rhagoletis pomonella*)寄主植物,通常含有黄酮类物质,而苹果实蝇却能抵抗根皮苷的毒性。因为成年苹果实蝇可以从叶面上获取各种营养,如蚜虫蜜露、叶面的微生物以及一些叶面渗出液,成分包含氨基酸、糖类,也包括根皮苷。这些摄取物质中包含的细菌在消化道内可以将根皮苷进行分解。实验证明,苹果实蝇摄取不包含微生物的根皮苷后,会在24小时内死亡[103]。

在一个将无菌蝗虫肠道移入特定微生物的实验中,发现一些细菌的次级代谢产物可以产生对蝗虫有益的酚类物质,其中一些次级代谢产物具选择抗菌性,可抑制外源微生物(如病原微生物)。当昆虫接种了某种特殊微生物时,经测量得出其肠道中细菌的多样性增加,因而昆虫防御系统的效率也可能增加[52]。

迄今为止,我们对共生体对植物—食物的影响的认识还非常有限。在昆虫植物相互作用过程中,共生体的作用非常关键[52]。追溯到动物进化的早期阶段,共生体可能为导致某些动物进化的一个因素,这足以体现它们的重要性。因此,蚜虫内共生菌(*Buchnera aphidicola*)为蚜虫共生体,基于长期严格共同进化模式,已经与寄主一起经历了2亿年的演变[55,119]。也许再次提醒我们:昆虫—植物体系与许多其他形式的生物有紧密的联系。

5.5 微生物影响寄主植物特性

共生体或植物病原微生物,都可能引起植物化学性质的变化,这也许会影响植食性昆虫,并且与植物间接的作用。

5.5.1 植物致病菌

在自然和农业生态系统中,植物病害的发生同样频繁。被感染的植物,除物理特征已发生改变外,在生物化学方面也不同于健康植物,对植食性昆虫的营养适应度也将改变。在感染植物中,对营养物质如糖、氨基酸、淀粉等和化感物质的吸收与分配通常都会显著地改变[16]。

当寄主植物被植物病原体感染时,植食性昆虫通常会受到消极的影响,尽管也有一些积极影响的报告[16,76]。叶甲(*Gastrophysa viridula*)是一种以酸模属植物(*Rumex* spp.)为食的专食性昆虫,相对于正常的植物,取食被锈菌(rust fungus)感染的植物后,叶甲(*G. viridula*)会有更高的幼虫死亡率,且会发育延迟,繁殖能力更低。

化学分析表明,那些被感染的叶片相对于正常的植物,氮元素含量降低,而草酸浓度升高,也许为对叶甲(*Gastrophysa viridula*)有害的原因。植物的这种反应在被损坏的叶片迅速扩展,同时,在一定范围内会转移到未被损害的叶片上[77]。然而,植物—病原菌—昆虫这三者间的关系非常复杂,昆虫受到锈菌为害,又可通过毁坏植物、诱发抗性来防御锈菌。

植物病原体有时可能对植食性昆虫有益,有一个典型的例子。生长在锈菌感染的欧洲白桦(*Betula pendula*)上的蚜虫(*Euceraphis betulae*)相对于未感染植株,种群增长率和胚胎发育率更高。蚜虫的性能和叶片一些的化学因素正相关,尤其是那些自由氨基酸基团(表5.6)。而植物受真菌感染后,在生理上产生的应答反应,与植物叶片在衰老阶段的反应类似。这些改变后叶片的化学成分为提高蚜虫性能的原因[90]。

表 5.6 欧洲白桦(*Betula pendula*)叶片的化学成分,及蚜虫在未感染叶片(AL)与被真菌感染叶片(FIL)上的表现(詹森等,2003)[90]

	AL	FIL
叶片成分		
酚类物质浓度	100	128
自由氨基酸浓度	100	239
蚜虫表现		
成虫数量	100	114
具有成熟胚胎的成虫数量	100	136
成虫体内的胚胎数量	100	127

注:所有值都为基于未感染叶片值的百分率。

偶尔发现,感染病毒的植物能够更好地为昆虫提供食物。那些将疾病传染给植物的昆虫(如同翅类、蓟马、螨虫、甲虫类等)能通过感染而在寄主上获益,实际上是对资源更好地开发利用。与健康的豌豆(*Phaseolus vulgaris*)相比,墨西哥豆瓢虫(*Epilachna varivestis*)更喜欢取食被感染的豌豆,而它们的幼虫也会在感染病毒的豌豆叶片上成长得更快。在甲虫与病毒之间明显存在互利关系,病毒可由昆虫传播,而昆虫在感染病害的植株上成长得更快。由植物病原体引起植物的化学变化最有可能提高昆虫的性能[120]。

5.5.2 内生真菌

已在几个实例中发现,由于内生真菌的存在,植物化学组成发生改变,继而对植食性昆虫的性能存在显著影响。虽然内生真菌的直接致病效果非常有限甚至完全丧失,但是它们依然可以抵抗植食性昆虫和植物病原体的入侵(图5.18)。这些内生菌对它们寄主植物的贡献主要表现为,能够产生生物碱或其他的化合物,从而使它的寄主可以预先对食草昆虫进行防御。内生真菌被定义为在健康的植物组织上存活的真菌,而植物不表现出症状。它在很多的植物上都存在并且对于农业有很大价值。近几年这方面的研究很多[38,47]。相对而言,许多稻科植物都缺乏防护性化学物质,植物内生真菌演化的动力,也许就在于它与寄主之间的互利共生关系。而耕种的稻科植物通常表现出较严重的内生菌感染,而天然的稻科植物通常由感染和未感染的植物混合而成。这也是最近内生菌的防护作用遭质疑的原因之一[61]。

植物内生真菌不仅局限于稻科植物,也有证据表明它们与许多被子植物也有关系[169],包括木本植物[140]以及一些其他的被子植物。当番茄植株被一种普遍而传播广泛的、由土壤携带的内生真菌(*Acremonium strictum*)感染时,杂食性棉铃虫(*Helicoverpa armigera*)幼虫的死亡率变高,生长发育时间也会变长。在这个实例中,负面影响也不能完全归因于植物中的生物碱或

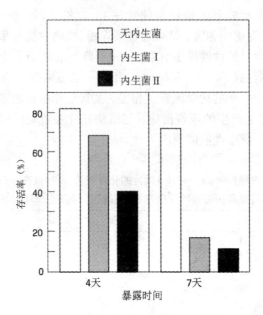

**图 5.18　谷物大害虫麦虱（*Blissus leucopterus hirtus*）在未感染和
接种内生菌的紫羊茅（*Festuca rubra*）的存活率**

注：两个内生菌品种在不同的稻科植物上获得。

其他有阻碍作用的化合物，也有可能是植物中其他真菌引发的生理学改变，这些改变影响了植食昆虫对食物的利用[89a]。

最近的一些研究表明，我们当前普遍认同的昆虫—植物的关系需要扩展到多重营养关系互作更广阔的角度。关于多重营养系统有一个有趣的例子，日本金龟子（*Popillia japonica*）的幼虫以牛毛草的根部为食。当牛毛草被内生真菌感染时，幼虫的摄食会因具有摄食抑制作用的生物碱的存在而降低。因而幼虫的活力下降，表现为它们更易受病原线虫的侵染[71]。尽管这个研究仅基于单一的植食昆虫与病原菌的关系，然而研究微生物共生体的影响可能对植物所属的食物链中更多成员的关系有所揭示。因此，对具有内寄生菌的意大利黑麦草（*Lolium multiflorum*）与蚜虫——寄生性天敌形成的食物链的比较显示：不同的蚜虫对于内寄生菌的有无存在不同的反应。而且，它们的寄生性天敌也随之发生改变，表明了在寄主植物中互利共生的微生物对多重营养关系的影响[43,124]。

内生真菌并不总为植物抵抗食草昆虫。正如一些研究显示，植物—寄生菌的互作关系也可以使食草昆虫受益[169]。然而，尽管有一些相反的结果，整体的文献报告均说明内生真菌对植物—食草昆虫关系起到相当大的作用。

微生物可以对植物的次级代谢物解毒并改变植物的化学成分，此类相关文献表明：昆虫—植物关系与其他有机物的相互作用是密不可分的。这部分内容将会在第 10 章中进行进一步讨论。

5.6　寄主植物影响植食性昆虫对病原体和杀虫剂的敏感性

目前已有证据表明，植物会影响昆虫对病原微生物如细菌[9,121]、病毒[144]、真菌[75]和线虫[15]的敏感性。这归因于植物对病原微生物可能存在抑制作用,也可能会增强病原微生物的繁殖力和毒性。

假设化感物质在营养级互作关系间起到决定性的作用,大多数研究主要集中在植物化合物对昆虫病原微生物的影响上。一些研究指出寄主植物对病原微生物致病性产生的化感物质的浓度不同,另一些研究通过将纯化合物与化感物质及病原菌加入人工饲料后,研究化感物质的效果。借助后一种方法,发现了芸香苷在一定浓度范围内,能保护拟尺蠖(*Trichoplusia ni*)幼虫不被苏云金胞杆菌(*Bacillus thuringiensis*)[101]产生的毒素伤害。已经发现几种化感物质可以影响病菌或病菌产生的毒素,但是也有其他一些叶片因素,如营养价值、年龄和水分等因素也能起到一定的影响。根据在测试前和在测试中提供的叶片不同,舞毒蛾的幼虫对杆状病毒的敏感性也不同。在接种正常剂量的病毒后,取食水解丹宁酸浓度较低叶片的幼虫死亡率高于取食较高水解丹宁酸浓度叶片的幼虫(表5.7)[94]。因为幼虫死亡率也与叶片组织中的酸性有关(会影响昆虫中肠的PH酸碱度)。由此看来,寄主植物和病毒对食草动物的敏感性之间的关系是由多因素影响的(可参照10.4部分)[95]。

表5.7 加入标准剂量的杆状病毒后,不同的寄主植物对舞毒蛾(*Lymantria dispar*)幼虫死亡率的影响(基廷等,1988)[94]

寄主植物	死亡率	水解丹宁(%叶片干重)
黑棕(*Quercus nigra*)	25	33.2
红橡木(*Q. rubra*)	47	36.6
美洲山杨(*Populus tremuloides*)	79	1.4
大齿杨(*P. grandidentata*)	86	1.2

另一种杂食性昆虫,白粉虱(*Trialeurodes vaporariorum*)对白僵菌(*Beauveria bassiana*)的敏感度因植物食物的不同存在很大的差别(图5.19)。白僵菌与杆状病毒相反,通过角质层入侵寄主。当昆虫以番茄为食时,这种真菌显现出较低的敏感度,可能由于角质层或血淋巴中存在的番茄碱是一种众所周知的抗菌物质[132]。

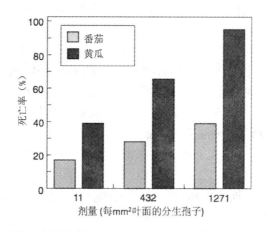

图5.19 在两种不同的寄主植物上,添加不同剂量的白僵菌(*Beauveria bassiana*)分生孢子,七天后三龄温室粉虱(*Trialeurodes vaporariorum*)的死亡率(普罗斯基等,2000)[132]

如上所示，杂食性昆虫对杀虫剂的敏感性会因它们取食不同的植物而发生变化。迁移蚱蜢取食燕麦时被溴氰菊酯杀死的概率是取食黑麦时的1/3[83]。根据寄主植物的不同，桃蚜（*Myzus persicae*）对杀虫剂的敏感度会显示出约200倍的变化[5]。即使在同一种植物中，昆虫对杀虫剂的敏感性也会随不同品种的不同生理变化而发生变化。桃蚜对杀虫剂显示出不同的敏感性，不仅体现在喂以不同品种的球芽甘蓝（*Brussels sprouts*），而且体现在施用不同氮肥的统一栽培品种的植物上。因而，寄主植物的状况也能极大地影响杀虫剂的耐受力[118]。

食草昆虫对病原体和杀虫剂的敏感性的生理机制至今仍存在许多未解之谜。然而，大多数研究都支持这样一种普遍假设：昆虫对病原微生物的敏感性与对寄主植物的适应性成反比。如昆虫以次优的寄主植物为食，将会对昆虫防御（如微生物感染）施加压力[114]。

5.7 植物食物质量与环境因素的关系

5.7.1 干旱

水是植物将矿物盐和其他物质摄入其体内的主要运输手段。植物通过根和茎将水分子运送到叶片，水分子在叶片中分解，得到的氢离子（H^+）用于光合作用。植物通常也以交换二氧化碳的形式流失水分子（交换速率可达每固定一个二氧化碳分子交换400个水分子），因此，植物的生长需要大量的水。

有许多研究揭示了植物水分与昆虫反应之间的关系，但是植物内水分胁迫影响昆虫性能的具体细节仍需要进一步阐明[73]。干旱对食草昆虫极为不利，尤其是吸食植物汁液的昆虫在这种连续的水分胁迫下受到的不利影响更为严重[87]。然而，也存在相反的情况，这种水分胁迫的影响可以忽略不计，甚至对食草昆虫的种群数量增加有益。异常温暖、干旱的气候往往伴随着森林和牧场中虫害的爆发。其原因至今没有得到很好的阐述，也许是干旱对植物抵御机制产生了负面的影响，而与此同时对昆虫的营养价值会提高[100]。干旱像其他压力一样，会导致植物叶片、内树皮、木质部汁液中可溶性糖类和氮水平的增加，如香脂冷杉的叶面糖含量可能会增加2.5倍。尽管叶面的氮浓度增加了，但由于细胞组织膨胀的减少和含水量的减少，许多食草昆虫对氮的摄入和利用都受到了影响。

同时发现，干旱能够扰乱许多木本及草本植物的氮代谢从而影响昆虫的生长和繁殖能力[113]。因此，植物缺乏水分往往导致内部自由氨基酸的累积。脯氨酸的含量如果积累达到一定水平，就能够反映水分缺失的情况[46,113]。在干旱的棉花中，自由脯氨酸的水平会增长到原来的50倍[151]。增加的脯氨酸（和其他自由氨基酸）水平可能会增加植物对虫害的易感性，这是由于对许多昆虫来说脯氨酸为一种进食刺激剂（第7章）。

5.7.2 空气污染

植物的结构特性，比如叶面形态和韧性，以及初级代谢产物和次生化合物的水平，都会受空气污染物影响，这已被休斯[88]和其他学者明确地提出[42,79]。

危害植物毒性最严重的空气污染物包括二氧化硫（SO_2）、臭氧、一氧化氮（NO）和二氧化氮（NO_2）。有关空气污染与植物虫害的变化两者之间关系的证据已经通过观测获得（在工业园附近的森林中虫害的爆发），而最近也已通过实验证实。使用控释二氧化硫进行的野外研究表明,蚜虫的摄食增强会导致额外产量损失[137]。许多蚜虫和其他吸食性昆虫取食暴露于空气污染物适度浓度的植物会生长得更好,而几种咀嚼式口器昆虫却表现为种群密度下降。

空气污染物会通过影响寄主植物质量或影响它们的天敌来影响植食性昆虫的种群密度。有证据表明,污染物对植物的营养质量的影响可能会有很大改变。在许多情况下自由氨基酸和还原糖的含量会增加,而叶片的蛋白含量可能增加也可能减少,这些变化都反映在氮与碳水化合物比例的改变上。

与人们预想的相反,将寄主植物暴露于SO_2中,对于植食性昆虫是有益的[182]。墨西哥豆瓢虫(*Epilachna varivestis*)更喜食用暴露于SO_2中的植物叶片,并且昆虫的生长速度和繁殖力会提高。应对氧化污染物及其他类型的压力,植物叶片产生常见的变化,还原型谷胱甘肽的数量发生很大增长。在大豆叶片中,发现谷胱甘肽浓度随着烟熏而增加,与此同时,昆虫的生长方式也发生了改变。当非熏蒸叶片增加谷胱甘肽后(离体的叶片通过叶柄吸收缩氨酸溶液),也会刺激昆虫生长,如同经SO_2处理的植物一样。表明这种化合物在污染—诱导效果中起重要作用(图 5.20)[88,89]。

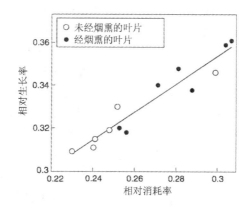

图 5.20 喂以烟熏的大豆叶片(24 小时 0.3 ppm 二氧化硫)或带有谷胱甘肽的未经烟熏的叶片的墨西哥豆瓢虫幼虫的摄食率和生长率(表现为相对摄食率和相对生长率)

空气污染物对植物和植食性昆虫存在影响,目前已无争议,但我们对空气污染物随后对昆虫种群密度发展的影响,或对生态系统功能的影响知之甚少。

酸雨是与空气污染密切相关的现象,主要由硫和氮的氧化物引起。酸雨可能不会直接对植物造成影响,在多数情况下通过间接改变土壤的性质和土壤中的微生物活动对植物产生消极的影响,特别是对缓冲能力差的土壤[88]。

尽管CO_2通常不被视为空气污染物,但它也许是由于人类活动对大气成分改动最大的部分。自 19 世纪中期以来,全球CO_2浓度上涨了近 30%,预计在接下来的 50~75 年翻倍。而二氧化碳含量的增加通常会影响植物生长及其物理和化学组成[50,127]。多数关于二氧化碳增加引发影响的研究表明,大部分植物叶片中的碳氮比率会增加[11]。观察中多数具咀嚼式口器的昆虫,为了弥补食物中较低的营养质量,相比取食正常CO_2含量的植物,要取食更多CO_2处理过的植物[181]。在比正常CO_2含量高出两倍的桦木上,叶片中的氮浓度减少了 23%,而浓缩丹宁酸的浓度为原来的 2 倍,淀粉的浓度变为原来的 3 倍。3 种鳞翅目昆虫取食这种叶片后,均表现为摄食量增加,生长停滞或下降,幼虫发育期延长[108]。

对于二氧化碳含量的增加，蚜虫的反应是多种多样的。在某些情况下，蚜虫的性能增强，而在其他方面也观察到消极的影响[86]。在野外实验中，观察到蚜虫的天敌密度在高浓度 CO_2 条件下比在对照条件下更大[127]。因此，预测自然群落应对大气中 CO_2 含量增加的变化仅依靠实验室中得出的结论仍然有限。同时，也应该认识到 CO_2 浓度增加可能会影响其他因素，如云量、降水和温度等，都可能对 CO_2—植物—昆虫之间的关系产生影响[11,178]。

5.8 结论

对昆虫来说，植物组织并非为最佳的食物来源。它是一种低氮和高化感物质的集合。食草昆虫，有自己特定的营养需求，不仅面临营养需求和所获食物中的化学成分间的差距，同时也面临植物的营养和抗性物质时空差异的问题。例如，北美落叶松的叶片单萜类化合物，在树内部的差异与树间的差异可能一样大。因此，单棵树可以表现为昆虫食物的化学嵌合体[133]。

当然，较大的营养差异可能对昆虫的性能产生很大的影响。在一棵桦树上，尺蠖 (*Epirrita autumnata*) 幼虫的生长差异可以高达30%。在植物内部的质量异质性很大程度上是由于植物内部导管的分布对资源的影响以及之前的损伤和营养、水、光照等因素的空间变化[125]。

植物的化学成分可发生很大的变化，是由许多环境因素引发的，如植物病原体、空气污染、营养和水资源利用以及其他一些因素。食草昆虫通过摄取大量的食物来应对营养不足，而增加了化感物质的含量引发中毒的风险，这是一个对于补偿进食的限制因素的例子。非特异性的和诱导解毒的有效机制可以中和有毒化合物，并且共生体可以帮助实现营养需求。

由植物提供的食物质量和昆虫所需的最低限度的食物质量之间的平衡是微妙的。此外，在更大程度上，这受一个复杂网络（例如病原体和共生微生物）中的其他成分的影响。

目前的文献大多都关注于昆虫植物互作的定量营养生态学。然而，大多数数据是基于实验室的研究，这些研究采用重量分析的技术，并使用离体的植物组织。斯兰斯基、斯奈伯[164]以及斯兰斯基和罗德里格兹[163]对这部分内容进行了综述。采用新的技术手段，在相关生态条件下，用进行光合作用的完整植物来喂养昆虫的代谢效率可以测量出来，这意味着在这一领域将取得真正的突破性进展[174]。

5.9 参考文献

1. Abisgold, J. D. and Simpson, S. J. (1987). The physiology of compensation by locusts for changes in dietary protein. *Journal of Experimental Biology*, 129, 329–46.

2. Ahmad, S. (1986). Enzymatic adaptations of herbivorous insects and mites to phytochemicals. *Journal of Chemical Ecology*, 12, 533–60.

3. Aidley, D. J. (1976). Increase in respiratory rate during feeding in larvae of the armyworm, *Spodoptera exempta*. *Physiological Entomology*, 1, 73–5.

4. Allen, S. E., Grimshaw, H. M., Parkinson, J. A., and Quarmby, C. (1974). *Chemical analysis of ecological materials*. Blackwell, London.

5. Ambrose, H. J. and Regupathy, A. (1992). Influence of host plants on the susceptibility of *Myzus persicae* (Sulz.) to certain insecticides. *Insect Science and its Application*, 13, 79–86.

6. Anantha Raman, K. V., Magadum, S. B., and Datta, R. K. (1994). Feed efficiency of the silkworm *Bombyx mori* L. hybrid (NB4D2 x KA). *Insect Science and its Application*, 15, 111 – 16.
7. Andersen, P. C., Brodbeck, B. V., and Mizell, R. F. (2003). Plant and insect characteristics in response to increasing density of *Homalodisca coagulata* on three host species: a quantification of assimilate extraction. *Entomologia Experimentalis et Applicata*, 107, 57 – 68.
8. Anderson, T. E. and Leppla, N. C. (1992). *Advances in insect rearing research and pest management*. Westview Press, Boulder, CO.
9. Appel, H. M. (1994). The chewing herbivore gut lumen: physicochemical conditions and their impact on plant nutrients, allelochemicals, and insect pathogens. In *Insect-plant interactions*, Vol. 5 (ed. E. A. Bernays), pp. 209 – 23. CRC Press, Boca Raton.
10. Axelsson, B. and Ågren, G. I. (1979). A correction for food respiration in balancing energy budgets. *Entomologia Experimentalis et Applicata*, 25, 260 – 6.
11. Ayres, M. P. (1993). Plant defense, herbivory, and climate change. In *Biotic interactions and global change* (ed. P. M. Kareiva, J. G. Kingsolver, and R. B. Huey), pp. 75 – 94. Sinauer, Sunderland, MA.
12. Barbehenn, R. V. (2003). Antioxidants in grass-hoppers: higher levels defend the midgut tissues of a polyphagous species than a graminivorous species. *Journal of Chemical Ecology*, 29, 683 – 702.
13. Barbehenn, R. V., Walker, A. C., and Uddin, F. (2003). Antioxidants in the midgut fluids of a tannin-tolerant and a tannin-sensitive caterpillar: effects of seasonal changes in tree leaves. *Journal of Chemical Ecology*, 29, 1099 – 116.
14. Barbehenn, R. V. and Martin, M. M. (1994). Tannin sensitivity in larvae of *Malacosoma disstria* (Lepidoptera): roles of the peritrophic envelope and midgut oxidation. *Journal of Chemical Ecology*, 20, 1985 – 2001.
15. Barbercheck, M. E., Wang, J., and Hirsh, I. S. (1995). Host plant effects on entomopathogenic nematodes. *Journal of Invertebrate Pathology*, 66, 169 – 77.
16. Barbosa, P. (1993). Lepidopteran foraging on plants in agrosystems: constraints and consequences. In *Caterpillars. Ecological and evolutionary constraints on foraging* (ed. N. E. Stamp and T. M. Casey), pp. 523 – 66. Chapman & Hall, New York.
17. Barbosa, P., Krischik, V. A., and Jones, O. G. (1991). *Microbial mediation of plant-herbivore interactions*. Wiley, New York.
18. Beck, S. D. and Reese, J. C. (1976). Insect-plant interactions: nutrition and metabolism. *Recent Advances in Phytochemistry*, 10, 41 – 92.
19. Behmer, S. T. and Nes, W. D. (2003). Insect sterol nutrition and physiology: a global overview. *Advances in Insect Physiology*, 31, 1 – 72.
20. Berenbaum, M. (1985). Brementown revisited: inter-actions among allelochemicals in plants. *Recent Advances in Phytochemistry*, 19, 139 – 69.
21. Berenbaum, M. R. (1986). Target site insensitivity in insect-plant interactions. In *Molecular*

aspects of insect-plant associations (ed. L. B. Brattsten and S. Ahmad) , pp. 257 – 72. Plenum Press, New York.

22. Berenbaum, M. R. (1995). Turnabout is fair play: secondary roles for primary compounds. *Journal of Chemical Ecology*, 21, 925 – 40.
23. Berenbaum, M. R. (2002). Postgenomic chemical ecology: from genetic code to ecological interactions. *Journal of Chemical Ecology*, 28, 873 – 96.
24. Berenbaum, M. and Neal, J. J. (1985). Synergism between myristicin and xanthotoxin, a naturally co-occurring plant toxicant. *Journal of Chemical Ecology*, 11, 1349 – 58.
25. Bernays, E. A. (1982). The insect on the plant—a closer look. In *Proceedings of the 5th international symposium on insect-plant relationships* (ed. J. H. Visser and A. K. Minks), pp. 3 – 17. Pudoc, Wageningen.
26. Bernays, E. A. (1986). Evolutionary contrasts in insects: nutritional advantages of holometabolous development. *Physiological Entomology*, 11, 377 – 82.
27. Bernays, E. A. (1997). Feeding by lepidopterous larvae is dangerous. *Ecological Entomology*, 22, 121 – 3.
28. Bernays, E. A. and Chapman, R. F. (1994). *Host-plant selection by phytophagous insects*. Chapman & Hall, New York.
29. Bernays, E. A. and Lewis, A. C. (1986). The effect of wilting on palatability of plants to *Schistocerca gregaria*, the desert locust. *Oecologia*, 70, 132 – 5.
30. Bernays, E. A., Cooper Driver, G., and Bilgener, M. (1989). Herbivores and plant tannins. *Advances in Ecological Research*, 19, 263 – 302.
31. Berry, R. E., Yu, S. J., and Terriere, L. C. (1980). Influence of host plants on insecticide metabolism and management of variegated cutworm. *Journal of Economic Entomology*, 73, 771 – 4.
32. Blau, P. A., Feeny, P., and Contardo, L. (1978). Allylglucosinolate and herbivorous caterpillars: a contrast in toxicity and tolerance. *Science*, 200, 1296 – 8.
33. Blua, M. J., Perring, T. M., and Madore, M. A. (1994). Plant virus-induced changes in aphid population development and temporal fluctuations in plant nutrients. *Journal of Chemical Ecology*, 20, 691 – 707.
34. Bown, D. P., Wilkinson, H. S., and Gatehouse, J. A. (2004). Regulation of expression of genes encoding digestive proteases in the gut of a polyphagous lepidopteran larva in response to dietary protease inhibitors. *Physiological Entomology*, 29, 278 – 90.
35. Brattsten, L. B. (1979). Biochemical defense mechanisms in herbivores against plant allelochemicals. In *Herbivores. Their interaction with secondary plant metabolites* (ed. G. A. Rosenthal and D. H. Janzen), pp. 200 – 70. Academic Press, New York.
36. Brattsten, L. B. (1988). Enzymic adaptations in leaffeeding insects to host-plant allelochemicals. *Journal of Chemical Ecology*, 14, 1919 – 39.
37. Brattsten, L. B. (1992). Metabolic defenses against plant allelochemicals. In *Molecular aspects of insect-plant associations* (ed. L. B. Brattsten and S. Ahmad), pp. 176 – 242. Plenum

Press, New York.

38. Breen, J. P. (1994). Acremonium endophyte interactions with enhanced plant resistance to insects. *Annual Review of Entomology*, 39, 401 – 23.

39. Brewer, J. W., O'Neill, K. M., and Deshon, R. E. (1987). Effects of artificially altered foliar nitrogen levels on development and survival of young instars of western spruce budworm, *Choristoneura occidentalis* Freeman. *Journal of Applied Entomology*, 104, 121 – 30.

40. Brodbeck, B. and Strong, D. (1987). Amino acid nutrition of herbivorous insects and stress to host plants. In *Insect outbreaks* (ed. P. Barbosa and J. C. Schultz), pp. 347 – 64. Academic Press, San Diego.

41. Brodbeck, B. V., Mizell, R. F., and Andersen, P. C. (1993). Physiological and behavioral adaptations of three species of leafhoppers in response to the dilute nutrient content of xylem fluid. *Journal of Insect Physiology*, 39, 73 – 81.

42. Brown, V. C. (1995). Insect herbivores and gaseous air pollutants—current knowledge and predictions. In *Insects in a changing environment* (ed. R. Harrington and N. E. Stork), pp. 219 – 49. Academic Press, London.

43. Bultman, T. L., McNeill, M. R., and Goldson, S. L. (2003). Isolate-dependent impacts of fungal endophytes in a multitrophic interaction. *Oikos*, 102, 491 – 6.

44. Campbell, B. C. (1989). On the role of microbial symbiotes in herbivorous insects. In *Insect-plant interactions*, Vol. 1 (ed. E. A. Bernays), pp. 1 – 44. CRC Press, Boca Raton.

45. Cianfrogna, J. A., Zangerl, A. R., and Berenbaum, M. R. (2002). Dietary and developmental influences on induced detoxification in an oligophage. *Journal of Chemical Ecology*, 28, 1349 – 64.

46. Claussen, W. (2005). Proline as a measure of stress in tomato plants. *Plant Science*, 168, 241 – 8.

47. Clay, K. and Schardl, C. (2002). Evolutionary origins and ecological consequences of endophyte symbiosis with grasses. *American Naturalist*, 160, S99 – S127.

48. Cohen, A. C. and Patana, R. (1984). Efficiency of food utilization by *Heliothis zea* (Lepidoptera: Noctuidae) fed artificial diets or green beans. *Canadian Entomologist*, 116, 139 – 46.

49. Cohen, A. C. (2004). *Insect diets: science and technology*. CRC Press, Boca Raton, FL.

50. Coley, D., Massa, M., Lovelock, E., and Winter, K. (2002). Effects of elevated CO_2 on foliar chemistry of saplings of nine species of tropical tree. *Oecologia*, 133, 62 – 9.

51. Dethier, V. G. and Schoonhoven, L. M. (1968). Evaluation of evaporation by cold and humidity receptors in caterpillars. *Journal of Insect Physiology*, 14, 1049 – 54.

52. Dillon, R. J. and Dillon, V. M. (2004). The gut bacteria of insects: nonpathogenic interactions. *Annual Review of Entomology*, 49, 71 – 92.

53. Dolezal, P., Bextine, B. R., Dolezalova, R., and Miller, T. A. (2004). Novel methods of monitoring the feeding behavior of *Homalodisca coagulata* (Say) (Hemiptera: Cicadellidae). *Annals of the Entomological Society of America*, 97, 1055 – 62.

54. Douglas, A. E. (1998). Nutritional interactions in insect-microbial symbioses: aphids and

their symbiotic bacteria *Buchnera*. *Annual Review of Entomology*, 43, 17–37.
55. Douglas, A. E. (2003a). *Buchnera* bacteria and other symbionts of aphids. In *Insect symbiosis* (ed. K. Bourtzis and T. Miller), pp. 23–38. CRC Press, Boca Raton.
56. Douglas, A. E. (2003b). The nutritional physiology of aphids. *Advances in Insect Physiology*, 31, 73–140.
57. Douglas, A. E., Minto, L. B., and Wilkinson, T. L (2001). Quantifying nutrient production by the microbial symbionts in an aphid. *Journal of Experimental Biology*, 204, 349–58.
58. Dowd, P. F. (1991). Symbiont-mediated detoxification in insect herbivores. In *Microbial mediation of plant-herbivore interactions* (ed. P. Barbosa, V. A. Krischik, and O. G. Jones), pp. 411–40. Wiley, New York.
59. Duffey, S. S., Bloem, K. A., and Campbell, B. C. (1986). Consequences of sequestration of plant natural products in plant-insect-parasitoid interactions. In *Interactions of plant resistance and parasitoids and predators of insects* (ed. D. J. Boethel and R. D. Eikenbary), pp. 31–60. Ellis Horwood, Chichester.
60. Duke, J. A. and Bogenschutz-Godwin, M. J. (1999). The synergy principle at work in plants, pathogens, insects, herbivores, and humans. In *Natural products from plants* (ed. P. B. Kaufman, L. J. Cseke, S. Warber, J. A. Duke, and H. L. Brielmann), pp. 183–205. CRC Press, Boca Raton.
61. Faeth, S. H. (2002). Are endophytic fungi defensive plant mutualists? *Oikos*, 98, 25–36.
62. Feeny, P. (1970). Seasonal changes in oak leaf tannins and nutrients as a cause of spring feeding by winter moth caterpillars. *Ecology*, 51, 565–81.
63. Feeny, P. (1992). The evolution of chemical ecology: contributions from the study of herbivorous insects. In *Herbivores. Their interactions with secondary plant metabolites*, Vol. 2 (2nd edn) (ed. G. A. Rosenthal and M. R. Berenbaum), pp. 1–44. Academic Press, San Diego.
64. Feyereisen, R. (1999). Insect P450 enzymes. *Annual Review of Entomology*, 44, 507–33.
65. Fiedler, K., Krug, E., and Proksch, P. (1993). Complete elimination of hostplant quinolizidine alkaloids by larvae of a polyphagous lycaenid butterfly *Callophrys rubi*. *Oecologia*, 94, 441–5.
66. Fischer, K. and Fiedler, K. (2000). Response of the copper butterfly *Lycaena tityrus* to increased leaf nitrogen in natural food plants: evidence against the nitrogen limitation hypothesis. *Oecologia*, 124
67. Forkner, R. E. and Hunter, M. D. (2000). What goes up must come down? Nutrient addition and predation pressure on oak herbivores. *Ecology*, 81, 1588–1600.
68. Gelman, D. B., Bell, R. A., Liska, L. J., and Hu, J. S. (2001). Artificial diets for rearing the Colorado potato beetle, *Leptinotarsa decemlineata*. *Journal of Insect Science*, 1(7), 1–11.
69. Gould, F. (1991). Arthropod behavior and the efficacy of plant protectants. *Annual Review of Entomology*, 36, 305–30.
70. Gouveia, S. M., Simpson, S. J., Raubenheimer, D., and Zanotto, F. P. (2000). Patterns of

respiration in *Locusta migratoria* nymphs when feeding. *Physiological Entomology*, 25, 88-93.

71. Grewal, S. K., Grewal, P. S., and Gaugler, R. (1995). Endophytes of fescue grasses enhance susceptibility of *Popillia japonica* larvae to an entomopathogenic nematode. *Entomologia Experimentalis et Applicata*, 74, 219-24.

72. Guthrie, W. D., Larvis, J. L., and Reed, G. L. (1984). Leaf-feeding damage by European corn borer (Lepidoptera: Pyralidae) larvae reared one generation each year on dent maize and eight generations each year on a meridic diet (for eight years) and then reared continuously on a meridic diet for eight additional years. *Journal of the Kansas Entomological Society*, 57, 352-4.

73. Haile, F. J. (2001). Drought stress, insects, and yield loss. In *Biotic stress and yield loss* (ed. R. K. D. Peterson and L. G. Higley), pp. 117-34. CRC Press, Boca Raton.

74. Hägele, B. F. and Rowell-Rahier, M. (1999). Dietary mixing in three generalist herbivores: nutrient complementation or toxin dilution. *Oecologia*, 119, 521-33.

75. Hajek, A. E. and St Leger, R. J. (1994). Interactions between fungal pathogens and insect hosts. *Annual Review of Entomology*, 39, 293-322.

76. Hammond, A. M. and Hardy, T. N. (1988). Quality of diseased plants as hosts for insects. In *Plant stressinsect interactions* (ed. E. A. Heinrichs), pp. 381-432. John Wiley, New York.

77. Hatcher, P. E. (1995). 3-Way interactions between plant-pathogenic fungi, herbivorous insects and their host plants. *Biological Reviews of the Cambridge Philosophical Society*, 70, 639-94.

78. Haukioja, E., Ossipov, V., and Lempa, K. (2002). Interactive effects of leaf maturation and phenolics on consumption and growth of a geometrid moth. *Entomologia Experimentalis et Applicata*, 104, 125-36.

79. Heliövaara, K. and Väisänen, R. (1993). *Insects and pollution*. CRC Press, Boca Raton.

80. Henn, M. (1999). The changes of polyphenols as a result of the passage through the gut of the gypsy moth *Lymantria dispar* (Lep., Lymantriidae): influence on the growth of the larvae. *Journal of Applied Entomology*, 123, 391-5.

81. Henn, M. W. and Schopf, R. (2001). Response of beech (*Fagus sylvatica*) to elevated CO_2 and N: influence on larval performance of the gypsy moth *Lymantria dispar* (Lep., Lymantriidae). *Journal of Applied Entomology*, 125, 501-5.

82. Henriksson, J., Haukioja, E., Ossipov, V., Ossipova, S., Sillanpää, S., Kapari, L., et al. (2003). Effects of host shading on consumption and growth of the geometrid *Epirrita autumnata*: interactive roles of water, primary and secondary compounds. *Oikos*, 103, 3-16.

83. Hinks, C. F. and Spurr, D. T. (1989). Effect of food plants on the susceptibility of the migratory grasshopper (Orthoptera: Acrididae) to deltamethrin and dimethoate. *Journal of Economic Entomology*, 82, 721-6.

84. Hiratsuka, E. (1920). Researches on the nutrition of the silkworm. *Bulletin of the Sericulture Experimental Station Japan*, 1, 257-315.

85. Hirayama, C., Konno, K., and Shinbo, H. (1996). Utilization of ammonia as a nitrogen source in the silkworm, *Bombyx mori*. *Journal of Insect Physiology*, 42,983 – 8.
86. Holopainen, J. K. (2002). Aphid response to elevated ozone and CO_2. *Entomologia Experimentalis et Applicata*, 104, 137 – 42.
87. Huberty, A. F. and Denno, R. F. (2004). Plant water stress and its consequences for herbivorous insects: a new synthesis. *Ecology*, 85, 1383 – 98.
88. Hughes, P. R. (1988). Insect populations on host plants subjected to air pollution. In *Plant stress-insect inter-actions* (ed. E. A. Heinrichs), pp. 249 – 319. John Wiley, New York.
89. Hughes, P. R. and Voland, M. L. (1988). Increase in feeding stimulants as the primary mechanism by which SO_2 enhances performance of Mexican bean beetle on soybean. *Entomologia Experimentalis et Applicata*, 48, 257 – 62.
89a. Jallow, M. F. A., Dugassa-Gobena, D., and Vidal, S. (2004). Indirect interaction between an unspecialized endophytic fungus and a polyphagous moth. *Basic and Applied Ecology*, 5, 183 – 91.
90. Johnson, S. N., Douglas, A. E., Woodward, S., and Hartley, S. E. (2003). Microbial impacts on plantherbivore interactions: the indirect effects of a birch pathogen on a birch aphid. *Oecologia*, 134, 388 – 96.
91. Kam, M. and Degen, A. A. (1997). Energy budget in free-living animals: a novel approach based on the doubly labelled water method. *American Journal of Physiology*, 272, 1336 – 43.
92. Karban, R. and Agrawal, A. A. (2002). Herbivore offense. *Annual Review of Ecology and Systematics*, 33,641 – 64.
93. Karowe, D. N. (1989). Differential effect of tannic acid on two tree-feeding Lepidoptera: implications for theories of plant anti-herbivore chemistry. *Oecologia*, 80, 507 – 12.
94. Keating, S. T., Yendol, W. G., and Schultz, J. C. (1988). Relationship between susceptibility of gypsy moth larvae (Lepidoptera; Lymantriidae) to a baculovirus and host plant foliage constituents. *Environmental Entomology*, 17, 952 – 8.
95. Keating, S. T., Schultz, J. C., and Yendol, W. G. (1990). The effect of diet on gypsy moth (*Lymantria dispar*) larval midgut pH, and its relationship with larval susceptibility to a baculovirus. *Journal of Invertebrate Pathology*, 56, 317 – 26.
96. Keena, M. A., Odell, T. M., and Tanner, J. A. (1995). Effects of diet ingredient source and preparation method on larval development of laboratory-reared gypsy moth (Lepidoptera: Lymantriidae). *Annals of the Entomological Society of America*, 88, 672 – 9.
97. Klekowski, R. Z. and Duncan, A. (1975). Physiological approach to ecological energetics. In *Methods for ecological bioenergetics* (ed. W. Grodzinski, R. Z. Klekowski, and A. Duncan), pp. 15 – 64. Blackwell, Oxford.
98. Kogan, M. (1986). Bioassays for measuring quality of insect food. In *Insect-plant interactions* (ed. J. R. Miller and T. A. Miller), pp. 155 – 89. Springer, New York.
99. Kopper, B. J., Jakobi, V. N., Osier, T. L., and Lindroth, L. R. (2002). *Effects of paper birch condensed tannin on whitemarked tussock moth (Lepidoptera: Lymantriidae) performance.*

Environmental Entomology, 31, 10 – 14.

100. Koricheva, J., Larsson, S., and Haukioja, E. (1998). Insect performance on experimentally stressed woody plants: a meta-analysis. *Annual Review of Entomology*, 43, 195 – 216.

101. Krischik, V. A. (1991). Specific or generalized plant defense: reciprocal interactions between herbivores and pathogens. In *Microbial mediation of plant-herbivore interactions* (ed. P. Barbosa, V. A. Krischik, and O. G. Jones), pp. 309 – 340. Wiley, New York.

102. Kukal, O. and Dawson, T. E. (1989). Temperature and food quality influences feeding behavior, assimilation efficiency and growth rate of arctic woolly-bear caterpillars. *Oecologia*, 79, 526 – 32.

103. Lauzon, C. R., Potter, S. E., and Prokopy, R. J. (2003). Degradation and detoxification of the dihydrochalcone phloridzin by *Enterobacter agglomerans*, a bacterium associated with the apple pest, *Rhagoletis pomonella* (Walsh) (Diptera: Tephritidae). *Environmental Entomology*, 32, 953 – 62.

104. Lavoie, B. and Oberhauser, K. S. (2004). Compensatory feeding in *Danaus plexippus* (Lepidoptera: Nymphalidae) in response to variation in host plant quality. *Environmental Entomology*, 33, 1062 – 9.

105. Lee, K. P., Raubenheimer, D., and Simpson, S. J. (2004). The effects of nutritional imbalance on compensatory feeding for cellulose-mediated dietary dilution in a generalist caterpillar. *Physiological Entomology*, 29, 108 – 17.

106. Li, X., Schuler, M. A., and Berenbaum, M. R. (2002). Jasmonate and salicylate induce expression of herbivore cytochrome P450 genes. *Nature*, 419, 712 – 15.

107. Lindroth, R. L. (1991). Differential toxicity of plant allelochemicals to insects: roles of enzymatic detoxification systems. In *Insect-plant interactions*, Vol. 3 (ed. E. A. Bernays), pp. 1 – 33. CRC Press, Boca Raton.

108. Lindroth, R. L., Arteel, G. E., and Kinney, K. K. (1995). Responses of three saturniid species to paper birch grown under enriched CO_2 atmospheres. *Functional Ecology*, 9, 306 – 11.

109. Loader, C. and Damman, H. (1991). Nitrogen content of food plants and vulnerability of *Pieris rapae* to natural enemies. *Ecology*, 72, 1586 – 90.

110. McEvoy, P. B. (1984). Increase in respiratory rate during feeding in larvae of the cinnabar moth *Tyria jacobaeae*. *Physiological Entomology*, 9, 191 – 5.

111. McNeill, S. and Southwood, T. R. E. (1978). The role of nitrogen in the development of insect plant relationships. In *Biochemical aspects of plant and animal coevolution* (ed. J. B. Harborne), pp. 77 – 98. Academic Press, London.

112. Mattson, W. J. (1980). Herbivory in relation to plant nitrogen content. *Annual Review of Ecology and Systematics*, 11, 119 – 61.

113. Mattson, W. J. and Haack, R. A. (1979). The role of drought stress in provoking outbreaks of phytophagous insects. In *Insect outbreaks* (ed. P. Barbosa and J. C. Schultz), pp. 365 – 407. Academic Press, San Diego.

114. Meade, T. and Hare, J. D. (1994). Effects of genetic and environmental host plant variation on the susceptibility of two noctuids to *Bacillus thuringiensis*. *Entomologia Experimentalis et Applicata*, 70, 165 –78.
115. Meisner, J., Ishaaya, I., Ascher, K. R. S., and Zur, M. (1978). Gossypol inhibits protease and amylase activity of Spodoptera littoralis Boisduval larvae. *Annals of the Entomological Society of America*, 71, 5 – 8.
116. Mellanby, K. and French, R. A. (1958). The importance of drinking water to larval insects. *Entomologia Experimentalis et Applicata*, 1, 116 – 24.
117. Mittler, T. E. (1970). Uptake rates of plant sap and synthetic diet by the aphid *Myzus persicae*. *Annals of the Entomological Society of America*, 63, 1701 – 5.
118. Mohamad, B. M. and van Emden, H. F. (1989). Host plant modification to insecticide susceptibility in *Myzus persicae* (Sulz.). *Insect Science and its Application*, 10, 699 – 703.
119. Moran, N. (2002). The ubiquitous and varied role of infection in the lives of animals and plants. *American Naturalist*, 160, S1 – S8.
120. Musser, R. O., Hum-Musser, S. M., Felton, G. W., and Gergerich, R. C. (2003). Increased larval growth and preference for virus-infected leaves by the Mexican bean beetle, *Epilachna varivestis* Mulsant, a plant virus vector. *Journal of Insect Behavior*, 16, 247 – 56.
121. Navon, A., Hare, J. D., and Federici, B. A. (1993). Interactions among *Heliothis virescens* larvae, cotton condensed tannin and the CryIA(c) δ-endotoxin of *Bacillus thuringiensis*. *Journal of Chemical Ecology*, 19, 2485 – 99.
122. Neal, J. J. (1987). Metabolic costs of mixed-function oxidase induction in *Heliothis zea*. *Entomologia Experimentalis et Applicata*, 43, 175 – 9.
123. Nelson, A. C. and Kursar, T. A. (1999). Interactions among plant defense compounds: a method for analysis. *Chemoecology*, 9, 81 – 92.
124. Omacini, M., Chaneton, E. J., Ghersa, C. M., and Müller, C. B. (2001). Symbiotic fungal endophytes control insect-parasite interaction webs. *Nature*, 409, 78 – 81.
125. Orians, C. M. and Jones, C. G. (2001). Plants as resource mosaics: a functional model for predicting patterns of within-plant resource heterogeneity to consumers based on vascular architecture and local environmental variability. *Oikos*, 94, 493 – 504.
126. Peng, Z. and Miles, P. W. (1988). Studies on the salivary physiology of plant bugs: function of the catechol oxidase of the rose aphid. *Journal of Insect Physiology*, 11, 1027 – 33.
127. Percy, K. E., Awmack, C. S., Lindroth, R. L., Kubiske, M. E., Kopper, B. J., Isebrands, J. G., et al. (2002). Altered performance of forest pests under atmospheres enriched by CO_2 and O_3. *Nature*, 420, 403 – 7.
128. Peterson, C. C., Walton, B. M., and Bennett, A. F. (1999). Metabolic costs of growth in free-living Garter snakes and the energy budget of ectotherms. *Functional Ecology*, 13, 500 – 7.
129. Peterson, R. K. D., Higley, L. G., Haile, J., and Barrigossi, J. A. F. (1998). Mexican bean beetle (Coleoptera: Coccinelidae) injury affects photosyn-thesis of *Glycine max* and

Phaseolus vulgaris. Environmental Entomology, 27, 373 – 81.

130. Petrusewicz, K. and MacFayden, A. (1970). *Productivity of terrestrial animals: principles and methods*, IBP Handbook 13. Blackwell, Oxford.

131. Phelan, P. L. and Stinner, B. R. (1992). Microbial mediation of plant-herbivore ecology. In *Herbivores. Their interactions with secondary plant metabolites*, Vol. 2 (2nd edn) (ed. G. A. Rosenthal and M. R. Berenbaum), pp. 279 – 315. Academic Press, San Diego.

132. Poprawski, T. J., Greenberg, S. M., and Ciomperlik, M. A. (2000). Effect of host plant on *Beauveria bassiana* and *Paecilomyces fumosoroseus*-induced mortality of *Trialeurodes vaporariorum* (Homoptera: Aleyrodidae). *Environmental Entomology*, 29, 1048 – 53.

133. Powell, J. S. and Raffa, K. F. (1999). Sources of variation in concentration and composition of foliar monoterpenes in tamarack (*Larix laricina*) seedlings: roles of nutrient availability, time of season, and plant architecture. *Journal of Chemical Ecology*, 25, 1771 – 97.

134. Raubenheimer, D. and Simpson, S. J. (1994). Ananalysis of nutrient budgets. *Functional Ecology*, 8, 783 – 91.

135. Rausher, M. D. (1982). Population differentiation in *Euphydras editha* butterflies, larval adaptation to different hosts. *Evolution*, 36, 581 – 90.

136. Reese, J. C. and Field, M. D. (1986). Defense against insect attack in susceptible plants: cutworm (Lepidoptera: Noctuidae) growth on corn seedlings and artificial diet. *Annals of the Entomological Society of America*, 79, 372 – 6.

137. Riemer, J. and Whittaker, J. B. (1989). Air pollution and insect herbivores: observed interactions and possible mechanisms. In *Insect-plant interactions*, Vol. 1 (ed. E. A. Bernays), pp. 73 – 105. CRC Press, Boca Raton.

138. Roessingh, P., Bernays, E. A., and Lewis, A. C. (1985). Physiological factors influencing preference for wet and dry food in *Schistocerca gregaria* nymphs. *Entomologia Experimentalis et Applicata*, 37, 89 – 94.

139. Rose, H. A. (1985). The relationship between feeding specialization and host plants to aldrin epoxidase activites of midgut homogenates in larval Lepidoptera. *Ecological Entomology*, 10, 455 – 67.

140. Saikkonen, N., Helander, M., Ranta, H., Neuvonen, S., Virtanen, T., Suomela, J., et al. (1996). Endophytemediated interactions between woody plants and insect herbivores? *Entomologia Experimentalis et Applicata*, 80, 269 – 71.

141. Schmidt, D. J. and Reese, J. C. (1986). Sources of error in nutritional index studies of insects on artificial diet. *Journal of Insect Physiology*, 32, 193 – 8.

142. Schroeder, L. A. (1981). Consumer growth efficiencies, their limits and relationships to ecological energetics. *Journal of Theoretical Biology*, 93, 805 – 28.

143. Schroeder, L. A. (1986). Protein limitation of a tree leaf feeding Lepidopteran. *Entomologia Experimentalis et Applicata*, 41, 115 – 20.

144. Schultz, J. C. and Keating, S. T. (1991). Host-plant-mediated interactions between the gypsy moth and a baculovirus. In *Microbial mediation of plant-herbivore interactions* (ed. P. Bar-

bosa, V. A. Krischik, and O. G. Jones), pp. 489–506. Wiley, New York.

145. Scott Brown, A. S., Simmonds, M. S. J., and Blaney, W. M. (2002). Relationship between nutritional composition of plant species and infestation levels of thrips. *Journal of Chemical Ecology*, 28, 2359–409.

146. Scriber, J. M. (1979). Effects of leaf-water supplementation upon post-ingestive nutritional indices of forb-, shrub-, vine-, and tree-feeding Lepidoptera. *Entomologia Experimentalis et Applicata*, 25, 240–55.

147. Scriber, J. M. (1984). Host-plant suitability. In *Chemical ecology of insects* (ed. W. J. Bell and R. T. Cardé), pp. 159–202. Chapman & Hall, London.

148. Scriber, J. M. and Slansky, F. (1981). The nutritional ecology of immature insects. *Annual Review of Entomology*, 26, 183–211.

149. Self, L. S., Guthrie, F. E., and Hodgson, E. (1964). Metabolism of nicotine by tobacco-feeding insects. *Nature*, 204, 300–1.

150. Shen, S. K. and Dowd, P. F. (1991). Detoxification spectrum of the cigarette beetle symbiont *Symbiotaphrina kochii* in culture. *Entomologia Experimentalis et Applicata*, 60, 51–9.

151. Showler, A. T. (2002). Effects of water deficit stress, shade, weed competition, and kaolin particle film on selected foliar free amino acid accumulations in cotton, *Gossypium hirsutum* (L.). *Journal of Chemical Ecology*, 28, 631–51.

152. Showler, A. T. and Moran, P. J. (2003). Effects of drought stressed cotton, *Gossypium hirsutum* L., on beet armyworm, *Spodoptera exigua* (Hübner), oviposition, and larval feeding preferences and growth. *Journal of Chemical Ecology*, 29, 1997–2011.

153. Simpson, S. J. and Abisgold, J. D. (1985). Compensation by locusts for changes in dietary nutrients: behavioural mechanisms. *Physiological Entomology*, 10, 443–52.

154. Simpson, S. J. and Raubenheimer, D. (1996). Feeding behaviour, sensory physiology and nutritiona feedback: a unifying model. *Entomologia Experimentalis et Applicata*, 80, 55–64.

155. Simpson, S. J. and Raubenheimer, D. (2001). The geometric analysis of nutrient-allelochemical interactions: a case study using locusts. *Ecology*, 82, 422–39.

156. Simpson, S. J. and Simpson, C. L. (1990). The mechanisms of nutritional compensation by phytophagous insects. In *Insect-plant interactions*, Vol. 2 (ed. E. A. Bernays), pp. 111–160. CRC Press, Boca Raton.

157. Simpson, S. J., Simmonds, M. S. J., and Blaney, W. M. (1988). A comparison of dietary selection behaviour in larval *Locusta migratoria* and *Spodoptera littoralis*. *Physiological Entomology*, 13, 225–38.

158. Singer, M. (2001). Determinants of polyphagy by a woolly bear caterpillar: a test of the physiological efficiency hypothesis. *Oikos*, 93, 194–204.

159. Singh, P. and Moore, R. F. (1985). *Handbook of insect rearing*, 2 Vols. Elsevier, Amsterdam.

160. Slansky, F. (1992). Allelochemical-nutrient interactions in herbivore nutritional ecology. In *Herbivores. Their interactions with secondary plant metabolites*, Vol. 2 (2nd edn) (ed. G. A.

Rosenthal and M. R. Berenbaum), pp. 135 – 74. Academic Press, San Diego.
161. Slansky, F. (1993). Nutritional ecology: the fundamental quest for nutrients. In *Caterpillars. Ecological and evolutionary constraints on foraging*, (ed. N. E. Stamp and T. M. Casey), pp. 29 – 91. Chapman & Hall, New York.
162. Slansky, F. and Feeny, P. (1977). Stabilization of the rate of nitrogen accumulation by larvae of the cabbage butterfly on wild and cultivated food plants. *Ecological Monographs*, 47, 209 – 28.
163. Slansky, F. and Rodriguez, J. G. (1987). *Nutritional ecology of insects, spiders, and related invertebrates*. John Wiley, New York.
164. Slansky, F. and Scriber, J. M. (1985). Food consumption and utilization. In *Comprehensive insect physiology, biochemistry, and pharmacology* (ed. G. A. Kerkut and L. I. Gilbert), Vol. 4, pp. 87 – 163. Pergamon, New York.
165. Slansky, F. and Wheeler, G. S. (1992). Caterpillars' compensatory feeding response to diluted nutrients leads to toxic allelochemical dose. *Entomologia Experimentalis et Applicata*, 65, 171 – 86.
166. Southwood, T. R. E. and Henderson, P. A. (2000). *Ecological methods* (3rd edn). Blackwell Science, Oxford.
167. Spencer, J. L., Mabry, T. R., and Vaughn, T. T. (2003). Use of transgenic plants to measure insect herbivore movement. *Journal of Economic Entomology*, 96, 1738 – 49.
168. Stamp, N. E. (1991). Stability of growth and consumption rates and food utilization efficiencies when insects are given an excess of food. *Annals of the Entomological Society of America*, 84, 58 – 60. 169.
169. Strauss, S. Y. and Irwin, R. E. (2004). Ecological and evolutionary consequences of multispecies plant-animal interactions. *Annual Review of Ecology, Evolution and Systemtics*, 35, 435 – 66.
170. Tabashnik, B. E. and Slansky, F. (1987). Nutritional ecology of forb foliage-chewing insects. In Nutri-tional ecology of insects, mites, spiders and related invertebrates (ed. F. Slansky and J. G. Rodriguez), pp. 71 – 103. John Wiley, New York.
171. Van Loon, J. J. A. (1988). A flow-through respi rometer for leaf chewing insects. *Entomologia Experimentalis et Applicata*, 49, 265 – 76.
172. Van Loon, J. J. A. (1991). Measuring food utilization in plant-feeding insects—toward a metabolic and dynamic approach. In *Insect-plant interactions*, Vol. 3 (ed. E. A. Bernays), pp. 79 – 124. CRC Press, Boca Raton.
173. Van Loon, J. J. A. (1993). Gravimetric vs. respirometric determination of metabolic efficiency in caterpillars of *Pieris brassicae*. *Entomologia Experimentalis et Applicata*, 67, 135 – 42.
174. Van Loon, J. J. A., Casas, J., and Pincebourde, S. (2005). Nutritional ecology of insect-plant interactions: persistent handicaps and the need for innovative approaches. *Oikos*, 108, 194 – 201.
175. Van't Hof, H. M. and Martin, M. M. (1989). The effect of diet water content on energy ex-

penditure by third-instar *Manduca sexta* larvae (Lepidoptera, Sphingidae). *Journal of Insect Physiology*, 35, 433-6.

176. Waldbauer, G. P. (1968). The consumption and utilization of food by insects. *Advances in Insect Physiology*, 5, 229-88.

177. Waldbauer, G. P. and Friedman, S. (1988). Dietary self-selection by insects. In *Endocrinological frontiers in physiological insect ecology* (ed. F. Sehnal, A. Zabza, and D. L. Denlinger), pp. 403-22. Wroclaw Technical University Press, Wroclaw.

178. Watt, A. D., Whittaker, J. B., Docherty, M., Brooks, G., Lindsay, E., and Salt, D. T. (1995). The impact of elevated atmospheric CO_2 on insect herbivores. In *Insects in a changing environment* (ed. R. Harrington and N. E. Stork), pp. 197-217. Academic Press, London.

179. Welter, S. C. (1989) Arthropod impact on plant gas exchange. In *Insect-plant interactions*, Vol. 1 (ed. E. A. Bernays), pp. 135-150. CRC Press, Boca Raton.

180. White, T. C. R. (1993). *The inadequate environment. Nitrogen and the abundance of animals*. Springer, Berlin.

181. Whittaker, J. B. (1999). Impacts and responses at population level of herbivorous insects to elevated CO_2. *European Journal of Entomology*, 96, 149-56.

182. Whittaker, J. B. and Warrington, S. (1995). Effects of atmospheric pollutants on interactions between insects and their food plants. In *Pests, pathogens and plant communities* (ed. J. J. Burdon and R. S. Leather), pp. 97-110. Blackwell, Oxford.

183. Wiegert, R. G. and Petersen, C. E. (1983). Energy transfer in insects. *Annual Review of Entomology*, 28, 455-86.

184. Wightman, J. A. (1981) Why insect energy budgets do not balance. *Oecologia*, 50, 166-9.

185. Wink, M. and Schneider, D. (1990). Fate of plant-derived secondary metabolites in three moth species (*Syntomis mogadorensis*, *Syntomeida epilais*, and *Creatonotus transiens*). *Journal of Comparative Physiology* B, 160, 389-400.

186. Wolf, T. J., Ellington, C. P., Davis, S., and Feltham, M. J. (1996). Validation of the doubly labelled water technique for bumblebees *Bombus terrestris* (L.) *Journal of Experimental Biology*, 199, 959-72.

187. Woods, H. A. and Bernays, E. A. (2000). Water homeostasis by wild larvae of *Manduca sexta*. *Physiological Entomology*, 25, 82-7.

188. Yu, S. J. (1986). Consequences of induction of foreign compound-metabolizing enzymes in insects. In *Molecular aspects of insect-plant associations* (ed. L. B. Brattsten and S. Ahmad), pp. 153-174. Plenum Press, New York.

189. Yue, Q., Johnson-Cicalese, J., Gianfagna, T. J., and Meyer, W. A. (2000). Alkaloid production and chinch bug resistance in endophyte-inoculated chewings and strong creeping red fescues. *Journal of Chemical Ecology*, 26, 279-92.

190. Zangerl, A. R., Hamilton, J. G., Miller, T. J., Crofts, A. R., Oxborough, K., Berenbaum, M. R., *et al.* (2002). Impact of folivory on photosynthesis is greater than the sum of

its holes. *Proceedings of the National Academy of Sciences of the USA*, 99: 1088 – 91.
191. Zanotto, F. P., Gouveia, S. M., Simpson, S. J., Raubenheimer, D., and Calder, P. C. (1997). Nutritional homeostasis in locusts: is there a mechanism for increased energy expenditure during carbohydrate overfeeding? *Journal of Experimental Biology*, 200, 2437 – 48.

第6章 选择寄主植物:如何发现寄主植物

6.1 术语
6.2 寄主植物的选择:一个链过程
6.3 搜寻机制
6.4 寄主植物的定位
　6.4.1 光学诱因与化学诱因
　6.4.2 对寄主植物特性的视觉反应
　6.4.3 对寄主植物的嗅觉反应
　6.4.4 飞蛾和步甲:嗅觉定向的两个例子
6.5 对寄主的气味检测的化学感觉基础
　6.5.1 嗅觉感受器的形态学研究
　6.5.2 嗅觉传导
　6.5.3 嗅觉电生理和敏感性
　6.5.4 嗅觉特征与编码
6.6 自然中对寄主植物的搜寻
6.7 结论
6.8 参考文献

　　前几章已经提到,植食昆虫最显著的特征之一是多数种类的昆虫是有选择性的摄食者,它们谨慎地选择植物以存放虫卵。近期对一些种类昆虫的研究表明它们不仅选择特定的植物种类,并且选择特定的植物器官。本章主要介绍昆虫的选择行为,在本章内容开始之前需要指出以下事实:某一种昆虫种类的寄主植物范围并不是在实验室实验条件下,行为可以接受或营养充分的所有的植物种类,在实际的自然环境中这些通常都会受到限制。而且,寄主选择行为或许会随着昆虫的发育阶段而改变,不同的生命阶段通常对寄主植物的偏好会有所不同。尽管初孵的昆虫幼体很小,能量储备有限,但如果它们认为作为它们孵化地的植物不合适时,会离开植株。

　　在几种情况下,植食昆虫必须寻找到寄主植物。土壤里越冬的蛹羽化时的位置有可能远离潜在食物或产卵的植物(如果为一年生植物)。另一种情况是经迁徙或扩散后到达到新的栖息地,而当地宿主植物已被耗尽。在自然环境下,寄主植物和非寄主植物一起成长形成混杂的植被环境。对于寄主植物专化性昆虫来说,在这种生境内寻找和辨认寄主植物的能力是很重要的,这种能力是本章和下章内容的重点。

6.1 术语

　　首先定义一些描述昆虫对寄主植物的选择行为的术语,对之后的介绍和分析是十分有益的。

1. 搜寻

每当昆虫与潜在的食物距离遥远时,需要搜寻并找到植株。为了正确定位寄主植物,昆虫需要向其所在方向移动并进行接触,或至少到达并停留于其附近,进一步验证它的特性。然而,在观察昆虫与植物的接触过程中,科学家们并没有得到建立这一接触机制的有效信息。"搜寻"这一术语的意思是在一个地方细心观察,努力找到某东西。"发现"(有时却被用作一个同义词[104])更确切地说是搜寻最终结果,所以才作为本章的副标题。因为搜寻暗含着一种方向性,这里需要重点指出的是,昆虫的运动方式多样,有可能是意外接触导致的随机运动,也有可能是强力定向运动。

2. 选择

严格意义上来说,"选择"意味着在各种选项中挑选。为了做到这一点,有必要对植物的感官知觉进行了解。从某种角度看,要在选择行为过程中证明对各选项的权衡是十分困难的,特别是当昆虫连续地与不同寄主植物进行接触时。连续接触不同寄主植物的情况比同时接触更为频繁,这也暗示了与之前的选项进行对比必须强化短期记忆。假设在最后接纳发生之前,要么保持一定距离接近后又避开,要么通过实际的接触实验,对所有选项进行评估。这种情况下可以使用"选择行为"这个术语。

3. 接纳

对寄主植物的接纳一般发生在持续摄食或者卵形态阶段。"接纳"并不包括"选择"这一术语中暗示的一些假设情况。例如,把一只甲虫放到单一栽培的豆科植物中间,它在爬上一棵豆科植株之后开始持续摄食,此时我们不能定论说甲虫选择豆科植株作为寄主,因为没有其他可供选择的植物。只能说是昆虫已经接纳了豆科植物。接纳受动机影响,也就是摄食或产卵的基本意愿,而这是昆虫内部生理状态参数的综合结果(例如,饱腹感和卵的成熟状况)。接纳这个术语不同于可接受性,后者是一种植物特性,被定义为某种特殊植物被选作摄食对象或产卵地的可能性。

4. 偏好

在双选或多选试验中,昆虫始终在一株选项植株上取食或产卵,说明比起其他植物,它更偏好这株植物。这种情况在田间也会出现,某种植物上的摄食和产卵程度高于根据相对生长情况预测出的程度。显然,偏好是一个相对的概念,仅适用于昆虫可实际获得的植物种类或者基因型。

5. 识别

这个术语常常与接纳一起出现。它意味着重新认识,隐含地指代一种由神经调控的过程。它暗示着昆虫有一个寻找植物内在的标准或者"形象"。这个形象以这样或那样的形式存在于昆虫的中枢神经系统。有关植物的传入感觉信息与储存的形象进行比较,如果能够充分匹配,植物就被识别为寄主。假定的形象是遗传性的,但可以通过经验修正到一个合理的程度(第8章)。

综上所述,搜寻、选择、偏好和识别这些术语似乎暗指复杂的行为过程和神经机制,对神经机制的讨论即将有所阐述。正确使用这些术语十分必要,它可以帮助避免混淆生态学者描述的联系模式(patterns of association)和行为学家的寻址机制(addressing mechanisms)[149]。这里我们把偏好作为一种昆虫特征,不受植物密度、分布或质量的影响(除非昆虫处于学习过程)(参见第 8 章)。

在这方面同样重要的是,把以上提到的行为术语与更改行为的化学物质联系起来。这些化学物质被统称为化学信息素(semiochemmicals)[116]或信息化学物质(infochemicals)[50]。为此,我们采用了德蒂尔等提出的术语[48],在表 6.1 中有所总结。化学信息素(semiochemmicals)和信息化学物质(infochemicals)专有名词中相应的术语有:利它素,对应引诱剂以及摄食和产卵的刺激物;利己素,对应防护剂和抑制性物质。花释放的能吸引传粉者的挥发物(第 12 章)对应的则是互利素。

化学信息素和信息化学物质的不同点在于,化学信息素的术语是以生产化学物质的起源而命名的,如利它素、利己素、互利素。而信息化学物质主要特指信息物质的传运。

表 6.1　一些有关昆虫行为的术语(德蒂尔等,1960)[48]

术语	功能
引诱剂(Attractant)	引起昆虫朝向味源运动的化学物质
忌避剂(Repellent)	引起昆虫远离味源运动的化学物质
抑制剂(Arrestant)	引起昆虫降低直线运动速率,而增加旋转速率的化学物质
取食或产卵刺激剂 (Feeding or ovipositional stimulant)	刺激昆虫取食或产卵的化学物质,取食刺激剂(feeding stimulant)与诱食剂同义
取食或产卵抑制剂 (Feeding or Ovipositional Deterrent)	抑制昆虫取食或产卵的化学物质,当此物质不存在时,昆虫开始恢复正常取食或产卵

6.2　寄主植物的选择:一个链过程

昆虫经常表现出程序化的行为和固定的可预知的行动顺序,即所谓的反应链[8]。这就意味着有一些行为元素会进入固定的顺序。对于特定的刺激,昆虫会表现出适当的反应(图 6.1)。

如果结果是为排斥某种植物或植物器官作为其摄食或产卵地,植食动物会"跳回"反应顺序中的较早步骤。由先前的经验导致其选择行为更改会致使更快做出决策或是改变原来的偏好,但这个顺序保持不变。在以下例子当中,我们将会发现整个过程以及每个阶段都可能相当的长和复杂。

在选择寄主植物过程中,两个主要连续的阶段会存在差别,这是由两个步骤间昆虫是否决定继续接触植物所界定的。这两个阶段为寻找和接触测试。第一个阶段也许以发现而结束;第二个阶段以接受或拒绝而结束。接受是一个重要的决定行为,因为这会导致摄入植物或者产下虫卵,同时也有可能对健康产生负面影响。图 6.2 中 A 所描绘的是寄主植物的选择过程。

在这个顺序中,植物提供给昆虫的线索的数量和强度不断增长,因此昆虫对植物的感觉信息强度不断增长。对寄主植物选择的标准顺序模式如下:

(1)昆虫不与植物进行身体接触,或休憩或随意四处活动,如爬行或飞行。
(2)感知植株散发的视觉或嗅觉线索。
(3)对线索做出反应,接近植株。

第6章 选择寄主植物:如何发现寄主植物

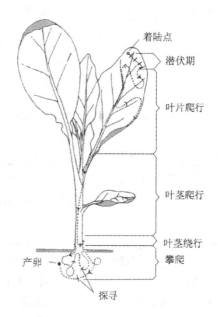

图6.1 昆虫对植物的探索过程

复杂的行为模式包括一系列的刺激和反应步骤,如寄生在卷心菜根部的甘蓝地种蝇(Delia radicum)的产卵活动所示。一直在空中飞翔的怀孕雌性蝇会对绿色叶片反射的黄绿光的波长(500~600 nm)做出反应。在"潜伏期"雌虫会沿着叶片爬行,时而停下来梳理,时而短暂性地飞行。下一阶段是"叶片爬行",雌虫在叶片上,通常沿着叶片边缘并且不断更换方向,不停地爬行。跗节上的味毛帮助它评估植株的适应性。如果它接触到合适的化学刺激物,就会爬到叶片的中脉或者叶茎,进入下个阶段——叶茎爬行。这时,雌虫头朝下围绕叶茎进行"叶茎绕行"。在"攀爬阶段"中,爬至卷心菜根茎部位。之后雌虫开始用产卵器展开"探寻"工作,测试土壤颗粒大小和水分含量。如果再次感受到合适的刺激物质,雌虫最终会将虫卵产在接近根茎部位的土壤中(卓然,1968)[189]

(4)发现植株,即触摸、攀爬或者降落在植株表面。
(5)通过接触测试检查植株表面(叶片表面的触碰检查)。
(6)试咬(常发生在咀嚼类昆虫身上)、探寻(常发生在吸食类昆虫身上)或者用产卵期穿孔,通过以上方式对植株进行破坏,组织内部成分得到释放。
(7)接受植株(表现为产下一个或多个虫卵或者持续摄食)或者排斥植株(表现为昆虫离开植株)。

在各个步骤之间,昆虫可能决定在接触植物之前或者之后离开。当它接近一批潜在寄主植物时,它可能重复同样的顺序来回应同种或不同种植物个体。最后它也许会返回去选择之前接触后离开的植物。

在这章和下面的章节中,将会用顺序理论框架讨论寄主植物的选择行为。讨论的重点是不同植物影响选择行为的诱因和感知诱因并影响选择行为的感觉器官。接受还是排斥这一重要决定不仅取决于植物诱因中的感觉信息,也取决于昆虫的生理状态(过饱、性成熟、卵成熟等[13])。中枢神经系统(CNS)[45]将以上两个变量连同之前储存在昆虫记忆中的经验信息整合起来。从本章内容来看,我们将首先做出以下假设:当昆虫不忙于迁徙或扩散活动时,其摄食和产卵动机高。

应该注意的是，不是所有的植食性动物都遵循以上的描述及表6.2中A总结的标准顺序。一些昆虫采取捷径，还有一些表现出更复杂的顺序。图6.2中B—E为一些研究得出的结论。

6.3 搜寻机制

为了理解食草类昆虫搜寻食物的方式，我们有必要对搜寻行为做出描述，同时讨论可能涉及的重要机制。

搜寻过程中行为步骤的顺序因昆虫种类不同而不同，并且还依靠有效的诱因。研究内容涵盖了从随机搜寻到高度定向搜寻等全部模式。在这个领域，随意的搜寻在多种昆虫身上均有所体现，例如，多食性的毛虫[46]、未成熟的和成熟的多食性的蝗虫[2,108]和成年的寡食性的马铃薯甲虫（$Leptinotarsa\ decemlineata$）。

在上述情况中，昆虫运动的频率、速率和方向与植物在感知范围内接受能力无关，感知范围指寄主植物的诱因被昆虫感觉系统所察觉的范围。随机运动的产生被解释为位于中枢神经系统（CNS）的"中心运动神经的程序"进行运作的结果。当一只昆虫受刺激去搜寻食物时，例如血液的海藻糖水平下降至低于某一水平（一个身体状态参数），这些程序便被激活，结果会引发昆虫的随机活动，内部储存（例如记忆）和固有感受的信息被使用[179]。当环境诱因未提供方向信息，或昆虫的感觉能力不足以感知必要的刺激物时，这个搜寻类型可能是最好的选择。在搜寻期间，搜寻运动有利于增加在一条路线上察觉资源的可能性，发生这种情况主要是因为这个路径是宽阔的。这在毛虫在地上运动来搜寻寄主植物上可以看出，毛虫抬起它们的头和胸部的第一节，并且左右地摇摆。

在随机搜寻期间，对于植物释放的刺激诱因，昆虫会表现出几种类型的定向反应。这种回应也许为非定向性或者定向性的。运动中非定向性的变化被归类为动性（kineses）[89,145]，昆虫也许会改变它的直线运动速度（直动态，orthokinesis）或改变转弯的速率和频率（斜动态，klinokinesis）。外界刺激物的强度（光强度、植物的气味、湿度等）和空间或时间异质性会决定上述反应的强度。一个（单方面的）感受器能通过感觉中枢神经系统对输入感觉信息进行时间对比以感知刺激物的浓度。这些动力反应经常导致区域限制性搜寻（在较小区域进行集中搜寻）和停滞。通常如果线性运动的速度降低，转弯的速率增加，它们大部分靠近或即将接触到寄主植物（表6.3和表6.4）[107]。

当寄主植物发出单一信号或和结合其他诱因发出时，正在搜寻的昆虫的感受系统能感知这些信号的方向，此时会发生定向性运动。这种情况的运动受外部诱因引导，同时也受中心运动神经程序的影响。对应某种外界刺激的定向性活动被定义为税金（taxes），可能接近来源（阳性）或远离来源（阴性）。许多昆虫种类常见于视觉和化学诱因以及二者形成的组合诱因导致的定向性。在几厘米的短距离内以及相对无干扰的静止空气中，昆虫会以积极的趋化现象来回应植物释放的气味成分。其中涉及来自嗅觉接收器的信息的时间对比（调转趋性），或者同时来源于一对双边性（嗅觉）接收器的感觉输入对比和同时来源于身体两侧刺激信息的感知（趋激性，对称定向性）。第三种定向性是补偿向性，保持感觉刺激的非对称分布以保证与刺激物的方向呈一个恒定的角度。

需要特别注意两种特殊的补偿向性：趋风性和趋光性。因为它们对植食昆虫的行为有调控作用。趋风性和趋光性即与主导风向或光向保持固定角度进行定向性运动。通过机械感受器感知气流或者通过光感受器感知光子流，以此感知风向或光向，例如连续反转运动使身体不

第6章 选择寄主植物:如何发现寄主植物

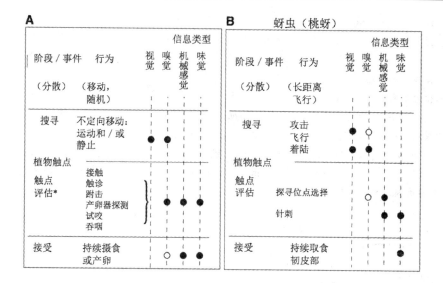

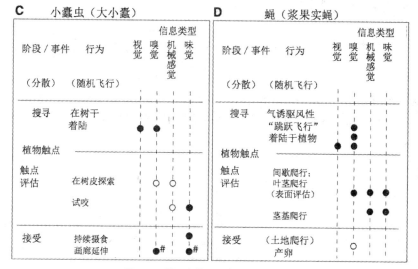

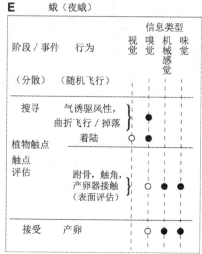

图6.2 不同昆虫对植物的探索过程不同

A 为植食昆虫选择寄主植物的一般顺序,左栏:行为阶段或事件;中间栏:一个行为阶段中常见的行为要素;右栏:影响选择行为的主要植物刺激物。黑点表示明确记载的作用于一些昆虫的植物诱因;白点表示暗示性的或可能的诱因;星号(*)表示许多种类昆虫表现出来的行为要素,但是并非体现在某种昆虫中所有的行为要素都是选择顺序中的必要部分。顶部的括号中,扩散代表之前的行为要素阶段(不属于寄主选择顺序)。B—E 举例说明寄主选择行为顺序中四个主要的行为顺序和特定要素。B 为具翅蚜虫(瘤蚜属,*Aphis* spp.);C 为小蠹虫成虫(大小蠹,*lps* spp.)。#表示如果没有抑制剂或忌避剂,昆虫会通过挖地道的形式扩大领地范围;D 为食草性蝇类成虫(*Delia*,*Rhagoletis* spp.),视觉诱因为地面状况,而机械感觉诱因为空气气流,以上二者均非源自植物方面的诱因;E 为夜蛾成虫(*Helicoverpa* spp., *Manduca sexta*),如 D 一样,也涉及视觉和机械感觉诱因(根据大量材料综合整理而成)

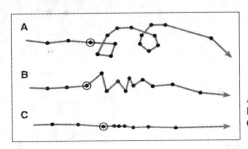

A——周期性增加转弯倾向,产生循环或转圈
B——交替旋转方向,产生曲线
C——在停顿之间调整运动长度,点代表降落,点排列成环时表示在寄主植物上产卵

图 6.3　昆虫的运动曲线

资源集中时使用的搜寻模式这种情况下,一旦发现寄主植物便对整个区域进行全部搜寻,这对于昆虫更加有利。这一策略增加了发现另一寄主植物的可能性(贝尔,1991)

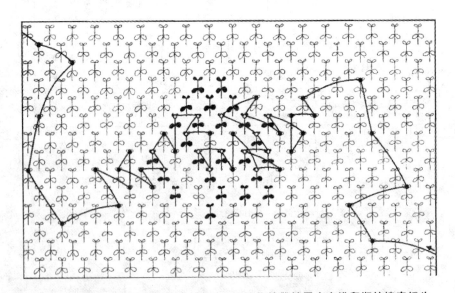

图 6.4　一种专食鼻花属(*Rhinanthus* spp.)植物的雌性昆虫在排卵期的搜索行为

在降落间的飞行时间较短,并且在靠近寄主植物时表现出更多的转弯飞行,从而增加了降落在寄主植物上的机会,更改飞行路线和决定降落(至少是后者)是由寄主植物的气味刺激所致,植物总数为 252,鼻花属植物数量为 25(10%),总降落次数与 45,降落在鼻花属植物上的次数为 15(33%)(道韦斯,1968)

断地向左向右摆动。爬行昆虫通过机械运动测定风向,而飞行昆虫则通过视觉检测风向。趋风性行为受植物气味影响,在实验室条件下对许多植食昆虫都有影响。与预料相反的是,当与植物的距离大于几厘米时,气味的诱因不会呈现出一个梯度趋势(见6.4.4)。在外界的空气运动是非常混乱的[111]。气味流形成的复杂湍流为不连续的气味分子团组成,这些分子团随风向运动。因此不存在浓度梯度(图6.5)。缩短发现气味来源的间隔时间的最佳办法是,垂直于空气流做横向运动以增加与气味分子团相遇的概率,并定位湍流的中位线。气味分子团最有可能产生于一个锥形空间的,锥顶端指向逆风方向。探索这个锥形空间的最佳途径是采用"之"字形的运动路线,直到遇到另一个气味分子团。这个探寻策略利用简单的运动行为规则,结合对空气运动的平均速度和方向认知。由此产生的抛物线式和"之"字式的运动模式与从理论推测的结果十分匹配(见6.4.4)。

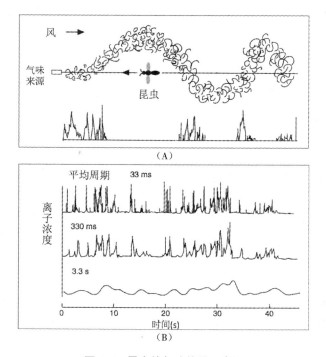

图6.5 昆虫的气味信号示意图

(A)一只昆虫朝向微弱气味源做逆风直线运动时,随时间推移遇到波动的气味湍流和气味信号示意图;(B)在不同的平均周期中,气味湍流中静止的离子探针产生的信号振幅。由大气湍流产生气味分子团,随后分子团被气味接收器感知。随着平均周期的增加,信号振幅的差异减少,进而导致嗅觉接收器对浓度的分辨率降低。然而,即使平均每3.3秒钟摄食一次,这个信号仍然是间断的,气味的来源仍然可以被清楚地区分开来(墨利斯,1986)[109]

横向趋光性(Photomenotaxis)是昆虫陆地行走的主要机制[145]。在野外很难检测出趋风性,因为缺乏风向的对照,以及普遍存在空气干扰,其中空气干扰了昆虫运动的方向,在土表边界层表现得尤为显著。但是对于趋光性的研究进展却相对顺利。其中一种方法是桑茨奇的"镜子测试"[145],另一种方法是证明趋光性的"转盘测试"[83]。

虽然对运动类型和植物衍生的诱因的描述可能用于表现不同的搜寻策略,但是对定位机制进行成功总结并记录在案的研究却数量有限。尤其是在野外条件下,不同的机制相互结合,而非单一机制发挥作用(见6.6)。一些文献中也记载了一些其他的搜索模式[16,177,179]。搜索模式暗示的可能与预期结果相反,随机爬行是一个非常有效的搜寻策略,并且随机运动的速率是决定非随机搜索成功与否的重要因素[107]。定向性运动通常被认为是一种适应性活动,因为它提高了搜索的效率,也就是说它会提高搜寻行为在单位时间和单位能量内的成功率。

6.4 寄主植物的定位

6.4.1 光学诱因与化学诱因

植物的视觉和味觉特征是植食昆虫定向运动的重要刺激信号。根据不同物种,两种特征的相对重要性略有不同,这种差异在昼夜物种间体现得尤为明显。两种刺激信号也通常被综合使用(见6.6章)。

光和化学作为植物诱因的性质在一些重要方面有很大不相同。光照可以根据强度、光谱成分、偏振等进行分类。光能量的单位——光子是以光速推进的。空气运动对植物的光谱反射模式影响不大,并且与植物距离的变化也不会改变相对稳定的反射模式。相反,植物释放的挥发性物质运动缓慢。挥发物在静止的空气中以扩散的形式向四周运动,在流动的空气中浓度不断发生变化。当接近植物后,气味浓度急剧上升。自然条件下,气流绝对静止和毫无气流干扰的情况十分罕见,但也并非完全没有。一般情况下,风速要比大多数有机分子的扩散速度大。在流动的空气中(正常情况),挥发物质由风带离植物,而后以气味分子团的形式随风飘散(见6.3)。

气味空间(odour-filled space)这个概念是以萨顿的扩散模型为基础的,在空气中形成半椭圆形的空间。然而,最近利用离子探测器,在较短的时间内可以明显发现,气味以分子团或蔓延开来的湍流中分子丝状物的形式进行随机漫游(图6.5)。在湍流边界之外,借用烟雾可以使之更为直观。当逆风运动时,昆虫会接触到分布在不同空间且浓度比其邻近植株略低的气味分子团。关于来自对昆虫体内性信息素气味空间分布信息的研究表明,其为一个点源。查普曼一直强调点源释放的气味湍流与植物群所释放的有所不同;显然,植物来源的气体会影响气味湍流的形状[33]。

总之,当考虑非生物因素时,不管分布、温度和风速如何,植物的光学特征是相对恒定的,但取决于光照强度。而植物释放的气味在空间分布和浓度变化的幅度较大,并在一定程度上依赖于风速、温度和光的强度。此外,植物挥发物的质量和数量根据植物的生理状态和是否受到植食动物的侵袭而发生变化[21,160](图4.7)。

除了这些非生物因素外,有关光和气味诱因的重要议题是它们的特异性和"活动空间",又称"有效区"或者有效的"吸引半径"[26]。

通常假定光不能作为识别寄主植物的因素,因为"所有的植物都是绿色的"(即占主导地位的反射—透光率色度为500~580纳米)。而形成明显反差的是,在一些植物种类中已经被发现其挥发化学物质或化学物质混合物,这些化学物质存在定性的(独特的混合物)或者定量的(特征比)差异[177]。因此研究更关注将气味作为搜索植物寄主的引导因素,尤其是在专门研究植食动物的情况中。然而,与叶片反射光谱的低可变度相反,反射光的强度在不同物种间具有显著不同,这是由于叶片表面存在蜡状晶体或毛状体,或者一些生物因素(年龄和营养状况)或非生物因素(密度和入射光强度)。

引诱昆虫到寄主植物的最大距离涉及一个重要概念——活跃空间(active space)。活跃空间被定义为引发植食动物反应的刺激物或诱因的强度临界值的空间范围。在缺乏视觉线索时,对一些寡食性物种而言,对植物气味的反应距离已经被证实为 5~30 m,葱蝇(*Delia antiqua*)的最大反应距离是 100 米(表6.2)。一些昆虫可以被吸引到气味陷阱的事实表明,挥发性植物化合物能在野外条件下吸引植食昆虫,有时在长距离下也可能发生。实蝇和叶甲可以被大量特定的花香吸引。这些也适用于一些杂食性昆虫,例如棉铃虫[77]和日本丽金龟。后者在空旷的地方能被最远距离达 400 米的陷阱吸引。在这种情况下,挥发饵的陷阱似乎对于控制昆虫密度十分有效和灵敏。

表6.2 植物诱因——味和光引起植食昆虫积极反应的距离

昆虫	距离(米)	参考文献
气味信号		
马铃薯甲虫(*Leptinotarsa decemlineata*)	0.6	83
荚象甲(*Ceutorhynchus assimilis*)	6	49
甘蓝地种蝇(*Delia radicum*)	20	58
大小蠹(*Dendroctonus* spp.)	24	63
草地贪夜蛾(*Pegomya betae*)	30	181
葱蝇(*Delia antiqua*)	50	138
	100	85
光信号		
甘蓝种蝇(*Delia brassicae*)	2	128
叶蝉(*Empoasca devastans*)	3.6	142
马铃薯叶甲(*Leptinotarsa decemlineata*)	8	171
苹果实蝇(*Rhagoletis pomonella*)	10	3

自然条件下有效空间的价值很大程度上取决于生物量和植被的复杂性,然而在昆虫搜寻寄主植物的野外研究中,目前对生物量和植被的复杂性这两个因素并没有开展广泛研究。所有刺激物(纯的物质)都产生于寄主植物个体或寄主植物群,被认为只在相对较短的距离内保留特性[162],即使在特殊情况下和其他植物的挥发物混合在一起时,仍然能够吸引昆虫。因此,交配后的雌性草地贪夜蛾(*Pegomya betae*)能够被远在 50 米以外的甜菜幼叶的气味所吸引,虽然这些气味已经混杂一些非寄主植物的气味[138]。而混合的植物群中的光在几米外就可被感知,在飞行类昆虫上体现得尤为明显。目前,关于活跃空间(active space)的感知是基于视觉信号还是气味信号,只存在极少的数据,而有人认为给出气味信号的活跃空间似乎比视觉信号的活跃空间更大,这样的结论似乎为时过早[19,128]。实际上,在野外条件下,视觉信号和气味信号总是同时发生作用,下文(见6.6)将揭示昆虫如何利用两种信号的结合来克服仅依靠其中一种信号所产生的弊端。

6.4.2 对寄主植物特性的视觉反应

植物的三个光学特征可能影响寄主选择行为:光谱特性、尺寸(大小)和样式(形状)[128]。昆虫的复眼对光谱的敏感度范围为 350 nm 到 650 nm 之间(接近紫光到红光的范围),并且包括人眼看不到的短波(图6.6)。昆虫的小眼作为昆虫复眼的光感受器和成像信息的单元有固定的焦点,这使得昆虫在很近的距离内能保持最佳视觉,然而在较远的距离对形状的感知较差。如想详细了解昆虫的光感受器和复杂的视觉系统的特点,可以查阅相关文献[24,153]。植物或植物器官的大小和形状在不同种类之间以及同一种类内部都表现出了巨大的差异性,这种差异性在近距离范围内可能有助于选择寄主植物。

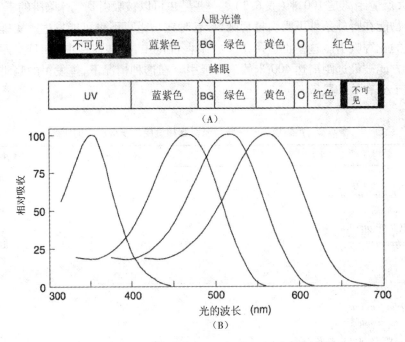

图6.6 人眼与蜂眼的光谱与光的波长和相对吸收的关系

(A) 人与欧洲粉蝶(*Pieris brassicae*)对可见光的波长范围(nm)对比(乔特尔和瓦泽,1997)[35];
(B) 一种夜蛾昆虫四色视觉系统(*Spodoptera* sp.)眼睛的光谱灵敏度曲线,每一种色素的吸收表示这种色素吸收的最大百分比(兰格等,1979)[97]

为了说明寄主利用视觉来选择植物的范围,下面将用昆虫对寄主植物的光诱因,如颜色和形状如何做出反应,这一实例具体说明。

1. 鳞翅目

对于日行性(day-foraging)蝴蝶对植物颜色做出反应这一方面的研究相对完善。当给受孕菜粉蝶(*Pieris brassicae* and *P. rapae*)提供用绿颜色的纸制作的人造树叶时,一些缺乏经验的个体会表现出着陆反应,尽管和真的卷心菜叶子相比着落的次数很少。它们一旦落到基质上,即使在寄主植物释放的化学物质缺失的情况下,也开始连续几秒钟地敲打基质。粉蝶(*P. brassicae*)表现出对真正颜色和特定波长的反应行为(图6.7),而另一种同源性粉蝶(*P. rapae*)则明显表现出着陆时对不同颜色的人工基质的偏好。在粉蝶(*P. brassicae*)和同源性粉蝶(*P. rapae*)中,二者的联系学习行为与不同色阶的绿色有关[169,174]。两种蝴蝶在着陆过程中改变了颜色偏好,从偏好绿色变为偏好花的黄色、蓝色、蓝紫色,与各自的产卵和取食需求有关。经证实凤蝶(*Battus philenor*)可以通过树叶的形状来辨别,这种蝴蝶把树叶的形状作为优先登陆宿主植物的一个信号[121]。粉蝶(*Eureka hecabe*)更加喜欢有长轮廓的人造叶子,这与它的寄主豆科植物的复叶形状是相一致的[79]。凤蝶(*Papilio aegeus*)专以芸香科植物为食,能对人造光源经偏振作用产生的光做出反应。

它的光感受器结合了对颜色和偏振的敏感度,而其他的昆虫,像蜜蜂对颜色和偏振的感知是在单独的小眼中完成的。由于植物叶子的方向在垂直或水平方向的变化会影响反射光的偏振,使叶片表面呈现光滑或灰绿的颜色。对人造叶片颜色的接受可能会引导昆虫接着对产卵地点进行选择[88]。如果黄色的中等大小的人工基板被垂直放置,夜蛾(*Mamestra brassicae*)会选择在上面着落。视觉目标和宿主气味相结合,会增加昆虫着陆的可能性[135]。

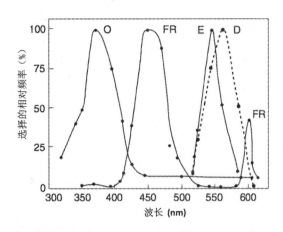

图 6.7 引起欧洲粉蝶(*Piers brssicae*)行为反应的不同波长

Y 轴为相对选择频率(%),最吸引昆虫的波长为 100%。开放空间效应(open-space reaction,O)不对植物做出反应以及没有增加飞行高度的倾向,它是由在紫外范围内的波长引起的。而摄食反应(feeding reaction,FR)最主要是由蓝光引发的,其次是黄光。此外,引发产卵(egg-laying,E)和嗡鸣(drumming,D)的波长稍微不同,但均在光谱的绿光部分(谢勒和库伯,1987)[143]

尽管毛虫侧单眼与蝴蝶成虫的复眼相比非常简单,但毛虫仍能够区别物体的尺寸和颜色,如果它们掉到地面上,仍能够确定植物的方向。

2. 双翅目

植食蝇类昆虫中的食蝇科(Tephritidae,果蝇)和花蝇科(Anthomyiidae,根蛆)的例子充分证明了视觉诱因对昆虫的作用[127]。一只雌性苹果实蝇(*Rhagoletis pomonella*)在寻找产卵地点(即苹果)的时候,它的视觉导向行为的发生过程可以被描述为一系列连贯的步骤。首先在 10 米远的距离,感知到有一棵树的轮廓与整个背景中其他植株构成对比。在这个阶段,色彩因素不太可能被感知,尤其是当这种昆虫迎着太阳光的时候,同样此时昆虫也不能感知形状等细节信息,因为它的视觉敏感度有限。其次,随着昆虫飞到距植物几米或更近的地方,至植物的前方、下方或树冠上方时,反射光的光谱和强度是引起昆虫飞落的主要诱因,例如植物的叶子、果实或树干。最后,昆虫更加接近植物(1 米以内),此时可以辨别基本的大小或形状(图 6.8)。

对甘蓝地种蝇(*Delia radicum*)而言,当对人造的叶片涂以模拟宿主植物叶片的颜色时,依据视觉反射,它们就会做出着陆的反应(图 6.9)。如果将三种模仿寄主植物真实叶片光谱的人造叶片与真实叶片同时提供给甘蓝地种蝇时,则不会表现出明显的降落偏好。它们会根据植物的年龄改变偏好。以萝卜为例,处于幼年时期的萝卜明显比成熟时期更受苍蝇青睐。在着落后期,叶片形状似乎不影响产卵,但是有柄的人造叶片与无柄的相比更受欢迎(图 6.10)。

当蝇类在不同大小的人造叶片中选择时,在四倍大小的叶片上的着陆率是正常尺寸叶片的四倍,而产卵量是正常尺寸的 2.5 倍[130,133]。果实蝇属(*Bactrocera*)昆虫中的杂食性类群和寡食性类群对颜色的偏好明显不同。杂食性的果实蝇(*B. tryoni*)偏爱蓝色人造球体,因为该球体能够反射紫外线(UV),而其他颜色的球体则不能。这种对紫外线的敏感性是有原因的,因为成熟果实表面有较厚的蜡质,会增加紫外线的反射。显然,寄主植物的颜色、形状和大小对植食飞行昆虫的选择有重要影响,因此也成为这一研究方向中常用的研究对象。而需要注意的是,视觉导向的同时也会受到气味感知的影响(见6.6)。

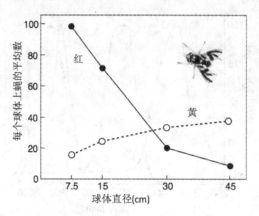

图 6.8　果园中苹果实蝇(*Rhagoletis pomonella*)对红色和黄色的无味黏性球随其直径的增加而做出的反应

依据球体上黏住的苹果实蝇的数量判断其视觉偏好,直径 7.5 cm 的球体的大小和颜色与成熟的苹果一致,更多的苹果实蝇黏着在直径更大的黄色球体上,这被认为是对绿叶的特殊替代刺激物的反应,因为这些苹果实蝇要在此处寻找蚜虫蜜露补充能量(普罗科皮,1968)[126]

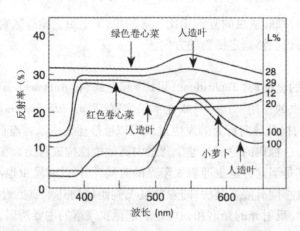

图 6.9　不同的十字花科寄主植物叶片的反射率特征和甘蓝地种蝇(*Delia radicum*)对真实叶片与人造高仿叶片的着陆回应

因为小萝卜是多项选择实验中苍蝇着陆最频繁的植株,因此着陆反应(L%)被表示为相对于小萝卜的百分比。其他可供选择的寄主植物有绿色卷心菜和红色卷心菜。对比甘蓝地种蝇在真实叶片和人造叶片的反应,二者着落的次数是相同的。纵轴表示入射光的反射率(由普罗科皮等重制,1983a)[129]

3. 同翅类

对于叶片颜色的吸引作用的研究,主要以蚜虫和粉虱作为研究对象[37,91,106]。这些昆虫肌肉力量很弱,风速超过了 1 米/秒时,它们便不能在逆风飞行时保持速度。然而,它们却能够控制地面爬行[81]。有翅型蚜虫能通过起飞或降落来控制它们的运动。决定降落的主要因素是对植物颜色的感知。因此甘蓝蚜(*Brevicoryne brassicae*)和桃蚜(*Myzus persicae*)优先选择反射长波的叶片,而很少或几乎不考虑植物分类属性。因为甜菜叶片比甘蓝叶片的长/短波反射率更高(图6.11),因此着落在甜菜叶片上的蚜虫的数量更多,尽管甜菜叶子原本不是它们的寄主。长/短波的反射率随叶片年龄和水分含量而改变。颜色的吸引使得对黄色敏感的蚜虫偏好落在黄色的植物叶片上,而不是落在它们平时的宿主植物上[91]。

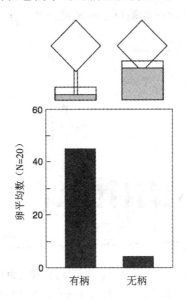

图 6.10　具柄人造叶片对甘蓝地种蝇(*Delia radicum*)产卵偏好的影响

有柄和无柄的人造叶片(绿色的 13×13 cm 纸片,浸在液状石蜡中且表面喷上甘蓝叶子的提取物)共同放在相同的实验环境中(罗斯塞和斯塔德勒,1990)

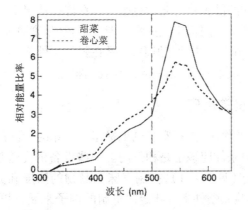

图 6.11　七月的阳光直射下,甜菜和甘蓝成熟叶片的上表面反射光的相对能量曲线

以 500 nm(长/短波反射率)为界,右侧曲线以及左侧虚线构成区域内相对能量比率为:甜菜 3.2,卷心菜 2.1(肯尼迪等,1961)

蚜虫(*Aphis fabae*)对接近甜菜叶片反射比率的深黄色较为偏爱,因此它飞落在甜菜(*Beta-*

Vulgaris)上的频率是落在芦苇(*Phragmites commuinis*)上的3倍(图6.12)。桃大尾蚜(*HyalopTerus pruni*)在夏季寄主芦苇(*Phragmites*)和冬季寄主李属(*Prunus* spp.)植物间表现出了寄主交替的现象(见8.4.1)。有翅的蚜虫在春天搜寻芦苇着落,因此在芦苇上降落的频率是附近的非甜菜寄主植物上的3倍[106]。这种情况下,芦苇叶片对黄色反射比的饱和度比甜菜叶片低。因此,不同的颜色和反射比强度引发的视觉导向反应依据物种不同而表现出差异性。粉虱避开短波(400 nm)光照的环境,但会落在绿光照射的范围(500 nm)。同蝴蝶和苍蝇一样,蚜虫对宿主植物的视觉导向性选择也会受到植物表面的蜡质的影响。换季初期,有翅的豌豆蚜(*Acythosiphon pisum*)落在蜡质含量低的豌豆上的密度比含标准蜡质的豌豆低[184]。

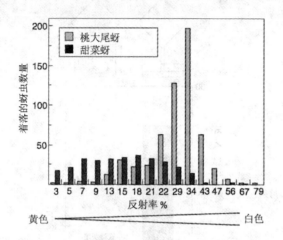

图6.12 两种蚜虫(桃大尾蚜虫和甜菜蚜)以视觉为导向的着落偏好
根据有翅蚜虫着落在16块色板上的数目判断蚜虫的偏好,16块色板组合成由黄色到白色的色阶,其短波波长的反射呈上升趋势(由左至右色彩饱和度递减)(摩力克,1969)[106]

鳞翅目、双翅目和同翅类以及其他类群的昆虫,都能利用植物或同一植株的不同叶片或不同器官反射率的差异来选择更有营养的组织。通常有营养的多为幼嫩组织,对黄光的反射能力较强。事实上,昼间活动的昆虫更易被黄色吸引。在许多情况下,黄色的叶表面作为一种特殊刺激,它们能像叶子一样散发同样带宽的峰值能量,而且强度更大。

尽管关于昆虫感光作用机理的研究很多,但是在野外条件下,关于植食昆虫视觉功效的研究,相对于其嗅觉研究来说成果有限,相关的研究我们将在之后的章节中继续讨论。

6.4.3 对寄主植物的嗅觉反应

若单独测验视觉刺激的作用,需将实验昆虫暴露于无味的、具可控光的物体之中。若单独测定寄主气味的作用,实验条件应该也是相似的。一些设备的完善使得定量分析气味的定向反应成为可能[61]。如上所述(见6.4.1),控制气味刺激物的浓度和分布比预想的更具不确定性。我们将讨论两个在实验室条件下气味定向机制的例子来展开更详细地论证,一个是飞行类昆虫,另一个是爬行类昆虫。

6.4.4 飞蛾和步甲:嗅觉定向的两个例子

一只雌性烟草天蛾(*Manduca sexta*)搜寻寄主植物时,表现出生物趋向性行为,即以顺风向为线索逆风飞行。位于触角的机械性感受器可以感受风向从而提供风向信息(通过调整趋性或趋激性,见6.3)。雌虫飞行的途径可以被描述为一个振幅有规律的"之"字(一系列反转)。

植物释放的气味何时开始发挥作用?寄主气味首先起到飞行激活剂(激发剂)的作用,诱使蛾类在休息或爬行状态下起飞。一旦起飞,飞蛾便可以接收到单个或许多寄主植物散发的气味流,然后在飞行中试图追击不让气味丢失。经过一段最小时间间隔后,嗅觉感受器细胞无法追踪到气味时,一种所谓的"放弃"反应随之发生。飞蛾会减慢速度,增大反转的振动幅度,从而增加横穿和顺风方向飞行的路径。"放弃"反应过程中,如果嗅觉感受器再次接收到气味分子,飞蛾会重新采用逆风"之"字形线路。这种行为的顺序可能不断重复直至最终接近寄主。飞蛾越接近气味源,飞行中反转的间隔越短。受气味引导(或受气味调节)的正趋风性为目前普遍认为的对寄主的搜寻机理。

实际上,雌蛾的寄主搜寻行为与雄蛾受气味调节进行逆风飞行以搜寻雌性的行为是一样的[10]。在后者的情况中,气味的信号是一种雌性散发的性信息素。目前关于控制这一行为的机制有如下观点:连续的反转运动受中枢神经系统中一种运动程序的控制,由嗅觉激发,随后开始自动运行(自我操控)[187]。然而,从"之"字形运动到"放弃"行为的转换却受嗅觉信息控制,在一定时间间隔(最短)内,活动的单调性可以导致"放弃"行为。视觉反馈也可使逆风飞行成为可能,也就是说周围主要是地面上,视觉图像通过反馈回路控制着运动反应。

在一定风速范围内,雌蛾的飞行参数(对地速度、角度轨迹)以及反转频率都会优先选用一些数值。趋风性(受气味调节)会帮助昆虫找到气味发源地,在本质上不同于受气味浓度梯度影响的趋化性定向。而事实上,在野外任何距离内都不存在这种气味梯度。目前,对雄性昆虫受性信息素引导寻找配偶的行为机制已经进行了较为完善的研究,如性信息素在不同时间和空间的作用等[94]。然而,有关野外环境下,植物气味的定向机制的研究仍十分有限[100,187]。在爬行昆虫用嗅觉定位寄主的研究实例中,有一个关于马铃薯甲虫(*Letinotarsa decemlineata*)的典型例子[177]。这类昆虫专以茄属植物为食且偏好马铃薯,是危害马铃薯最大的害虫之一。在马铃薯甲虫成虫羽化后的前7天内,为发育与飞行相关的肌肉,需要大量摄食,首先它们会通过爬行确定寄主植物的位置。在实验中,为了将它们的爬行行为进行量化,将风洞与"运动补偿器(locomotion compensator)结合使用。这种仪器可以自动记录行进足迹,同时避免昆虫接触任何障碍物。

当饥饿的甲虫感觉到有清洁的空气流吹过时,会表现出趋风性反应,与风向保持一个相对恒定的角度(图6.13)。其爬行足迹显示出甲虫旋转360度形成一个环状路线。当气流带来未受损的马铃薯植株的气味时,路线中的直线部分会显著增长。此时环形路线消失,爬行的平均速度加快,甲虫会花更长的时间逆风爬行,即为正向的趋风性(气味调节)反应。当气流中有非寄主气味时,如卷心菜,此时的足迹参数与在清洁空气中记录的参数相似。如果把土豆与卷心菜的气味混合,甲虫的爬行足迹与在清洁空气中的足迹无显著差异(图6.13)。

在另一种茄科植物中也发现了类似的效果,多毛番茄(*Lycopersicon hirsutum*)为一种不适合甲虫寄生的植物。番茄和土豆挥发物的混合物也不能吸引甲虫。混合物中由于番茄气味阻止甲虫飞向它们寄主植物的现象被称为"气味遮盖"[163]。在混合耕作系统中,这一现象被认为可以起到减少植食昆虫数量的作用(第13章)。

174　　　　　　　　　　　　　　　昆虫—植物生物学

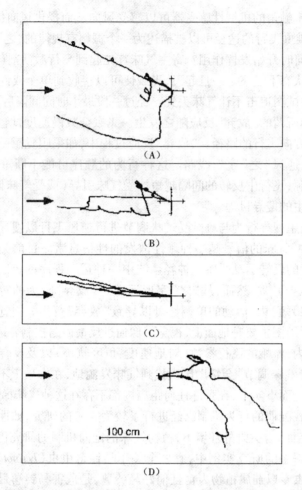

图6.13　在4个连续10分钟间隔内(A—D)，一只雌性马铃薯甲虫的爬行足迹

刺激气流分别为：(A)清洁的气流；(B)含有卷心菜气味的气流；(C)含有马铃薯(甲虫最喜爱的寄主植物)气味的气流；(D)含有卷心菜和马铃薯混合气味的气流。箭头表示气流方向。甲虫爬行至一定最大距离时，绘图仪会把甲虫的位置重置回原点(十字图标中心)。图(C)中，总爬行距离和直线足迹比其他三种情况下更多，而其他三种情况下，并无明显差异(蒂里和维瑟，1986)[162]

表6.3　四个主要目的植食性昆虫成虫对于植物气味的行为反应

昆虫	食性	气味源	实验环境类型	数据来源	参考文献
半翅目					
忽布疣额蚜(*Phorodon humuli*)	M	G	L/F	SCR	28
瘤蚜(*Cryptomyzus korschelti*)	O	HP	L		180
蚜虫(*Cavariella aegopodii*)	O	G	F(T)		34
菜缢管蚜(*Lipaphis erysimi*)	O	S	L	SCR	118
菜蚜(*Brevicoryne brassicae*)	O	S	L(F_)	SCR	118, 123
稻麦蚜(*Rhopalosiphum padi*)	O	G	L(F_)		122
甜菜蚜(*Aphis fabae*)	P	HP	L(F_)	SCR	90,118
棉蚜(*Aphis gossypii*)	P	HP	F		125
鞘翅目					
马铃薯甲虫(*Leptinotarsa decemlineata*)	O	HP/G	L(F_)	EAG/SCR	98, 162, 176

续表

昆虫	食性	气味源	实验环境类型	数据来源	参考文献
棉铃象(Anthonomus grandis)	O	G	L/F	EAG/SCR	52, 53, 54
云杉八齿小蠹(Ips typographus)	O	G	L/F	SCR	105, 168
条跳甲(Phyllotreta spp.)	O	S	L/F		124
白菜籽龟象(Ceutorhynchus assimilis)	O	S/HP	L/F	SCR	22, 58
日本丽金龟(Popillia japonica)	P	G	F		1
蔬菜叶象甲(Listroderes obliquus)	P	G/S*	L		99
宽肩叶甲(Oreina cacaliae)	O	HP/HPE	L		86
金龟子(Phyllopertha diversa)	P	G	L	SCR	74
松树皮象(Hylobius abietis)	O	HPE	L	SCR	186
双翅目					
茎蝇(Psila rosae)	M	S	L/F	EAG	70, 71, 117
葱地种蝇(Delia antique)	O	S	L/F	EAG/SCR	70, 80, 85
甘蓝地种蝇(Delia radicum)	O	S	L/F	EAG	40, 70, 117
苹果实蝇(Rhagoletis pomonella)	O	G	L/F	EAG	60, 65, 114
柑橘小实蝇(Daces dorsalis)	P	G	L/F	EAG	102
鳞翅目					
棉铃虫(Helicoverpa subflexa)	M	HPE	L		165
邻菜夜蛾(Acrolepiopsis assectella)	M	S	L		161
小菜蛾(Plutella xylostella)	O	HPE	L	EAG	120
烟草天蛾(Manduca secta)	O	G/HP/HPE	L	EAG	164, 100
凤蝶(Papilio polyxenus)	O	G	L	EAG	15
棉铃虫(Helicoverpa virescens)	P	HPE	L	SCR	82, 137, 166
粉纹夜蛾(Trichoplusia ni)	P	HP	L		96
玉米螟(Ostrinia nubilalis)	P	HP/G	L	EAG	29, 170
灰翅夜蛾(Spodoptera littoralis)	P	HP/G	L	SCR	84, 141
甘蓝夜蛾(Mamestra brassicae)	P	HP/HPE/S/G	L	EAG	134, 136
苹果蠹蛾(Cydia pomonella)	O	G/S	L/F	EAG	6, 38, 76

其中,M 为单食性;O 为寡食性;P 为多食性;HP 为完整的寄主植物;HPE 为寄主植物提取物;G 为一般的绿叶挥发物;S 为特定植物的挥发物;L 为使用嗅觉仪或风洞在实验室内的行为测试;F 为野外测试、陷阱捕获(F(T))或直接观察;(F_)为在实验室条件下表现对气味的嗅觉趋向反应,在野外并没有表现出来;EAG 为应用触角电位仪对气味源感知的记录;SCR 为单细胞记录法对气味源感知的记录;* 表示挥发物为异硫氰酸酯,为十字花科(寄主植物之一)植物特有。

由嗅觉引导引发随后的视觉定向被认为是植食昆虫(包括专食性和杂食性昆虫)搜寻寄主植物中主要使用的机制[154,177]。另外,已在几种毛虫及多种植食昆虫中得到证实,在距离寄主几厘米的范围内可以发生趋化反应[93,115]。表格6.3总结了四个目的植食性昆虫成虫对植物气味的行为反应。可以发现,在每一个目中,专食性昆虫只对其寄主植物释放出的气味做出反应。

植食昆虫普遍能够利用植物挥发物作为信号,了解关于植物情况的信息,从而对寄主植物进行优化选择。经卷叶病毒感染的马铃薯能更强烈地吸引杂食性的桃蚜(Myzus persicae),因为被感染的植株比未感染的植株更适合成为桃蚜的寄主[57]。然而,杂食性铃夜蛾(Helicoverpa virescens)会避免在被同种毛虫啃食过的植物上产卵。在夜间,被取食过的植物会释放特殊的挥发物,这些物质强烈排斥夜行性雌性蛾类搜寻产卵地[2]。植食昆虫的许多天敌(包括节肢动物)也会利用植物挥发物获取化学信息(第10章)[51,157]。

6.5 对寄主的气味检测的化学感觉基础

昆虫在搜寻食物、产卵地点、配偶以及社交时,主要依靠化学感受器接收信息。事实上,昆虫生活在一个化学世界里。化学感受指嗅觉(嗅觉,检测挥发性化学刺激物的器官)和味觉(味觉,接触性化学感受器以检测溶解或固体的化学物质,第7章)两者之间并无绝对差别,因为有时昆虫的味觉感受器会对气味做出反应[156],而且味觉受体蛋白也在触角中表达并且具有嗅觉功能[183]。

6.5.1 嗅觉感受器的形态学研究

嗅觉感受细胞与所谓感受器(单一的:感受器),包含神经元的器官、副卫细胞和表皮组织相连(图6.14)。

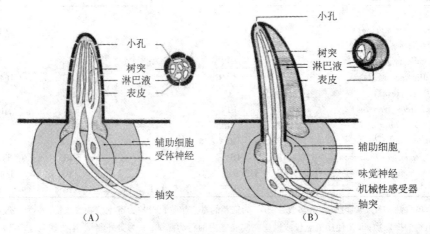

图6.14 昆虫嗅觉感受器

(A)和(B)是昆虫味觉感受器的纵切和横切的图示。嗅觉感受器由内含的两个化学感受器双向神经元控制,味觉感受器由内含的两个化学感受器和一个机械性感受器控制

神经元的细胞体(核周体,perikarya)与表皮组织紧密相连。树突通常位于特化的表皮组织上,根据外部形态的差异分属不同类别,其中包括毛发状的(毛形感器)、钉状和锥状的(锥形感受器通常为对植物气味的感器)。通常情况下,嗅觉感受器有2~5个神经元[32,87,112],但在蝗虫体内,一个锥型感受器中分布的神经元高达50个。意大利蜜蜂(*Apis mellifera*)的板型感受器受30个神经元控制[69]。化学感觉神经元大部分为双极神经元,它们的轴突不需要通过中间突触就可以连接到神经中枢。树突为神经元细胞的丝状延伸,深入到感受器的凹陷中,对化学刺激物引发神经产生的电位差(感受电位)做出反应。当这种电位高出一定的临界值时就会引发一系列的反应。

嗅觉感受器和味觉感受器在结构上有一些重要差别。嗅觉感受器是多极的,整个感受器壁上有成千上万个小孔(直径大约为10~50 nm)并且树突经常有分支[158]。与之相反,味觉感受器为单孔结构,气孔(直径200~400 nm)大多位于感受器(钉状、毛发状或乳突状)顶部

（图 6.14）。这两种感受器的树突末端都接近气孔。淋巴液由感受器基部的膜原细胞和毛原细胞分泌入感受器管腔，因而树突在这种干燥机制的作用下免受淋巴液的影响。

嗅觉感受器主要存在于触角，但在昆虫产卵期间也可能会位于上颚和下唇须。嗅觉感受器和相关嗅觉受体细胞的数量在不同物种间存在很大差异。完全变态的昆虫，其幼虫的嗅觉细胞数量很少（如甲虫幼虫的嗅细胞少于 10 个，蝇蛆大概有 100 个嗅觉细胞[183,188]）。而半翅目雌成虫则有高达数千个嗅觉细胞[32]。带有嗅觉受体的细胞感器可能为多受体式，也就是说，它们还可能包含热受体、湿受体和机械性受体[47,147]。

6.5.2 嗅觉转导

转导过程是将化学刺激物质（质量和数量）转化为受体电位，最终转换成的动作电位的过程，其中包括一系列步骤。最近对嗅觉传导的分子基础的了解有了相当大的进展。当前公认的模式如图 6.15 所示。首先，挥发性刺激物分子通过感受器上的孔道进入感受器管腔，并且与这些分子结合形成小的水溶性嗅觉结合蛋白（OBPs），这样挥发性刺激物分子（配体）就被结合到树突膜受体上了。昆虫的嗅觉受体（OR）为 G 蛋白偶联（GPC，包含 7 个跨膜蛋白）受体，通过第二信使，如环磷酸腺苷（cAMP）和参与树突膜离子通道开放的 1,4,5－三磷酸肌醇的激活而发挥功效。离子通道的开放导致树突膜的去极化。当去极化的感受器电位幅度超过阈值，会导致动作电位的产生，从轴突膜向中枢神经系统的触角神经叶小球传导。而刺激物分子的活动最有可能被淋巴液中的气味降解酶终止[182]。

自 2000 年来，随着果蝇（*Drosophila melanogaster*）的全基因组序列测序完成，揭开了关于气味检测分子的遗传机理，果蝇（*Drosophila melanogaster*）也成了在昆虫这一类群中此类研究的模式生物。这种昆虫有 1300 个嗅觉神经元与 43 个触角神经叶小球相连。目前，已有 25～60 个 OBP 基因和 61 个七跨膜 GPC-OR 基因被报道[182,183]。虽然它们共享一些碱基序列，该序列多样性很高，据显示只有 17%～26% 为保守序列，并且与其他动物类群的 OR 基因没有明显的同源性序列。对果蝇（*D. melanogaster*）基因的推测应该也同样适用于其他类群的昆虫。目前，关于昆虫嗅觉传导方面的研究难题为阐释嗅觉特异性中 OBPs 的功能和对参与气味识别的嗅觉受体配体特异性描述[190]。利用果蝇（*D. melanogaster*）的基因组信息和其他物种 OR 基因序列同源性，在基因沉默技术的帮助下，我们可以展开 OR 基因对独特气味识别的研究[64]。

6.5.3 嗅觉电生理和敏感性

目前主要采用两种电生理技术对昆虫嗅觉系统的敏感性和特异性进行研究。一种是记录触角嗅觉神经元产生的全部的受体电位反射，也就是所谓的触角电位反应（EAG）。第二种是对单独的感受器进行记录（即所谓的单感器或单细胞记录，SCR）获得动作电位，也称为峰电活动（spike activity）。这是在中枢神经系统处理的真正传输信号。这两种方法各有其优势和局限。EAG 体现了整个嗅觉神经元群的反应，但其灵敏度有限。SCR 对嗅觉检测灵敏度高，但只能记录触角神经上少量的细胞样本[185]。

同大多数感觉细胞一样，化学感受器对刺激强度的变化（如化学物质浓度的变化）尤为敏感。主要有两种类型：激发，动作电位随着刺激气味的增加而增加；抑制，非刺激状态导致激发率下降，同时发生峰电现象（图 6.16）。

嗅觉细胞每秒可处理多达 33 个气味脉冲[14,110]，因此可以对瞬时出现的气味湍流中的气味

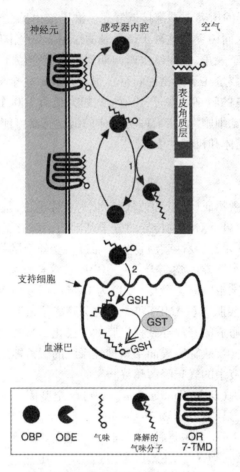

图 6.15 气味识别的生化路径简图

疏水性气味分子穿过表皮上的小孔到达嗅感器内腔,与亲水性气味结合蛋白(OBPs,包含 7 个跨膜结构蛋白)结合,并且把气味分子传送到位于神经元细胞膜上的受体蛋白——气味受体(ORs)上。嗅感器内腔中的气味分子降解酶(ODEs)(途径 1)可将气味分子降解。神经细胞体周围的支持细胞细胞质中,含有灭活酶,如谷胱甘肽-S-转移酶(GST),可能有助于气味分子失去活性(途径 2)。GSH 代表谷胱甘肽;7-TMD 代表 7 个跨膜域(沃格特,2003)[182]

信息进行反应(图 6.5)。

在触角电位以及单细胞记录中显示:浓度与反应之间的关系呈 S 形曲线(图 6.17、图 6.18)。随着气味浓度的增加,触角电位的波幅值和动作电位的频率值通常的增幅可高达 1.5~3 倍,直至到达饱和浓度时,不会发生进一步的增幅反应。反应浓度与饱和浓度之间的浓度差是易于辨别的(即曲线上升阶段的剂量值)(图 6.17、图 6.18)。在理论上,这能够帮助我们了解昆虫可能会做出刺激趋性行为所感知的气味梯度(见 6.4.3)。味觉受体起到浓度探测器的作用,与之相反,通过在一段时间内追踪分子量,嗅觉受体可以起到通量探测器的作用[167]。

神经的收敛使得检测的灵敏度大为提高。嗅觉受体细胞上的轴突与后大脑的嗅觉神经叶上有限的中间神经元进行突触结合,即发生神经收敛[78]。此处的中间神经元同时接收来自许多受体细胞的信息,在较低的浓度就开始去极化,而非必须达到一定的浓度阈值。聚合提高了信号的信噪比(signal-to-noise ratio),噪声是周边嗅觉系统引发的自发背景活动。例如,在测量

第6章 选择寄主植物：如何发现寄主植物　　179

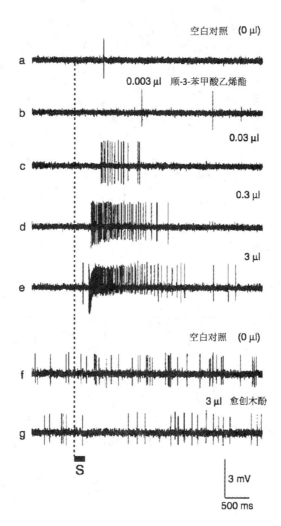

图 6.16　雌性烟草天蛾不同 A 型毛状感器的电生理记录

　　上图为一只雌性烟草天蛾(Manduca sexta)触角上两种不同的 A 型毛状感器上得到的电生理记录：b—e 显示兴奋反应，g 显示抑制反应。将含有 30 ms 溶有气味源的矿物油的滤纸置入嗅觉测量管，并向触角吹气以刺激触角。a—e 表示一个感觉器中的一个嗅觉受体神经元(ORN)的反应：(a)矿物油(对照)；(b)0.003 ml；(c)0.03 l；(d)0.3 ml；(e)顺 - 3 - 苯甲酸乙烯酯，一种芳香酯，3 ml；(f—g)表示在不同感觉器中的一个嗅觉受体神经元(ORN)的反应；(f)矿物油；(g)愈创木酚 3 ml，刺激条(S) = 200 毫秒(谢尔德和希尔德布兰德，2001)[148]

绿色叶挥发物对马铃薯甲虫中间神经元对触角的刺激作用上，其浓度为触角受体阈值的千分之一至百分之一[41]。在触角神经叶中，嗅觉神经元轴突与球状神经纤维上进行突触连接，这种球状神经纤维被称为神经球。神经球是嗅觉神经细胞、中间神经元和投射神经元以突触结合的小的缠绕团(图 6.19)[7]。在一些蛾类和蝶类中，有 60~70 个神经球，据记载蜜蜂有 166 个神经球样结构，而蝗虫有 1 000 个。一些投射神经元在神经球部位分叉并且其轴突连接到大脑中心，如蕈形体，如果将投射神经元和触角的嗅觉细胞的数量进行比较，可以计算出收敛比率，它的值大概与敏感度呈正相关。东亚飞蝗(Locusta migratoria)的收敛比为 150∶1，意蜂

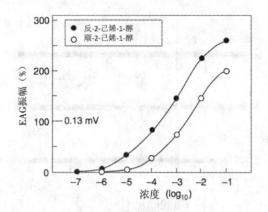

图 6.17 两种绿色叶片挥发物浓度和马铃薯雌性甲虫的触角电位反应(EAG)强度之间的关系

浓度表示为在液状石蜡中的稀释度(V/V),触角电位反应为对一种绿色叶片挥发物,标准剂量的顺-3-己烯醇-(1)(10^{-3} 或 1 ml/ml)的反应,反式化合物发生引发反应的浓度值为顺式化合物的1/10(维德,1976)[176]

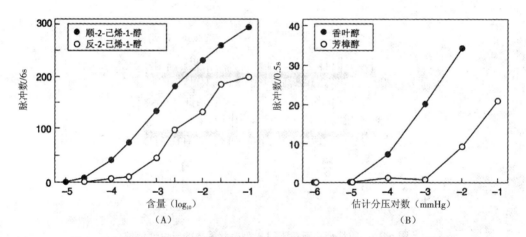

图 6.18 植物挥发物浓度和两种昆虫嗅觉受体的单个神经元反应之间的关系

(A)为单一的支配雌性马铃薯甲虫感觉器的嗅觉神经元在受到两种绿色叶挥发物刺激时剂量与反应的关系,浓度表示为在液状石蜡中的稀释度(v/v)(马和维德,1978)[98];(B)烟草夜蛾雌虫触角上仅能感知萜类化合物气味的三种不同嗅觉细胞体现出的剂量与反应之间的关系,气味的蒸汽压力也被纳入考虑因素(谢尔德和希尔德布兰德,2001)[148]

(*Apis mellifera*)的收敛比为650∶1,而烟草天蛾(*Manduca sexta*)的收敛比为330∶1。烟草天蛾(*M. sexta*)的触角神经叶输出的神经元可以将几毫秒内气味强度的变化进行精细编码[175]。

6.5.4 嗅觉特征与编码

昆虫的嗅觉受体如何对外界挥发性化学刺激信息进行编码以增加寻找寄主植物的机会?分析嗅觉特异性需要记录单细胞的反应。目前,可通过测试植物化学物质单一成分或混合物研究其诱发化学感应活动(对嗅觉神经元激发或抑制)的效果。嗅觉系统具有滤器的功能,因为嗅觉受体神经元只对环境中的一定类别的挥发性化学物质有敏感性。以往的研究认为嗅觉神经元和味觉神经元(第7章)主要分为两类:"特异性"和"非特异性"受体神经元。根据定

第6章 选择寄主植物：如何发现寄主植物

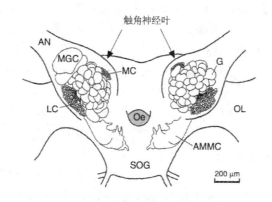

图6.19 雄性烟草天蛾大脑（Manduca sexta）的正面图

显示了中脑的两个神经纤维网、触角叶以及触角机械感觉和运动中心（AMMC）。大血管小球复合体（MGC）仅存在于雄性天蛾中。触角叶中间神经元的大部分细胞体都集中在两个细胞群：中间群（MC）和侧面群（LC）。其中，AN，触角神经；G，神经球；SOG，食管下神经节；Oe，食管；OL，视神经叶（安东和霍姆伯格，1999）[7]

义，特异性神经元只对少数与结构相关的化合物有反应，而非特异性神经元则对大量与结构不相关的化合物有反应。在昆虫的嗅觉感受器中，性信息素感受器是典型的特异性受体[114]。

在过去的10年中，我们对于嗅觉神经元特异性的认识在逐渐发生变化。在对昆虫行为的研究中，越来越多的研究表明，特异性嗅觉神经元细胞对一般的植物挥发物也有反应，如甲虫、蛾类对绿叶挥发物和萜类物质均有反应[5,74,159]。另外，也发现了特异性嗅觉神经元对寄主植物特殊的挥发物反应，例如鞘翅目[22]、鳞翅目幼虫[178]和蚜虫[118]。现在看来，非特异性神经元受体是比较少见的。这种观点的变化可以用如下事实解释：以往在实验室内刺激物的浓度往往过高（相对于当时的环境中的普遍浓度），而且缺乏对特异性嗅觉神经元关键刺激物的了解[74]。

最近的研究结果表明，寡食性和杂食性昆虫的嗅觉感受神经元具有高敏感性和高选择性，其选择对象往往为许多植物中常见的化学物质以及某些植物群落特有的化学物质[113]。因此，杂食性夜蛾的大多数（80%）嗅觉受体神经元细胞对某种物质显示选择性和敏感性反应，因为这种化合物能刺激其产卵。然而，由于同时还有其他的挥发性化合物释放，因此这一特定化合物的浓度在昆虫产卵方面传递什么样的信息，目前还不得而知。

目前使用分子技术分析嗅觉神经元特异性的范例为：一个嗅觉神经元代表一个受体蛋白[72]。单独的受体蛋白可能与挥发性配体（结构不同）相互作用。一些配体不能激活受体，而一些配体可以激活多种受体。一种受体受不同的配体影响，可能会产生兴奋或抑制反应。通常情况下，根据最低阈值的分子来决定神经元细胞的特异性，此时特异性与分子类型（类别）相一致[167]。

嗅觉受体神经元可分为不同的反应类型。在烟草天蛾（Manduca sexta）[148]和桉树天牛（Phoracacantha semipunctata）[12]中发现了3种反应类型。马铃薯甲虫[98]和菜粉蝶[43]的触角感受器有5种反应型，象鼻虫（Pissodes notatus）[20]有12种，而果蝇（Drosophila melanogaster）[39]有16种。这些反应类型的数量取决于气味测试和触角神经元的取样。备受关注的是，周围嗅觉系统的组织特点是能在同一感受器内不同特性的嗅觉神经元被区域划分。这样划分优势为，与神经元作用的挥发性混合化合物能以更准确的方式接收和传播[167]。

目前,对于嗅觉编码方面的研究,已从传统的对嗅觉神经元的反应类型方面扩展到对触角叶和前脑对嗅觉信息的处理过程的研究[31,78,113]。利用扫描显微镜和钙敏感荧光染料可以在单一成分化合物或混合物接触触角感受器的时候,对神经球的激活模式进行实时监控[67,68,151]。在此基础上,可以构造神经球三维图像[18,95],并可以确定气味的空间性表征[18,66,140]。目前的研究发现,具相同功能的触角嗅觉神经元将其轴突与神经球相连,神经球可以起到独立处理单元的作用[31,113]。投射神经元将嗅觉信息从单一神经球传输到覃形体和其他前脑中心,经电生理分析发现,每个神经球都有一个特定的分子接受范围[92],并且对于单一成分的植物化合物,会有不止一个神经球参与到处理过程中[31,140]。

在自然环境下,与行为有关的气味信号常常是混合的。昆虫的嗅觉系统依照植物气味的质量[31]、数量(浓度,比率)、空间分布等信息进行编码,而后将其转化为决策行为。

在寡食性昆虫中,编码的形式为"线性标记"(Labelled-line),嗅觉神经元的活性可能引发动力反应,或经气味诱导引起的趋风性行为,既可能为积极反应,也可能为消极反应。在杂食性昆虫中,普遍存在"交叉纤维模式(Across-fibre pattern)"代码,嗅觉神经元通过对一系列有混杂的、有重叠但不完全相同的物质进行反应。参与"交叉纤维模式(Across-fibre pattern)"过程的细胞属于中间神经元细胞和触角叶的投射神经元细胞。虽然这两个编码模式有时被认为是相互排斥的,但它们更可能代表了一个连续过程的两个极端[152]。许多植物都能释放出成分复杂的混合物质,其中包括绿叶挥发物和萜类化合物质(第4章)。由于对这些物质缺乏定性分类,因此对这些释放物质依据其数量比例进行神经编码则显得尤为重要,因为这些比率可能包含植物的分类信息[177]。因此,"交叉纤维模式(Across-fibre pattern)更适合这种目的,并且需要较少的受体就能完成这项任务[23]。

在行为方面,一般的绿叶挥发物可能互相协调,也可能与特定的挥发物或信息素相作用[27,38,60]。同样,在嗅觉受体方面,已被证明相互作用会发生在寄主植物的气味之间或气味和信息素之间[73,119,173,180]。

目前,关于嗅觉特异性,编码原则和处理植物气味信息的中枢神经系统等领域的研究正在迅速发展。这些研究主要关注于一些代表物种:意大利蜜蜂(*Apis mellifera*)[67]、夜蛾(*Spodoptera* spp.)[140]、铃夜蛾(*Helicoverpa* spp.)[151]以及烟草天蛾(*Manduca sexta*)[148]。

6.6 自然中对寄主植物的搜寻

植食昆虫在野外寻找寄主植物时,会遇到分布不均却种类繁多的刺激物。而野外的环境缺少对一些刺激条件(影响昆虫行为反应的)和非生物参数的控制。因此,在野外条件下,很难对光学和气味这两种主要刺激方式的重要性进行评估。在一些昆虫中,已经证明主要刺激方式是交互作用的。在搜寻食物或产卵地的过程中,不同类别的刺激方式的重要性可能会随着与植物距离的改变而改变。刺激方式的交互作用可能是表6.3中差异的原因之一,因为在实验室中观察到的与气味有关的行为反应不能在野外进行验证。

例如,马铃薯甲虫在定向行为中只根据气味做出反应,并对气味显示正向趋向反应,这在实验室研究中已经被证明[162,172]。然而,通过野外条件对昆虫搜寻寄主植物行为的观察,因寄主植物的位置不同,气味对于寄主植物的定位有不同的结果。瓦尔德发现,野外条件下,距离马铃薯植物(未说明大小)小于6米时,会出现逆风趋向反应[49]。然而,杰米和同事在野外发现

第6章 选择寄主植物:如何发现寄主植物

只有很少一部分甲虫逆风移动,即使在这种情况下,它们爬行的轨迹也没有显示出朝向马铃薯的定向运动[83]。甲虫表现出了明显的趋光性,但并没有表现出气味诱导的趋风性。而在马铃薯植株附近,甲虫的直线运动轨迹发生中断,同时转弯的频率增加。杰米和同事估计,基于嗅觉线索、视觉线索或二者的结合,爬行甲虫可以检测到马铃薯植株的最大距离约为60厘米。然而,进入到这个检测半径的甲虫中,只有1/2能被吸引到植株。由于在自然植被中存在复杂的气味掩蔽现象,这很可能是昆虫进行小半径检测的原因之一。我们可以得出以下结论,自然条件下,马铃薯植株可能随机的分散于各种非寄主植物之间,如果甲虫从60厘米以外的距离搜寻马铃薯植株,对于寄主植物的发现是一个偶然事件[83]。这些研究发现与摩尔和凯瑞华提出的"交替随机和非随机搜寻策略"相符合[107]。

在被研究过的对寄主搜寻行为的植食性昆虫中,对苹果实蝇(*Rhagoletis pomonella*)和甘蓝地种蝇(*Delia radicum*)的研究最为深入[3,4,62,139]。苹果实蝇以视觉导向搜寻寄主的行为如上文所述(见6.4.2)。它们可对特定视觉刺激做出强烈反应,但只有在被苹果的气味"激活"后才如此。依据物体的大小和它们的动机状态,它们显示出对黄色或红色的偏好(见6.2)。如果苹果实蝇正在寻找产卵地点,会比较偏好相对较小的球形红色物体。为了获得碳水化合物,果蝇以苹果树叶上的蚜虫蜜露为食。当对碳水化合物的需求较高时,较大黄色球体比红色球体更受苹果实蝇青睐。黄色已经成为苹果叶片绿色色调中一种超常的替代刺激。苹果的气味吸引苹果实蝇逆风飞行,并且气味诱导的趋风性帮助果蝇在树木群中可以成功定位唯一结果的苹果树。以同样的方式,它们也可以定位人工合成的气味源。一旦树结出苹果,就要依靠视觉对果实的大小或颜色进行判断来决定。然而,如果树上的果实数量太少或者还是青色而非红色时,就无法同叶片区别,此时还要用气味线索来进行协助选择(图6.20)。

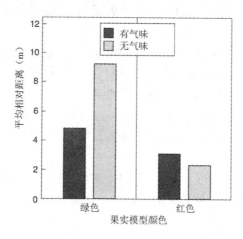

图6.20 苹果实蝇(*Rhagoletis pomonella*)在对寄主的选择行为中,
嗅觉信息和视觉信息对其的影响

寄主果实的气味(经人工合成的6种酯类物质混合物)以500毫克/小时的速度释放,对苹果实蝇发现树木上的绿色或红色果实模型(共16个模型)的影响(阿鲁杰和普罗科皮,1993)[3]

芬奇和科利尔提出一个新的假说(已有7个假说),来解释植物多样性对昆虫选择行为的影响[62]。这是基于对甘蓝地种蝇和其他昆虫(与十字花科植物相关)大量详细的行为观察而提出的。芬奇和科利尔认为选择过程分3个阶段,第一阶段受植物挥发物的控制,第二阶段受视觉因素控制,最后阶段主要由植物非挥发性的化学物质影响(第7章)。而第二阶段在很大程度上被忽略,并且被认为昆虫是随意选择正确的(寄主)植物或错误的(非寄主)植物。在第一阶段,嗅觉(植物挥发物)引导昆虫靠近植物,但没有提供确切的方向信息(见6.3)。在第二阶段,反复接触寄主植物是至关重要的,因为可以强化刺激信息,使昆虫进入第三阶段。在第三个阶段,是否决定产卵或摄食主要基于接触信息,尽管甘蓝地种蝇为一个特例[40]。

对昆虫搜寻寄主植物的研究主要集中在体积较大的昆虫中,因为这些昆虫的特征更加直观醒目,并且可以更方便地观察到一段时间内它们在植物之间的运动。因此我们对蝴蝶的产卵行为进行了一系列较为深入的研究[9,150]。得出的结论为,视觉以及包括光学诱因、接触化学感受诱因等因素均对昆虫产卵行为起到十分重要的作用,可以在寄主选择行为中实现时间和精力的最优化。

与果蝇的上述情况相反,在森林生态系统中,小蠹虫的寄主选择行为与上述苹果实蝇相反,主要受化学诱因来控制。寄主植物气味之间复杂的相互作用、小蠹虫或相关微生物释放的聚集信息素,以及种间的抑制信息素形成了高度复杂的化学通信系统[25,103,131]。

6.7 结论

虽然我们对影响昆虫搜寻的寄主植物特征,以及植食昆虫检测植物特征的方法有了一定的了解,但总体而言,还只是对少数物种的研究较为深入。显然,在野外测定距离对昆虫搜寻植物的影响很难进行。许多寡食性昆虫距离寄主植物较远时,并没有明显表现出定向行为,也就是说它们遇到合适的寄主植物是一种巧合。那么,搜寻本质上为一个随机过程,遇到寄主植物在很大程度上是由空间因素决定的[30,44,83]。然而,已观察到一些昆虫在自然条件下能整合不同线索的信息,而这些整合信息仅通过搜寻行为表现出来,要比通过各种感官反应表现得更为复杂[17,33,75]。

在过去10年中,在对气味检测的相关领域,由于分子生物学和神经生理学等方面有了迅速发展,大大增进了我们对有关机制的理解。目前面临的挑战是,行为和进化的意义(与利用气味信息相关)与分子生物学和生理学知识如何联系起来。植物顶空挥发性物质成分在定性和定量方面都十分复杂,但在调查具体化合物对昆虫搜寻寄主植物的影响时,必须要了解自然条件下挥发物的浓度。目前,在少数物种中已能对具吸引作用的挥发性物质的最小混合体进行鉴定[102,155]。

可以预见,在不久的将来,分子生物学、神经生物学、行为学和植物化学等学科的融合将大大推进我们对于昆虫搜寻寄主植物机制的了解[113]。此外,对植物光和气味诱因方面也取得了一定的成果[62]。

6.8 参考文献

1. Ahmad, S. (1982). Host location by the Japanese beetle: evidence for a key role for olfaction in a highly polyphagous insect. *Journal of Experimental Zoology*, 220, 117–20.

2. Aikman, D. and Hewitt, G. (1972). An experimental investigation of the rate and form of dispersal in grasshoppers. *Journal of Applied Ecology*, 9, 807–17.

3. Aluja, M. and Prokopy, R. (1993). Host odor and visual stimulus interactions during intratree

hostfinding behavior of *Rhagoletis pomonella* flies. *Journal of Chemical Ecology*, 19, 2671 – 96.
4. Aluja, M., Prokopy, R. J., Buonaccorsi, J. P., and Cardé, R. T. (1993). Wind tunnel assays of olfactory responses of female *Rhagoletis pomonella* flies to apple volatiles: effect of wind speed and odour release rate. *Entomologia Experimentalis et Applicata*, 68, 99 – 108.
5. Anderson, P., Hansson, B. S., and Löfquist, J. (1995). Plant-volatile receptor neurons on the antennae of female and male *Spodoptera littoralis*. *Physiological Entomology*, 20, 189 – 98.
6. Ansebo, L., Coracini, M. D. A., Bengtsson, M., Liblikas, I., Ramírez, M., Borg-Karlson, A. - K., *et al.* (2004). Antennal and behavioural response of codling moth *Cydia pomonella* to plant volatiles. *Journal of Applied Entomology*, 128, 488 – 93.
7. Anton, S. and Homberg, U. (1999). Antennal lobe structure. In *Insect olfaction* (ed. B. S. Hansson), pp. 97 – 124. Springer, Berlin.
8. Atkins, M. D. (1980). *Introduction to insect behavior.* Macmillan, New York.
9. Auckland, J. N., Debinski, D. M., and Clark, W. R. (2004). Survival, movement, and resource use of the butterfly *Parnassius clodius*. *Ecological Entomology*, 29, 139 – 49.
10. Baker, T. C. (1988). Pheromones and flight behavior. In *Insect flight* (ed. G. Goldsworthy and C. Wheeler), pp. 231 – 55. CRC Press, Boca Raton, FL.
11. Balkovsky, E. and Shraiman, B. I. (2002). Olfactory search at high Reynolds numbers. *Proceedings of the National Academy of Sciences of the USA*, 99, 12589 – 93.
12. Barata, E. N., Mustaparta, H., Pickett, J. A., Wadhams, L. J., and Araujo, J. (2002). Encoding of host and non-host plant odours by receptor neurones in the eucalyptus woodborer, *Phoracantha semipunctata* (Coleoptera: Cerambycidae). *Journal of Comparative Physiology A*, 188, 121 – 33.
13. Barton Browne, L. (1993). Physiologically induced changes in resource oriented behavior. *Annual Review of Entomology*, 38, 1 – 25.
14. Bau, J., Justus, K. A., and Cardé, R. T. (2002). Antennal resolution of pulsed pheromone plumes in three moth species. *Journal of Insect Physiology*, 48, 433 – 42.
15. Baur, R., Feeny, P., and Städler, E. (1993). Oviposition stimulants for the black swallowtail butterfly: identification of electrophysiologically active compounds in carrot volatiles. *Journal of Chemical Ecology*, 19, 919 – 37.
16. Bell, W. J. (1984). Chemo-orientation in walking insects. In *Chemical ecology of insects* (ed. W. J. Bell and R. T. Cardé), pp. 93 – 109. Chapman & Hall, London.
17. Bell, W. J. (1991). *Searching behaviour. The behavioural ecology of finding resources.* Chapman & Hall, London.
18. Berg, B. G., Galizia, C. G., Brandt, R., and Mustaparta, H. (2002). Digital atlases of the antennal lobe in two species of tobacco budworm moths, the oriental *Helicoverpa assulta* (male) and the American *Heliothis virescens* (male and female). *Journal of Comparative Neurology*, 446, 123 – 34.
19. Bernays, E. A. and Chapman, R. F. (1994). *Host-plant selection by phytophagous insects.* Chapman & Hall, New York.

20. Bichao, H., Borg-Karlson, A. K., Araujo, J., and Mustaparta, H. (2003). Identification of plant odours activating receptor neurones in the weevil *Pissodes notatus* F. (Coleoptera, Curculionidae). *Journal of Comparative Physiology A*, 189, 203 – 12.
21. Blaakmeer, A., Geervliet, J. B. F., Van Loon, J. J. A., Posthumus, M. A., Van Beek, T. A., and De Groot, A. E. (1994). Comparative headspace analysis of cabbage plants damaged by two species of *Pieris* caterpillars: consequences for inflight host location by *Cotesia* parasitoids. *Entomologia Experimentalis et Applicata*, 73, 175 – 82.
22. Blight, M. M., Pickett, J. A., Wadhams, L. J., and Woodcock, C. M. (1989). Antennal responses of *Ceutorhynchus assimilis* and *Psylliodes chrysocephala* to volatiles from oilseed rape. *Aspects of Applied Biology*, 329 – 34.
23. Boeckh, J. and Ernst, K. D. (1983). Olfactory food and mate recognition. In *Neuroethology and behavioral physiology* (ed. F. Huber and H. Markl), pp. 78 – 94. Springer, Berlin.
24. Briscoe, A. D. and Chittka, L. (2001). The evolution of color vision in insects. *Annual Review of Entomology*, 46, 471 – 510.
25. Byers, J. A. (1995). Host-tree chemistry affecting colonization of bark beetles. In *Chemical ecology of insects*, Vol. 2 (ed. R. T. Cardé and W. J. Bell), pp. 154 – 213. Chapman & Hall, New York.
26. Byers, J. A., Anderbrant, O., and Löfqvist, J. (1989). Effective attraction radius: a method for comparing species attractants and determining densities of flying insects. *Journal of Chemical Ecology*, 15, 749 – 65.
27. Byers, J. A., Birgersson, G., Löfqvist, J., Appelgren, M., and Bergström, G. (1990). Isolation of pheromone synergists of bark beetle, *Pityogenes chalcographus*, from complex insect-plant odors by fractionation and subtractive-combination bioassay. *Journal of Chemical Ecology*, 16, 861 – 76.
28. Campbell, C. A. M., Pettersson, J., Pickett, J. A., Wadhams, L. J., and Woodcock, C. M. (1993). Spring migration of damsonhop aphid *Phorodon humuli* (Homoptera, Aphididae), and summer host plant-derived semiochemicals released on feeding. *Journal of Chemical Ecology*, 19, 1569 – 76.
29. Cantelo, W. W. and Jacobson, M. (1979). Corn silk volatiles attract many pest species of moths. *Journal of Environmental Science and Health A*, 14, 695 – 707.
30. Cappuccino, N. and Kareiva, P. M. (1985). Coping with capricious environment: a population study of a rare pierid butterfly. *Ecology*, 66, 152 – 61.
31. Carlsson, M. A. and Hansson, B. S. (2003). Plasticity and coding mechanisms in the insect antennal lobe. In *Insect pheromone biochemistry and molecular biology* (ed. G. J. Blomquist and R. G. Vogt), pp. 699 – 728. Elsevier, Amsterdam.
32. Chapman, R. F. (1982). Chemoreception: the significance of sensillum numbers. *Advances in Insect Physiology*, 16, 247 – 356.
33. Chapman, R. F. (1988). Odors and the feeding behavior of insects. *ISI Atlas of Science: Animal and Plant Sciences*, pp. 208 – 12.

34. Chapman, R. F., Bernays, E. A., and Simpson, S. J. (1981). Attraction and repulsion of the aphid, *Cavariella aegopodii*, by plant odors. *Journal of Chemical Ecology*, 7, 881 – 8.
35. Chittka, L. and Waser, N. M. (1997). Why red flowers are not invisible to bees. *Israel Journal of Plant Sciences*, 45, 169 – 83.
36. Clyne, P. J., Warr, C. G., and Carlson, J. R. (2000). Candidate taste receptors in *Drosophila*. *Science*, 287, 1830 – 4.
37. Coombe, P. E. (1982). Visual behaviour of the greenhouse whitefly *Trialeurodes vaporariorum*. *Physiological Entomology*, 7, 243 – 51.
38. Coracini, M., Bengtsson, M., Liblikas, I., and Witzgall, P. (2004). Attraction of codling moth males to apple volatiles. *Entomologia Experimentalis et Applicata*, 110, 1 – 10.
39. De Bruyne, M., Foster, K., and Carlson, J. R. (2001). Odor coding in the *Drosophila* antenna. *Neuron*, 30, 537 – 52.
40. De Jong, R. and Städler, E. (1999). The influence of odour on the oviposition behaviour of the cabbage root fly. *Chemoecology*, 9, 151 – 4.
41. De Jong, R. and Visser, J. H. (1988). Integration of olfactory information in the Colorado potato beetlebrain. *Brain Research*, 447, 10 – 17.
42. De Moraes, C. M., Mescher, M. C., and Tumlinson, J. H. (2001). Caterpillar-induced nocturnal plant volatilcs repel conspecific females. *Nature*, 410, 577 – 80.
43. Den Otter, C. J., Behan, M., and Maes, F. W. (1980). Single cell responses in female *Pieris brassicae* (Lep.: Pieridae) to plant volatiles and conspecific egg odours. *Journal of Insect Physiology*, 26, 465 – 72.
44. Dethier, V. G. (1959). Foodplant distribution and density and larval dispersal as factors affecting insect populations. *Canadian Entomologist*, 91, 581 – 96.
45. Dethier, V. G. (1982). Mechanisms of host plant recognition. *Entomologia Experimentalis et Applicata*, 31, 49 – 56.
46. Dethier, V. G. (1989). Patterns of locomotion of polyphagous arctiid caterpillars in relation to foraging. *Ecological Entomology*, 14, 375 – 86.
47. Dethier, V. G. and Schoonhoven, L. M. (1968). Evaluation of evaporation by cold and humidity receptors in caterpillars. *Journal of Insect Physiology*, 14, 1049 – 54.
48. Dethier, V. G., Barton Browne, L., and Smith, C. N. (1960). The designation of chemicals in terms of the responses they elicit from insects. *Journal of Economic Entomology*, 53, 134 – 6.
49. De Wilde, J. (1976). The olfactory component in hostplant selection in the adult Colorado beetle (*Leptinotarsa decemlineata* Say). *Symposia Biologica Hungarica*, 16, 291 – 300.
50. Dicke, M. and Sabelis, M. W. (1988). Infochemical terminolgy: based on cost-benefit analysis rather thanorigin of compounds? *Functional Ecology*, 2, 131 – 9.
51. Dicke, M. and Van Loon, J. J. A. (2000). Multitrophic effects of herbivore-induced plant volatiles in an evolutionary context. *Entomologia Experimentalis et Applicata*, 97, 237 – 49.
52. Dickens, J. C. (1984). Olfaction in the boll weevil, *Anthonomus grandis* Boh. (Coleoptera:

Curculionidae); electroantennogram studies. *Journal of Chemical Ecology*, 10, 1759–85.
53. Dickens, J. C. (1986). Orientation of boll weevil *Anthonomus grandis* Boh. (Coleoptera: Curculionidae), to pheromone and volatile host compound in the laboratory. *Journal of Chemical Ecology*, 12, 91–8.
54. Dickens, J. C. (1990). Specialized receptor neurons forpheromones and host plant odors in the boll weevil, *Anthonomus grandis* Boh. (Coleoptera: Curculionidae). *Chemical Senses*, 15, 311–31.
55. Douwes, P. (1968). Host selection and host finding in the egg-laying female *Cidaria albulata* L. (Lep. Geometridae). *Opuscula Entomologica*, 33, 233–79.
56. Drew, R. A. I., Prokopy, R. J., and Romig, M. C. (2003). Attraction of fruit flies of the genus *Bactrocera* tocolored mimics of host fruit. *Entomologia Experimentalis et Applicata*, 107, 39–45.
57. Eigenbrode, S. D., Ding, H., Shiel, P., and Berger, P. H. (2002). Volatiles from potato plants infected withpotato leafroll virus attract and arrest the virus vector, *Myzus persicae* (Homoptera: Aphididae). *Proceedings of the Royal Society of London*, Series B, 269, 455–60.
58. Evans, K. A. and Allen-Williams, L. J. (1993). Distantol factory response of the cabbage seed weevil, *Ceutorhynchus assimilis*, to oilseed rape odour in the field. *Physiological Entomology*, 18, 251–6.
59. Fadamiro, H. Y., Wyatt, T. D., and Birch, M. C. (1998). Flying beetles respond as moths predict: optomotor anemotaxis to pheromone plumes at different heights. Journal of Insect Behavior, 11, 549–57.
60. Fein, B. L., Reissig, W. H., and Roelofs, W. L. (1982). Identification of apple volatiles attractive to the apple maggot, *Rhagoletis pomonella*. *Journal of Chemical Ecology*, 8, 1473–87.
61. Finch, S. (1986). Assessing hostplant finding by insects. In *Insect-plant interactions* (ed. J. R. Miller and T. A. Miller), pp. 23–63. Springer, New York.
62. Finch, S. and Collier, R. H. (2001). Host-plant selection by insects—a theory based on 'appropriate/inappropriate landings' by pest insects of cruciferous plants. *Entomologia Experimentalis et Applicata*, 96, 91–102.
63. Finch, S. and Skinner, G. (1982). Upwind flight bythe cabbage root fly, *Delia radicum*. *Physiological Entomology*, 7, 387–99.
64. Fire, A. (1999). RNA-triggered gene silencing. *Trends in Genetics*, 15, 358–63.
65. Frey, J. E. and Bush, G. L. (1990). Rhagoletis sibling species and host races differ in host odor recognition. *Entomologia Experimentalis et Applicata*, 57, 123–31.
66. Galizia, C. G. and Menzel, R. (2000). Odour perception in honeybees: coding information in glomerular patterns. *Current Opinion in Neurobiology*, 10, 504–10.
67. Galizia, C. G. and Menzel, R. (2001). The role of glomeruli in the neural representation of odours: results from optical recording studies. *Journal of Insect Physiology*, 47, 115–30.
68. Galizia, C. G., Sachse, S., and Mustaparta, H. (2000). Calcium responses to pheromones

and plant odours in the antennal lobe of the male and female moth *Heliothis virescens*. *Journal of Comparative Physiology A*, 186, 1049 – 63.

69. Getz, W. M. and Akers, R. P. (1994). Honeybee olfactory sensilla behave as integrated processing units. *Behavioral and Neural Biology*, 61, 191 – 5.

70. Guerin, P. M. and Städler, E. (1982). Host odour perception in three phytophagous Diptera—a comparative study. In *Proceedings of the 5th international symposium on insect-plant relationships, Wageningen,1982* (ed. J. H. Visser and A. K. Minks), pp. 95 – 105. Pudoc, Wageningen.

71. Guerin, P. M. and Städler, E. (1984). Carrot fly cultivar preferences: some influencing factors. *Ecological Entomology*, 9, 413 – 20.

72. Hallem, E. A., Ho, M. G., and Carlson, J. R. (2004). The molecular basis of odor coding in the *Drosophila* antenna. *Cell*, 117, 965 – 79.

73. Hansson, B. S., Van der Pers, J. N. C., and Löfqvist, J. (1989). Comparison of male and female olfactory cell response to pheromone compounds and plant volatiles in the turnip moth, *Agrotis segetum*. *Physiological Entomology*, 14, 147 – 55.

74. Hansson, B. S., Larsson, M. C., and Leal, W. S. (1999). Green leaf volatile-detecting olfactory receptor neurons display very high sensitivity and specificity in a scarab beetle. *Physiological Entomology*, 24, 121 – 6.

75. Harris, M. O. and Foster, S. P. (1995). Behavior and Integration. In *Chemical ecology of insects*, Vol. 2 (ed. R. T. Cardé and W. J. Bell), pp. 3 – 46. Chapman & Hall, New York.

76. Hern, A. and Dorn, S. (2004). A female-specific attractant for the codling moth, *Cydia pomonella*, from apple fruit volatiles. *Naturwissenschaften*, 91, 77 – 80.

77. Hesler, L. S., Lance, D. R., and Sutter, G. R. (1994). Attractancy of volatile non-pheromonal semiochemicals to northern corn rootworm beetles (Coleoptera: Chrysomelidae) in eastern South Dakota. *Journal of the Kansas Entomological Society*, 67, 186 – 92.

78. Hildebrand, J. G. (1996). Olfactory control of behavior in moths: central processing of odor information and the functional significance of olfactory glomeruli. *Journal of Comparative Physiology A*, 178, 5 – 19.

79. Hirota, T. and Kato, Y. (1999). Influence of visual stimuli on host location in the butterfly, *Eurema hecabe*. *Entomologia Experimentalis et Applicata*, 101, 199 – 206.

80. Honda, I., Ishikawa, Y. and Matsumoto, Y. (1987). Electrophysiological studies on the antennal olfactory cells of the onion fly, *Hylemya antiqua* Meigen (Diptera: Anthomyiidae). *Applied Entomology and Zoology*, 22, 417 – 23.

81. Isaacs, R., Willis, M. A., and Byrne, D. N. (1999). Modulation of whitefly take-off and flight orientation by wind speed and visual cues. *Physiological Entomology*, 24, 311 – 18.

82. Jackson, D. M., Severson, R. F., Johnson, A. W., and Herzog, G. A. (1986). Effects of cuticular duvane diterpenes from green tobacco leaves on tobacco budworm (Lepidoptera: Noctuidae). *Journal of Chemical Ecology*, 12, 1349 – 59.

83. Jermy, T., Szentesi, Á., and Horváth, J. (1988). Host plant finding in phytophagous in-

sects: the case of the Colorado potato beetle. Entomologia Experimentalis etApplicata, 49, 83 -98.
84. Jönsson, M. and Anderson, P. (1999). Electrophysiologicalresponse to herbivore-induced host-plantvolatiles in the moth Spodoptera littoralis. PhysiologicalEntomology, 24, 377-85.
85. Judd, J. G. R. and Borden, J. (1989). Distant olfactoryresponse of the onion fly, Delia antiqua, to hostplantodour in the field. Physiological Entomology, 14,429-41.
86. Kalberer, N. M., Turlings, T. C. J., and Rahier, M. (2001). Attraction of a leaf beetle (Oreina cacaliae) todamaged host plants. Journal of Chemical Ecology, 27, 647-61.
87. Keil, T. A. (1999). Morphology and development ofthe peripheral olfactory organs. In Insect olfaction (ed. B. S. Hansson), pp. 5-47. Springer, Berlin.
88. Kelber, A. (1999). Why 'false' colours are seen bybutterflies. Nature, 402, 251.
89. Kennedy, J. S. (1986). Some current issues in orientation to odour sources. In *Mechanisms in insect olfaction* (ed. T. L. Payne, M. C. Birch, and C. E. J. Kennedy), pp. 11-25. Oxford University Press, Oxford.
90. Kennedy, J. S., Booth, C. O., and Kershaw, W. J. S. (1959). Host finding by aphids in the field. II. *Aphis fabae* Scop. (gynoparae) and *Brevicoryne brassicae* L., with a reappraisal of the role of host finding behaviour in virus spread. *Annals of Applied Biology*, 47, 424-44.
91. Kennedy, J. S., Booth, C. O., and Kershaw, W. J. S. (1961). Host finding by aphids in the field. III. Visual attraction. *Annals of Applied Biology*, 49, 1-21.
92. King, J. R., Christensen, T., and Hildebrand, J. G. (2000). Response characteristics of an identified, sexually dimorphic olfactory glomerulus. *Journal of Neuroscience*, 20, 2391-9.
93. Klingler, J. (1958). Die Bedeutung der Kohlendioxyd Ausscheidung der Wurzeln für die Orientierung der Larven von *Otiorhynchus sulcatus* F. und anderer bodenbewohnender phytophager Insektenarten. *Mitteilungen der Schweizerischen Entomologischen Gesellschaft*, 31, 205-69.
94. Kuenen, L. P. S. and Cardé, R. T. (1994). Strategies for recontacting a lost pheromone plume: casting and upwind flight in the male gypsy moth. *Physiological Entomology*, 19, 15-29.
95. Laissue, P. P., Reiter, C., Hiesinger, P. R., Halter, S., Fischbach, K. F., and Stocker, R. F. (1999). Threedimensional reconstruction of the antennal lobe in *Drosophila melanogaster*. *Journal of Comparative Neurology*, 405, 543-52.
96. Landolt, P. J. (1989). Attraction of the cabbage looper to host plants and host plant odor in the laboratory. *Entomologia Experimentalis et Applicata*, 53, 117-24.
97. Langer, H., Hamann, B., and Meinecke, C. C. (1979). Tetrachromatic visual system in the moth *Spodoptera exempta* (Insecta: Noctuidae). *Journal of Comparative Physiology*, 129, 235-9.
98. Ma, W. C. and Visser, J. H. (1978). Single unit analysis of odour quality coding by the olfactory antennal receptor system of the Colorado beetle. *Entomologia Experimentalis et Applicata*, 24, 520-33.
99. Matsumoto, Y. (1970). Volatile organic sulfur compounds as insect attractants with special

reference to host selection. In *Control of insect behavior by natural products* (ed. D. L. Wood, R. M. Silverstein, and M. Nakajima), pp. 133-160. Academic Press, New York. 100.

100. Mechaber, W. L., Capaldo, C. T., and Hildebrand, J. G. (2002). Behavioral responses of adult female tobacco hornworms, *Manduca sexta*, to hostplant volatiles change with age and mating status. *Journal of Insect Science*, 2(5), 1-8.

101. Meisner, J. and Ascher, K. R. S. (1973). Attraction of *Spodoptera littoralis* larvae to colours. *Nature*, 242, 332-4.

102. Metcalf, R. L. and Metcalf, E. R. (1992). *Plant kairomones in insect ecology and control*. Chapman & Hall, New York.

103. Miller, D. R. and Borden, J. H. (2000). Dose-dependent and species-specific responses of pine bark beetles (Coleoptera: Scolytidae) to monoterpenes in association with pheromones. *Canadian Entomologist*, 132, 183-95.

104. Miller, J. R. and Strickler, K. L. (1984). Finding and accepting host plants. In *Chemical ecology of insects* (ed. W. J. Bell and R. T. Cardé), pp. 127-57. Chapman & Hall, London.

105. Moeckh, H. A. (1981). Host selection behavior of bark beetles (Col: Scolytidae) attacking *Pinus ponderosa*, with special emphasis on the western pine beetle, *Dendroctonus brevicomis*. *Journal of Chemical Ecology*, 7, 49-83.

106. Moericke, V. (1969). Hostplant specific colourbehaviour by *Hyalopterus pruni* (Aphididae). *Entomologia Experimentalis et Applicata*, 12, 524-34.

107. Morris, W. F. and Kareiva, P. M. (1991). How insect herbivores find suitable host plants: the interplay between random and nonrandom movement. In *Insect-plant interactions*, Vol. 3 (ed. E. A. Bernays), pp. 175-208. CRC Press, Boca Raton.

108. Mulkern, G. B. (1969). Behavioral influences on food selection in grasshoppers (Orthoptera: Acrididae). *Entomologia Experimentalis et Applicata*, 12, 509-23.

109. Murlis, J. (1986). The structure of odour plumes. In *Mechanisms in insect olfaction* (ed. T. L. Payne, M. C. Birch, and C. E. J. Kennedy), pp. 27-38. Oxford University Press, Oxford.

110. Murlis, J., Elkinton, J. S., and Cardé, R. T. (1992). Odor plumes and how insects use them. *Annual Review of Entomology*, 37, 505-32.

111. Murlis, J., Willis, M. A., and Cardé, R. T. (2000). Spatial and temporal structures of pheromone plumes in fields and forests. *Physiological Entomology*, 25, 211-22.

112. Mustaparta, H. (1984). Olfaction. In *Chemical ecology of insects* (ed. W. J. Bell and R. T. Cardé), pp. 37-70. Chapman & Hall, London.

113. Mustaparta, H. (2002). Encoding of plant odour information in insects: peripheral and central mechanisms. *Entomologia Experimentalis et Applicata*, 104, 1-13.

114. Nojima, S., Linn, C., and Roelofs, W. (2003). Identification of host fruit volatiles from flowering dogwood (*Cornus florida*) attractive to dogwood-origin *Rhagoletis pomonella* flies. *Journal of Chemical Ecology*, 29, 2347-57.

115. Nordenhem, H. and Nordlander, G. (1994). Olfactory oriented migration through soil by

root-living *Hylobius abietis* (L.) larvae (Col., Curculionidae). *Journal of Applied Entomology*, 117, 457-62.

116. Nordlund, D. A. and Lewis, W. J. (1976). Terminologyof specific releasing stimuli in intraspecific and interspecific interactions. *Journal of Chemical Ecology*, 2, 211-20.

117. Nottingham, S. F. (1987). Effects of nonhost-plant odors on anemotactic response to host-plant odor in female cabbage root fly, *Delia radicum*, and carrot rust fly, *Psila rosae*. *Journal of Chemical Ecology*, 13, 1313-18.

118. Nottingham, S. F., Hardie, J., Dawson, G. W., Hick, A. J., Pickett, J. A., Wadhams, L. J., et al. (1991). Behavioral and electrophysiological responses of aphids to host and nonhost plant volatiles. *Journal of Chemical Ecology*, 17, 1231-42.

119. Ochieng, S. A., Park, K. C., and Baker, T. C. (2002). Host plant volatiles synergize responses of sex pheromone-specific olfactory receptor neurons in male *Helicoverpa zea*. *Journal of Comparative Physiology A*, 188, 325-33.

120. Paliniswamy, P. and Gillot, C. (1986). Attraction of diamondback moths, *Plutella xylostella* (L.) (Lepidoptera: Plutellidae), by volatile compounds of canola, white mustard, and faba bean. *Canadian Entomologist*, 118, 1279-85.

121. Papaj, D. R. and Prokopy, R. J. (1989). Ecological and evolutionary aspects of learning in phytophagous insects. *Annual Review of Entomology*, 34, 315-50.

122. Pettersson, J. (1970). Studies on *Rhopalosiphum padi* (L.). I. Laboratory studies on olfactometric responses to the winter host *Prunus padus* L. *Lantbrukshoegskolan Annaler*, 36, 381-99.

123. Pettersson, J. (1973). Olfactory reactions of *Brevicoryne brassicae* (L.) (Hom.: Aph.). *Swedish Journal of Agricultural Research*, 3, 95-103.

124. Pivnick, K. A., Lamb, R. J., and Reed, D. (1992). Responses of flea beetles, *Phyllotreta* spp., to mustard oils and nitriles in field trapping experiments. *Journal of Chemical Ecology*, 18, 863-73.

125. Pospisil, J. (1972). Olfactory orientation of certain phytophagous insects in Cuba. *Acta Entomologica Bohemoslovaca*, 69, 7-17.

126. Prokopy, R. J. (1968). Visual responses of applemaggot flies, *Rhagoletis pomonella* (Diptera: Tephritidae): orchard studies. *Entomologia Experimentalis et Applicata*, 11, 403-22.

127. Prokopy, R. J. (1986). Visual and olfactory stimulus interaction in resource finding by insects. In *Mechanisms in insect olfaction* (ed. T. L. Payne, M. C. Birch, and C. E. J. Kennedy), pp. 81-9. Oxford University Press, Oxford.

128. Prokopy, R. J. and Owens, E. D. (1983). Visual detection of plants by herbivorous insects. *Annual Review of Entomology*, 28, 337-64.

129. Prokopy, R. J., Collier, R., and Finch, S. (1983a). Leaf color used by cabbage root flies to distinguish among host plants. *Science*, 221, 190-2.

130. Prokopy, R. J., Collier, R., and Finch, S. (1983b). Visual detection of host plants by cabbage root flies. *Entomologia Experimentalis et Applicata*, 34, 85-9.

131. Reddy, G. V. P. and Guerrero, A. (2004). Interactions of insect pheromones and plant semiochemicals. *Trends in Plant Science*, 9, 253–61.
132. Roden, D. B., Miller, J. R., and Simmons, G. A. (1992). Visual stimuli influencing orientation by larval gypsy moth *Lymantria dispar* (L.). *Canadian Entomologist*, 124, 287–304.
133. Roessingh, P. and Städler, E. (1990). Foliar form, colour and surface characteristics influence oviposition behaviour in the cabbage root fly *Delia radicum*. *Entomologia Experimentalis et Applicata*, 57, 93–100.
134. Rojas, J. C. (1999). Electrophysiological and behavioral responses of the cabbage moth to plant volatiles. *Journal of Chemical Ecology*, 25, 1867–83.
135. Rojas, J. C. and Wyatt, T. D. (1999). Role of visual cues and interaction with host odour during the host-finding behaviour of the cabbage moth. *Entomologia Experimentalis et Applicata*, 91, 59–65.
136. Rojas, J. C., Wyatt, T. D., and Birch, M. C. (2000). Flight and oviposition behavior toward different host plant species by the cabbage moth, *Mamestra brassicae* (L.) (Lepidoptera: Noctuidae). *Journal of Insect Behavior*, 13, 247–54.
137. Røstelien, T., Borg-Karlson, A. K., and Mustaparta, H. (2000). Selective receptor neurone responses to E-β-ocimene, β-myrcene, E,E-α-farnesene and *homo*farnesene in the moth *Heliothis virescens*, identified by gas chromatography linked to electrophysiology. *Journal of Comparative Physiology A*, 186, 833–47.
138. Röttger, U. (1979). Untersuchungen zur Wirtswahl der Rübenfliege *Pegomya betae* Curt. (Diptera, Anthomyidae). I. Olfaktorische Orientierung zur Wirtspflanze. *Zeitschrift für Angewandte Entomologie*, 87, 337–48.
139. Rull, J. and Prokopy, R. J. (2000). Attraction of apple maggot flies, *Rhagoletis pomonella* (Diptera: Tephritidae) of different physiological states to odourbaited traps in the presence and absence of food. *Bulletin of Entomological Research*, 90, 77–88.
140. Sadek, M. M., Hansson, B. S., Rospars, J.-P., and Anton, S. (2002). Glomerular representation of plant volatiles and sex pheromone components in the antennal lobe of the female *Spodoptera littoralis*. *Journal of Experimental Biology*, 205, 1363–76.
141. Salama, H. S., Rizk, A. F., and Sharaby, A. (1984). Chemical stimuli in flowers and leaves of cotton that affect behaviour in the cotton moth, *Spodoptera littoralis* (Lepidoptera: Noctuidae). *Entomologia Generalis*, 10, 27–34.
142. Saxena, K. N. and Khattar, P. (1977). Orientation of *Papilio demoleus* larvae in relation to size, distance, and combination pattern of visual stimuli. *Journal of Insect Physiology*, 23, 1421–8.
143. Scherer, C. and Kolb, G. (1987). Behavioral experiments on the visual processing of color stimuli in *Pieris brassicae* L. (Lepidoptera). *Journal of Comparative Physiology A*, 160, 645–56.
144. Schneider, D. (1992). 100 years with pheromone research: an assay on Lepidoptera. *Naturwissenschaften*, 79, 241–50.

145. Schöne, H. (1984). *Spatial orientation: the spatial control of behavior in animals and man*. Princeton University Press, Princeton, NJ.
146. Schoonhoven, L. M. (1987). What makes a caterpillar eat? The sensory code underlying feeding behavior. In *Perspectives in chemoreception and behavior* (ed. R. F. Chapman, E. A. Bernays, and J. G. Stoffolano), pp. 69–97. Springer, New York.
147. Shields, V. D. C. and Hildebrand, J. G. (1999). Fine structure of antennal sensilla of the female sphinx moth, *Manduca sexta* (Lepidoptera: Sphingidae). II. Auriculate, coeloconic, and styliform complex sensilla. *Canadian Journal of Zoology*, 77, 302–13.
148. Shields, V. D. C. and Hildebrand, J. G. (2001). Responses of a population of antennal olfactoryreceptor cells in the female moth *Manduca sexta* to plant-associated volatile organic compounds. *Journal of Comparative Physiology A*, 186, 1135–51.
149. Singer, M. C. (2000). Reducing ambiguity in describing plant-insect interactions: 'preference', 'acceptability' and 'electivity'. *Ecology Letters*, 3, 159–62.
150. Singer, M. C., Stefanescu, C., and Pen, I. (2002). When random sampling does not work: standard design falsely indicates maladaptive host preferencesin a butterfly. *Ecolog Letters*, 5, 1–6.
151. Skiri, H. T., Galizia, C. G., and Mustaparta, H. (2004). Representation of primary plant odorants in the antennal lobe of the moth *Heliothis virescens* using calcium imaging. *Chemical Senses*, 29, 253–67.
152. Smith, B. H. and Getz, W. M. (1994). Nonpheromonal olfactory processing in insects. *Annual Review of Entomology*, 39, 351–75.
153. Srinivasan, M. V., Poteser, M., and Kral, K. (1999). Motion detection in insect orientation and navigation. *Vision Research*, 39, 2749–66.
154. Städler, E. (1992). Behavioral responses of insects toplant secondary compounds. In *Herbivores: their interactions with secondary plant metabolites*, Vol. II (2nd edn) (ed. G. A. Rosenthal and M. R. Berenbaum), pp. 45–87. Academic Press, New York.
155. Städler, E. (2002). Plant chemical cues important for egg deposition by herbivorous insects. In *Chemoecology of insect eggs and egg deposition* (ed. M. Hilkerand T. Meiners), pp. 171–204. Blackwell, Berlin.
156. Städler, E. and Hanson, F. E. (1975). Olfactory capabilities of the 'gustatory' chemoreceptors of the tobacco hornworm larvae. *Journal of Comparative Physiology*, 104, 97–102.
157. Steidle, J. L. M. and Van Loon, J. J. A. (2003). Dietary specialization and infochemical use in carnivorous arthropods: testing a concept. *Entomologia Experimentalis et Applicata*, 108, 133–48.
158. Steinbrecht, R. A. (1997). Pore structures in insect olfactory sensilla: a review of data and concepts. *International Journal of Insect Morphology and Embryology*, 26, 229–45.
159. Stranden, M., Liblikas, I., Koenig, W. A., Almaas, T. J., Borg-Karlson, A. K., and Mustaparta, H. (2003). (−)-Germacrene D receptor neurones in three species of heliothine moths: structure-activity relationships. *Journal of Comparative Physiology A*, 189, 563–77.

160. Takabayashi, J., Dicke, M., and Posthumus, M. A. (1994). Volatile herbivore-induced terpenoids in plant-mite interactions: variation caused by biotic and abiotic factors. *Journal of Chemical Ecology*, 20, 1329 – 54.

161. Thibout, E., Auger, J., and Lecomte, C. (1982). Host plant chemicals responsible for attraction and oviposition in *Acrolepiopsis assectella*. In *Proceedings of the 5th international symposium on insect-plant relationships*, Wageningen, 1982 (ed. J. H. Visser and A. K. Minks), pp. 107 – 15. Pudoc, Wageningen.

162. Thiéry, D. and Visser, J. H. (1986). Masking of host plant odour in the olfactory orientation of the Colorado potato beetle. *Entomologia Experimentalis et Applicata*, 41, 165 – 172.

163. Thiéry, D. and Visser, J. H. (1987). Misleading the Colorado potato beetle with an odor blend. *Journal of Chemical Ecology*, 13, 1139 – 46.

164. Tichenor, L. H. and Seigler, D. S. (1980). Electroantennogram and oviposition responses of *Manduca sexta* to volatile components of tobacco and tomato. *Journal of Insect Physiology*, 26, 309 – 14.

165. Tingle, F. C., Heath, R. R., and Mitchell, E. R. (1989). Flight response of *Heliothis subflexa* (Gn.) females (Lepidoptera: Noctuidae) to an attractant from groundcherry, *Physalis angulata* L. *Journal of Chemical Ecology*, 15, 221 – 31.

166. Tingle, F. C. and Mitchell, E. R. (1992). Attraction of *Heliothis virescens* (F.) (Lepidoptera: Noctuidae) to volatiles from extracts of cotton flowers. *Journal of Chemical Ecology*, 18, 907 – 14.

167. Todd, J. L. and Baker, T. C. (1999). Function of peripheral olfactory organs. In *Insect olfaction* (ed. B. S. Hansson), pp. 67 – 97. Springer, Berlin.

168. Tommerås, B. A. and Mustaparta, H. (1987). Chemoreception of host volatiles in the bark beetle *Ips typographus*. *Journal of Comparative Physiology A*, 161, 705 – 10.

169. Traynier, R. M. M. (1986). Visual learning in assays of sinigrin solution as an oviposition releaser for the cabbage butterfly, *Pieris rapae*. *Entomologia Experimentalis et Applicata*, 40, 25 – 33.

170. Valterova, I., Bolgar, T. S., Kalinova, B., Kovalev, B. G., and Vrkoc, J. (1990). Host plant components from maize tassel and electroantennogramme responses of *Ostrinia nubilalis* to the identified compounds and their analogues. *Acta Entomologica Bohemoslovaca*, 87, 435 – 44.

171. Van der Ent, L. J. and Visser, J. H. (1991). The visual world of the Colorado potato beetle. *Proceedings of the Section Experimental and Applied Entomology of the Netherlands Entomological Society*, 2, 80 – 5.

172. Van der Pers, J. N. C. (1981). Comparison of electroantennogram response spectra to plant volatiles in seven species of *Yponomeuta* and in the tortricid *Adoxophyes orana*. *Entomologia Experimentalis et Applicata*, 30, 181 – 92.

173. Van der Pers, J. N. C., Thomas, G., and Den Otter, C. J. (1980). Interactions between plant odors and pheromone reception in small ermine moths (Lepidoptera: Yponomeutidae). *Chemical Senses*, 5, 367 – 71.

174. Van Loon, J. J. A., Everaarts, T. C., and Smallegange, R. C. (1992). Associative learning

in host-findingby female *Pieris brassicae* butterflies: relearning preferences. In *Proceedings of the 8th International Symposium on Insect-Plant Relationships* (ed. S. B. J. Menken, J. H. Visser, and P. Harrewijn), pp. 162–4. Kluwer Academic Publishers, Dordrecht.

175. Vickers, N. J., Christensen, T. A., Baker, T. C., and Hildebrand, J. G. (2001). Odour-plume dynamics influence the brain's olfactory code. *Nature*, 410, 466–70.
176. Visser, J. H. (1976). Electroantennogram responses of the Colorado beetle, *Leptinotarsa decemlineata*, to plant volatiles. *Entomologia Experimentalis et Applicata*, 25, 86–97.
177. Visser, J. H. (1986). Host odor perception in phytophagous insects. *Annual Review of Entomology*, 31, 121–44.
178. Visser, J. H. (1987). Cited in Ref. 146
179. Visser, J. H. (1988). Host-plant finding by insects: orientation, sensory input and search patterns. *Journal of Insect Physiology*, 34, 259–68.
180. Visser, J. H. and De Jong, R. (1988). Olfactory coding in the perception of semiochemicals. *Journal of Chemical Ecology*, 14, 2005–18.
181. Vité, J. P. and Gara, R. I. (1962). Volatile attractants from ponderosa pine attacked by bark beetles. (Coleoptera: Scolytidae). *Contributions of the Boyce Thompson Institute*, 21, 251–73.
182. Vogt, R. G. (2003). Biochemical diversity of odor detection: OBPs, ODEs and SMNPs. In *Insect pheromone biochemistry and molecular biology* (ed. G. J. Blomquist and R. G. Vogt), pp. 391–445. Elsevier, Amsterdam.
183. Vosshall, L. B. (2003). Diversity and expression of odorant receptors in *Drosophila*. In *Insect pheromone biochemistry and molecular biology* (ed. G. J. Blomquistand R. G. Vogt), pp. 567–91. Elsevier, Amsterdam.
184. White, C. and Eigenbrode, S. D. (2000). Effects of surfacewax variation in *Pisum sativum* on herbivorous and entomophagous insects in the field. *Environmental Entomology*, 29, 773–80.
185. Wibe, A. (2004). How the choice of method influenceon the results in electrophysiological studies of insect olfaction. *Journal of Insect Physiology*, 50, 497–503.
186. Wibe, A., Borg-Karlson, A. K., Norin, T., and Mustaparta, H. (1997). Identification of plant volatiles activating single receptor neurons in thepine weevil (*Hylobius abietis*). *Journal of Comparative Physiology A*, 180, 585–95.
187. Willis, M. A. and Arbas, E. A. (1991). Odor-modulated upwind flight of the sphinx moth, *Manduca sexta* L. *Journal of Comparative Physiology A*, 169, 427–40.
188. Zacharuk, R. Y. and Shields, V. D. (1991). Sensilla ofimmature insects. *Annual Review of Entomology*, 36, 331–54.
189. Zohren, E. (1968). Laboruntersuchungen zu Massenanzucht, Lebensweise, Eiablage und Eiablageverhalten der Kohlfliege, *Chortophila brassicae* (Bouché) (Diptera: Anthomyiidae). *Zeitschrift für Angewandte Entomologie*, 62, 139–88.
190. Zwiebel, L. J. (2003). The biochemistry of odour detection and its future prospects. In *Insect pheromone biochemistry and molecular biology* (ed. G. J. Blomquist and R. G. Vogt), pp. 371–90. Elsevier, Amsterdam.

第 7 章　选择寄主植物：何时接受植物

7.1　接触期：对植物特征的详细评价
7.2　接触期内自然界植物的物理特征
　7.2.1　毛状体
　7.2.2　叶面结构
7.3　植物化合：接触化感作用评估
7.4　植物化学对于寄主植物选择的重要性：一段历史性的间奏
7.5　摄食与产卵的刺激
　7.5.1　初级植物的代谢物
　7.5.2　植物次级代谢物促进接受：信号刺激
　7.5.3　通常产生的植物次生代谢物作为刺激剂
7.6　摄食与产卵的抑制
　7.6.1　抑制是寄主识别过程中的一个普遍原则
　7.6.2　寄主标记是避免植食性昆虫相互竞争的机制
7.7　植物可接受性：刺激与抑制之间的平衡
7.8　基于接触化学感受器的寄主植物的选择行为
　7.8.1　化学感受器
　7.8.2　味觉编码
　7.8.3　毛虫成为编码原则的模型
　7.8.4　信号刺激受体：卓越的专家
　7.8.5　糖和氨基酸的受体：养分的探测器
　7.8.6　抑制受体：全面感受神经元
　7.8.7　外围化学物质的相互作用
　7.8.8　刺吸式昆虫的寄主植物选择
　7.8.9　产卵偏好
　7.8.10　寄主植物的选择：三级系统
7.9　化学感应系统与寄主植物偏好的演变
7.10　结论
7.11　参考文献

寻找寄主植物的植食性昆虫一旦接触到植物后，它们就会进入到对寄主植物选择的"接触期"。这个时期包含了一系列行为要素，这些行为可以对植物物理与化学特征进行详细评价。

7.1 接触期:对植物特征的详细评价

在最初与植物的接触之后,昆虫的运动往往会突然中止。这种行为被称为"限制性行为";昆虫倾向于将自己的活动范围限制在一个小的区域。例如,一只昆虫在第一次着陆之后可能会立刻飞走,此后再落到相同或相邻的叶子上。一只爬行昆虫开始可能沿植物的茎爬行,以小圆环的形式移动。毛虫类常常摇动它们的头部,可能是为了更好地对气味进行定向。一些植物结构例如叶边缘、叶脉或者茎在这个阶段似乎在引导昆虫爬行移动。在爬行的过程中,昆虫们时常对所处环境进行评价监测,表现为它们反复用足、触角、口器或产卵器触碰植物表面;经常可以观察到昆虫用跗节、触角、触须进行剐蹭和发出嗡鸣,以及产卵器拖拽行为。这些行为是对植物提供的物理和化学上的接触信息所做的直接反应。同时,叶片的边界层中含有相对高浓度且易挥发的植物化合物,也能影响昆虫的行为[1,6,205]。值得注意的是,许多昆虫对于植物的选择,包括任何继续试探评价或者放弃触碰的植物个体或者器官,都基于植物表面的物理或者化学特征[11,59,207]。

在检测行为的下一步中,昆虫可能会损害植物,从而使植物释放出内部化合物,这是包含了复杂的初级和次生代谢产物的混合物。损害常常是由昆虫的口器造成的,这一步被定义为试咬,就刺吸式昆虫而言,则定义为试吸。试咬通常比正常的取食对植物的创伤面积更小,并且植物材料在口腔中的存留时间可能比正常摄取的植物材料更长。在接触期内收集的信息由中枢神经系统做出判断、接受,并且在做出最后决定后,开始食物摄入或产卵过程。在整个过程中,聚集的感觉信息达到了最大值。食物接受量通常表示为一定食物摄入的最小量。

对于产卵的接受可以通过基质上一个或多个卵显示出来。应当指出的是,实际的食物摄入量或产卵数量的范围是可变的,并不仅取决于昆虫感官评价的结果,同时也取决于昆虫个体当时的生理状态和经验(第8章)。从进化的角度来看,接受可视为在寄主植物的选择行为中做出的重要决定,因为它对于后续的养分和能量的获取,以及产卵的过程和对后代生存都有直接的影响。

7.2 接触期内自然界植物的物理特征

昆虫一旦和植物接触,通过接触(机械感觉)和化学(取食或味觉)的刺激,就会获得许多前期阶段所不能获得的有关植物特性的额外信息。植物器官或组织的物理特性可以对昆虫的选择行为产生深远的影响。正如第3章所讨论的,植物表面的毛状体以及晶体结构、叶片的厚度和韧性、硅的含量等因素均可能导致昆虫的回避行为,这些植物特性往往被认为与防御功能相关(表7.1)。

表7.1 植物的物理特性对鳞翅目、半翅目和鞘翅目昆虫选择寄主植物的影响

	植物种类	昆虫种类	幼虫或成虫	参考文献
毛状体				
无腺体	水豆	棉铃虫 (*Helicoverpa armigera*) (鳞翅目)	幼虫	173
	棉花	西方草盲蝽 (*Lygus hesperis*) (半翅目)	幼虫+成虫	8

续表

	植物种类	昆虫种类	幼虫或成虫	参考文献
具腺体	大豆	豆叶甲（*Cerotoma trifurcata*）（鞘翅目）	成虫	108
	野生土豆	马铃薯块茎蛾（*Phthorimaea operculella*）（鳞翅目）	成虫	116
	苜蓿	马铃薯小绿叶蝉（*Empoasca fabae*）（半翅目）	幼虫+成虫	160
	曼陀罗	烟草跳甲（*Epitrix hirtipennis*）（鞘翅目）	成虫	78
组织厚度				
豆荚	黄豆	豆荚蛀虫（*Grapholita glycinivorella*）（鳞翅目）	幼虫	148
茎	番茄	马铃薯蚜虫（*Macrosipum euphorbiae*）（半翅目）	成虫	158
叶	芥末	芥末甲虫（*Phaedon cochleariae*）（鞘翅目）	幼虫	214
蜡层				
	卷心菜	菜粉蝶（*Pieris rapae*）（鳞翅目）	幼虫	210
	树莓	覆盆子蚜虫（*Amphorophora rubi*）（半翅目）	成虫	113
	芥末	芥末甲虫（*Phaedon cochleariae*）（鞘翅目）	成虫	211

昆虫在身体的许多部位都配备了机械感受器[120]，这些感受器可能与植物表面的结构和质地相对应。以植物的特征为出发点，本节列举的几个例子将更详细地表明植物的物理特征能在多大程度上影响昆虫对寄主植物的选择。昆虫与寄主植物在接触期内相互作用的主要界面为植物表面：在植物表面被破坏之前，植物不会受到损伤，因此我们将首先观察植物表面的特点。

7.2.1 毛状体

植物表面通常被许多毛状体覆盖，这些毛状体有些具有腺体，有些没有腺体。这些结构可能对昆虫的运动和取食造成障碍，尤其是体型较小的昆虫和螨类。植物种内毛状体的种类或密度的变异已经被研究应用于抵抗一些害虫的滋生。在很多情况下，毛状体的韧性是由一到两个基因决定的，这就使得选择相对简单[152,200]。

具腺体的毛状体（黏液毛）被认为在形态结构上和化学成分上都可以阻止昆虫在植物上定居（见4.7）。由于昆虫的爬行，可能会致使毛状体中的腺体产生机械损伤，或者持续分泌。腺体的分泌物也许对昆虫有趋避作用甚至有毒，可以将饥饿的小型昆虫黏到植物表面[72]。而个体较大的昆虫能避开那些带有能释放化感物质的腺毛的植物或栽培品系。举个例子来说，马铃薯甲虫在野生种马铃薯（*Solanum berthaultii*）和栽培品种（*Solanum tuberosum*）二者中进行选择时，会倾向选择栽培品系。而当栽培的马铃薯嫩叶紧靠野生嫩叶时，两者都不被选择，这暗示着有化学物质从野生种马铃薯（*S. berthaultii*）毛状体流出，而当野生叶片去除毛状体后则能被甲虫接受（图7.1）[72,237]。当用丙酮冲洗野生种马铃薯（*S. berthaultii*）叶片后，将所得液体应用于叶蝶（*S. tuberosum*），对马铃薯甲虫仍具有很大的威慑作用。其中包含了几种活性化合物，但是它们的确切成分至今仍然未知。专食一种茄属（*Solanum*）植物的金花虫科甲虫（*Gratiana spadicea*），其跗爪为等长生长，相对于其他幼体异速生长的特征，可以看作对寄主植物上不同毛状体类型的形态适应[123]。

7.2.2 叶面结构

对于寻找产卵场所的雌性昆虫而言，叶表面形态结构可能具有重要作用。相比较平滑的人造叶面，小菜蛾（*Plutella xylostella*）更喜欢粗糙的人造叶面（图7.2），雌性主要沿着叶子的叶脉、小叶和茎腔产卵。甘蓝地种蝇(*Delia radicum*)在具垂直褶皱的人造叶片基部的产卵量是在具水平褶皱人造叶片上产卵量的2.5倍。另外,从对叶片的探索转接到对茎的探索行为更可能发生在垂直褶皱的叶片上(图7.3)[165]。甘蓝地种蝇(*D. radicum*)为一种寡食性昆虫,取食葱属(*Allium* spp.)植物,研究表明,人造植物的大小、形状和方向均对甘蓝地种蝇的选择产生影响。当在植物模型上喷洒寄主植物挥发物质后,人造植物基部的产卵量与之协同增加[80]。

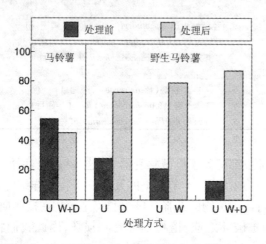

图7.1　马铃薯甲虫对经不同方式处理过的普通马铃薯与野生马铃薯的选择率

去除野生种马铃薯(*Solanum berthaultii*)的毛状体,通过浸置于(D,95% 乙醇),或擦刷(W,软毛刷)或浸置于擦刷结合(W+D)。在处理组与非处理组(U)中,马铃薯甲虫成虫对其叶片进行选择。从实验结果得出,除W+D组合对于马铃薯甲虫没有明显影响外,其余的三组去除毛状体后,相对于对照组,野生种马铃薯(*S. berthaultii*)显示出很强的倾向性(叶奈仇和廷吉,1994)[237]

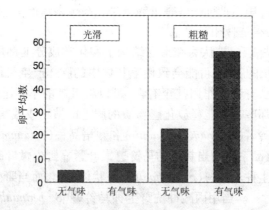

图7.2　触觉和嗅觉的结合因素对于小菜蛾(*Plutella xylostella*)产卵的影响

平滑或粗糙的塑料盖子作为产卵基质,使用或不使用10 ppm异硫氰酸烯丙酯作为味源物质(这种化合物主要为十字花科植物释放)进行实验对比。带有味道的粗糙基质被证明最具引诱性;粗糙基质比带有味道的光滑基质的引诱性更强(吉普塔和索尔斯坦森,1960)[75]

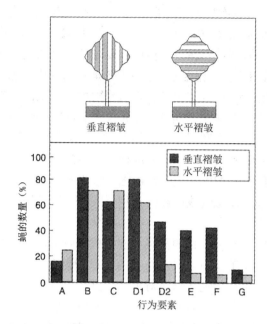

图 7.3 叶片纸模型(水平或垂直褶皱)对甘蓝地种蝇(*Delia radicum*)产卵行为的触觉影响

对每个行为要素 A—G,甘蓝地种蝇完成每个行为要素以百分比的形式展示出来:A,短期访问,对叶片无探索行为;B,休息,梳理;C,探索叶片表面;D1,沿着叶片边界或叶脉直线运动;D2,对茎的直接趋向运动;E,绕茎做水平环形运动,朝向地面;F,从茎到地面运动,探索沙地表面;G,试图产卵。在水平褶皱叶片上完成整个行为过程的甘蓝地种蝇比在垂直褶皱叶片上的少。这种差别可能与对叶子的探索到茎的探索间的转化有关(D1—D2)。相比较而言,能够完成 F 和 G 行为的雌性甘蓝地种蝇数量明显比雄性少(罗斯塞弗和斯塔德勒,1990)[165]

7.3 植物化合:接触化感作用评估

之前的章节表明,植物的物理特性在很大程度上可以影响昆虫的寄主选择行为。在此,我们仍要关注植食性昆虫食性高度专一的问题(第 2 章),然而显而易见的是,单纯对植物物理特性的行为反应,不足以对此专一化提供令人满意的解释。主要原因是这类物理和形态特征的模式目前仍然缺失,这与目前植物化学模式的状况形成鲜明的对比[93]。许多植物科的特征是产生其他科不能产生的次级代谢产物(第 4 章)。甚至科内,不同属间,其次级代谢产物也存在定性和定量的不同。这种植物的化学分类模式为植食性昆虫的寄主植物的专一性提供线索,并且现已被确定下来,而且对这种模式的认知利用程度已经很高。在本章的后续部分,我们将继续阐述这种模式。

7.4 植物化学对寄主植物选择的重要性:一段历史性的间奏

一直以来,大多数植食性昆虫多表现出寄主植物特异性机制,对生物学家而言,是一个挑战。约 200 年前,瑞士植物学家康多勒[41]提出,植物化学是昆虫选择寄主植物的决定因素。之后法布尔[60]提出了"植物感觉"(botanical sense)这个术语,指的是行为专化性的感觉基础[184]。

荷兰植物学家维夏菲尔特[231]首次揭示了昆虫的选择机制,他证明了芥子油苷是十字花科植物的主要化学分类特征,是欧洲粉蝶(*Pieris brassicae*)和菜粉蝶(*P. rapae*)幼虫接受植物的决定性因素[185]。而对这种行为的化感基础的发现时间,要比位于幼虫的下颚且对糖苷类物质特别敏感的味觉细胞的发现晚得多[176]。德蒂尔[47]证明了伞形科精油中的萜类化合物决定了黑凤蝶(*Papilio polyxenes*)幼虫(专食伞形科植物)对寄主植物的接受。弗伦克尔在一篇题为《次生植物物质存在的理由》的文章中收集的数据表明,专食性昆虫对于食物的选择仅基于次生代谢产物的存在与否,一些寡食性昆虫利用特定植物类群的次生代谢产物作为其识别刺激剂,而这些次生代谢产物对其他昆虫而言,却成了防御屏障[63]。德蒂尔使用了"刺激信号"(token stimuli)这一术语,来指代被食草昆虫用来作为寄主识别的植物次生性物质[48]。杰米更关注阻碍物的作用,即植物次生性物质抑制摄食或产卵的作用,他认为昆虫对寄主植物的选择主要基于避免非寄主植物的阻碍物[94,95,98]。为了平衡对植物次生化合物的各种观点,肯尼迪和布什提出,在对寄主植物的选择过程中,要将植物的初级和次级代谢物质的重要性结合在一起,即"双重区分"[103]。这些概念对我们了解昆虫对寄主植物的选择行为有很大的帮助。这些概念包括初生性和次生性植物化合物,也包括这些物质对植食性昆虫行为的刺激或抑制作用。

下面我们将要讨论昆虫利用化学物质寻找寄主植物的机制。在本章中,我们更关注被味觉感受器感知到的非挥发性(sapid)化合物。目前,对存在于取食地点或其附近的气味物质的作用的研究较少,但有迹象表明,即使在接触阶段,挥发性物质仍可能起到一定的作用。

许多植物化学物质往往只局限于细胞内或细胞外的封闭部位(见4.11)。而外部的封闭部位特指植物角质层,每次对寄主植物选择机制的讨论都会特别涉及。如前所述,存在于植物表面的化合物可能会先于细胞内容物(经损伤后才能释放)对选择行为产生影响,这是昆虫与生俱来的反应或是基于其先前经验的结果[36,106]。一些非极性的表皮化合物,如长链烷烃和酯类,大概只存在于表面上[59,93]。糖类、氨基酸和次生代谢产物,不论极性或者非极性的,也只存在于植物表面(表7.3)。在下文中,将要对已发现的表面产生的化合物引发的行为反应进行讨论。

7.5 取食和产卵的刺激

7.5.1 初级植物的代谢物

受光合作用的影响,所有植物初级代谢物都包含碳水化合物和氨基酸。大量证据表明,大部分植食性昆虫使用碳水化合物作为取食刺激剂(表7.2)。在大多数被研究的昆虫中,二糖蔗糖和单糖果糖、葡萄糖都是有效的刺激剂。这些糖类在绿叶中浓度很高(2%~10%干重,大约相当于10~50 mmol/l),在果实和花蜜中浓度更高(达0.25 mol/l)。作为取食刺激剂,它们对于刺激昆虫取食存在剂量依赖性(图7.4)。这些物质也是合成机体组织并且提供能量的重要营养物质(第5章)。

表 7.2　各种糖类对植食性昆虫刺激作用的比较(伯奈斯和辛普森 1982)[14]

	蝗虫类		甲虫类		毛虫类	
	东亚飞蝗 (*Locusta migratoria*)	沙漠蝗 (*Schistocerca gregaria*)	苜蓿叶象甲 (*Hypera postica*)	马铃薯甲虫 (*Leptinotarsa decemlineata*)	欧洲粉蝶 (*Pieris brassicae*)	灰翅夜蛾属 (*Spodoptera* spp.)
戊糖						
L-阿戊糖	+	·	—	—	—	—
L-鼠李糖	—	—	·	·	·	—
D-核糖	—	—	·	·	·	—
D-木糖	—	—	·	·	·	—
己糖						
D-果糖	+ + + + +	+ + + + +	+ + + +	+	·	+ + + + +
D-半乳糖	+ +	+	·	+	—	+ +
D-葡萄糖	+ + +	+ + + +	+	+	+ +	+ +
D-甘露糖	—	+	+ +	·	—	+
L-山梨糖	·	·	·	·	·	·
二糖						
D-纤维二糖	·	·	·	·	·	·
D-乳糖	—	—	·	·	·	+
D-麦芽糖	+ + + + +	+ + + +	+ +	—	—	+ + +
D-蜜二糖	+ + +	+ + +	·	—	—	+ +
D-蔗糖	+ + + + +	+ + + + +	+ + + + +	+ + + + +	+ + + + +	+ + + + +
D-海藻糖	+	+ +	·	—	—	·
三糖						
D-蜜三糖	+ + + +	+	+ + +	+ +	·	+ +
D-棉籽糖	+ + + +	+ + +	·	—	—	+ + +
醇类						
肌醇	+	·	·	—	—	—
山梨糖醇	+	+	·	—	—	—
甘露醇	+	+	·	—	·	—

注：+ + + + +，高刺激性；+，弱刺激性；—，无影响；·未测试。

虽然植物蛋白质是动物生长的必需成分，但是能够刺激植食性昆虫取食的蛋白质分子至今仍未被发现。然而，仅有很少的研究涉及此方面。最近从一只寄生蜂上发现了有趣的现象，寄主蛋白质可以作为利它素被其味觉感知[7]。尽管蛋白质并不能直接刺激摄食行为，但是它的组成部分——氨基酸，却充当了很多物种的进食刺激剂[14]。然而，甚至在亲缘关系密切的物种之间，20 种天然氨基酸的刺激作用在感觉水平上也会呈现出显著的不同[190,227]。一般而言，对于昆虫有 10 种必需氨基酸，但这些必需氨基酸并不一定比其他非必需氨基酸的刺激作用更强，也并非能刺激更多的物种。

在昆虫中都发现有糖和氨基酸的味觉感受器细胞的存在，而且由糖和氨基酸引发的化学感应的反应强度通常与它们的行为效果有一致的关系(潘祖托和艾伯特)[114,130,153]。参与植物主要代谢的其他物质，如糖醇[70]、磷脂和核苷酸、一些矿物质和维生素等，尽管没有深入研究，但是众所周知，这些物质均能影响食物的吸收[14,87]。

在植物不同部分的糖和氨基酸的浓度变化很大，这可能是昆虫在选择进食部位时的重要线索(第 4 章)。测试并量化糖和氨基酸作为摄食刺激物的作用，可通过基于琼脂的基质进行试验。这种方法可以对刺激的相对效果进行评价。而目前仅对少数物种使用了这种测试方法[87]。在无选择的情况下,在人工基质上以植物的标准蔗糖含量没有进一步增加人工基质的基础上,即可诱导昆虫产生最大摄食率。然而,对摄食率和植物组织的关联未进行研究,因此,以上结果不能直接说明在寄主选择行为中糖的作用。例如,寡食性和多食性的毛虫类,即使在人工培养基上培养了四个龄期,在双选条件下,依然首选植物组织。当完整的植株与人工基质中的某种化学成分进行比较时,会产生很多问题。

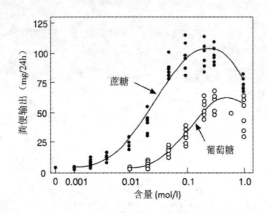

图 7.4 欧洲粉蝶(*Pieris brassicae*)幼虫对蔗糖和葡萄糖做出的行为反应

这两种糖混合于琼脂(琼脂、水、纤维素的混合物)中,纵坐标上的参数是在 24 小时内 6 个幼虫的排泄物的干重,这能够反映食物的摄入量。在血糖水平较低时,与葡萄糖相比,蔗糖是一种相当强的进食刺激物(马,1972)[144]

首先,由于二者在触感上存在显著差异,因此在技术上不能排除对这种差异的偏好。其次,在这类研究中,人工基质中通常只含有一种糖类物质和一到两个附加化合物,因此可能营养不足。当通过几个小时后排泄物重量或基质的消耗量来测得摄食率时,和植物组织进行摄食率的对比是存在问题的。在营养不良的情况下,摄食率可能受到血淋巴应对低营养水平而产生的积极的生理反馈的影响(见 5.3.3)。

研究表明,糖类物质能促进昆虫产卵,例如,多食性的欧洲玉米螟(*Ostrinia nubilalis*)[44,46]。与大多数产卵昆虫一样,雌蛾似乎不会损伤植物组织,并且产卵反应必须基于对叶表面糖含量的感知。除此之外,叶表面的亲酯性化合物(烷烃、酯类、脂肪酸)为代谢主要的产物,它们同时还能促进许多昆虫(如蚜虫和蝗虫)试咬或探测的过程,以及随后的取食和产卵过程(伯奈斯和查普曼[11],艾根博德[58],艾根博和艾利斯[59])。

植物表面的主要成分,特别是糖和氨基酸,会影响昆虫对寄主植物的选择(表 7.3),而在植物内部,这些物质的浓度会随植物发育阶段、年龄、生理状况和环境因素产生很大的变化,因此很难仅靠这些物质解释昆虫对寄主植物选择的特异性。因此,我们不得不考虑植物次生代谢化学物质的作用。

表7.3 能影响昆虫行为的叶表面化学物质

化学物质	植物物种	参考文献
果糖,葡萄糖,蔗糖	玉米,向日葵	45
氨基酸	蚕豆,甜菜	99
氨基酸	玉米,向日葵	45
脂类	卷心菜及其他的种类	59
半乳糖醇(糖醇)	卫矛	100
p-羟基苯甲醛	高粱	236
芸苔葡糖硫苷(硫代葡萄糖苷)	卷心菜	73, 228
各种芥子油苷	油菜	118
根皮苷(酚醛)	苹果	105
蒽醌(酚醛)	多年生黑麦草	2
木樨草素,反绿原酸(酚醛)	胡萝卜	61
法卡林二醇(聚乙炔)	胡萝卜	206
倍半萜	野生番茄	101
醋酸三萜	甘薯	149
α-和β-二醇,饱和烃	烟草	92
酪胺(生物碱),反绿原酸	防风草	33
柚苷,橘皮苷(黄酮),奎宁酸	柑橘	85
马兜铃酸	马铃薯	147
吡咯生物碱	千里光	232
各种醇	胡杨	110
α-托可醌	胡杨	110

7.5.2 植物次级代谢物促进接受:信号刺激

如第4章提到的,植物为植食性昆虫提供种类丰富的次生代谢物。通常,一类化学物质只存在于一个或几个相近的植物科中。然而,也有一些次生代谢物会广泛分布于多个植物科中,例如酚类和黄酮类物质。

在有些情况下,特定的次生代谢物可作为单食性或寡食性昆虫的摄食或产卵刺激剂,自维夏菲尔特提出以来[231],此类研究不断增加。表7.4列出了由植物次生物质控制的摄食或产卵活动的例子。在有些情况下,可以通过类比法发现活性化合物(对取食相同植物上的其他昆虫有刺激作用);在其他情况下,可以采取以生物测定的方法进行识别。对于产卵行为,经研究发现,一种或几种特定化合物对于产卵的刺激程度与寄主植物的完整提取物甚至是完整植株的刺激程度相似。这些物质作为"刺激信号"是非常好的例证,并且只限定在特定的植物类群中。昆虫对这些物质的化学感受能使它们进行精确地寄主识别。对此方面研究得较为深入的类群存在于鳞翅目、双翅目、鞘翅目以及十字花科、伞形科和葱科之中[203]。

表 7.4 寄主植物、刺激信号、化学物质类别及刺激信号受体的情况

昆虫	寄主植物	刺激信号	化学物质类别	参考文献	受体确定情况	参考文献
鳞翅目——取食						
粉蝶 (*Pieris* spp.)	芸苔 (*Brassica* spp.)	黑芥子苷	芥子油苷	231	是	176
家蚕 (*Bombyx mori*)	桑 (*Morus* spp.)	桑色素	类黄酮	178		
蛱蝶 (*Euphydryas chalcedona*)	车前 (*Plantago*)	梓醇	倍半萜烯	32		
小菜蛾 (*Plutella xylostella*)	芸苔 (*Brassica* spp.)	黑芥子苷 + 黄酮醇	芥子苷,类黄酮	230	是	230
辰砂飞蛾 (*Tyria jacobaeae*)	千里光 (*Senecio jacobaea*)	千里光菲啉及氧化物	吡啶生物碱	20	是	20
烟草天蛾 (*Manduca sexta*)	茄属 (*Solanum* spp.)	刺天茄甙 D	甾类糖苷	43	是	43
鳞翅目——产卵						
粉蝶 (*Pieris* spp.)	芸苔 (*Brassica* spp.)	芸苔葡糖硫苷	芥子苷	164, 227	是	57,209
黑凤蝶 (*Papilio polyxenes*)	胡萝卜 (*Daucus carota*)	木樨草素-糖苷	类黄酮	61	是	166
蓝闪荆凤蝶 (*Battus philenor*)	马兜铃 (*Aristolochia*)	马兜铃酸	环烯醚萜苷	175		
鹿眼蛱蝶 (*Junonia coenia*)	车前 (*Plantago*)	珊瑚木苷 + 梓醇	香豆素,倍半萜烯	154		
凤蝶 (*Eurytides marcellus*)	巴婆树 (*Asimina triloba*)	3-咖啡酰-粘液酰-奎尼酸	酚酸衍生物	79		
鞘翅目——取食						
黄曲条跳甲 (*Phyllotreta armoraciae*)	芸苔 (*Brassica* spp.)	黑芥子苷 + 黄酮苷	芥子苷,类黄酮	144		
榆蓝叶甲 (*Plagioderma versicolora*)	柳属 (*Salix* spp.)	水杨苷	酚醛树脂	122		
叶甲 (*Chrysolina brunsvicensis*)	金丝桃属 (*Hypericum*)	金丝桃素	醌	161	是	161
玉米根虫 (*Diabrotica* spp.)	南瓜属 (*Cucurbita* spp.)	葫芦素	类固醇(皂苷)	128	是	141
膜翅目——产卵						
叶蜂 (*Euura lasiolepis*)	柳属 (*Salix* spp.)	特里杨甙	酚苷	171		
双翅目——产卵						
甘蓝地种蝇 (*Delia radicum*)	芸苔 (*Brassica* spp.)	芸苔葡糖硫苷 + 'CIF'	芥子苷 吲哚衍生物	142 168	是 是	201 168
胡萝卜茎蝇 (*Psila rosae*)	胡萝卜 (*Daucus* spp.)	法卡林二醇 + 香柠檬烯,等	聚乙炔 呋喃并香豆素	202	是	207
呋喃香豆素葱蝇 (*Delia antiqua*)	葱属 (*Allium* spp.)	正丙基二硫醚	二硫化物	121	是	207
黑森瘿蚊 (*Mayetiola destructor*)	中国春小麦 (*Triticum aestivum*)	苯并恶唑啉酮(MBOA) 1-廿八烷醛	异羟肟酸 叶蜡醛	139		

续表

昆虫	寄主植物	刺激信号	化学物质类别	参考文献	受体确定情况	参考文献
同翅类——取食						
甘蓝蚜 (*Brevicoryne brassicae*)	芸苔 (*Brassica* spp.)	黑芥子苷	芥子苷	234		
苹果蚜 (*Aphis pomi*)	苹果属 (*Malus*)	根皮苷	查尔酮	137		
豌豆蚜 (*Acyrthosiphon spartii*)	金雀花属 (*Cytisus*)	金雀花碱	生物碱	199		
粗尾修尾蚜 (*Megoura crassicauda*)	蚕豆属 (*Vicia* spp.)	酰基化黄酮苷	类黄酮	213		

注:CIF——卷心菜识别因子;MBOA——6-甲氧基苯苯唑啉。

含有次级代谢物的刺激信号的作用随物质的不同而变化极大。例如,在两种菜粉蝶(*Pieris* spp.)中,当在人造叶片或两种菜粉蝶的非寄主叶片金甲豆(*Phaseolus lunatus*)上喷洒一种芥子苷后[164,228],能引起强烈的产卵行为。而其他的一些芥子油苷却不能引发产卵行为(图7.5)。在燕尾蝶(*Papilio* spp.)中,这种情况更为复杂,在一系列的化学混合物中,只有寄主植物的某些特异性物质能引发其完整的行为反应(图7.6)[61,86,146]。

表7.4表示,对于取食相同植物的不同寡食性昆虫而言,刺激信号可能各不相同。这方面的例证有:以胡萝卜为食的胡萝卜根蝇(*Psila rosae*)和黑凤蝶(*Papilio polyxenes*);以卷心菜为食的跳甲(*Phyllotreta armoraciae*)、菜蛾(*Plutella*)和粉蝶(*Pieris*)的幼虫;以葱科为食的葱谷蛾(*Acrolepiopsis assectella*)和葱蝇(*Delia antiqua*)。当特定的化合物被证实为刺激信号后,通常没有对其他物质进行深入的研究,尽管目前对于完整植株的整套行为反应仍未获得。举一个有趣的例子,卷心菜根蝇(*Delia radicum*),芥子油苷作为寄主植物主要的特异性物质,是卷心菜根蝇特异性的产卵刺激剂[167]。而在后续对叶片抽提物进行的生物测定研究中,发现了一种非芥子油苷物质是一种更为强烈的刺激剂CIF,它所引发的刺激强度比芥子油苷高100倍(图7.7)[90,169]。CIF可以通过跗节的感毛能刺激比对芥子油苷敏感的神经元之外的神经元[117]。

近40年来,对马铃薯甲虫(*Leptinotarsa decemlineata*)、烟草天蛾(*Manduca sexta*)与茄属植物的研究否定了茄属植物的次生代谢物质是昆虫接受寄主的刺激信号[97]。由此,一个可供选择的寄主识别机制被提出:寄主植物可以被接受,是因为缺乏了(至少是可感受到的部分)抑制取食的物质,而非寄主植物因为存在抑制取食的物质而不被接受[95,97]。由于高效液相色谱、核磁共振和质谱分析方法的使用,使得在刺激信号方面的研究有了很大的进展[43,140]。对于马铃薯甲虫,一种至今仍未鉴定出来的小甾体生物碱以及对于烟草天蛾,一种人参激素醣剌天茄苷D,被证实是其刺激信号。这两种化合物已被证实存在于马铃薯植物中。关于昆虫—植物关系中的刺激信号方面的研究,近年来一直在稳定增加(表7.4)。

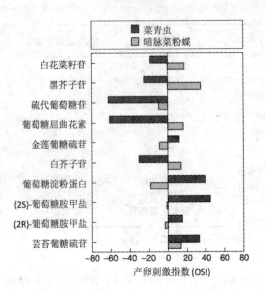

图7.5 不同物质作用下的菜青虫和暗脉菜粉蝶的产卵刺激指数的不同

在菜青虫(*Pieris rapae*)和暗脉菜粉蝶(*P. napi oleracea*)的非寄主植物青豆上喷洒不同芥子油苷(2 ml,0.1 mmol/l)后,对二者产卵的刺激作用产卵刺激指数(OSI)表示在双选实验中的偏好(与喷洒 2 ml,0.1 g 甘蓝提取物的青豆做对照);甘蓝提取物的主要芥子苷物质为芸苔葡糖硫苷(glucobrassicin);负值的 OSI 表示雌性昆虫倾向于选择喷洒甘蓝提取物的青豆。对于每个物种芥子油苷对于产卵刺激所起的作用都是不同的,每个物种都有其选择的倾向性

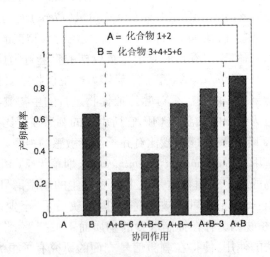

图7.6 雌性蓝凤蝶产卵概率

以蓝凤蝶(*Papilio protenor*)寄主植物蜜柑(*Citrus unshui*)提取物中的不同化合物组合,喷洒于滤纸碟上,观察雌性蓝凤蝶的产卵概率。供试化合物为:(1)柚皮苷,0.1%;(2)橙皮苷,0.05%;(3)脯氨酸,0.2%;(4)辛弗林,0.1%;(5)水苏碱,0.2%;(6)奎尼酸,0.2%。化合物1和2(A)的混合无效;A+B 可协同作用。删除化合物4(i.e. A+B-4)、5或者6导致了刺激反应显著减少(本田,1990)[84]

在源于植物表面的化合物中(表7.3),当味觉感受器接触完整植株表面时,它们的实际浓

度其实是未知的。基于植物化学提取物的浓度（假设全部提取）和植物表面化合物的数量，可以被表示为每单位表面积的微摩尔数。但是，其中能与味觉感受器接触的比例与浓度仍然未知。许多的昆虫被植物表皮的极性化合物刺激[59]。昆虫味感器也许能以某种未知机制将极性化合物从上表皮蜡质中释放出来，或者味感器能够通过气孔接触到叶片内部。对这些问题进行更深入的研究将是非常有意义的。

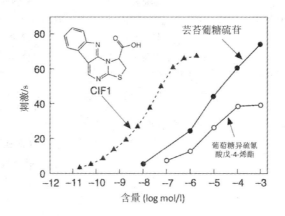

图7.7　不同物质的含量与刺激的关系

甘蓝地种蝇（*Delia radicum*）跗节的感毛与芥子油苷（芸苔糖硫苷和葡萄糖异硫氰酸戊-4-烯酯）接触后神经活动的反应曲线（与感器尖端接触后第一秒内的动作电位的数量）引起甘蓝地种蝇产卵反应最强烈的刺激的物质为CIF1，是发现于甘蓝叶表面抽提物中的化合物，比以上两种物质对于产卵的刺激作用更强，但是这种物质与以上两种物质分属不同的化学类别，能刺激对芥子油苷敏感的神经元之外的细胞（罗斯塞弗，1992b[168]；赫特，1999[90]）

基于溶剂的方法通常从植物表面进行抽提[111,207,232]。这种方法最近备受争议，因为用这种方法无法将存在于角质蜡层中的化合物抽提出来[162]。

7.5.3　通常将产生的植物次生代谢产物作为刺激剂

由无亲缘关系的、不同科的植物释放的次生代谢产物对昆虫的取食有刺激作用，这方面的发现在不断增加。尤其是对于酚酸和类黄酮类（表7.5）化合物。例如，咖啡酸和绿原酸都能刺激家蚕（*Bombyx mori*）取食桑科植物。而绿原酸也能够刺激马铃薯甲虫对茄属植物的取食[87]。家蚕和棉铃象甲（*Anthonomus grandis*）都能被槲皮素所刺激[178]。多食性昆虫也可能会受到它们食物中的类黄酮类物质的刺激。普遍存在的槲皮素芸香糖苷已被证明是南美沙漠蝗（*Schistocerca americana*）[15]和烟芽夜蛾（*Helicoverpa virescens*）幼虫的取食刺激剂[178]。考虑次生代谢化合物和营养化学物质的普遍存在，很难设想专食性昆虫对寄主植物的接受方式，因为这些化合物可以形成明确的信号。

表7.5 作为昆虫摄食刺激剂的不同黄酮类物质(哈本和格瑞,1994)

类黄酮	取食刺激剂	寄主植物	昆虫	参考文献
黄酮醇 O-糖苷	异槲皮苷,桑色素	桑树 (*Morus alba*) (桑科)	家蚕 (*Bombyx mori*) (鳞翅目)	77
	异槲皮苷	陆地棉 (*Gossypium hirsutum*) (锦葵科)	棉铃象甲 (*Anthonomus grandis*) (鞘翅目)	81
	山奈酚木糖基半乳糖苷	马萝卜 (*Armoracia rusticana*) (十字花科)	辣根跳甲 (*Phyllotreta armoraciae*) (鞘翅目)	144
	芸香苷	许多物种	南美沙漠蝗 (*Schistocerca Americana*) (直翅目)	16
	芸香苷	许多物种	玉米夜蛾 (*Helicoverpa zea*) (鳞翅目)	74
黄酮 O-糖苷	广寄生甙,金丝桃苷,芸香苷, 槲皮苷,异槲皮苷	荞麦 叶甲 (*Fagopyrum esculentum*) (蓼科)	叶甲 (*Galerucella vittaticollis*) (鞘翅目)	151
	7-α-L-鼠李糖-6- 甲氧基木樨草素	跳甲 (*Alternanthera phylloxeroide*) (苋科)	叶甲 (*Agasicles* sp.) (鞘翅目)	238
	木樨草素-7-葡糖苷	柳属 (*Salix*) (杨柳科)	叶甲(*Lochmea capreae*) (鞘翅目)	119
黄酮 C-糖苷	8个C-糖基化黄酮	稻 (*Oryza sativa*) (禾本科)	褐飞虱(*Nilaparvata lugens*) 白背飞虱(*Sogatella furcifera*) 灰飞虱(*Laodelphax striatellus*) (同翅类)	21
二氢黄酮醇和 二氢黄酮	花旗松素,二氢山奈酚, 生松素	李属 (*Prunus* spp.) (蔷薇科)	小蠹 (*Scolytus mediterraneus*) (鞘翅目)	109
二氢查尔酮 O-糖苷	根皮苷	苹果属 (*Malus* spp.) (蔷薇科)	苹果蚜(*Aphis pomi*) 苹草缢管蚜(*Rhopalosiphum insertum*) (同翅类)	105
黄烷醇 O-糖苷	儿茶素7-O-木糖苷	榆树(*Ulmus americanus*) (榆科)	欧洲大榆小蠹(*Scolytus multistriatus*) (鞘翅目)	55
类黄酮	异荭草甙,苜蓿素, 苜蓿素7-O-葡糖苷	苞茅(*Hyparrhenia hirta*) (禾本科)	飞蝗(*Locusta migratoria*) 沙漠蝗(*Schistocerca gregaria*) (直翅目)	31

7.6 摄食和产卵的抑制

弗兰克指出,对大多数的植食性昆虫而言,植物的次生性物质是抑制昆虫取食的防御性物质,但一些特殊的昆虫除外,这些昆虫利用植物的次生性物质作为其接受寄主的刺激信号[63]。目前,对于寄主植物防御昆虫的机制尚缺乏系统的研究。杰米表明,昆虫对非寄主植物的排斥是由于取食抑制剂的存在(拒食剂)[94,95]。有一种"三明治"法实验,是将测试植物夹在寄主植物的两片叶叠间。这种方法可以避免由于摄食刺激剂的缺乏或对非寄主植物的低偏好性造成的排斥反应。一项关于两种蝗虫,即专性取食禾本科植物的东亚飞蝗(*Locusta migratoria*)和多食性的沙漠蝗(*Schistocerca gregaria*)的研究取得了相同的结论。对植物的接受,是判定该植物为寄主植物的一个标准,而取食量是另一个标准,由此,使得对植物接受等级更细微的辨别得以实现。以人造基质上的物质的被取食量来衡量昆虫对植物的接收程度时,东亚飞蝗(*Locusta migratoria*)会将禾本科食物完全取食,而只取食少部分非寄主植物。所有的非寄主植物都包含抑制剂,就像禾本科植物只接受少量的昆虫物种取食。而沙漠蝗对不同植物的取食表现了巨大的取食量差异,然而,即使取食少量的植物,仍然有抑制剂的存在[10]。

7.6.1 抑制是寄主识别过程中的一个普遍原则

目前,对许多植食性昆虫的研究揭示了一些对于摄食抑制剂反应的普遍原则。第一,非寄主通常含有抑制剂。第二,通常单食性和寡食性昆虫对非寄主植物抑制剂比多食性昆虫敏感(表格7.6)。这已在蝗虫[10,11]和几种毛虫上都得到证明[28]。第三,抑制剂不仅在非寄主中存在,在一些可接受昆虫的植物中也存在。在这些例子中,抑制剂的作用因为刺激剂的同时存在而呈现出明显的中和性[38,88,95]。一些单食性和寡食性昆虫的寄主植物刺激剂已被鉴定出来。由于缺乏刺激及存在可能的抑制作用,这为非寄主植物对昆虫的排斥提供了一种解释。通过对非寄主植物注入或涂抹刺激剂,一些非寄主植物也能显示出被接受性,并且其可能存在的(弱性的)抑制剂会被掩盖[115,156,231]。

在植物次级代谢物抑制昆虫摄食方面有大量的文献[138]。这方面资料的积累促进了植物源化合物的鉴定工作,也为农作物防虫方面提供了依据[65]。而相对于取食抑制,在产卵抑制剂方面进行的研究相对较少[163]。基于目前得到的资料表明,在识别食物方面,抑制是许多昆虫选择寄主植物时的一个重要机制。

7.6.2 寄主标记是避免植食性昆虫相互竞争的机制

对于一个寻找产卵地的受孕雌性昆虫,当其着陆后,不仅受植物外部的化学组成影响,并且还要受先前到访昆虫留下的化合物影响。一些雌性蝶类、甲虫、蝇类,伴随其产卵而产生的一些物质,含有抑制同种雌性的产卵并且抑制后来到访的雌性产卵的作用[83,183]。这些物质被称为"寄主标记激素"或"疏散信息素"。从这种激素化合物的化学结构的少数例子中得知,它们的化学结构变化很大。

对寄主植物进行标记是一种非常普遍的现象,例如在许多果蝇中,如雌性樱桃绕实绳(*Rhagoletis cerasi*),将卵产入樱桃的表皮下,之后在樱桃表面拖拽其产卵器,此时,一些标记物质就被涂抹到樱桃表面。而其他雌性,在落到此樱桃表面后,会用跗节上的化感器察觉到这些物质。用人工合成的天然产物类似物进行化感实验,结果显示,在标记化合物的识别过程中存在与其他受体不同的结构[208],由此表明,这种标记信息素会刺激一种特殊的受体。

表 7.6 一些植物次生性化合物的主要类别对一种寡食性(O)的鳞翅目昆虫和一种杂食性(P)的鳞翅目或同翅类昆虫的抑制作用

化合物	类别	昆虫	昆虫食性	有效浓度(ppm)	抑制率(%)	参考文献
黑芥子苷	芥子苷	黑凤蝶(*Papilio polyxenes*)	O	900	66	29
		蓓带夜蛾(*Mamestra configurata*)	P	3 100	50	192
亚麻苦苷	氰苷	夜蛾 *Heliothis subflexa*	O	1 235	40	18
		烟夜蛾(*Heliothis virescen*)	P	12 350	40	18
绿原酸	酚醛酸	夜蛾 *Heliothis subflexa*	O	3 540	45	18
		烟夜蛾(*Heliothis virescen*)	P	35 400	50	18
根皮苷	类黄酮	麦二叉蚜 *Schizaphis graminum*	O	200	50	56
		桃蚜(*Mmyzus persicae*)	P	4 360	100	187
士的宁	生物碱	欧洲粉蝶(*Pieris brassicae*)	O	30	100	114
		甘蓝夜蛾(*Mamestra brassicae*)	P	3 900	75	30
咖啡因	生物碱	美洲夜蛾(*Heliothis subflexa*)	O	≤0.2	30	18
		烟夜蛾(*Heliothis virescen*)	P	1	20	18
散白草素	双萜	莎草黏虫(*Spodoptera exempta*)	O	100	Thr	107
		灰翅夜蛾(*Spodoptera littoralis*)	P	300	Thr	107
印楝素	三萜	欧洲粉蝶(*Pieris brassicae*)	O	7	50	112
		草地贪夜蛾(*Spodoptera frugiperda*)	P	315	50	159

在两种菜粉蝶,即欧洲粉蝶(*Pieris brassicae*)和菜粉蝶(*P. rapae*)的例子中,发现无论是种内还是种间卵液,都能对雌性产卵行为产生强烈的抑制作用。这表明标记化学物质引起了其他个体的回避行为[174]。研究发现从卵液中分离出的一些燕麦蒽酰胺生物碱具有这种作用。这类化合物只在粉蝶(*Pieris*)的卵中发现,而粉蝶科的另外两种昆虫的卵中则不存在,在其他5种非粉蝶科的鳞翅目昆虫的卵中也不存在[22]。研究发现,粉蝶(*Pieris*)在叶片的背光面产卵后并无产卵器拖拽行为。而这些带有卵块的叶片不会再接受其他雌性产卵,即使落地后叶面朝上。由此开展了对假定标记物质的移动方面的调查。然而,进一步的研究未能揭示卵液中有效成分的迁移。有趣的是,曾经带有卵的叶表面提取物能阻止产卵,但是却不包括卵中所含的生物碱[23]。樱桃绕实绳的标记物仅由昆虫产生,与之相反,在粉蝶(*Pieris*)的例子中植物也会起到一定作用。显然,植物表面与粉蝶(*Pieris*)的卵接触后,诱导发生了化学变化,一些至今仍未知的物质产生了,该物质对即将产卵的雌性具强大的抑制作用。在其他一些昆虫—植物的例子中,也发现了植食性昆虫卵诱导寄主植物产生的化学反应,继而可以对卵寄生蜂产生影响[82]。

最近有一些综述对阻碍产卵信息素的行为和化学生态学进行了详细的讨论[3,150]。

7.7 植物可接受性:刺激和抑制之间的平衡

植物的化学物质(主要的或者次要的)在昆虫选择寄主植物方面所起的刺激或抑制性作用相互平衡,并且这些作用最终的平衡将决定决策的结果:抑制或以不同程度接受,表现为选择时的偏好[11,51,129]。若从寄主植物专化性的不同类别考虑,在理解选择行为时,这种"平衡模式"是一个非常有用的概念。

在杂食性昆虫中,许多植物中普遍存在的一些主要代谢产物足以刺激昆虫摄食,而只有产生大量或高效抑制剂的植物,才会导致摄食刺激受到抑制,因此植物会被拒绝取食。而在寡食性昆虫中,对于缺乏特定刺激信号的寄主植物也以类似的原则进行寄主识别(例如7.5.2中所讨论到的)。此外,有一类寡食性和单食性的昆虫必须感受到刺激信号才会接受寄主植物(表7.4),对于这种类别,刺激信号通常是某种分类单元特定的次生代谢产物,由专一的味觉受体感知(见7.8.4)。

近年来有种观点认为,出现不同类别的寄主选择机制,其实是一个程度问题,而不明确定义或明确分类。在上述最后的例子中,昆虫与特定植物类群结合,其感官形成了专性识别,是所有类别中最重要的、明确的信号。而起抑制和刺激作用的化学物质通常是不平衡的。换而言之,对摄食的抑制只在一定程度上能与摄食刺激平衡。抑制作用到一定程度后,没有刺激剂能引起摄食反应。这已经通过"三明治"实验得以证实,绝大多数非寄主植物具有的拒食作用不会因为寄主植物叶的存在而发生中和。

7.8 基于接触化学感受器的寄主植物的选择行为

7.8.1 化学感受器

上述对植物物质做出的行为反应主要是基于味觉神经元的检测。与嗅觉细胞一样,味觉细胞的细胞体位于角质层以下,并将树突伸到毛状、锥形或乳突状、尖端有孔的感受器内(图6.14)。味觉感受器主要位于口前腔(例如内唇感受器)和口器、跗节、产卵管与触角上(图7.8),当昆虫与植物进行接触时,这些感受器可以与植物表面或植物细胞内容物进行简短而间歇性的接触。接触化感器的数量在不同物种间以及同一物种的不同阶段间存在显著差异;尤其在完全变态的类群中,幼虫的化感器数量比成虫少很多[34]。在蝗虫中,存在着这样的趋势,食性越专化的类群,其感受器的数量越少[37]。而一些单食性类群中,其寄主植物对其他昆虫有高度抑制性,这些单食性类群的感受器数量最少[24]。在通常情况下,三到五个味觉神经元通常与一个味觉感受器相连,大多数感受器还包含一个机械感受神经元(图6.14)。

7.8.2 味觉编码

昆虫的味觉感受器与嗅觉感受器相似(第6章),通过将植物化合物质的信息转导为动作电位(或"峰值"),以电信号来传递神经信息。每单位时间内动作电位的数量和峰电位序列的及时细节(temporal details)(如峰值间隔的分布)所包含的信息将以编码的形式(不通过间歇突触)传递到第一个中转站(the first relay station),位于食管下或局部体神经节,再传递到中枢神经系统[104,134,170]。食管下神经节内包含有支配下颚肌肉的运动神经元,最终将控制摄食活动[25]。如植物汁液等的复合性刺激往往能同时刺激相同或不同的感受器细胞,并且它们的轴突在成对神经节(segmental ganglia)上集合。通过从外周感受器,如机械力受体或内受器传入的信息或者从大脑其他部位输入的信息,在这里发生整合。在整合过程发生后(这个过程可能只需要几分之一秒的时间),摄食可能发生,也可能不会发生。复杂的是,传达到大脑的感觉信息并不是固定的,而是随着年龄、当日的时间、生理状态和其他生物与非生物因素的变化而变化[27]。相对于嗅觉信息(第6章),我们关于接触—化学感受信息的处理过程了解甚少,尽管它在对寄主植物选择中起主导作用[170]。嗅觉信息通过触角和口器上的受体传递到神经小球(位于第二脑节的神经纤维网),而从分布更为广泛的味觉受体得到的信息似乎并未聚合于中枢神经系统的特定区域。

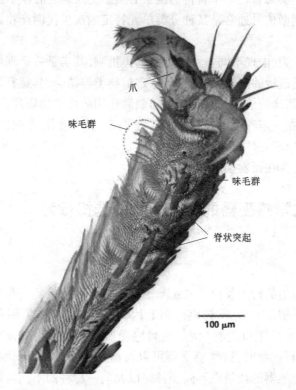

图7.8 雌性菜粉蝶(*Pieris rapae*)前足跗节末端腹面的扫描电镜显微照片

通过分析,所谓的"输入输出"关系是一种破译感觉编码的方法:输入(动作电位系列)是一种通过刺激味觉感受器发生的生理学现象,而输出是在基于食物消耗的绝对数量或对于不同基质的取食或产卵偏好程度的量化。以输入和输出关系为基础,推导出了编码原则。在这类研究中,感受器并非被细胞识别,而通常被认为是神经生理学的反应单位。目前有一个分析方法的基本原理为:由标准的尖端记录方法获得细胞外记录数据,需分离很多的味觉神经元产生细胞外记录峰电位,这一过程仍然存在技术上的困难。第二个原因是,仅在少数情况下,能对支配感觉器的特异性神经元进行详细的分析,例如识别"最好的糖""最好的盐"或者"水"的神经元[50]。事实上,对这种特异性的研究非常艰巨,目前有一些针对毛虫上颚味毛(含8个味觉细胞)的味觉系统的初步研究[190],并对粉蝶(*Pieris*)[57,209]和葱蝇(*Delia*)[167,196]的跗节感器进行了研究。目前得到的大多数数据是关于毛虫的,并且这些研究显示,即使在亲缘关系密切的物种之间,味觉的特异性也存在很大的差异[190]。理论上,没有必要为推导味觉编码在细节方面了解这些特异性[53]。这个概念对于最经常讨论的两种化学感受编码模式进行了定义:单线标记和交叉纤维模式,下面将要分别讨论。

7.8.3 毛虫成为编码原则的模型

许多毛虫都是非常专一的取食者,因此,毛虫类已成为感觉编码和行为研究最有利的模型。这是因为经切除实验证实,许多物种仅需上颚的两根毛(每根含有四个味觉细胞)即可完成完整的寄主植物识别行为过程(图 7.9)。这种八细胞味觉神经元,约占化学感受器总数的 10%。关于化学感受编码的主要问题之一,是在选择行为时采取的极端决定(接受或拒绝)的编码之间是否存在明显的差异。德蒂尔对 7 种毛虫(同属但亲缘关系较远)的研究得出的结论为:对于接受或拒绝行为,在感官模式间并无显著的差异[49]。这表明神经系统是通过几个味觉神经元以同步交叉阅读模式进行输入,并由此输出,继而影响了行为决定。之后基于这个想法,在脊椎动物中正式形成了"交叉纤维"味觉编码模式[51]。在一项早期的研究中,已对 7 种昆虫上颚味觉神经元的敏感范围进行了研究,但几乎没有证据能证明存在专一的味觉神经元[54]。

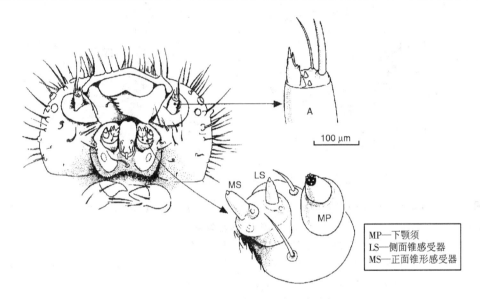

图 7.9 从下方观察毛虫的头部以及放大的触角(A)和下颚

在寡食性的烟草天蛾(*Manduca sexta*)与多食性夜蛾(*Spodoptera*)和铃夜蛾(*Helicoverpa*)幼虫中,位于上颚锥形感受器侧面和中间的纤维比率与可接受性相关[188,194]。在烟草天蛾(*Manduca sexta*)中,可能性最大的编码机制为交叉纤维模式[53,188],而目前还没有关于其味觉细胞特异性的详细信息。明显地,是由两个上颚锥形感受器的联合输入(形成交叉纤维模式)决定这些毛虫对寄主植物非常微妙的偏好行为[182,194]。一项关于烟草天蛾(*Manduca sexta*)对三个茄属植物的偏好编码研究表明,时间模式为叠加于交叉纤维模式之上的第三种编码模式。由于味觉细胞的适应速率不同,不同细胞的交叉纤维适应速率随时间发生改变,因此,把行为反应和相关的感觉反应时间域联系起来是非常重要的[181]。

大部分对毛虫的化学感应生理及行为的研究主要集中于上颚的八细胞味觉神经元。其他的味觉器官位于口前腔内。许多毛虫类在上唇内表面有两个膜状感受器,每一个感受器都有三个化学受体神经元,在吞咽反应中会涉及这些感受器[182]。马铃薯甲虫的成虫和幼虫也具有这种内唇感受器[127,131],而蝗虫在内唇表面和咽下部有好几组[34]。

最近的研究显示从内唇、触角及上颚须感受器的输入也有助于昆虫对植物食物的辨别[40,69,230]。显然，这些器官应该受到更多的关注。

成虫的感受器和味觉神经元数量比幼虫多得多[34]，尤其在鳞翅目和鞘翅目昆虫中，成虫与幼虫间的差别至少为10倍。这些增加的受体数量有可能涉及成虫更为复杂的行为任务。对于幼虫而言，仅需要取食到适合的食物，而对于分散的成虫而言，它们除找到食物之外还要找到交配对象，雌虫还需要找到产卵场所。尽管目前在处理大量受体方面的技术仍不完善，但对甲虫[76,136]和蛾类成虫的食物偏好编码已经成功解析[26]。通过记录马铃薯甲虫外颚叶感受器的反应得出的结论为三个寄主植物的汁液比非寄主植物对味觉神经元引起的反应更为持续。对于不同茄属植物的选择可能主要基于交叉纤维的编码模式，但也有使用单线标记编码的迹象（图7.10）。

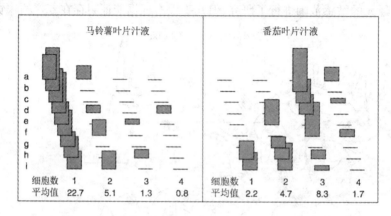

图7.10　9个马铃薯甲虫(a-i)对马铃薯(*Solanum tuberosum*)和番茄(*Lycopersicon esculentum*)叶片汁液反应的交叉纤维模式

成虫外颚叶味觉感受器的4个细胞的活动水平以条形表示(9个个体的平均值在底部表示)。对马铃薯和番茄反应的主要区别为细胞1的活性低或缺失，而2和3对番茄汁液具有较高的细胞活性，由此为昆虫对两种植物的分辨提供了基础(施伯林和米切尔,1991)[76]

7.8.4　信号刺激受体：卓越的专家

在研究昆虫对寄主植物化学感受专化性时，有一个非常重要事件，即发现欧洲粉蝶(*Pieris brassicae*)的味觉神经元对十字花科植物的次生性物质具有高度敏感性[176]。分别位于左右上颚外颚叶锥形感受器中的细胞，能对一定浓度的十字花科特有物质——芥子油甙产生反应。而这两类细胞在光谱上有重叠，但是并不完全相同。在这些细胞内，需要最低水平的代谢活动来标记对植物材料的接收情况。这种化学感受器细胞可以被定义为"标记线"，即将信息传输到大脑，而这些信息的数量与行为反应强度相关。然而，这些"标记线"型受体对刺激信号的影响可以被中和或被抑制，如生物碱或酚酸，是所谓的受体抑制剂[186,224]。图7.11为一种简单的刺激模型，但目前还不清楚这个模型是否对天然的(复杂的)刺激适用，例如植物汁液[170]。

至此之后，有更多的例子表明，味觉神经元对植物次生代谢物非常敏感。而这类化学感受细胞似乎是植食昆虫特有的，因为在其他动物类群中从未见过记载，如脊椎动物，对其味觉系统已经进行了广泛深入的研究。而这与植食昆虫对寄主植物的专化性一样，并不都适用于其他植食性动物(包括脊椎动物)。在一些单食性或寡食性昆虫中，刺激信号是通过将植物化学与行为调查结合而发现的,经电生理分析揭示了刺激受体神经元的存在。对这些细胞进行刺激将给大脑传递一个信号:接受这种食物或接受这个产卵场所。这是到目前为止,对所有已记录的,这种专化的细胞检测刺激化学物质的情况。在多食性(*Estigmene acrea*)幼虫中也发现了这种情况,它对吡咯生物碱表现出极端的敏感性,吡咯生物碱可以用于其防卫和信息素的产生[19]。目前已经发现了一种专化的抑制性神经元(图7.11)[226]。

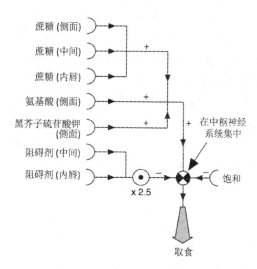

图7.11 源自口器不同部位化学感受器的输入信号在中枢神经系统(CNS)整合并调节欧洲粉蝶(*Pieris brassicae*)幼虫的摄食行为方式

对蔗糖、氨基酸和芥子苷产生脉冲的感应细胞位于内唇的(epi)及外颚叶侧面(lat)和中间(med)的锥形感器上,这种脉冲将对刺激昆虫取食产生积极影响(+),而源自抑制细胞的输入信号则会对取食产生抑制影响(-)。饱感是一项生理参数,会在昆虫饱腹的时候抑制取食;"取食"或"不取食"取决于积极和消极间的输入比例(如神经脉冲频率)(斯洪霍芬,1987)[182]

应该指出的是,交叉纤维模式和线性标记模式并不互相排斥,这两种模式可以合并到交叉纤维模式中(即许多细胞,每个都有不同但又重叠的光谱范围)参与复杂编码的刺激(例如植物汁液)。然而,一些光谱范围较窄且灵敏度良好的细胞(线性标记细胞)可能会起到更主要的影响,甚至在行为决策方面起到决定性作用。同样,抑制细胞也可能在决策过程中发挥主导作用。而即使存在一个或多个主要信息渠道,也不能排除其他味觉神经元的功能,后者对植物化学提供更精细的感觉评价,将有助于最终的决策。

7.8.5 糖和氨基酸的受体:养分的探测器

在7.5.1中,我们讨论了主要代谢产物作为摄食刺激剂的重要意义。在毛虫类中,一些味觉神经元对植物的主要代谢产物(如糖)敏感,这些味觉神经元将会被特定的摄食刺激激活,如它们可以被糖类物质激活,但不能被氨基酸或植物次级代谢物激活[190]。在粉蝶(*Pieris*)幼虫的上颚锥形感受器的八个味觉神经元中,两个是光谱敏感范围有重叠但不相同的"感糖"细胞[114]。刺激这些细胞对于诱导摄食率充分实现是非常必要的。在各种昆虫类群中,也发现了对氨基酸敏感的味觉神经元(表7.7)。

有些时候,对糖和氨基酸的感知过程会通过相同的细胞产生。众所周知,在马铃薯甲虫成虫中，一个对糖敏感的上颚神经元也对两种氨基酸，即 γ-氨基丁酸（GABA）和丙氨酸敏感[133]。此外，红萝卜甲虫（*Entomoscelis americana*）幼虫的感糖细胞，对一些糖类（如蔗糖、麦芽糖）和一些氨基酸有反应[132]。然而，奇怪的是，在成虫中这种感糖细胞似乎对氨基酸不起反应[212]。显然，不同的物种其味觉神经元的光谱敏感范围不同，甚至在同一物种的不同发展阶段也可能会有所不同。经深入研究，许多腐食性双翅目成虫的感糖细胞位于其喙上。虽然目前已普遍认同单独受体结合位点的存在，但这些感糖细胞通常对糖和氨基酸都具敏感性[141]。而在许多（但并非全部）鳞翅目昆虫中，往往通过单独的细胞来接受糖和氨基酸的信息[190,227]。

另一类能与一般化合物发生反应的细胞为"肌醇细胞"。很多毛虫类具有专化的糖醇受体细胞，糖醇为一类取食刺激剂，如肌醇[190]。而令人不解的是，为什么在大多数供试毛虫的八个上颚化学感受器神经元中，有一个甚至两个经常是专化的肌醇感神经元，这似乎是数量比例相对较高的神经元。肌醇也许是一个植物一般性质的指标，如年龄和/或蛋白质含量[143]。在巢蛾（*Yponomeuta*）中已发现不同的味觉神经元。两个立体异构糖醇，己六醇和山梨醇，它们是强大的幼虫摄食刺激剂。巢蛾（*Yponomeuta*）为卫矛科植物专食性类群，高浓度糖醇通常存在于卫矛科植物中[155]，通过将蔷薇科（非寄主植物）李属（*Prunus*）叶片浸置于己六醇中后，巢蛾可接受李属植物[155]。

7.8.6 抑制受体：全面感受神经元

在许多毛虫类中，已经发现存在一个或多个味觉神经元，能对非寄主植物的次生植物物质做出反应。这些神经元细胞被称为"抑制受体"。将可以完全被昆虫接受的寄主植物用这些次生化合物处理后，会导致受体细胞激活产生抑制力，最终导致对这种植物的排斥反应[191]。这类味觉神经元能对多种类别的次级植物化合物有反应，因此它们可以被看作全面的味觉神经元。当然，术语"全面"并不意味着它们能应对一切物质（如糖）或所有的次级植物化合物。对于这种细胞类型，不同的幼虫显示不同的敏感性[181,190]。目前仍不太清楚的是，抑制细胞如何表达这种敏感性，但是基于电生理学和基因组的研究，有证据表明这涉及不同受体位点的调整[71]。

表7.7　12种鳞翅目幼虫及甲虫（一种幼虫和一种成虫）的氨基酸感受器的光谱敏感范围

	毛虫类												甲虫	
	P.b.	P.r.	H.z.	E.a.	M.a.	D.p.	P.p.	L.d.	C.e.	A.o.	C.f.+	G.g.	L.d.	E.a.
	L	L	L	M	M	L	L	L	L	L	L	L/M		
参考文献	177, 227	54, 227	54	54	54	54	54	54	54	179	153	13	135	132
精氨酸*	o	o	o	o	+	o	o	o	++	o	+	+	+	+
组氨酸*	+++	+	o	o	o	+	o	o	+	+	+++	+		+
异亮氨酸*	++	+++	++	o	o	o	o	o	+++	++	++	+		
亮氨酸*													++	
赖氨酸*	o	o	o	o	o	o	o	o	o	o	o	o		
蛋氨酸*	++	++	++	o	o	o	o	o	+	+	+	+	+	+
苯基丙氨酸*	+++	+	o	o	o	o	o	o	o	+	+	+		
苏氨酸*	++	o	o	+	+	++	++	o	o	+	+	+		
色氨酸*	++	+	o	+	o	o	o	o	+	+	+	+		
缬胺酸*	++	++	++	−	o	++	++	o	+	+	+	+		

续表

	毛虫类												甲虫	
	P.b.	P.r.	H.z.	E.a.	M.a.	D.p.	P.p.	L.d.	C.e.	A.o.	C.f.+	G.g.	L.d.	E.a.
	L	L	L	M	M	L	L	L	L	L	L	L/M		
丙氨酸	++	++	o	+++	++	++	o	o	-		+	+	+++	++
天门冬酰胺	++	++									+++			
天冬氨酸	o	o	o	o	+	o	o	o	++		+	+		++
半胱氨酸	+	o		++	o		o	-	o	++				
胱氨酸			++	o		o	o	o						
伽马氨基丁酸	++	++									+		+++	+
谷氨酸	o	o	o	++	+	o	o	o	o		++	+		++
甘氨酸	+	o	o	++	+	o	o	+	o		+		+	++
脯氨酸	++	++	o	++	++	+++	o	++	o		+		++	++
丝氨酸	++	++	+	+++	+	o	o	o	+	o	+		++	++
络氨酸	o	o	+	+	o	o	o	o		+	o			

注:+++,强烈的反应;++,中度反应;+轻微反应;0,无反应;-与对照相比的抑制;L/M,侧面的/中间的锥形感器。P.b.,欧洲粉蝶(*Pieris brassicae*);P.r.,菜粉蝶(*Pieris rapae*);H.z.,玉米夜蛾(*Helicoverpa zea*);E.a.,毛毛虫(*Ecrisia acrea*);M.a.,东部天幕毛虫(*Malacosoma americana*);D.p.,君主斑蝶(*Danaus plexippus*);P.p.,黑凤蝶(*Papilio polyxenes*);L.d.,舞毒蛾(*Lymantria dispar*);C.e.,卷叶螟(*Calpodes ethlius*);A.o.卷叶蛾(*Adoxophyes orana*);C.f.,云杉卷叶蛾(*Choristoneura fumiferana*);G.g.,毛熊灯蛾(*Grammia geneura*);L.d.马铃薯叶甲(*Leptinotarsa decemlineata*);E.a.,油菜叶甲(*Entomoscelis Americana*);*,必需氨基酸;+,在不同的浓度下,检验不同的化合物。

在欧洲粉蝶(*Pieris brassicae*)和菜粉蝶(*P. rapae*)幼虫的上颚味毛中,都有通用和专化的抑制细胞存在[224]。位于感受器侧面的专化性细胞(图7.9)是一个"感强心甾"受体,因为它对强心甾极为敏感(阈值大约为 10^{-8} mol/l)。强心甾存在于寄主植物特定成员(十字花科)中,这些化合物是甾族类物质的强烈抑制剂,能使同科的植物不被接受。同样的抑制细胞也对酚酸和黄酮类有反应,但是仅在浓度为1000多倍的情况下发生。在其他毛状感受器中存在着普适抑制神经元,也会受强心甾刺激,但仅在浓度超过10倍时[226]。目前唯一的关于专化性抑制细胞的例子是仅对强心甾敏感的细胞。可以假设,专化性抑制细胞是由普适细胞通过丢失对其他抑制物(如生物碱)的感受位点而演变来的(图7.12)。

最近的一些研究表明,毛虫类所谓的抑制神经元是以"单线标记"的形式发挥作用的:与未经处理的对照组相比,将可接受的食物外面涂有特定的抑制化合物,导致昆虫拒食,与几个毛虫类抑制感受器的激活速率相符(图7.13)[126,155,197]。

我们试图以化学感受器神经元刺激范围为基础来解释昆虫对食物的选择行为,会对理解昆虫是否决定在特定的植物上取食有所帮助。然而,也有学者指出,应该根据活动的行为效果对味觉神经元进行分类,而非根据引发活动的化学物质类型进行分类[13]。按照这种观点,刺激取食和抑制神经元都应被认为是味觉系统的基本标记线。

7.8.7 外围化学物质的相互作用

从以上关于刺激和抑制受体的讨论围绕一个模式,即摄食刺激和摄食抑制剂的相关信息是由独立的化学感受器神经元检测,并将信息分别传送到大脑,随后输入信息会根据数学规则进行权衡。相对简单的数学规则可以从人工饲料喂养的粉蝶(*Pieris*)和夜蛾(*Mamestra*)幼虫

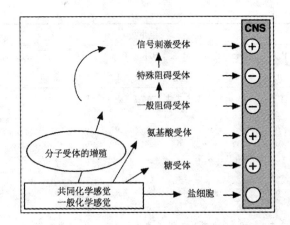

图7.12 假想的专食性植食昆虫味觉感受器类型进化途径

中枢神经系统(CNS)中圆圈含有加号或减号代表了与一阶中间神经元的兴奋或抑制传导(斯洪霍芬和范隆,2002)[190]

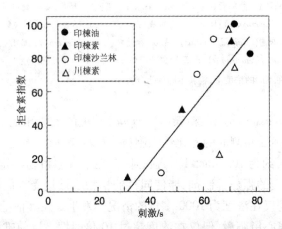

图7.13 欧洲粉蝶(*Pieris brassicae*)幼虫的锥形感受器的抑制受体细胞的峰值频率与拒食指数(由双选实验决定)之间的关系

根据相同化合物的等摩尔浓度下,对不同浓度的印楝制剂(Margosan-O)、印楝素、印楝制剂(salannin)以及川楝素做出反应的脉冲频率与拒食指数,发现抑制细胞反应的强度和拒食指数之间存在一个显著的相关性

中推导得出[186]。对于其他毛虫类、甲虫、蚱蜢的电生理研究,揭示了外围化学物质的相互作用,而这是不符合线性算法的:感受器将一个或几个味觉神经元的反应进行混合,而这种混合并非是对单个成分反应的简单相加(图7.14)。对于感糖味觉神经元的抑制化合物的作用已进行了较深入的研究[35,64,191],在不同的物种中,化合物间的相互作用结果不同[195]。一个例子为花青素对粉蝶(*Pieris*)感糖细胞的影响。黄酮类化合物不仅能刺激味毛侧面和中间的抑制细胞,而且能抑制这两个感受器上的蔗糖敏感细胞(图7.15)。当刺激剂与抑制受体反应时,会产生相反的效果[193]。

在目前讨论的例子中,在感官水平的相互作用并非全是抑制的,也可能是协同作用的。例如,在具带灯蛾(*Isia isabella*)幼虫的感黑芥子苷细胞中由于蔗糖的存在而敏感性加强,而蔗糖单独存在时只能刺激感糖细胞(图7.16)[54]。

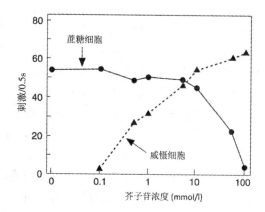

图 7.14 实夜蛾(*Heliothis subflexa*)幼虫在受到 5 mmol/l 蔗糖与不同浓度的黑介子苷酸混合液刺激时，外侧的锥形感器上的感糖和抑制细胞的脉冲频率

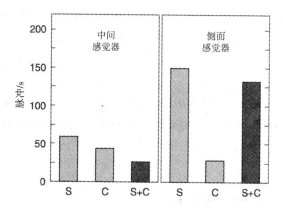

图 7.15 花青素苷(一种花色素苷)对欧洲粉蝶(*Pieris brassicae*)幼虫的两个上颚的栓锥形感器上对糖反应的抑制作用

当用 15 mmol/L 的蔗糖(S)和 2.5mmol/L 的花青素苷(C)与这两种刺激的混合(S + C)分别来刺激时，反应以总的脉冲频率表现，两个感受器对混合液的反应神经活动比单一的化合物刺激的神经活动要低(范隆,1990)[224]

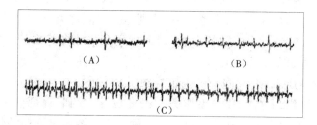

图 7.16 具带灯蛾(*Isia isabella*)幼虫上颚中间锥形感器的协同受体反应

(A)对 0.001 mol/升黑芥子苷的反应;(B)对 0.1 mol/l 蔗糖的反应;(C)对黑芥子硫苷酸钾和蔗糖的混合物的反应。实验表明对黑芥子苷有反应的细胞对混合液的反应更强(德蒂尔和库奇,1971)[54]

这两种化合物都刺激同一细胞,但是如果将两种化合物结合会产生更强的刺激。欧洲松

毛虫(*Dendrolimus pini*)幼虫上颚的味觉神经元会对大量糖类物质产生反应。当此神经元受到葡萄糖和肌醇的混合物刺激时,会产生比由单独的葡萄糖或肌醇刺激更强的反应[182]。

很多例子中都证实了这种外围化合物相互作用的存在,这是由于在此领域已由对纯化合物的研究转移到对两种化学物质的混合物的研究,甚至转移到对自然植物汁液(无法在化学成分上定义)的复杂刺激反应的研究。显然,关于对植物汁液反应的知识,对于理解昆虫选择不同寄主植物的化学感受机制是非常重要的。而对两种化合物相互作用的研究,并不能代表叶片汁液复杂刺激情况。三萜类化合物川楝素对欧洲粉蝶(*Pieris brassicae*)幼虫具有很强的抑制摄食作用。它能刺激中间的抑制神经元并抑制感糖和芥子苷神经元,而这两种物质都可以刺激取食[189]。三萜类印楝素也激活中间的抑制细胞,只是程度相对较轻,并没有影响刺激受体细胞的反应[112]。当对比川楝素和印楝素的抑制作用时,采用寄主植物的叶片用叶蝶法进行生物测定,抑制细胞的反应表示对其的抑制水平,并且假定川楝素对感受器的抑制或刺激作用很小。因此,研究外围化合物相互作用的发生和重要性,应尽可能地接近在摄食或产卵阶段出现的刺激情况[225]。

对于不同物种的外围化合物相互作用机制,目前仍然未知。竞争或变构性的相互作用也许发生在膜受体结合位点[64,141],但至今仍缺乏直接的证据证明。外围化合物相互作用的一个额外机制可能是味觉神经元之间的电偶联(electrotonic coupling),在此方面有一些来自电生理学和超微结构的证据[91,235]。

当抑制化合物对兴奋受体产生负面影响时,将对抑制作用的神经编码有所助益。这种额外的抑制编码机制已被发现,如抑制物对蔗糖敏感的神经元产生抑制作用。最近对于这个主题,有关于各种味觉系统编码原则的系统讨论(弗雷泽[64],斯洪霍芬和范隆[190],罗杰和纽兰[170])。

7.8.8 刺吸式昆虫的寄主植物选择

需要引起我们关注的有两种主要摄食模式:咀嚼式和刺吸式,而我们了解到的相关化学知识也应与这两种摄食方式结合起来。刺吸式昆虫专食植物的组织液和细胞液。为了鉴定出在选择特定的植物组织或细胞时,刺吸式昆虫使用的化学线索,需要对特定隔室(compartments)内的数据进行化学分析,这在技术上是非常困难的。如第3章所述,半翅目是一类典型的刺吸式昆虫,将其下颚和上颚针刺入到植物组织的表皮下。这与咀嚼式昆虫在这个过程中将全部组织浸软且使细胞破裂的行为不同,半翅昆虫,特别是一些同翅类,如蚜虫、粉虱及其他一些韧皮部取食者,将口针巧妙地刺入植物的组织,似乎能完全避免细胞损伤。两个上颚的口针联结,形成双管道,一个管道负责吸取食物,另一个负责传递唾液(图3.2)。

口针刺入植物的表层后,沿细胞间的通道穿过叶肉细胞间的细胞壁,一直通向维管束细胞。口针一旦进入植物组织之中,就会调整方向来寻找合适的取食部位(图7.17)。昆虫会控制口针朝向维管束,有时会转180度。目前有一种假说试图通过蔗糖的化学浓度梯度或pH值(两者在韧皮部的值均高于周围组织)的方式对韧皮部细胞进行定位。使用刺吸电位技术(EPG)已经对蚜虫口针的刺入行为进行了较深入的研究[218]。口针在植物体内不断进行穿刺,以监测口针尖端的电压。与咀嚼式昆虫的情况不同,蚜虫对叶片上细胞内(外)的物质的化学感受评价依靠位于咽上部和下部的味觉器官上的化学感受器(包含了约100个味觉神经元)。正是由于其体积极小以及位置特殊,这些化学感应器官的特性和敏感性无法通过常规的电生理学方法进行解释。

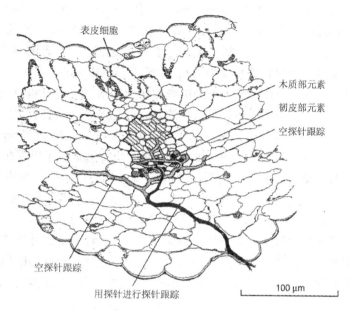

图7.17　甜菜蚜(*Aphis fabae*)的口针取食蚕豆叶脉筛管组织的吸食路径

显示为很多分支,代表了早期对韧皮部定位过程中的搜索行为。这些空的分支为唾液鞘的成分,在口针抽回时仍然可见(恰林和艾斯齐,1993)[219]

EPG测定的顺序大体可以分为三个阶段,即通路阶段、木质部阶段和韧皮部阶段。通路阶段先于韧皮和木质阶段,最少持续大约10分钟,通过表皮和外围组织进行机械穿刺,同时唾液开始分泌。口针从植物细胞壁间穿过,并进行周期性的机械运动,并分泌一种胶形唾液包裹口针,称为唾液鞘。

这种唾液鞘留在植物组织中,表明口针顶端曾经到过此处(图7.17)。蚜虫通过简短的细胞穿刺过程(持续5~10秒)试食细胞的内容物,这些物质在1秒内就能被运输到咽部的味觉器官,而从角质层到韧皮部的大部分路径都是位于细胞外的[219]。

当蚜虫口针处于水的压力下时,在木质部时期可通过EPG进行检测。在此期间,蚜虫使用其活跃的肌肉驱动的吮吸机制来吸取水分,因而木质部通常呈现负静压状态。到第三步时,口器尖端找到了目标营养物质,也就是韧皮部细胞。这时会发生两个次过程,第一个次过程为水性唾液的分泌,大约持续1分钟,随后或许还有对韧皮细胞物质的被动吸收。寻找合适的筛管从而获取食物是一个烦琐的过程,并且在实际取食之前,要对一些韧皮部筛管细胞进行取样。然而对于韧皮部筛管细胞的选择基于何种因素目前仍然未知。对蚜虫而言,一般需要平均2~7小时开始首次接触到韧皮部,而时间的长短要依据蚜虫—植物的具体组合而定[220]。一旦定位成功,它们就可以持续几小时或者几天吸食一个简单的筛管组织,有时还可以长达10天[217]。

蚜虫及其他刺吸式口器的昆虫与咀嚼式口器昆虫的一个重要不同之处在于,在取食过程中,到达目标组织沿途的细胞不被损坏,并且细胞质和液泡的物质不发生混合。许多植物表皮层和叶肉细胞中的次生物质以一种糖基化的形式被储存起来,而且只有先转变成为糖苷配基的形式,才能够被活化进行积极的防御(见4.11),而刺吸式口器昆虫能够有效地规避这种活化。然而,蚜虫摄食导致了植物体转录组大规模的改变。在一个对攻击拟南芥(*Arabidopsi*)相互作用的全基因组芯片研究中,桃蚜(*Myzus persicae*)取食导致约 830 个基因上调,而经咀嚼式口器菜粉蝶(*Pieris rapae*)幼虫取食,仅有约 130 个基因上调,刺吸式口器的西花蓟马(*Frankliniella occidentalis*)取食约有 170 个基因上调。有趣的是,通过桃蚜的摄食会导致拟南芥 1 350 个基因下调,而这一数字在菜粉蝶中为 60,在西花蓟马中为 30[42]。

由于刺吸式昆虫是基于植株细胞水平的机械和化学因素来决定对植物的接受或拒绝,因而对此详细的线索仍知之甚少。而对刺激信号的研究似乎涉及一些物种,如专食十字花科的甘蓝蚜(*Brevicoryne brassicae*)等。在甘蓝蚜的一种寄主植物——白芥子(*Sinapis alba*)中,发现主要的芥子苷白芥子硫苷出现在花茎的表皮细胞上的情况要远远多于出现在叶子的表皮细胞上。甘蓝蚜虫更喜欢以花茎为食而非叶片[56]。EPG 记录表明,在叶片上,昆虫所做的持续探测不足 2 分钟,这个时间刚能刺透表皮。相反,在花茎上的第一次探测,大多数情况下可以持续多于 10 分钟的时间,并导致对韧皮部的摄食[66]。拒食可能基于对植物表面化感物质的感知,通过触角或跗节上的接触化学感应器完成。在通路阶段,以对表皮或叶肉细胞(或基于韧皮部的物质)的取样进行判断[67,223]。因而仅在很少的情况下,这种抑制性化感物质才能被检测出,例如一种抑制剂(DIMBOA),出现在玉米和小麦中,主要贮存在维管束鞘细胞中,在韧皮部汁液浓度很低[68]。

7.8.9 产卵偏好

当雌性成虫选择一种植物作为其产卵场所时,其决定对幼虫的生存至关重要,因为初孵幼虫的活动性和能量储备都极其有限,因此依靠幼虫自身寻到适合位置的概率极小。寡食性地种蝇(*Delia*)属(双翅目:花蝇科)的两种昆虫取食十字花科植物,当雌性接触芥子油苷时会引发产卵行为。而不同种类的雌性对不同类别的芥子油苷有不同的偏好。与芥子油苷发生神经反应的化学感受器位于跗节感觉毛上,被各种芥子油苷物质激活后,引导机体产生相应的行为反应(图 7.18)。从以上结果可得出的结论为,跗节上的感受器在寄主植物识别方面起到十分重要的作用(即使不是决定性的)[204]。

菜粉蝶(*Pieris rapae*)与暗脉菜粉蝶(*P. napi oleracea*)对不同的芥子油苷表现出不同的偏好(图 7.5)。对跗节味觉感受器的电生理研究表明,昆虫倾向选择的化学物质恰是最能激发芥子油苷的受体细胞活性的物质[209]。基于目前的研究,发现这种输入和输出之间的关系非常奇特,因为感觉输入(记录的细胞数目相对于全部数目的味觉神经元)仅占雌性 2 100 个跗节感觉受体的 1%~2%。这些发现,同以上毛虫类的发现相似,在亲缘关系很近的蝴蝶之间,感觉特性呈现很大不同。也许每个物种的感觉系统都要随对特定寄主植物选择而演化。

在同一物种(如在亚种之间)间,也能观察到感觉反应的显著不同,这表明了系统在进化上的灵活性。暗脉菜粉蝶(*Pieris napi*)的两个亚种,对芥子油苷(图 7.19)的反应存在显著不同[57]。对于幼虫取食有抑制作用的强心苷,对两种雌性成虫也有着强大的产卵抑制作用[38]。强心苷不会影响"感芥子油苷"细胞。对于芥子油苷的偏好程度由"感芥子油苷"神经元(与偏好正相关)和"感强心苷"细胞(与偏好负相关)所决定,这是有两个单线标记编码模式相互平衡且最基本的交叉纤维模式。这一例子也再次清楚地表明了,单线标记模式和交叉纤维模式间存在连续性。当一只雌性成虫落在表面上有芥子油苷和强心苷混合物的十字花科植物上时,两种神经元都被刺激,而且在决定接受和拒绝之间进行平衡。

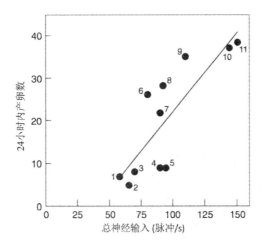

图 7.18 来自萝卜地种蝇(*Delia floralis*)足和唇瓣的两个不同的感受器的所有神经输入（发生刺激后第一秒的神经冲动）和产卵行为（无选择的条件下，在 24 小时内产卵数量）之间的关系

11 种不同的芥子油苷以 10^{-2} mol/l 的浓度喷洒在一片人造叶子上，在神经输入和行为输出间发现了显著的相关性。其中，1 为芝麻菜苷；2 为屈曲花苷；3 为 2-羟基-3-丁烯基硫苷式；4 为白芥子硫苷；5 为新葡萄糖芸苔素；6 为黑芥子苷；7 为葡萄糖芜菁芥素；8 为金莲葡萄糖硫苷；9 为豆瓣菜苷；10 为葡萄糖异硫氰酸戊-4-烯酯；11 为芸苔葡萄糖硫苷（西蒙等，1994）[196]

7.8.10 寄主植物的选择：三级系统

对寄主植物选择主要有三个主要因素：

(1) 外部化学感应系统，对很多化学刺激敏感，包括刺激剂和抑制剂。

(2) 中枢神经系统(CNS)，识别感觉模式，一些模式被识别为可被接受，它们可以引起摄食和产卵行为（可能受"动力中心"调节起增效作用；其他一些模式会引起拒绝行为。而最后的决定有可能是由咽下神经节做出，然而，这一过程可能不止在一个位置上发生[170]。应用"锁和钥匙"的概念到这个机制中，将是一个非常有用的简化模型。专食性昆虫的感觉模型，必须与中枢神经系统的一个基准匹配，目的是为了引发摄食行为，这是相较于广食性昆虫而言的。换言之，对于广食性昆虫，有许多不同的感觉受体活动的"钥匙"能适合于中枢神经系统模型中的"锁"；而对于专食性昆虫，"锁"会更精挑细选"钥匙"（图 7.20）[182]。

(3) 第三个决定对潜在的植物食物接受或拒绝的原因，涉及对内在化学敏感系统的贡献。

当食物营养成分与生理需求的差异较大时，这个系统就会预警中枢神经系统，并导致对营养选择的改变（见 5.3.3）。

在寄主植物选择的三级系统中，其感觉受体、中枢神经系统和营养反馈各成分之间相互影响，这个系统并非一个封闭的系统，而是不断地与许多生态因素相互作用[184]。

7.9 化学感应系统与寄主植物偏好的演化

在之前的几节中，我们讨论昆虫的化学感应系统对寄主植物接受和排斥行为起到的关键作用。基于观察（第 2 章）发现，专食性昆虫在数量上极大地超越了杂食性物种，许多学者一

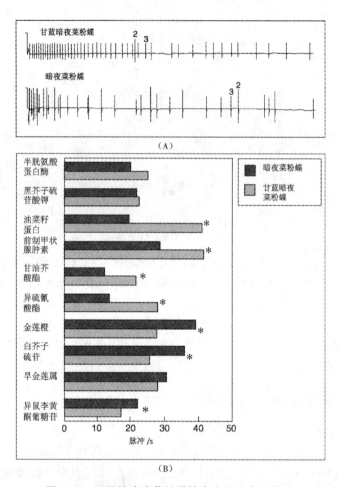

图 7.19 不同的暗脉菜粉蝶的电生理反应记录

(A)暗脉菜粉蝶(*Pieris napi*)的两个雌性亚种暗夜菜粉蝶(*Pieris napi oleracea*)和甘蓝暗夜菜粉蝶(*P. napi napi*)的跗节感觉毛对 10 mg/ml 的芥子油苷的电生理反应记录。在一秒内粉蝶亚种比亚种感觉细胞激发的频率(标为"2")更为频繁;(B)对 10 种不同的芥子油苷的反应,反应强度表现为在第一秒的刺激内,一种脉冲的数目,在(A)中被标为"3",七种混合物(被标为 *)对两亚种产生了显著不同的反应

直关注这样的假说,即昆虫与植物的关系的演化实际上取决于昆虫神经系统(包括外周及中枢水平)的进化[9,96]。在这一假说中,化学感应系统应该先于寄主植物变化前发生改变,随后可能会导致新的昆虫与植物关系的产生。而选择应在昆虫对植物的识别系统在基因上发生改变之后,因为一个新的基因组将可能会有新的植物选择偏好,如果能容忍一些生理因素(如植物毒素,见 5.4 和 11.7)和植物的一些生态特性(如天敌,见 11.7)[9],这种新建立的选择就会成功。一些对于昆虫神经系统进化的限制因素,可能对新的专食性物种的出现起重要作用,但也有可能不是必需的。

这一设想涉及植食性昆虫化学感受器的遗传基础。决定了对寄主植物特性和偏好的基因数越少,那些特征进化就越快的特点。根据凤蝶(Papilio)的杂交实验,在很少的遗传位点上发生改变,可能会对这些凤蝶的寄主偏好行为产生很大的影响[215]。对于昆虫化学感受器功能和

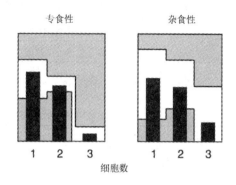

图7.20　专食性和杂食性昆虫的中枢神经系统对感觉输入的处理模式

黑框为当可食植物刺激时,三个化学感应受体(1~3)的动作电位频率。封闭的白色空间表示为,在食物的可接受的范围内感觉输入允许发生的变化,细胞3是一个抑制感受器(斯洪霍芬,1987)[182]

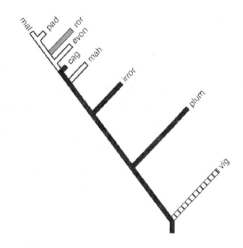

图7.21　基于寄主植物等位酶数据和植物状态的九种巢蛾(Yponomeuta)的系统发育树

巢蛾种类:*cag*, *cagnagellus*; *evon*, *evonymellus*; *irror*, *irrorellus*; *mah*, *mahalebellus*; *mal*, *malinellus*; *pad*, *padellus*; *plum*, *plumbellus*; *ror*, *rorellus*; *vig*, *vigintipunctatus*。寄主植物:黑色,卫矛科;白色,蔷薇科;阴影,杨柳科;(*Y. vigintipunctatus*)取食景天科植物,但它的姊妹群(*Y. yamagawanus*)以卫矛属(卫矛科)植物为食(门肯等,1992)[125]

基因的研究表明,基因单一突变能够将昆虫由单食性改变为杂食性,而反之亦然[52]。通过两种巢蛾(Yponomeuta)的种间杂交实验表明,对一种摄食抑制剂(查耳酮甙)的敏感性仅由一个显性基因遗传[222]。而转向抑制作用减弱的寄主植物,可能是巢蛾进化的一个重要因素(表7.8)[124]。对于这一属昆虫的系统发育重建表明,卫矛科是其祖先的寄主植物,后来转变为蔷薇科(图7.21)。苹果巢蛾(*Yponomeuta malinellus*)取食蔷薇科苹果(*Malus*)属,另一种昆虫巢蛾(*Yponomeuta rorellus*)取食杨柳科柳属(*Salix*)植物,而这两种昆虫的化学感受器对于这两属特有的化合物都缺乏敏感性,而这些物质对研究中涉及的其他物种能起到抑制作用(表7.8)[124]。也可能出现相反的情况,即寄主范围变窄。当抑制神经元对特定抑制物缺乏敏感性时,也可能导致寄主范围变宽。有一个例子可以证明这种变化,家蚕(*Bombyx mori*)基因的一些位点产生突变,会致使其取食原来拒绝的植物[5],然而仍需要对受体数量和特性进行更深入的研究。

表 7.8　四种巢蛾属(*Yponomeuta*)昆虫锥形感受器的味觉受体的化学感应敏感度

昆虫	寄主植物(科)	锥形感受器侧面/中间的特异性味觉受体的反应			
		卫矛醇	山梨醇	根皮苷	水杨苷
巢蛾	卫矛科(*Euonymus europaeus*)	+/+ *	-/-	-/+	n.t./+
巢蛾	蔷薇科(*Prunus/Crataegus* spp.)	+/- [+]	+/- *	-/+	+/+
巢蛾	蔷薇科(*Malus* spp.)	-/-	+/-	-/- *	+/+
巢蛾	杨柳科(*Salix* spp.)	-/-	-/-	-/+	-/+ *

注：+，感受器敏感；-，感受器不敏感；n.t.，未检测；*，寄主植物中存在的化合物；[+]，卫矛醇以低浓度的形式存在于一些蔷薇科寄主植物中(在卫矛科中浓度为10%)。

如果一个原本编码抑制感受器的基因因某种突变表达为对糖类兴奋的敏感神经元,这也许可以解释信号敏感受体的起源(图 7.12)。事实上,这种现象已经在黑腹果蝇(*Drosophila melanogaster*)的味觉突变体中发现[4]。对果蝇的基因组研究已揭示了一个由 60 个基因组成的编码七跨膜蛋白质(味觉受体蛋白)的基因家族[39]。研究植食性昆虫的受体蛋白的配体特异性及同源器官很有可能会提高我们对寄主植物的味觉识别和进化的知识。除了不同的受体位点,不同的细胞内传导机制也会导致对不同抑制剂的感觉差别。由于相同的抑制神经元可产生不同的传导途径,因此烟草天蛾(*M. sexta*)幼虫能够辨别水杨苷与马兜铃酸[71]。

专食性昆虫的化学感受器,将植物特异的化学物质作为刺激信号,这相对于多种竞争性信号而言是高度明确的刺激信号。这种系统显然存在适应优势。由基因所决定的对寄主植物的偏好的改变仅能遗传七代[198],之后就会转变为与宿主植物关系紧密或无关的种类,其寄主植物的范围可能扩大也可能缩小,目前已有一些相关的报道(第 11 章)。这种行为上的改变可能起源于受体蛋白基因的随机的改变。这些改变能否被固定主要取决于选择优势以及基因与发展的限制[96]。

7.10　结论

昆虫一旦与潜在的寄主植物发生联系,就会随之对其进行详尽的评价行为,在此期间,昆虫利用机械感觉和由植物提供的化学感应(主要是味觉)刺激而对寄主植物的选择,在很大程度上是由中枢神经对植物产生的众多味觉刺激信号进行评价。我们目前对于这类反应的知识会受到水溶性化合物的影响,并且事实上,我们在叶片表面的非极性植物化学物质对于昆虫的影响方面仍一无所知。

当昆虫接收到植物化学信号后,昆虫就会通过专一性不同(从高度专一性如接受刺激信号的受体,到广适的如抑制物受体)的味觉神经元对这些信号进行编码。在行为水平上,已被证明,对于植物的接受与否,是由刺激的和抑制的化学物质的平衡决定的。直到最近,人们才发现这种平衡的过程可以被追踪,或至少部分追踪,通过对刺激和抑制信号输入的比例,检测化学感觉的活动而实现。而这个比例似乎可以直接决定倾向性(接受或拒绝)的级别。然而,在其他一些情况下,密码仍未被破译,至今还没有发现昆虫对植物复杂辨别能力的生理基础,这将是一个很大的挑战。目前已有越来越多关于对外周化学物质相互作用的研究,通过以植物汁液(自然的)的刺激研究化学感应活动,是解释野外条件下产生强烈相互作用的最好方法。

从目前对昆虫味觉受体已有的了解中我们可以得出这样的结论,植食性昆虫拥有一个高度敏感的系统,因而能够分辨植物之间及植物各部分之间化学物质的细微差别。另一个重要的结论是,每个物种甚至每个生物型都具有特异的感官系统,可以很准确地区分寄主植物和非寄主植物,当然还能将许多不同的寄主植物进行精细区分。

在一些专食性取食者中,存在高度专一性的味觉受体,而广适性的抑制神经元包含许多由单基因决定的受体位点,说明了专食性的进化趋势以及寄主转移的可能性。受体的活动是昆虫做出接受或拒绝决定的基础,受体水平上的变化将会影响昆虫的行为。如受体的某一位点发生变异,导致对抑制剂的敏感性丢失,因此发生了寄主转移(图 7.12)。有时含抑制剂的不可接受的植物会变得可接受,寄主的范围由此被扩大。鳞翅目巢蛾属(*Yponomeuta*)是说明以上问题的一个很好的例子。

对于植食性昆虫而言,其植物—食物专化性的进化在很大程度上是由感官或中枢神经的限制决定的[225]。

7.11 参考文献

1. Ahmad, S. (1982). Host location by the Japanese beetle: evidence for a key role for olfaction in a highly polyphagous insect. *Journal of Experimental Zooloyy*, 220, 117 – 20.

2. Allebone, J. E., Hamilton, R. J., Bryce, T. A., and Kelly, W. (1971). Anthraquinone in plant surface waxes. *Experientia*, 27, 13 – 14.

3. Anderson, P. (2002). Oviposition pheromones in herbivorous and carnivorous insects. In *Chemoecology of insect eggs and egg deposition* (ed. M. Hilker and T. Meiners), pp. 235 – 63. Blackwell, Berlin.

4. Arora, K., Rodrigues, V., Joshi, S., Shanbhad, S., and Siddiqi, O. (1987). A gene affecting the specificity of the chemosensory neurons of *Drosophila*. *Nature*, 330, 62 – 3.

5. Asaoka, K. (2000). Deficiency of gustatory sensitivity to some deterrent compounds in 'polyphagous' mutant strains of the silkworm, *Bombyx mori*. *Journal of Comparative Physiology A*, 186, 1011 – 18.

6. Baur, R., Feeny, P., and Städler, E, (1993). Oviposition stimulants for the black swallowtail butterfly: identification of electrophysiologically active compounds in carrot volatiles. *Journal of Chemical Ecoloyg*, 19, 919 – 37.

7. Benedet, F., Leroy, T., Gauthier, N., Thibaudeau, C., Thibout, E., and Renault, S. (2002). Gustatory sensilla sensitive to protein kairomones trigger host acceptance by an endoparasitoid. *Proceedings of the Royal society*, *Biological Sciences*, *Series B*, 269, 1879 – 86.

8. Benedict, J. H., Leigh, T. F., and Hyer, A. H. (1983). *Lygus hesperus* (Heteroptera: Miridae) oviposition behavior, growth and survival in relation to cotton trichome density. *Environmental Entomology*, 12, 331 – 5.

9. Bernays, E. A. (2001). Neural limitations in phytophagous insects: impolications for diet breadth and evolution of host affiliation. *Annual Review of Entomology*, 46, 703 – 27.

10. Bernays, E. A. and Chapman, R. F. (1978). Plant chemistry and acridoid feeding behaviour.

In *Biochemical aspects of plant and animal coevolution* (ed. J. B. Harborne), pp. 91 – 141. Academic Press, London.

11. Bernay, E. A. and Chapman, R. F. (1994). *Host-plant selection behaviour of phytophagous insects*. Chapman & hall, New York.
12. Bernays, E. A. and Chapman, R. F. (2000). A neurophysiological study of sensitivity to a feeding deterrent in two sister species of *Heliothis* with different diet breadths. *Journal of Insect Physiology*, 46, 905 – 12.
13. Bernays, E. A. and Chapman, R. F. (2011). Taste cell responses in the polyphagous arctiid, *Grammia geneura*: towards a general pattern for caterpillars. *Journal of Insect Physiology*, 47, 1029 – 43.
14. Bernays, E. A. and Simpson, S. J. (1982). Control of food intake. *Advances in Insect Physiology*, 16, 59 – 188.
15. Bernays, E. A., Howard, J. J., Champagne, D., and Estesen, B. J. (1991). Rutin: a phagostimulant for the polyphagous acridid *Schistocerca americana*. *Entomologia Experimentalis et Applicata*, 60, 19 – 28.
16. Bernay, E. A., Howard, J. J., Champagne, D., and Estesen, B. J. (1991). Rutin: a phagostimulant for the polyphagous acridid *Schistocerca americana*. *Entomologia Experimentalis et Applicata*, 60, 19 – 28.
17. Bernays, E. A., Chapman, R. F., and Singer, M. S. (2000). Sensitivity to chemically diverse phagostimulants in a single gustatory neuron of a polyphagous caterpillar. *Journal of Comparative Physiology A*, 186, 13 – 19.
18. Bernays, E. A., Oppenheim, S., Chapman, R. F., Kwon, H., and Gould, F. (2000). Taste sensitivity of insect herbivores to deterrents is greater in specialists than in generalists: a behavioral test of the hypothesis with two closely related caterpillars. *Journal of Chemical Ecology*, 26, 547 – 63.
19. Bernays, E. A., Chapman, R. F., and Hartmann, T. (2002). A highly sensitive taste receptor cell for pyrrolizidine alkaloids in the lateral galeal sensillum of a polyphagous caterpillar, *Estigmene acraea*. *Journal of Comparative Physiology A*, 188, 715 – 23.
20. Bernays, E. A., Hartmann, T., and Chapman, R. F. (2004). Gustatory responsiveness to pyrrolizidine alkaloids in the *Senecio* specialist, *Tyria jacobaeae* (Lepidoptera, Arctiidae). *Physiological Entomology*, 29, 67 – 72.
21. Besson, E., Dellamonica, G., Chopin, J., Markham K. R., Kim, M. Koh, H. S., et al. (1985). C-Glycosylflavones from *Oryza sativa*. *Phytochemistry*, 24, 1061 – 4.
22. Blaakmeer, A., Stork, A., Van Veldhuizen, A., Van Beek, T. A., De Groot, Æ van Loon, J. J. A., et al. (1994a). Isolation, identification, and synthesis of miriamides, new hostmarkers from eggs of *Pieris brassicae*. *Journal of Natural Products*, 57, 90 – 9.
23. Blaakmeer, A., Hagenbeek, D., Van Beek, T. A., De Groot, Æ Schoonhoven, L. M., and van Loon, J. J. A. (1994b). Plant response to eggs *vs.* host marking pheromone as factors inhibiting oviposition by *Pieris brassicae*. *Journal of Chemical Ecology*, 20, 1657 – 65.

24. Bland, R. G. (1989). Antennal sensilla of Acrididae (Orthoptera) in relation to subfamily and food preference. *Annals of the Entomological Society of America*, 82, 368-84.
25. Blaney, W. M. and Simmonds, M. S. J. (1987). Control of mouthparts by the subesophageal ganglion. In *Arthropod brain* (ed. A. P. Gupta), pp. 303-22. Wiley, New York.
26. Blaney, W. M. and Simmonds, M. S. J. (1990). A behavioural and electrophysiological study of therde of tarsal chemoreceptors in feeding by adults of *Spodoptera*, *Heliothis virescens* and *Helicoverpa armigera*. *Journal of Insect Physiology*, 36, 743-56.
27. Blaney, W. M., Schoonhoven, L. M. and Simmonds, M. S. J. (1986). Sensitivity variations in insect chemoreceptors: a review. *Experientia*, 42, 13-19.
28. Blaney, W. M., Simmonds, M. S. J., Ley, S. V., and Katz, R. B. (1987). An electrophysiological and behavioural study of insect antifeedant properties of natural and synthetic drimane-related compounds. *Physiological Entomology*, 12, 281-91.
29. Blau, P. A., Feeny, p., and Contardo, L. (1978). Allylglucosinolate and herbivorous caterpillars: a contrast in toxicity and tolerance. *Science*, 200, 1296-8.
30. Blom, F. (1978). Sensory activity and food intake: a study of input-output relationships in two phytophagous insects. *Netherlands Journal of Zoology*, 28, 277-340.
31. Bouaziz, M., Simmonds, M. S. J., Grayer, R. J., Kite, G. C., and Damak, M. (2001). Flavonoids from *Hyparrhenia hirta* Stapf (Poaceae) growing in Tunisia. *Biochemical Systematics and Ecology*, 29, 849-51.
32. Bowers, M. D. (1983). The role of iridoid glycosides in host-plant specificity of checkerspot butterflies. *Journal of Chemical Ecology*, 9, 475-94.
33. Carter, M., Sachdev-Gupta, K., and Feeny, P. (1994). Tyramine, an oviposition stimulant for the black swallowtail butterfly from the leaves of wild parsnip. *Abstracts of the 11th Annual Meeting of the International Society of Chemical Ecology*, Syracuse, 55.
34. Chapman, R. F. (1982). Chemoreception: the significance of receptor numbers. *Advances in Insect Physiology*, 16, 247-356.
35. Chapman, R. F. (2003). Contact chemoreception in feeding by phytophagous insects. *Annual Review of Entomology*, 48, 455-84.
36. Chapman, R. F. and Bernays, E. A. (1989). Insect behavior at the leaf surface and learning as aspects of host plant selection. *Experientia*, 45, 215-22.
37. Chapman, R. F. and Fraser, J. (1989). The chemosensory system of the monophagous grasshopper, *Bootettix argentatus* Bruner (Orthoptera: Acrididae). *International Journal of Insect Morphology and Embryology*, 18, 111-18.
38. Chew, F. S. and Renwick, J. A. A. (1995). Chemical ecology of hostplant choice in *Pieris* butterflies. In *Chemical ecology of insects* (2nd edn) (ed. R. T. Cardé and W. J. Bell), pp. 214-38. Chapman & Hall, NewYork.
39. Clyne, P. J., Warr, C. G., and Carlson J. R. (2000). Candidate taste receptors in *Drosophila*. *Science*, 287, 1830-4.
40. De Boer, G. (1993). Plasticity in food preference and diet-induced differential weighting of

chemosensory information in larval *Manduca sexta*. *Journal of Insect Physiology*, 39, 17 – 24.

41. De Candolle, A. P. (1804). *Essai sur les propriétés médicinales des plantes, comparées avec leurs formes extérieures et leur classification naturelle*. Didot Jeune, Paris.

42. De Vos, M., Van Oosten, V. R., Van Pelt, J. A., vanLoon, L. C., Dicke, M., and Pieterse, C. M. J. (2004). Herbivore-induced resistance: differential effectiveness against a set of microbial pathogens in *Arabidopsis thaliana*. In *Biology of plant-microbe interactions*, Vol. 4 (ed. I. Tikhonovich, B. Lugtenberg, andN. Provorov), pp. 40 – 3. International Society for Molecular Plant-Microbe Interactions, St Paul, MN.

43. Del Campo, M. L., Miles, C. I., Schroeder, F. C., Müller, C., Booker, R., and Renwick, J. A. A. (2001). Host recognition by the tobacco hornworm is mediated by a host plant compound. *Nature*, 411, 186 – 9.

44. Derridj, S., Gregoire, V., Boutin, J. P., and Fiala, V. (1989). Plant growth stages in the interspecific oviposition preference of the European corn borer and relations with chemicals present on the leaf surfaces. *Entomologia Experimentalis et Applicata*, 53, 267 – 76.

45. Derridj, S., Fiala, V., and Boutin, J. P. (1991). Host plant oviposition preference of the European corn borer (*Ostrinia nubilalis* Hbn.). A biochemical explanation. *Symposia Biologica Hungarica*, 39, 455 – 6.

46. Derridj, S., Wu, B. R., Stammitti, L., Garrec, J. P., and Derrien, A. (1996). Chemicals on the leaf surface, information about the plant available to insects. *Entomologia Experimentalis et Applicata*, 80, 197 – 201.

47. Dethier, V. G. (1941). Chemical factors determining the choice of food by *Papilio* larvae. *American Naturalist*, 75, 61 – 73.

48. Dethier, V. G. (1947). *Chemical insect attractants and repellents*. Blakiston, Philadelphia.

49. Dethier, V. G. (1973). Electrophysiological studies of gustation in lepidopterous larvae. II. Taste spectra in relation to food-plant discrimination. *Journal of Comparative Physiology*, 82, 103 – 33.

50. Dethier, V. G. (1976). The *hungry fly*. Harvard University Press, Cambridge, MA.

51. Dethier, V. G. (1982). Mechanisms of host plant recognition. *Entomologia Experimentalis et Applicata*, 31, 49 – 56.

52. Dethier, V. G. (1987). Analyzing proximate causes of behavior. In *Evolutionary genetics of invertebrate behavior* (ed. M. D. Huettel), pp. 319 – 28. Plenum Press, New York.

53. Dethier, V. G. and Crnjar, R. M. (1982). Candidate codes in the gustatory system of caterpillars. *Journal of General Physiology*, 79, 549 – 69.

54. Dethier, V. G. and Kuch, J. H. (1971). Electrophysiological studies of gustation in lepidopterous larvae. I. Comparative sensitivity to sugars, amino acids, and glycosides. *Zeitschrift für Vergleichende Physiologie*, 72, 343 – 63.

55. Doskotch, R. W., Mikhail, A. A., and Chatterjee, S. K. (1973). Structure of the water-soluble feeding stimulant for *Scolytus multistriatus*: a revision. *Phytochemistry*, 12, 1153 – 5.

56. Dreyer, D. L. and Jones, K. C. (1981). Feeding deterrency of flavonoids and related phenol-

ics towards *Schizaphis graminum* and *Myzus persicae*: aphid feeding deterrents in wheat. *Phytochemistry*, 20, 2489–93.

57. Du, Y.-J., Van Loon, J. J. A., and Renwick, J. A. A. (1995). Contact chemoreception of oviposition stimulating glucosinolates and an oviposition deterrent cardenolide in two subspecies of *Pieris napi*. *Physiological Entomology*, 20, 164–74.
58. Eigenbrode, S. D. (1996). Plant surface waxes and insect behaviour. In *Plant cuticles* (ed. G. Kerstiens), pp. 201–21. Bios Scientific Publishers, Oxford.
59. Eigenbrode, S. D. and Espelie, K. E. (1995). Effects of plant epicuticular lipids on insect herbivores. *Annual Review of Entomology*, 40, 171–94.
60. Fabre, J. H. (1886). *Souvenirs entomologiques*, Vol. 3. Delagrave, Paris.
61. Feeny, P., Sachdev-Gupta, K., Rosenberry, L., and Carter, M. (1988). Luteolin 7-O-(6"-O-malonyl)-β-D-glucoside and *trans*-chlorogenic acid: oviposition stimulants for the black swallowtail butterfly. *Phytochemistry*, 27, 3439–48.
62. Deleted.
63. Fraenkel, G. S. (1959). The raison d'être of secondary plant substances. *Science*, 129, 1466–70.
64. Frazier, J. L. (1992). How animals perceive secondary plant compounds. In *Herbivores: their interactions with secondary plant metabolites*, Vol. 2 (2nd edn) (ed. G. A. Rosenthal and M. R. Berenbaum), pp. 89–133. Academic Press, New York.
65. Frazier, J. L. and Chyb, S. (1995). Use of feeding inhibitors in insect control. In *Regulatory mechanisms in insect feeding* (ed. R. F. Chapman and G. de Boer), pp. 364–81. Chapman & Hall, New York.
66. Gabrys, B., Tjallingii, W. F., and Van Beek, T. A. (1997). Analysis of EPG recorded probing by cabbage aphid on host plant parts with different glucosinolate contents. *Journal of Chemical Ecology*, 23, 1661–73.
67. Garzo, E., Soria, C., Gomez-Guillamon, M. L., and Fereres, A. (2002). Feeding behavior of *Aphis gossypii* resistant accessions on different melon genotypes (*Cucumis melo*). *Phytoparasitica*, 30, 129–40.
68. Givovich, A. and Niemeyer, H. M. (1991). Hydroxamic acids affecting barley yellow dwarf virus transmission by the aphid *Rhopalosiphum padi*. *Entomologia Experimentalis et Applicata*, 59, 79–85.
69. Glendinning, J. I., Valcic, S., and Timmermann, B. N. (1998). Maxillary palps can mediate taste rejection of plant allelochemicals by caterpillars. *Journal of Comparative Physiology A*, 183, 35–43.
70. Glendinning, J. I., Nelson, N. M., and Bernays, E. A. (2000). How do inositol and glucose modulate feeding in *Manduca sexta* caterpillars? *Journal of Experimental Biology*, 203, 1299–315.
71. Glendinning J. I., Davis, A., and Ramaswamy, S. (2002). Contribution of different taste cells and signaling pathways to the discrimination of 'bitter' taste stimuli by an insect. *Journal*

of Neuroscience, 22, 7281–7.

72. Gregory, P., Avé, D. A., Bouthyette, P. J., and Tingey, W. M. (1986). Insect-defensive chemistry of potato glandular trichomes. In *Insects and the plant surface* (ed. B. Juniper and T. R. E. Southwood), pp. 173–83. Edward Arnold, London.
73. Griffiths, D. W., Deighton, N., Birch, A. N. E., Patrian, B., Baur, R., and Städler, E. (2001). Identification of glucosinolates on the leaf surface of plants from the Cruciferae and other closely related species. *Phytochemistry*, 57, 693–700.
74. Guerra, A. A. and Shaver, T. N. (1969). Feeding stimulants from plants for larvae of the tobacco budworm and bolworm. *Journal of Economic Entomology*, 62, 98–100.
75. Gupta, P. D. and Thorsteinson, A. J. (1960). Food plant relationships of the diamond-back moth (*Plutella maculipennis*). II. Sensory regulation of oviposition of the adult female. *Entomologia Experimentalis et Applicata*, 3, 241–50.
76. Haley Sperling, J. L. and Mitchell, B. K. (1990). A comparative study of host recognition and the sense of taste in *Leptinotarsa*. *Journal of Experimental Biology*, 157, 439–59.
77. Hamamura, Y., Hayashiya, K., Naito, K., Matsuura, K., and Nishida, J. (1962). Food selection by silkworm larvae. *Nature*, 194, 754–5.
77a. Harborne, J. B. and Grayer, R. J. (1994). Flavonoids and insects. In *The flavonoids: advances in research since* 1986 (ed. J. B. Harborne), pp. 589–618. Chapman & Hall, London.
78. Hare, J. D. and Elle, E. (2002). Variable impact of diverse insect herbivores on dimorphic *Datura wrightii*. *Ecology*, 83, 2711–20.
79. Haribal, M. and Feeny, P. (1998). Oviposition stimulant for the zebra swallowtail butterfly, *Eurytides marcellus*, from the foliage of pawpaw, *Asimina triloba*. *Chemoecology*, 8, 99–110.
80. Harris, M. O. and Miller, J. R. (1984). Foliar form influences ovipositional behaviour of the onion fly. *Physiological Entomology*, 9, 145–55.
81. Hedin, P. A., Thompson, A. C., and Minyard, J. P. (1966). Constituents of the cotton bud. III. Factors that stimulate feeding by the cotton weevil. *Journal of Economic Entomology*, 59, 181–5.
82. Hilker, M., Rohfritsch, O., and Meiners, T. (2002). The plant's response towards insect egg deposition. In *Chemoecology of insect eggs and egg deposition* (ed. M. Hilker and T. Meiners), pp. 205–33. Blackwell, Berlin.
83. Hoffmeister, T. S. and Roitberg, B. D. (2002). Evolutionary ecology of oviposition marking pheromones. In *Chemoecology of insect eggs and egg deposition* (ed. M. Hilker and T. Meiners), pp. 319–47. Blackwell, Berlin.
84. Honda, K. (1990). Identification of host-plant chemicals stimulating oviposition by swallowtail butterfly, *Papilio protenor*. *Journal of Chemical Ecology*, 16, 325–37.
85. Honda, K. (1995). Chemical basis of differential oviposition by lepidopterous insects. *Archives of Insect Biochemistry and Physiology*, 30, 1–23.
86. Honda, K., Tada, A., Hayashi, N., Abe, F., and Yamauchi, T. (1995). Alkaloidal ovi-

position stimulants for a danaid butterfly, *Ideopsis similis* L., from a host plant, *Tylophora tanakae* (*Asclepiadaceae*). *Experientia*, 51, 753-6.

87. Hsiao, T. H. (1985). Feeding behavior. In *Comprehensive insect physiology, biochemistry & pharmacology*, Vol. 9 (ed. G. A. Kerkut and L. I. Gilbert), pp. 497-512. Pergamon Press, New York.

88. Hsiao, T. H. (1988). Host specificity, seasonality and bionomics of *Leptinotarsa* beetles. In *Biology of Chrysomelidae* (ed. P. Jolivet, E. Petitpierre, and T. H. Hsiao), pp. 581-99. KluwerAcademic, Dordrecht.

89. Huang, X. and Renwick, J. A. A. (1993). Differential selection of host plants by two *Pieris* species: the role of oviposition stimulants and deterrents. *Entomologia Experimentalis et Applicata*, 68, 59-69.

90. Hurter, J., Ramp, T., Patrian, B., Städler, E., Roessingh, P., Baur, R., et al. (1999). Oviposition stimulants for the cabbage root fly: isolation from cabbage leaves. *Phytochemistry*, 51, 377-82.

91. Isidoro, N., Solinas, M., Baur, R., Roessingh, P., and Städler, E. (1994). Ultrastructure of a tarsal sensillum of *Delia radicum* L. (Diptera: Anthomyiidae) sensitive to important host-plant compounds. *International Journal of Insect Morphology and Embryology*, 23, 115-25.

92. Jackson, D. M., Severson, R. F., Johnson, A. W., Chaplin, J. F., and Stephenson, M. G. (1984). Ovipositional response of tobacco budworm moths (Lepidoptera: Noctuidae) to cuticular chemical isolates from tobacco leaves. *Environmental Entomology*, 13, 1023-30.

93. Jeffree, C. E. (1986). The cuticle, epicuticular waxes and trichomes of plants, with reference to their structure, functions and evolution. In *Insects and the plant surface* (ed. B. Juniper and T. R. E. Southwood), pp. 23-64. Edward Arnold, London.

94. Jermy, T. (1958). Untersuchungen über Auffinden und Wahl der Nahrung beim Kartoffelkäfer (*Leptinotarsa decemlineata* Say). *Entomologia Experimentalis et Applicata*, 1, 179-208.

95. Jermy, T. (1966). Feeding inhibitors and food preference in chewing phytophagous insects. *Entomologia Experimentalis et Applicata*, 9, 1-12.

96. Jermy, T. (1993). Evolution of insect-plant relationships—a devil's advocate approach. *Entomologia Experimentalis et Applicata*, 66, 3-12.

97. Jermy, T. (1994). Hypotheses on oligophagy: how far the case of the Colorado potato beetle supports them. In *Novel aspects of the biology of Chrysomelidae* (ed. P. H. Jolivet, M. L. Cox, and E. Petitpierre), pp. 127-39. Kluwer Academic, Dordrecht.

98. Jermy, T. and Szentesi, Á. (1978). The role of inhibitory stimuli in the choice of oviposition site by phytophagous insects. *Entomologia Experimentalis et Applicata*, 24, 458-71.

99. Jördens-Röttger, D. (1979). Das Verhalten der schwarzen Bohnenblattlaus *Aphis fabae* Scop. gegenüber chemischen Reizen von Pflanzenoberflächen. *Zeitschrift für Angewandte Entomologie*, 88, 158-66.

100. Juniper, B. E. and Jeffree, C. E. (1983). *Plant surfaces*. Edward Arnold, London.

101. Juvik, J. A., Babka, B. A., and Timmermann, E. A. (1988). Influence of trichome exu-

dates from species of *Lycopersicon* on oviposition behavior of *Heliothis zea* (Boddie). *Journal of Chemical Ecology*, 14, 1261 – 78.

102. Kennedy, J. S. (1987). Animal motivation: the beginning of the end? In *Advances in chemoreception and behaviour* (ed. R. F. Chapman, E. A. Bernays, and J. G. Stoffolano), pp. 17 – 31. Springer, New York.

103. Kennedy, J. S. and Booth, C. O. (1951). Host alternation in *Aphis fabae* Scop. I. Feeding preferences and fecundity in relation to the age and kind of leaves. *Annals of Applied Biology*, 38, 25 – 65.

104. Kent, K. S. and Hildebrand, J. G. (1987). Cephalic sensory pathways in the central nervous system of *Manduca sexta* (Lepidoptera: Sphingidae). *Philosophical Transactions of the Royal Society, London*, B315, 3 – 33.

105. Klingauf, F. (1971). Die Wirkung des Glucosids Phlorizin auf das Wirtswahlverhalten von Rhopalosiphum insertum (Walk.) und *Aphis pomi* de Geer (Homoptera: Aphididae). *Zeitschrift für Angewandte Entomologie*, 68, 41 – 55.

106. Klingauf, F., Nöcker-Wenzel, K., and Röttger, U. (1978). Die Rolle peripherer Pflanzenwachse für den Befall durch phytophage Insekten. *Zeitschrift für Pflanzenkrankheiten und Pflanzenschutz*, 85, 228 – 37.

107. Kubo, I., Lee, Y. - W., Bligh-Nair, V., Nakanishi, K., and Chapeau, A. (1976). Structure of ajugarins. *Journal of the Chemical Society—Chemical Communications*, 1976, 949 – 50.

108. Lam, W. - K. F. and Pedigo, L. P. (2001). Effect of trichome density on soybean pod feeding by adult bean leaf beetles (Coleoptera: Chrysomelidae). *Journal of Economic Entomology*, 94, 1459 – 63.

109. Levy, E. C., Ishaaya, I., Gurevitz, E., Cooper, R., and Lavie, D. (1974). Isolation and identification of host compounds eliciting attraction and bite stimuli in the fruit bark beetle, *Scolytus mediterraneus* (Col., Scolytidae). *Journal of Agricultural and Food Chemistry*, 22, 376 – 9.

110. Lin, S., Binder, B. F., and Hart, E. R. (1998). Insect feeding stimulants from the surface of Populus. *Journal of Chemical Ecology*, 24, 1781 – 90.

111. Lombarkia, N. and Derridj, S. (2002). Incidence of apple fruit and leaf surface metabolites on *Cydia pomonella* oviposition. *Entomologia Experimentalis et Applicata*, 104, 79 – 86.

112. Luo, L. - E., van Loon, J. J. A., and Schoonhoven, L. M. (1995). Behavioural and sensory responses to some neem compounds by *Pieris brassicae* larvae. *Physiological Entomology*, 20, 134 – 40.

113. Lupton, F. G. H. (1967). The use of resistant varieties in crop protection. *World Review of Pest Control*, 6, 47 – 58.

114. Ma, W. - C. (1972). Dynamics of feeding responses in *Pieris brassicae* Linn as a function of chemosensory input: a behavioural, ultrastructural and electro-physiological study. *Mededelingen Landbouwhoge-school Wageningen*, 72 – 11, 1 – 162.

115. Ma, W. -C. and Schoonhoven, L. M. (1973). Tarsal chemosensory hairs of the large white butterfly *Pieris brassicae* and their possible role in oviposition behaviour. *Entomologia Experimentalis et Applicata*, 16, 343-57.

116. Malakar, R. and Tingey, W. M. (2000). Glandular trichomes of *Solanum berthaultii* and its hybrids with potato deter oviposition and impair growth of potato tuber moth. *Entomologia Experimentalis et Applicata*, 94, 249-57.

117. Marazzi, C. and Städler, E. (2004a). *Arabidopsis thaliana* leaf-surface extracts are detected by the cabbage root fly (*Delia radicum*) and stimulate oviposition. *Physiological Entomology*, 29, 192-8.

118. Marazzi, C., Patrian, B., and Städler, E. (2004b). Secondary metabolites of the leaf surface affected by sulphur fertilisation and perceived by the diamondback moth. *Chemoecology*, 14, 81-6.

119. Matsuda, K. and Matsuo, H. (1985). A flavonoid, luteolin-7-glucoside, as well as salicin and populin, stimulating the feeding of leaf beetles attacking salicaceous plants. *Applied Entomology and Zoology*, 20, 305-13.

120. McIver, S. B. (1985). Mechanoreception. In *Comprehensive insect physiology, biochemistry & pharmacology*, Vol. 6 (ed. G. A. Kerkut & L. I. Gilbert), pp. 71-132. Pergamon Press, New York.

121. Matsumoto, Y. and Thorsteinson, A. J. (1968). Effect of organic sulfur comounds on oviposition in onion maggot *Hylemya antiqua* Meigen (Diptera, Anthomyiidae). *Applied Entomology and Zoology*, 3, 5-12.

122. Matsumoto, Y. (1969). Some plant chemicals influencing the insect behaviors. *Proceedings of the* 11th *International Congress of Botany*, Seattle, p. 143.

123. Medeiros, L. and Moreira Gilson, R. P. (2002). Moving on hairy surfaces: modifications of *Gratiana spadicea* larval legs to attach on its host plant *Solanum sisymbriifolium*. *Entomologia Experimentalis et Applicata*, 102, 295-305.

124. Menken, S. B. J. (1996). Pattern and process in the evolution of insect-plant associations: *Yponomeuta* as an example. *Entomologia Experimentalis et Applicata*, 80, 297-305.

125. Menken, S. B. J., Herrebout, W. M., and Wiebes, J. T. (1992). Small ermine moths (*Yponomeuta*): their host relations and evolution. *Annual Review of Entomology*, 37, 41-66.

126. Messchendorp, L., Van Loon, J. J. A., and Gols, G. J. Z. (1996). Behavioural and sensory responses to drimane antifeedants in *Pieris brassicae* larvae. *Entomologia Experimentalis et Applicata*, 79, 195-202.

127. Messchendorp, L., Smid, H., and Van Loon, J. J. A. (1998). Role of an epipharyngeal sensillum in the perception of deterrents by Colorado potato beetle larvae. *Journal of Comparative Physiology A*, 183, 255-64.

128. Metcalf, R. L., Metcalf, R. A., and Rhodes, A. M. (1980). Cucurbitacins as kairomones for diabroticite beetles. *Proceedings of the National Academy of Sciences of the USA*, 77,

3769–72.

129. Miller, J. R. and Strickler, K. L. (1984). Finding and accepting host plants. In *Chemical ecology of insects* (ed. W. J. Bell and R. T. Cardé), pp. 127–57. Chapman & Hall, London.

130. Mitchell, B. K. (1974). Behavioural and electro-physiological investigations on the responses of larvae of the Colorado potato beetle (*Leptinotarsa decemlineata*) to amino acids. *Entomologia Experimentalis et Applicata*, 17, 255–64.

131. Mitchell, B. K. (1988). Adult leaf beetles as models for exploring the chemical basis of host-plant recognition. *Journal of Insect Phsyiology*, 34, 213–25.

132. Mitchell, B. K. and Gregory, P. (1979). Physiology of the maxillary sugar sensitive cell in the red turnip beetle, *Entomoscelis americana*. *Journal of Comparative Physiology*, 132, 167–78.

133. Mitchell, B. K. and Harrison, G. D. (1984). Characterization of galeal chemosensilla in the adult Colorado beetle, *Leptinotarsa decemlineata*. *Physiological Entomology*, 9, 49–56.

134. Mitchell, B. K. and Itagaki, H. (1992). Interneurons of the subesophageal ganglion of *Sarcophaga bullata* responding to gustatory and mechanosensory stimuli. *Journal of Comparative Physiology A*, 171, 213–30.

135. Mitchell, B. K. and Schoonhoven, L. M. (1974). Taste receptors in Colorado beetle larvae. *Journal of Insect Physiology*, 20, 1787–93.

136. Mitchell, B. K., Rolseth, B. M., and McCashin, B. G. (1990). Differential responses of galeal gustatory sensilla of the adult Colorado potato beetle, *Leptinotarsa decemlineata* (Say), to leaf saps from host and non-host plants. *Physiological Entomology*, 15, 61–72.

137. Montgomery, M. E. and Arn, H. (1974). Feeding response of *Aphis pomi*, *Myzus persicae* and *Amphorophora agathonica* to phlorizin. *Journal of Insect Physiology*, 20, 413–21.

138. Morgan, E. D. and Mandava, N. B. (1990). *CRC Handbook of natural pesticides*, Vol. 6, *Insect attractants and repellents*. CRC Press, Boca Raton.

139. Morris, B. D., Foster, S. P., and Harris, M. O. (2000). Identification of 1-octacosanal and 6-methoxy-2-benzoxazolinone from wheat as ovipositional stimulants for Hessian fly, *Mayetiola destructor*. *Journal of Chemical Ecology*, 26, 859–73.

140. Müller, C., and Renwick, J. A. A. (2001). Different phagostimulants in potato foliage for *Manduca sexta* and *Leptinotarsa decemlineata*. *Chemoecology*, 11, 37–41.

141. Mullin, C. A., Chyb, S., Eichenseer, H., Hollister, B., and Frazier, J. L. (1994). Neuroreceptor mechanisms in insect gustation: a pharmacological approach. *Journal of Insect Physiology*, 40, 913–31.

142. Nair, K. S. S. and McEwen, F. L. (1976). Host selection by the adult cabbage maggot, *Hylemya brassicae* (Diptera: Anthomyiidae): effect of glucosinolates and common nutrients on oviposition. *Canadian Entomologist*, 108, 1021–30.

143. Nelson, N. and Bernays, E. A. (1998). Inositol in two host plants of *Manduca sexta*. *Entomologia Experimentalis et Applicata*, 88, 189–93.

144. Nielsen, J. K., Larsen, L. M., and Sørensen, H. (1979). Host plant selection of the horse-radish flea beetle, *Phyllotreta armoraciae* (Coleoptera: Chrysomelidae): identification of two flavonol glycosides stimulating feeding in combination with glucosinolates. *Entomologia Experimentalis et Applicata*, 26, 40-8.

145. Deleted.

146. Nishida, R. (1995). Oviposition stimulants of swallowtail butterflies. In *Swallowtail butterflies: their ecology and evolutionary biology* (ed. J. M. Scriber, Y. Tsubaki, and R. C. Lederhouse), pp. 17-26. Scientific Publishers, Gainesville, FL.

147. Nishida, R. and Fukami, H. (1989). Oviposition stimulants of an Aristolochiceae-feeding swallowtail butterfly, *Atrophaneura alcinous*. *Journal of Chemical Ecology*, 15, 2565-75.

148. Nishijima, Y. (1960). Host plant preference of the soybean pod borer, *Grapholita glicinivorella* (Matsumura) (Lep., Encosmidae). *Entomologia Experimentalis et Applicata*, 3, 38-47.

149. Nottingham, S. F., Son, K. C., Wilson, D. D., Severson, R. F., and Kays, S. J. (1989). Feeding and oviposition preferences of sweet potato weevils, *Cyclas formicarius elegantulus* (Summers), on storage roots of sweet potato cultivars with differing surface chemistries. *Journal of Chemical Ecology*, 15, 895-903.

150. Nufio, C. R. and Papaj, D. R. (2001). Host marking behavior in phytophagous insects and parasitoids. *Entomologia Experimentalis et Applicata*, 99, 273-93.

151. Ohta, K. C., Matsuda, K., and Matsumoto, Y. (1998). Feeding stimulation of strawberry leaf beetle, *Galerucella vittaticollis* Baly (Coleoptera: Chrysomelidae) by quercetin glycosides in polygonaceous plants. *Japanese Journal of Applied Entomology and Zoology*, 42, 45-9.

152. Panda, N. and Khush, G. S. (1995). *Host plant resistance to insects*. CAB International, London.

153. Panzuto, M. and Albert, P. J. (1998). Chemoreception of amino acids by female fourth and sixth-instar larvae of the spruce budworm. *Entomologia Experimentalis et Applicata*, 86, 89-96.

154. Pereyra, P. C. and Bowers, M. D. (1988). Iridoid glycosides as oviposition stimulants for the buckeye butterfly, *Junonia coenia* (Nympalidae). *Journal of Chemical Ecology*, 14, 917-28.

155. Peterson, S. C., Herrebout, W. M., and Kooi, R. E. (1990). Chemosensory basis of host colonization by small ermine moth larvae. *Proceedings of the Koninklijke Nederlandse Akademie van Wetenschappen*, 93, 287-94.

156. Peterson, S. C., Hanson, F. E., and Warthen, J. D. (1993). Deterrence coding by a larval *Manduca* chemosensory neurone mediating rejection of a nonhost plant, *Canna generalis* L. *Physiological Entomology*, 18, 285-95.

157. Deleted.

158. Quiras, C. F., Stevens, M. A., Rick, C. M., and Kok-Yokomi, M. K. (1977). Resistance in tomato to the pink form of the potato aphid, *Macrosiphum euphorbiae* (Thomas): the role

of anatomy, epidermal hairs and foliage composition. *Journal of the American Society of Horticultural Science*, 102, 166 – 71.

159. Raffa, K. (1987). Influence of host plant on deterrence by azadirachtin of feeding by fall armyworm larvae (Lepidoptera: Noctuidae). *Journal of Economic Entomology*, 80, 384 – 7.

160. Ranger, C. M., Backus, E. A., Winter, R. E. K., Rottinghaus, G. E., Ellersieck, M. R., and Johnson, D. W. (2004). Glandular trichome extracts from *Medicago sativa* deter settling by the potato leafhopper *Empoasca fabae*. *Journal of Chemical Ecology*, 30, 927 – 43.

161. Rees, C. J. C. (1969). Chemoreceptor specificity asso-ciated with choice of feeding site by the beetle, *Chrysolina brunsvicensis* on its foodplant, *Hypericum hirsutum*. *Entomologia Experimentalis et Applicata*, 12, 565 – 83.

162. Reifenrath, K., Riederer, M., and Müller, C. (2005). Leaf surfaces of Brassicaceae lack feeding stimulants for *Phaedon cochleariae*. *Entomologia Experimentalis et Applicata*, 115, 41 – 50.

163. Renwick, J. A. A. (1990). Oviposition stimulants and deterrents. In *Handbook of natural pesticides*, Vol. 6, *Insect attractants and repellents* (ed. E. D. Morgan and N. B. Mandava), pp. 151 – 60. CRC Press, Boca Raton, FL.

164. Renwick, J. A. A., Radke, C. D., Sachdev-Gupta, K., and Städler, E. (1992). Leaf surface chemicals stimulating oviposition by *Pieris rapae* (Lepidoptera: Pieridae). *Chemoecology*, 3, 33 – 8.

165. Roessingh, P. and Städler, E. (1990). Foliar form, colour and surface characteristics influence oviposition behaviour in the cabbage root fly, *Delia radicum*. *Entomologia Experimentalis et Applicata*, 57, 93 – 100.

166. Roessingh, P., Städler, E., Schöni, R., and Feeny, P. (1991). Tarsal contact chemoreceptors of the black swallowtail butterfly *Papilio polyxenes*: responses to phytochemicals from host and non-host plants. *Physiological Entomology*, 16, 485 – 95.

167. Roessingh, P., Städler, E., Fenwick, G. R., Lewis, J. A., Nielsen, J. K., Hurter, J., et al. (1992a). Oviposition and tarsal chemoreceptors of the cabbage root fly are stimulated by glucosinolates and host plant extracts. *Entomologia Experimentalis et Applicata*, 65, 267 – 82.

168. Roessingh, P., Städler, E., Hurter, J., and Ramp, T. (1992b). Oviposition stimulant for the cabbage root fly: important new cabbage leaf surface compound and specific tarsal receptors. In *Proceedings of the 8th International Symposium on Insect-Plant Relationships* (ed. S. B. J. Menken, J. H. Visser, and P. Harrewijn), pp. 141 – 2. Kluwer Academic, Dordecht.

169. Roessingh, P., Städler, E., Baur, R., Hurter, J., and Ramp, T. (1997). Tarsal chemoreceptors and oviposition behaviour of the cabbage root fly (*Delia radicum*) sensitive to fractions and new compounds of host-leaf surface extracts. *Physiological Entomology*, 22, 140 – 8.

170. Rogers, S. M. and Newland, P. L. (2003). The neurobiology of taste in insects. *Advances in Insect Physiology*, 31, 141 – 204.

171. Roininen, H., Price, P. W., Julkunen-Tiitto, R., Tahvanainen, J., and Ikonen, A.

(1999). Oviposition stimulant for a gall-inducing sawfly, *Euura lasiolepis*, on willow is a phenolic glucoside. *Journal of Chemical Ecology*, 25, 943–53.

172. Roitberg, B. D. and Prokopy, R. J. (1987). Insects that mark host plants. *BioScience*, 37, 400–6.
173. Romeis, J., Shanower, T. G., and Peter, A. J. (1999). Trichomes on pigeonpea (*Cajanus cajan* (L.) Millsp.) and two wild *Cajanus* spp. *Crop Science*, 39, 564–9.
174. Rothschild, M. and Schoonhoven, L. M. (1977). Assessment of egg load by *Pieris brassicae* (Lepidoptera: Pieridae). *Nature*, 266, 352–5.
175. Sachdev-Gupta, K., Feeny, P., and Carter, M. (1993). Oviposition stimulants for the pipevine swallowtail, Battus philenor, from an *Aristolochia* host plant: synergism between inositols, aristolochic acids and a monogalactosyl diglyceride. *Chemoecology*, 4, 19–28.
176. Schoonhoven, L. M. (1967). Chemoreception of mustard oil glucosides in larvae of *Pieris brassicae*. *Proceedings of the Koninklijke Nederlandse Akademie van Wetenschappen*, C70, 556–68.
177. Schoonhoven, L. M. (1969). Amino-acid receptor in larvae of *Pieris brassicae*. *Nature*, 221, 1268.
178. Schoonhoven, L. M. (1972). Secondary plant substances and insects. *Recent Advances in Phytochemistry*, 5, 197–224.
179. Schoonhoven, L. M. (1973). Plant recognition by lepidopterous larvae. *Symposia of the Royal Entomological Society of London*, 6, 87–99.
180. Schoonhoven, L. M. (1981). Chemical mediators between plants and phytophagous insects. In *Semiochemicals: their role in pest control* (ed. D. A. Nordlund, R. L. Jones, and W. J. Lewis), pp. 31–50. John Wiley, New York.
181. Schoonhoven, L. M. (1982). Biological aspects of antifeedants. *Entomologia Experimentalis et Applicata*, 31, 57–69.
182. Schoonhoven, L. M. (1987). What makes a caterpillar eat? The sensory code underlying feeding behaviour. In *Advances in chemoreception and behaviour* (R. F. Chapman, E. A. Bernays, and J. G. Stoffolano), pp. 69–97. Springer, New York.
183. Schoonhoven, L. M. (1990). Host-marking pheromones in Lepidoptera, with special reference to *Pieris* spp. *Journal of Chemical Ecology*, 16, 3043–52.
184. Schoonhoven, L. M. (1991). 100 years of botanical instinct. *Symposia Biologica Hungarica*, 39, 3–14.
185. Schoonhoven, L. M. (1997). Verschaffelt 1910: a founding paper in the field of insect-plant relationships. *Proceedings Koninklijke Nederlandse Akademie van Wetenschappen*, 100, 355–61.
186. Schoonhoven, L. M. and Blom, F. (1988). Chemoreception and feeding behaviour in a caterpillar: towards a model of brain functioning in insects. *Entomologia Experimentalis et Applicata*, 49, 123–9.
187. Schoonhoven, L. M. and Derksen-Koppers, I. (1976). Effects of some allelochemics on food

uptake and survival of a polyphagous aphid, *Myzus persicae*. *Entomologia Experimentalis et Applicata*, 19, 52 – 6.

188. Schoonhoven, L. M. and Dethier, V. G. (1966). Sensory aspects of host-plant discrimination by lepidopterous larvae. *Archives Néerlandais de Zoologie*, 16, 497 – 530.
189. Schoonhoven, L. M. and Luo, L. E. (1994). Multiple mode of action of the feeding deterrent, toosendanin, on the sense of taste in *Pieris brassicae* larvae. *Journal of Comparative Physiology A*, 175, 519 – 24.
190. Schoonhoven, L. M., and van Loon, J. J. A. (2002). An inventory of taste in caterpillars: each species its own key. *Acta Zoologica Academiae Scientiarium Hungaricae*, 48 (Suppl 1), 215 – 63.
191. Schoonhoven, L. M., Blaney, W. M., and Simmonds, M. S. J. (1992). Sensory coding of feeding deterrents in phytophagous insects. In *Insect-plant interactions*, Vol. 4 (ed. E. A. Bernays), pp. 59 – 79. CRC Press, BocaRaton.
192. Shields, V. D. C. and Mitchell, B. K. (1995a). Sinigrin as a feeding deterrent in two crucifer feeding, polyphagous lepidopterous species and the effects of feeding stimulant mixtures on deterrency. *Philosophical Transactions of the Royal Society*, London, B347, 439 – 46.
193. Shields, V. D. C. and Mitchell, B. K. (1995b). The effect of phagostimulant mixtures on deterrent receptor(s) in two crucifer-feeding lepidopterous species. *Philosophical Transactions of the Royal Society*, London, B347, 459 – 64.
194. Simmonds, M. S. J. and Blaney, W. M. (1991). Gustatory codes in lepidopterous larvae. *Symposia Biologica Hungarica*, 39, 17 – 27.
195. Simmonds, M. S. J. and Blaney, W. M. (1996). Azadirachtin: advances in understanding its activity as an antifeedant. *Entomologia Experimentalis et Applicata*, 80, 23 – 6.
196. Simmonds, M. S. J., Blaney, W. M., Mithen, R., Birch, A. N., and Fenwick, R. (1994). Behavioural and chemosensory responses of the turnip root fly (*Delia floralis*) to glucosinolates. *Entomologia Experimentalis et Applicata*, 71, 41 – 57.
197. Simmonds, M. S. J., Blaney, W. M., Ley, S. V., Anderson, J. C., Banteli, R., Denholm, A. A., et al. (1995). Behavioural and neurophysiological responses of *Spodoptera littoralis* to azadirachtin and a range of synthetic analogues. *Entomologia Experimentalis et Applicata*, 77, 69 – 80.
198. Singer, M. C., Thomas, C. D., and Parmesan, C. (1993). Rapid human-induced evolution of insecthost associations. *Nature*, 366, 861 – 3.
199. Smith, B. D. (1966). Effect of the plant alkaloid sparteine on the distribution of the aphid *Acyrtosiphon spartii* (Koch). Nature, 212, 213 – 14. 200. Smith, C. M. (1989). Plant resistance to insects. A fundamental approach. John Wiley, New York.
200. Städler, E. (1978). Chemoreception of host plant chemicals by ovipositing females of *Delia* (*Hylemya*) *brassicae*. *Entomologia Experimentalis et Applicata*, 24, 711 – 20.
201. Städler, E. (1986). Oviposition and feeding stimuli in leaf surface waxes. In *Insects and the plant surface* (ed. B. Juniper and T. R. E. Southwood), pp. 105 – 21. Edward Arnold, Lon-

don.

202. Städler, E. (1992). Behavioral responses of insects to plant secondary compounds. In *Herbivores: their interactions with secondary plant metabolites*, Vol. 2 (2nd edn) (ed. G. A. Rosenthal and M. R. Berenbaum), pp. 45–88. Academic Press, New York.

203. Städler, E. (2002). Plant chemical cues important for egg deposition by herbivorous insects. In *Chemoecology of insect eggs and egg deposition* (ed. M. Hilker and T. Meiners), pp. 171–204. Blackwell, Berlin.

204. Städler, E. and Buser, H. R. (1984). Defense chemicals in leaf surface wax synergistically stimulate oviposition by a phytophagous insect. *Experientia*, 40, 1157–9.

205. Städler, E. and Buser, H. R. (1984). Defense chemicals in leaf surface wax synergistically stimulate oviposition by a phytophagous insect. *Experientia*, 40, 1157–9.

206. Städler, E. and Roessingh, P. (1991). Perception of surface chemicals by feeding and ovipositing insects. *Symposia Biologica Hungarica*, 39, 71–86.

207. Städler, E., Ernst, B., Hurter, J., and Boller, E. (1994). Tarsal contact chemoreceptor for host marking pheromone of the cherry fruit fly, *Rhagoletis cerasi*: responses to natural and synthetic compounds. *Physiological Entomology*, 19, 139–51.

208. Städler, E., Renwick, J. A. A., Radke, C. D., and Sachdev-Gupta, K. (1995). Tarsal contact chemoreceptor response to glucosinolates and cardenolides mediating oviposition in *Pieris rapae*. *Physiological Entomology*, 20, 175–87.

209. Stoner, K. A. (1990). Glossy leaf wax and plant resistance to insects in *Brassica oleracea* under natural infestation. *Environmental Entomology*, 19, 730–9.

210. Stork, N. E. (1980). Role of waxblooms in preventing attachment to *Brassicas* by the mustard beetle, *Phaedon cochleariae*. *Entomologia Experimentalis et Applicata*, 28, 100–7.

211. Sutcliffe, J. F. and Mitchell, B. K. (1982). Characterization of galeal sugar and glucosinolate-sensitive cells in *Entomoscelis americana* adults. *Journal of Comparative Physiology A*, 146, 393–400.

212. Takemura, M., Nishida, R., Mori, N., and Kuwahara, Y. (2002). Acylated flavonol glycosides as probing stimulants of a bean aphid, *Megoura crassicauda*, from *Vicia angustifolia*. *Phytochemistry*, 61, 135–40.

213. Tanton, M. T. (1962). The effect of leaf toughness on the feeding of larvae of the mustard beetle, *Phaedon cochleariae* Fab. *Entomologia Experimentalis et Applicata*, 5, 74–8.

214. Thompson, J. N. (1994). *The coevolutionary process*. University of Chicago Press, Chicago.

215. Thorsteinson, A. J. (1958). The chemotactic influence of plant constituents on feeding by phytophagous insects. *Entomologia Experimentalis et Applicata*, 1, 23–7.

216. Tjallingii, W. F. (1995). Regulation of phloem sap feeding by aphids. In *Regulatory mechanisms in insect feeding* (ed. R. F. Chapman and G. de Boer), pp. 190–209. Chapman & Hall, New York.

217. Tjallingii, W. F. (2000). Comparison of AC and DC systems for electronic monitoring of stylet penetration activities by homopterans. In *Principles and applications of electronic monitoring*

and other techniques in the study of homopteran feeding behavior (ed. G. P. Walker and E. A. Backus), pp. 41 – 69. Thomas Say Publications in Entomology, Entomological Society of America, Lanham, MD.

218. Tjallingii, W. F. and Hogen Esch, T. (1993). Fine structure of aphid stylet routes in plant tissues in correlation with EPG signals. *Physiological Entomology*, 18, 317 – 28.

219. Tjallingii, W. F. and Mayoral, A. M. (1992). Criteria for host-plant acceptance by aphids. In *Proceedings of the 8th International Symposium on Insect-Plant Relationships* (ed. S. B. J. Menken, J. H. Visser, and P. Harrewijn), pp. 280 – 2. Kluwer Academics, Dordrecht.

220. Van Drongelen, W. (1979). Contact chemoreception of host plant specific chemicals in larvae of various *Yponomeuta* species (Lepidoptera). *Journal of Comparative Physiology*, 134, 265 – 2.

221. Van Drongelen, W. and Van Loon, J. J. A. (1980). Inheritance of gustatory sensitivity in F1 progeny of crosses between *Yponomeuta cagnagellus* and *Y. malinellus* (Lepidoptera). *Entomologia Experimentalis et Applicata*, 28, 199 – 203.

222. Van Helden, M. and Tjallingii, W. F. (1993). The resistance of lettuce (*Lactuca sativa* L.) to *Nasonovia ribisnigri*: the use of electrical penetration graphs to locate the resistance factor (s) in the plant. *Entomologia Experimentalis et Applicata*, 66, 53 – 8.

223. Van Loon, J. J. A. (1990). Chemoreception of phenolic acids and flavonoids in larvae of two species of *Pieris*. *Journal of Comparative Physiology A*, 166, 889 – 99.

224. Van Loon, J. J. A. (1996). Chemosensory basis of feeding and oviposition behaviour in herbivorous insects: a glance at the periphery. *Entomologia Experimentalis et Applicata*, 80, 7 – 13.

225. Van Loon, J. J. A. and Schoonhoven, L. M. (1999). Specialist deterrent chemoreceptors enable *Pieris* caterpillars to discriminate between chemically different deterrents. *Entomologia Experimentalis et Applicata*, 91, 29 – 35.

226. Van Loon, J. J. A. and Van Eeuwijk, F. A. (1989). Chemoreception of amino acids in larvae of two species of *Pieris*. *Physiological Entomology*, 14, 459 – 69.

227. Van Loon, J. J. A., Blaakmeer, A., Griepink, F. C., Van Beek, T. A., Schoonhoven, L. M., and De Groot, Æ. (1992). Leaf surface compound from *Brassica oleracea* (Cruciferae) induces oviposition by *Pieris brassicae* (Lepidoptera: Pieridae). *Chemoecology*, 3, 39 – 44.

228. Deleted.

229. Van Loon, J. J. A., Wang, C. Z., Nielsen, J. K., Gols, R., and Qiu, Y. T. (2002). Flavonoids from cabbage are feeding stimulants for diamondback moth larvae additional to glucosinolates: chemoreception and behaviour. *Entomologia Experimentalis et Applicata*, 104, 27 – 34.

230. Verschaffelt, E. (1910). The causes determining the selection of food in some herbivorous insects. *Proceedings of the Koninklijke Nederlandse Akademie van Wetenschappen*, 13, 536 – 42.

231. Vrieling, K. and Derridj, S. (2003). Pyrrolizidine alkaloids in and on the leaf surface of *Senecio jacobaea* L. *Phytochemistry*, 64, 1223 – 8.

232. Deleted.
233. Wensler, R. J. D. (1962). Mode of host selection by an aphid. *Nature*, 195, 830 –1.
234. White, P. R., Chapman, R. F., and Ascoli-Christensen, A. (1990). Interactions between two neurons in contact chemosensilla of thegrasshopper *Schistocera americana*. *Journal of Comparative Physiology*, A, 167, 431 –6.
235. Woodhead, S., Galeffi, C., and Marini Betollo, G. B. (1982). *p*-Hydroxybenzaldehyde is a major constituent of the epicuticular wax of seedling *Sorghum bicolor*. *Phytochemistry*, 21, 455 –6.
236. Yencho, C. G. and Tingey, W. M. (1994). Glandular trichomes of *Solanum berthaultii* alter host preference of the Colorado potato beetle, *Leptinotarsa decemlineata*. *Entomologia Experimentalis et Applicata*, 70, 217 –25.
237. Zielske, A. F., Simons, J. N., and Silverstein, R. M. (1972). A flavone feeding stimulant in alligatorweed. *Phytochemistry*, 11, 393 –6.

第8章 寄主植物的选择:变化即规律

8.1 地理变异
8.2 相同地区的种群间差异
8.3 个体间的差异
8.4 对寄主植物偏好引起改变的环境因素
　8.4.1 季节性
　8.4.2 温度
　8.4.3 捕食风险
8.5 引起寄主植物偏好改变的内部因素
　8.5.1 发育阶段
　8.5.2 昆虫性别影响食物选择
8.6 经验诱导寄主植物偏好的改变
　8.6.1 非联系式改变
　8.6.2 联系式改变
8.7 幼虫和早期成虫的经验
8.8 经验诱导寄主植物偏好改变的适应意义
8.9 结论
8.10 参考文献

　　生命的基本特征是不断变化的。明显的变化存在于界、门、目、种等类别之间,但是即使是种内也存在着变异。而这种变异对于植食性昆虫对寄主的选择尤为重要。一种昆虫的寄主范围并非一成不变。研究表明,某种昆虫的一些个体甚至整个种群都可能拒食这种昆虫正常寄主范围内的某些植物。在昆虫种内或种间的寄主植物范围和偏好程度似乎是经常变化的,而且先前认为的"植食性昆虫具有固定的寄主偏好"现在认为是不客观的。对于昆虫寻找特定类型的寄主植物并接受的偏好差异,可能取决于基因或先前的经验(包括联结式学习及其他一些类型的学习)。大量文献表明,表型变异(phenotypic variation)是植食性昆虫(个体或种群)选择寄主植物普遍采用的利用因素。而现在关于寄主偏好的遗传分化现象却鲜有报道,尽管这种现象可能并不少见[49,59]。基于目前有限的信息,假定昆虫的寄主选择行为和行动的变化,通常涉及基因型和经验的因素。每一个昆虫作为一个个体,这些个体具有各自的取食偏好。从这个角度看,"异常"的行为是不存在的。本章的重点在寄主植物选择的表型方面,寄主植物选择行为的基因型变异将在第11章中进行讨论(见11.3)。

8.1 地理变异

　　多数情况下,位于不同分布区的昆虫会呈现不同的寄主偏好。在北美,斑幕潜叶蛾(*Phyl-*

lonorycter blancardella)仅取食苹果属(*Malus*)植物。而在其原产地欧洲,它的寄主范围相当广泛,并且在蔷薇科七个属的植物上都能生长得非常好[85]。这与叶蝉科(*Graphocephala ennahi*)的情况恰好相反。在20世纪初,通过研究证实,新北区的本土物种仅以杜鹃花(*Rhododendron*)植物为食,而在欧洲却成为取食13个不同科植物的杂食性物种。对此的一种解释为,当昆虫叶蝉(*G. ennahi*)被引入新的区域后,这个物种即开始在新的生态条件扩充其寄主范围。然而,对上述斑幕潜叶蛾(*P. blancardella*)的例子却无法用这个理论进行解释[120]。至于为什么两个物种表现出的反应会相反,目前仍未明确。但无论什么原因,这两个例子至少表明当一种昆虫引入新的区域后,寄主范围可能会发生变化。

不同种群的马铃薯甲虫,对寄主植物选择存在巨大的偏好性。在北美南部,这种昆虫只取食刺萼龙葵(*Solanum rostratum*)和茄属(*S. augustifolium*),如果以银叶茄(*S. elaeagnifolium*)或栽培马铃薯(*S. tuberosum*)喂养,则几乎不能存活。然而位于亚利桑那州的种群却能(也是唯一的)取食银叶茄(*S. elaeagnifolium*)。位于美国北部,可以取食栽培马铃薯(*S. tuberosum*)的种群,却不能在银叶茄(*S. elaeagnifolium*)上存活[63]。北卡罗来纳州的种群在刺龙葵(*S. rostratum*)上可以繁荣生长,但更北区域的种群在这种寄主上却表现了很低的存活率。而在刺龙葵(*S. rostratum*)上的存活与否,实际上反映了马铃薯甲虫种群内或种群间的遗传变异[61]。最近的研究表明,马铃薯甲虫寄主范围的扩大现象涉及其中重要基因的改变[82]。

以上这种在不同地理种群间寄主范围的变异现象,在主要的昆虫类群中都会出现,以下为两个蝴蝶和蝗虫的例子。北美黑条黄凤蝶(*Papilio glaucus*)是一种真正的杂食性昆虫,遍布北美大陆的大部分区域,取食至少七个科的植物,但在任何一个地理区域内,北美黑条黄凤蝶(*P. glaucus*)的取食植物范围却非常有限。佛罗里达州的凤蝶(*P. glaucus*)仅以木兰属(*Magnolia virginiana*)植物为食,而俄亥俄州的种群则为杂食性的[24]。基于这种对食物的偏好,及对不同寄主植物的选择差异,已发现了一些亚种[25,118]。沙漠蝗(*Schistocerca emarginata*)是一类杂食性蝗虫,也存在一些明显为单食性的种群。种群间取食习性的差异,并非仅由不同种群所在的地区和可供取食的食物差异造成的,也有可能是基于种群间的遗传分化。像这种可能取决于基因隔离程度的寄主—关联的种群,可以表示为寄主族(host races)。从寄主族开始,基因的遗传差异可能会逐步积累,以致形成同域不同物种(见11.2)[131]。

还有更多的例子是关于不同昆虫种群由于地理区域不同,其寄主发生变异的例子,包括鳞翅目、同翅类、鞘翅目、双翅目等类群。所有的例子均表明,尽管一个物种在它的所有分布范围内总体而言是杂食性的,但位于局部区域的种群则有可能是专食性的[54,136]。

不同地区的种群对寄主的偏好性,事实上通常反映了对本地条件的适应。本地的一些因素,如存在食物竞争者,可能会对当地种群施加选择压力,从而导致寄主植物专一性。这通过两种亲缘关系密切的象甲(*Larinus sturnus*)和象甲(*L. jaceae*)对寄主的偏好得到了证明。象甲(*L. sturnus*)和象甲(*L. jaceae*)都取食蓟和矢车菊的顶端部分,当这两种昆虫同时存在时,尽管这两个物种几乎有相同的食物生态位,但为了避免幼虫对食物和空间的竞争,它们选择不同的蓟属植物作为其繁殖寄主,表现的是一种生态转移现象(表8.1)[144]。

表 8.1　两个近缘种象甲和象甲的寄主植物的地区差异(托沃尔费,1970)[144]

地区	象甲(L. sturnus)	象甲(L. jaceae)
瑞士,侏罗	大矢车菊(Centaurea scabiosa)	垂花飞廉(Carduus nutans)
瑞士,苏格兰	大矢车菊(Centaurea scabiosa)	
德国,法尔兹		大矢车菊(Centaurea scabiosa)
法国,阿尔萨斯	垂花飞廉(Carduus nutans)	

地理分隔造成种群间的遗传差异,可能是形成寄主偏好性的原因。搜集不同地区的象甲(L. sturnus)进行寄主选择实验,可以揭示它们的祖先起源地,由此也暗示了寄主植物选择差异的遗传基础。这些昆虫与植物可能形成了一种镶嵌模式,这种模式继续发展就形成了现今的昆虫—植物联合体[144,145]。在其他的一些实例中,不同地区植物的性能差别也是寄主选择差异的一项重要因素。

对于寄主植物选择差异的最直接原因,可能要归因于不同种群的感受器所受到的不同调节,形成不同的感觉输入,最终导致行为差异。芥子油苷是在寄主识别过程中发挥重要作用的一类化合物,来自不同地区(欧洲和北美)的暗脉粉蝶(Pieris napi)对各种芥子油苷的反应差异很大(图 8.1)。这些差异可能与两个大陆间寄主植物的差异有关[142]。

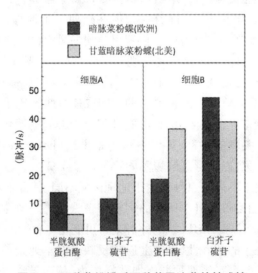

图 8.1　两种菜粉蝶对两种芥子油苷的敏感性

暗脉粉蝶(Pieris napi)的两个亚种,粉蝶采集于欧洲,粉蝶采集于北美,测试其跗节的化学感受器对两种芥子油苷的敏感性。当用 100 mg/ml 的白花菜苷或白芥子硫苷进行刺激时,两种感芥子油苷细胞(A 和 B)的神经冲动频率如图所示,对不同的刺激两个亚种均表现出显著的差异($P < 0.05$)(杜等,1995)[51]

除了研究种内的寄主植物偏好性差异外,这种差异的范围(如昆虫与寄主关系保守性的方面)也是非常重要的,且有待研究。这方面有一个关于马铃薯甲虫的例子,研究来自北美的两个马铃薯甲虫种群和欧洲的三个种群,对于寄主植物(茄属)及非寄主植物行为反应的异同之处。尽管北美和欧洲的种群已分别独立进化了约 70 年(约 150 代),但是经研究发现,位于相同大陆的种群间的取食行为差异更为显著,表明了两个大陆显著的生态差异(如可供选择的寄主植物、自然天敌等),到目前为止并未引起显著的行为改变[134]。而在20世纪中期,在美国发现了一种不同于以往染色体类型的种群,该种群迅速扩散至整个大陆并与原始种群杂交,而这个新出现的种群至今未在欧洲发现[63]。

8.2 相同地区的种群间差异

蛱蝶(*Euphydryas editha*)的成虫可以在加利福尼亚玄参科的五种不同植物上产卵。一些种群是严格专食性的,而一些其他种群通常为专食性,但是也会偶尔选择其他寄主植物。一些少数种群的雌虫可以在多达四种不同属寄主植物上产卵。而这些种群间对寄主植物选择的差异,并非仅由可利用寄主的差异引起,因为这些种群的生境的寄主植物几乎相同。更为有趣的是,尽管在寡食性种群中的个体间仍存在差异,一些个体是单食性的,而一些个体甚至可以接受所有可能的寄主。在这样的种群中,这种对寄主植物的专食性程度,似乎呈一种连续变化的方式取食[125]。尽管在实验室条件下,这种蝴蝶的寄主植物偏好性不如自然环境下明显,但是在自然环境观察到的种群间的差异也是被保持的,因此也表明了在寄主利用方面存在广泛生态型变化的遗传基础[126]。如上所述,不同的种群表现出不同的寄主偏好行为,部分原因是植物的种群间变异[55]。蛱蝶伴随(*E. editha*)蝴蝶种群的基因变异,它们的寄主植物种群也产生了基因变异[127,128]。

位于不同地区的同种植物,可能会受到不同昆虫的攻击,这种植食性昆虫对寄主植物偏好的差异是由基因决定的。这种地域性差异的种群,无论是昆虫还是植物,都是新物种形成的中间阶段。

8.3 个体间的差异

研究昆虫对不同寄主植物的取食反应时,不时会遇到行为异常且不遵从"正常"寄主范围偏好的一些个体。然而,这种情况发生得太过频繁,以致不能仅把其看作异常情况。许多研究者深受传统哲学的影响,忽略了这种行为或生理的变化范围,然而这种"中庸之道"也会忽略一些生命的基本原则[11]。

在观察水蜡树天蛾(*Sphinx ligustri*)幼虫的一个实验中,发现了一种"不正常"行为。正常情况下,水蜡树天蛾(*S. ligustri*)幼虫仅取食水蜡树和相关的一些木樨科植物。将十只水蜡树天蛾(*S. ligustri*)幼虫从它们正常寄主植物上移除,并让其饥饿一小时,然后将幼虫放到取自相同植株的夹竹桃(*Nerium oleander*)叶片上。开始时和预想的一样,在四小时内,幼虫都没有取食,或最多只进行尝试性的取食。然而,有一个个体例外,这只幼虫很快便开始取食,在观察实验的过程中,消耗掉3.8 cm²的叶片。竹桃属植物富含高毒性的强心苷,而显然并未对水蜡树天蛾(*S. ligustri*)这个特殊的个体造成行为障碍。在所有植食性昆虫中,生长和发育并未受不寻常食物的显著影响,这种情况并不罕见(图8.2)[112]。同样,在正常寄主范围之外的植物上"错误"产卵,在植食性昆虫中已有非常丰富的记录,需要引起更多的关注[90,119]。这种个体对寄主选择的偏离情况,实际上反映了由突变导致的随机变异。这种变异是由自然选择保留的,从而使得一个物种能够应对环境条件的改变。

图 8.2　在四个小时内观察五只被置于非寄主植物（蒲公英）叶片（所有的叶片都采集于同一植株）上的烟草天蛾（*Manduca sexta*）幼虫的摄食能力

其中的三只幼虫表现出小口试咬但拒绝取食的行为，而另两只幼虫则立即开始取食，并消耗了一定的叶片组织，白色叶区表示消耗的部分

经观察发现，广食性物种比专食性物种产卵所犯的"错误"更多[117]，这可能是由于广食性雌虫要处理更多的信息。它们的神经处理信息的能力有限，可能导致它们做出误差决策[14]。

8.4　对寄主植物偏好引起改变的环境因素

8.4.1　季节性

在一个种群的不同个体间，甚至同一个体对寄主植物的偏好都会随时间发生变化。在蚜虫类群中，这种季节性变化尤为显著。蚜虫在其连续世代中，其寄主植物会发生一种强制性改变（图 8.3）。

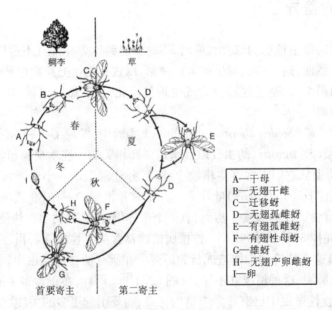

图 8.3　更换寄主的禾谷缢管蚜（*Rhopalosiphum padi*）的生活史

而这是现生蚜虫约10%种群的典型特性,尽管现生的许多不更换寄主的物种被认为源自更换寄主的祖先[77,78,121]。寄主交换物种的生命周期简述如下:春季,有翅的雌性蚜虫离开其第一寄主;夏季,它们的后代生活在第二寄主植物(快速生长的草本植物)内;当蚜虫离开第一寄主而未找到合适的取食植物时,会处于巨大的风险中,然而似乎可以通过第二寄主有较高食物质量得以补偿。研究表明,吸食草本植物汁液的蚜虫比木本的物种有更快的生长率和更高的繁殖率(平均差异达160%)[81]。

在草本植物上更高的生长繁殖率,可能是由于草本植物韧皮部汁液中可用的氨氮数量较高造成的[48]。夏季的蚜虫都是无翅、快速成熟、孤雌生殖的个体,这些个体能够快速连续繁殖新的个体。秋季初期,蚜虫开始形成翅,成为母性,飞回第一寄主进行产卵。

夏季和冬季的寄主植物,通常分别属于差别很大的不同科植物(表8.2)。例如,在温暖的季节,桃大尾蚜(*Hyalopterus pruni*)会取食芦苇(*Phragmites communis*),而到冬季,就会转移到李属(*Prunus*)植物上。造成蚜虫在夏末更换寄主的因素是什么呢?可能是由于一些季节因素,如光周期、温度及涉及植物生理条件的一些诱因,控制着这种更换寄主的现象[84,87],而最根本的原因,可能为利用木本植物转到草本植物,实现对其生长模式的补充[77]。

表8.2 隶属于三个属的蚜虫更换的一级和二级寄主植物科

蚜虫属	一级寄主植物	二级寄主植物
缢管蚜(*Rhopalosiphum*)	蔷薇科	禾本科
绵蚜(*Pemphigus*)	杨柳科	菊科 禾本科,伞形科
卷叶棉蚜(*Prociphilus*)	忍冬科 木樨科,蔷薇科	松科

这种寄主偏好性随季节发生改变的现象同样也发生在其他昆虫类群中。西欧的叶蝉(*Muellerianella fairmairei*)是一种二化性昆虫。春季,它在绒毛草(*Holcus lanatus*,禾本科)上产卵,但其二代雌虫仅在灯心草(*Juncus effusus*,灯芯草科)上产卵[50]。在鳞翅目的一些二化性类群中也发现了类似的例子,在其不同的世代中,它们会选择完全不同的寄主植物。因此,尺蠖(*Tephroclystis virgaureata*)春季代的幼虫取食一些菊科植物,如一枝黄花(*Solidago*)和狗舌草(*Senecio*);而夏季代幼虫则取食蔷薇科植物,如山楂(*Crataegus*)和李子(*Prunus*)[75]。对两个世代幼虫产卵位置的选择因素进行研究,将是非常有趣的,但目前仅有很少的研究涉及此内容。由于季节的变化,可能导致原寄主植物的化学和/或营养水平发生改变,这种改变积累到一定程度,昆虫就会从一种植物转移到另一种植物。当然,昆虫固有的偏好也可能发生改变。

在鳞翅目昆虫中,马兜铃燕尾蝶(*Battus philenor*)对寄主植物的季节性偏好,可能是由于植物的部分形态发生变化引起的。雌虫主要基于叶片的形状来搜寻寄主,而在寄主植物上的产卵数发生变化会导致其寄主偏好发生改变。在一定时间内,雌虫更喜欢飞落到窄叶植物上并忽略宽叶植物,而在另外的时间却相反。马兜铃燕尾蝶(*B. philenor*)采用的这种"图像搜寻"的方法,与鸟类捕食特定类群所采用的方法一样[138]。经行为学研究表明,蝴蝶的这种搜寻行为的改变,可能与成虫的经验有关[100]。

8.4.2 温度

昆虫对植物取食的偏好,有时会随温度变化而变化。在马铃薯和蜀羊泉(*Solanum dulca-*

mara)间,马铃薯甲虫在通常情况下都会表现出对马铃薯的偏好,而当温度为25 ℃或更高时,则会表现出蜀羊泉的偏好(图8.4)。

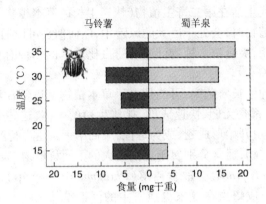

图8.4   在不同温度条件下马铃薯甲虫的取食偏好

食物的消耗量(每只甲虫在四小时内消耗食物的毫克数)随温度而变化。在20 ℃到25 ℃之间,对食物的偏好从马铃薯(*Solanum tuberosum*)转移到蜀羊泉(*S. dulcamara*)(邦格,1970)[23]

是否由于两种植物(或其中一种)的化学成分随温度改变而改变,从而影响了对昆虫的感知,亦或温度的变化致使昆虫化学感受器发生了变化,从而影响了它们的行为?目前仍然不得而知。

另一个温度影响摄食的例子是关于苜蓿蚜虫(*Terioaphis maculata*)的。苜蓿栽培品系通常对苜蓿蚜虫存在抗性,然而当温度降低到10 ℃时,则容易感染这种昆虫。也许是由于在低温条件下,苜蓿蚜虫的代谢水平降低,其迁移率也随之降低[109]。但是在这个例子中,我们同样不清楚温度改变的是昆虫还是植物,或者两者都被改变了。在其他的类群,如其他蚜虫、异翅亚目、麦瘿蝇(*Mayetiola destructor*)中也发现了类似的情况,但其机制仍然未知[89]。

8.4.3   捕食风险

影响植食性昆虫对食物选择的另一个外因,可能是捕食者的存在与否。在没有捕食性蜘蛛存在时,红腿蝗(*Melanoplus femurrubrum*)主要取食稻科植物。然而,当蜘蛛存在时,它们则明显偏好草本(非稻科)植物(图8.5)。结构更为复杂的草本植物可能为它们逃避捕食者提供了庇护。

因此,这种行为上的改变,并非其食物偏好发生改变,而使其生境的选择发生了改变[10]。其他的一些关于躲避天敌行为的例子,认为这种选择变化主要是基于学习行为的结果[45]。

8.5   引起寄主植物偏好改变的内部因素

一个昆虫个体对植物取食的偏好并非恒定不变,如在个体发育过程中,当营养需求改变时,其取食偏好也会发生改变。

8.5.1   发育阶段

许多植食性昆虫在幼虫的不同发育阶段,其取食范围会变宽或变窄。高龄期幼虫通常比

第 8 章 寄主植物的选择:变化即规律

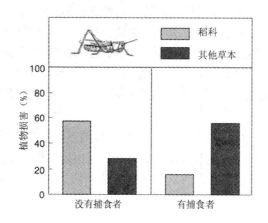

图 8.5 在捕食性蜘蛛存在(或缺失)时红腿蝗(*Melanoplus femurrubrum*)对不同植物的取食情况

当遭遇捕食风险时,昆虫会降低取食稻科植物的比例,而转向其他草本植物(贝克尔曼等,1997)[10]

低龄期的幼虫能接受更多的植物。例如,在豹灯蛾(*Arctia caja*)幼虫中,存在非常显著的取食范围扩大的现象。低龄期幼虫的取食范围很窄,而老熟幼虫几乎不拒绝任何植物[86]。许多文献也记载了一些相反的情况——高龄期幼虫比初孵幼虫更挑剔。这可能是基于学习行为的一种"偏好感应",这部分内容将在 8.6.2 中进行讨论。

不同发育阶段的取食偏好改变可能是营养需求发生变化的一种体现。目前没有任何证据表明,昆虫的营养需求在其个体发育过程中始终恒定。一些研究表明,昆虫的取食偏好可能随时间发生改变。如连续用两种人工饲料喂养舞毒蛾(*Lymantria dispar*)幼虫,一种饲料的脂肪含量低于最佳水平,而另一种饲料的蛋白含量低于最佳水平。从低龄幼虫到老熟幼虫,取食偏好明显由低脂、高蛋白食物转向高脂、低蛋白食物。这种食物选择的改变可能反映了对营养需求的改变[129]。在野外的观察中发现,在舞毒蛾幼虫的不同发育阶段会取食多种不同种植物,这被认为是其实现营养物摄取最优化,或与营养需求改变相关的选择行为[5]。飞蝗(*Locusta migratoria*)的不同发育阶段对蛋白质及碳水化合物的需求也发生相应的改变。其下颚须的神经反应与蛋白质和糖摄入量的变化相一致,表明了营养需求在感受器的敏感性方面存在反馈[124]。

一些幼虫在其生长过程中,会在寄主植物的不同部位进行取食。目前对取食位置转变的原因尚不清楚。也许由于营养因素,或随幼虫体形不断变大而增加的捕食风险变化可能对昆虫施加选择压力,从而导致取食位置的转移[102]。而植物的季节性变化也许也有一定的影响,但同样反映了昆虫对取食位置的选择会受到体形大小的影响。

暗脉粉蝶(*Pieris napi*)在低龄幼虫阶段拒食十字花科植物蒜芥(*Alliaria petiolata*),由于丙烯氰的存在会引起摄食后的毒副反应。而高龄幼虫对这种化合物不敏感,但会被同一植株的另一种糖苷所抑制[106]。这个例子表明,在幼虫发展阶段,对植物防御性化合物的敏感性可能会发生改变。

灰蝶科幼虫在不同发育阶段,其取食习惯会发生十分惊人的改变。最初这些昆虫是植食性的,但是在其发育过程的某个阶段中,幼虫会停止取食并掉落到地上。它们可能会被蚂蚁带回蚁巢,这些"喜蚁的"幼虫利用其特殊腺体分泌一些化合物,这些化合物大约含 20% 的糖,有时也含少量氨基酸。这些物质可以满足蚂蚁的营养需求,而作为回报,蚂蚁可以保护灰蝶幼虫免受捕食者捕食或拟寄生物的寄生。而这些灰蝶幼虫一旦进入蚁巢,则会吞食蚂蚁的幼体作为其食物[9,36]。这种从植食性到肉食性的奇异改变,是许多灰蝶科昆虫的一种普通模式,而这与某些"同类相食"的习性有相似之处。在其他的一些植食性昆虫中也存在这种模式[53],如黑粉虫(*Tribolium castaneum*)[143]、西南玉米螟(*Diatraea grandiosella*)[135]和乳草叶甲(*Labodomera clivicollis*)[46]。

在昆虫的幼虫或成虫阶段,其食性都可能发生巨大的差异。西方玉米根虫(*Diabrotica virgifera virgifera*)的幼虫为严格意义的单食性,其幼虫仅取食玉米根部,而其成虫却是杂食性的[31]。与其他生物学特性一样,取食行为从幼虫到成虫阶段,可能由于营养需求及环境条件的改变而发生大幅度的改变。

8.5.2 昆虫性别影响食物选择

在舞毒蛾幼虫的食物选择实验中,雌雄幼虫(包括低龄幼虫和高龄幼虫)对不同食物表现出非常显著的偏好。相对于雌性幼虫,雄性幼虫会更倾向于低蛋白高脂肪含量的食物,而雌性则倾向于高蛋白低脂肪含量的食物。在这个类群中,只有雄性有翅能飞行,而脂肪可为飞行运动储备能量;雌性为了卵的发育需要更多的蛋白质[129]。因此这种性别间的取食偏好,可能反映了各自生理需求的适应。雌雄幼虫激素的差异,可能会精确地控制脂肪与蛋白质的摄入水平。在另一项研究中发现,为热带竹节虫(*Lamponius portoricensis*)提供四种寄主植物的叶片,雌雄昆虫对各寄主植物的消耗量明显不同,和舞毒蛾一样,显示出不同性别的取食差别[107]。

在野外的观察也表明,许多蝗虫中存在雌雄取食的偏好性。然而在小车蝗(*Oedaleus senegalensis*)中发现,雄性(任何发育阶段)及未成熟的雌性倾向于取食谷类植物的叶片,而成熟雌性则会显示出对谷类植物谷穗的偏好,从而满足在卵形成时对高蛋白的需求[26]。在另一项关于14种蝗虫的野外研究中报道,两性的食物成分和取食偏好存在显著性差异。这些发现提出了一个有趣的假设:昆虫的雌雄基因型占据不同食物生态位[141]。

8.6 经验诱导寄主植物偏好的改变

相对于脊椎动物,植食性昆虫的行为(无论是一般行为或是与食物相关的行为)主要由遗传决定。然而,经过学习也能显著地改变其取食和产卵行为。昆虫的学习行为主要分两种类型:联系式学习(associative learning)和非联系式学习(non-associative learning)。

联系式学习是指在不同的事件中发现某种时间关联性的行为。当昆虫先暴露于无效刺激(条件刺激,CS)中,接着又暴露于有效刺激(无条件刺激,US)中,会相继产生消极或积极的反应,而后用CS单独刺激时,它可能会产生两种反应。

非联系式学习是一种更为简单的类型,不涉及条件刺激和非条件刺激。昆虫的习惯化(Habituation)和诱导食物偏好属于非联系式学习,而厌食、选择食物以及经验诱导的产卵行为改变都属于联系式学习[13,133]。

8.6.1 非联系式改变

1. 对阻碍物的习惯化

第8章 寄主植物的选择:变化即规律

习惯化(或减敏现象)是对一个反复呈现的刺激反应逐渐减弱的过程[80],被认为是最简单的学习形式。因为植食性昆虫对植物的接受主要由抑制摄食(见7.6.1)的植物次生代谢物质决定,在对这些物质抑制作用的习惯化方面,已经进行了相当深入的研究。例如,将两种蝗虫(即沙漠蝗虫(*Schistocerca gregaria*)和东亚飞蝗(*Locusta migratoria*))刚蜕皮的五龄幼虫分成两组。一组以未处理的高粱叶片饲养(自然组),另一组(有经历组)先用尼古丁酒石酸氢盐(抑制剂)处理过高粱叶片饲养19个小时,然后再用未处理的高粱叶片喂养5小时。之后的每天用抑制剂处理过的高粱叶片喂养10只"自然组"幼虫和10只"有经历组"幼虫,检测19小时内叶片的消耗量。图8.6中A表明"有经历组"中的沙漠飞蝗(杂食性)已经习惯了这种抑制物,它们比"自然组"幼虫消耗更多的叶片;而B表明东亚飞蝗的幼虫(寡食性)也同样表现出一些习惯化行为,虽然程度相对较轻[68]。

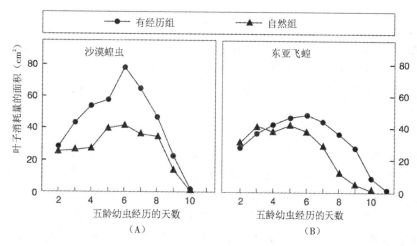

图8.6 "有经历组"和"自然组"的五龄幼虫在这一龄期内每天消耗的经抑制物处理的高粱叶片数量

(A)杂食性沙漠飞蝗(*Schistocerca gregaria*);(B)寡食性东亚飞蝗(*Locusta migratoria*)(杰米等,1982)[69]

一些类别的化学物质可能会引起习惯化反应,然也存在着一些绝不会被接受的化学物质。例如,烟草天蛾幼虫会迅速适应食物中的水杨苷,但却不能适应食物中掺入的少量马兜铃酸[58]。同样,蝗虫可以习惯相对较弱的刺激,而对那些可以抑制取食超过12小时的化学刺激不能适应[69,132]。

昆虫不会对非寄主植物习惯化,即使它们仅低于可接受的水平[67]。然而,对于一些边缘的寄主植物的接受水平,可能会随对一定抑制物的习惯化得到提高[64]。在专食性昆虫的非寄主植物中,可能存在非常复杂的抑制刺激,这种刺激可能阻碍了昆虫对这些植物抑制效果的习惯化。通过对几种鳞翅目幼虫的观察,这种解释得到了证实,经反复暴露,这些幼虫容易对单一成分的抑制物(印楝素)习惯化,但不能对含有多种抑制物的植物提取物习惯化[2,22]。

这种习惯化是在中枢神经系统发生的一个过程?还是外周化学感受系统对一种抑制物接受性的改变?研究发现,在烟草天蛾(*Manduca sexta*)幼虫的食物中混合水杨苷抑制物后,其抑制物感受器细胞的敏感性会降低,这表明化学感受器参与了学习过程。其影响就是幼虫更容易接受水杨苷处理过的植物,而正常情况下这是被拒绝的[111]。这些研究一致认为,除了中枢系统,外周化学感受系统也参与了这种学习的过程[57,113]。然而,研究沙漠飞蝗(*Schistocerca gregaria*)幼虫对于抑制物尼古丁酒石酸氢盐(NHT)的反应,却没有发现任何外周感受器的变化,表明在这种物种中仅涉及中枢反应。在一个设计精细的实验中,将小的尼龙管固定在昆虫的上颚须周围,使上颚须完全伸入到尼龙管中。上颚须中几乎包含了昆虫口器上30%的化学感受器。对实验组昆虫,在尼龙管中添加尼古丁酒石氰酸盐(NHT)溶液,每天保持几个小时,而在对照组尼龙管中添加蒸馏水。四天后,将尼龙管保留在原处,但都已被清空,将用NHT处理的叶子作为食物喂养两组幼虫。尽管此时幼虫上颚须感受器中直接的化学感受器被尼龙管阻断了,但是实验组幼虫消耗处理过的叶片比对照组更显著。这些结果清晰地表明,习惯化是通过中枢神经系统调节的。上颚须感受器所提供的关于NHT的信息,在预实验时已被储存在中枢神经系统,在实验中当昆虫取食NHT处理过的叶片时会将其他口器感受器提供的信息与储存信息进行比较[132]。

为什么有能力可惯化的昆虫不能对所有抑制化合物习惯化?答案可能是,有些化合物虽然具抑制作用,但在生理方面是无毒的,而其他抑制物则是有毒的。对于后者,昆虫除了必须降低味觉系统(外围或中枢)的敏感性外,还要激活其摄食后(post-ingestive)机制来降低对有毒化合物的生理敏感性。在解毒系统无法应对有害化学物质的情况下,昆虫就不会产生习惯化而导致死亡[57]。

2. 诱导取食偏好

在早期的昆虫学专著中,柯比和斯宾塞提出,尽管昆虫可能取食各种植物,但在接触某种植物后,仍会表现出其取食偏好性[74]。需要指出的是,当一些个体在一种植物上取食一段时间后,它们宁愿饿死也不会再取食其他植物,而这些植物可能是它们一开始完全可以接受的。这种现象反映了植食性昆虫的某种学习行为,而在很长一段时间里,人们认为这种学习仅能表现先天固有的和简单的刺激——应答行为模式。而直到杰米和他的同事设计了步骤严密的实验[68],之后才引起了广泛的关注。用人工饲料饲养棉铃虫(*Helicoverpa zea*)直至四龄,当蜕皮成五龄幼虫后,将它们分为四组,分别用人工饲料、天竺葵(*Pelargonium hortorum*)、蒲公英(*Taraxacum officinale*)、甘蓝(*Brassica oleracea*)喂养。当刚蜕皮的六龄幼虫面对食物选择时(图8.7)它们更偏好在五龄时所取食的植物。用人工饲料饲养的幼虫因没有取食某种植物的经历,因此表现出与其他三个组明显不同的取食偏好模式(图8.8)。先前取食经历的影响被称为"诱导偏好"[68]。目前对其性质和影响仍然了解很少,而且也不符合通常的学习行为类别。

用缺乏植物化学特性的人工饲料喂养的昆虫能保持刚孵化时的本性。它们可以欣然接受所提供的任何寄主植物甚至一些非寄主植物。而在取食植物1~2天内,它们对植物的无差异感消失,而且形成了自身的取食经验[110]。初孵幼虫在孵化后开始取食的1~2天,就会拒绝后期的其他食物(图8.9)。以取食经验为基础的偏好,可能是由于对一些植物成分的味觉反馈造成的,这些成分作为其适合或不适合的信号[103,105]。

以上是这种幼虫偏好取食一种寄主植物的例子,与脊椎动物幼年时的"铭记"有惊人的相似之处。术语"铭记"被定义为一种不可逆转的学习行为,发生在一个非常短暂的关键时期,即动物个体的早期生命阶段。而昆虫的诱导取食偏好在高龄幼虫中也可能发生,可能通过新的经验进行调整,因此能忽略掉早期的"铭记"影响。

诱导取食偏好持续的最短时间为4小时,这可以通过欧洲粉蝶(*Pieris brassicae*)幼虫取食金莲花(*Tropaeolum majus*)得以证明[83]。在不同的昆虫类群中,诱导偏好的持续时间存在很大

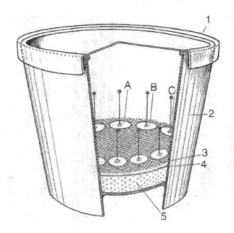

图 8.7　昆虫对不同植物的偏好性实验

A,B,C 是三种不同植物的叶碟,用大头针固定;1,用培养皿覆盖;2,食物容器纸杯;3,金属网筛;4,湿润的过滤纸;5,石蜡层(杰米,1968)[68]

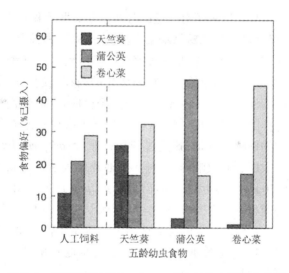

图 8.8　谷实夜蛾 (*Helicoverpa zea*) 六龄幼虫的取食偏好

从孵化到四龄幼虫末都喂养人工饲料,然后在五龄幼虫期间分别喂养人工饲料、天竺葵、蒲公英和卷心菜(杰米等,1968)[68]

的差异,这也取决于昆虫取食的植物。

在一些情况下,诱导偏好性是非常极端的。在达尔文的著作中,引用了米歇尔的观察数据[37],在自然的状态下金星蚕(*Bombyx Hesperus*)幼虫取食金缕梅(*Hamamelis virginiana*)叶片,但是用经臭椿(*Ailanthus*)喂养后,它们就不会再取食金缕梅,最后甚至因饥饿而死。当用卷心菜喂养欧洲粉蝶(*Pieris brassicae*)后,将五龄幼虫转移到另一寄主植物金莲花(*Tropaeolum majus*)上时,幼虫同样表现为拒食并且最终饿死[83]。在其他的鳞翅目类群中也观察到了这种极端的诱导[115]。这种诱导偏好的极端形式被称为"丰盛宴席上饿死"的现象(参照"罗马将领卢库勒斯的丰盛晚宴"[67])。

目前,在直翅目、竹节虫目、异翅亚目、同翅类、膜翅目、鞘翅目和鳞翅目中的一些类群中,

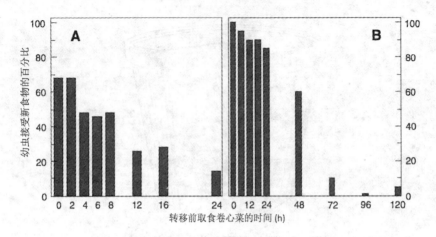

图8.9 幼虫接受新食物的百分比与转移前取食卷心菜的时间的关系

(A)菜粉蝶(*Pieris rapae*)一龄幼虫取食卷心菜的时间对旱金莲接受程度的影响(伦威克和黄,1995)[104];(B)取食卷心菜的时间对人工饲料接受程度的改变(斯洪霍芬,1977)[112]

已经有了诱导偏好现象的发现。这种相对广泛的分布现象表明这种行为改变是相对普遍发生的,而改变最显著的例子仅来自一些鳞翅目幼虫[133]。同样,在一些鳞翅目其他类群的例子中,没有这种行为改变的现象[62,67]。因此不能将这种普遍性扩展到整个植食性昆虫领域。

目前,已经对植物诱导偏好及相应的一些行为变化的内在机制进行了研究[17],但对其所涉及的生理过程的研究仍有待深入。至于诱导偏好的神经反应过程,我们尚不清楚外周感受器或中枢神经系统发挥了怎样的作用。在一些情况下,当昆虫取食一种植物时,它们的化学感受器对这种植物特定化合物的敏感性增加,这通过灰翅夜蛾(*Spodoptera littoralis*)幼虫的观察得到了证实。当灰翅夜蛾幼虫以卷心菜喂养后,对芥子油苷会表现出比缺少这种化合物的人工饲料更高的敏感性(图8.10)[114]。最近对于盐泽灯蛾(*Estigmene acrea*)幼虫的观察也得出了相似的结论。当以含烷生物碱(PAs)的植物饲养盐泽灯蛾幼虫后,幼虫的感PAs受体细胞会对PAs的敏感性增强。当幼虫在缺乏PAs的合成食物中存活几代后,其敏感性会持续下降,而又可以通过取食含有纯的PAs的食物增加[30]。因此,取食之前吃过的植物很有可能比取食新的植物能产生更强的神经刺激信号[39]。

在昆虫化学感觉系统中,存在着这种"外周学习"现象,但这并不意味着中枢神经系统发挥的作用较小,事实上偏好行为主要发生在中枢神经系统中。

昆虫的诱导偏好行为证明昆虫不仅能区分寄主与非寄主植物,还能够区分不同的寄主植物。因此,对每一种寄主植物,植食性昆虫都能详细地感知其"化学感觉轮廓"[38],或是"化学完形"(chemical gestalt)[76]。这个概念强调了在以味觉和嗅觉为基础的寄主识别过程中所涉及的刺激信号是非常精细而复杂的。

在野外环境下,引起昆虫取食行为的自然刺激是非常复杂的,但有些时候也可以鉴定出某单一成分在取食诱导偏好中发挥了主要作用。例如,刺天茄苷D为一种甾体类化合物,出现在茄属植物的叶片中,对烟草天蛾幼虫的诱导偏好性有重要作用[39]。通过在对烟草天蛾(*Manduca sexta*)幼虫(图8.11)[108]和灰翅夜蛾(*Spodoptera littoralis*)幼虫[28]的观察表明,挥发性化合物可以作为诱导偏好食物识别的重要线索。

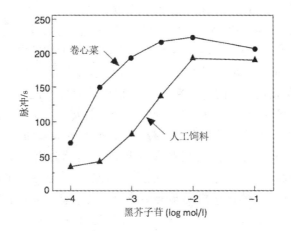

图 8.10　当受到不同浓度的芥子油苷(黑芥子苷)的刺激时,灰翅夜蛾(*Spodoptera littoralis*)
五龄幼虫上颚须(锥形感器)的神经反应(脉冲/秒)

昆虫以卷心菜叶或人工饲料饲养,以卷心菜为食的个体对芥子油苷表现出比以人工饲料为食的个体更高的敏感性(斯洪霍金等,1987)[114]

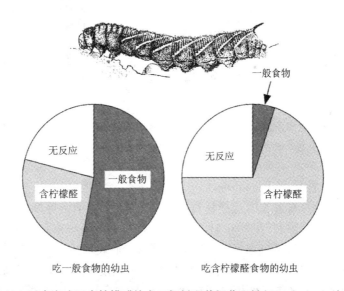

图 8.11　以含有或不含柠檬醛的人工饲料喂养烟草天蛾(*Manduca sexta*)幼虫
其五龄幼虫面对选择时的嗅觉定位

两种饲料在距昆虫很近的距离处以选择的方式呈现,嗅觉定位反应以幼虫向含有或不含柠檬醛的饲料移动的百分比呈现(萨克赛纳和斯洪塞金,1978)[108]

8.6.2　联系式改变

1. 厌食学习

德蒂尔提出,植食性昆虫的厌食学习行为是由植物引起的暂时不适而导致的一种后天行为[41]。这种在脊椎动物(包括人类)中被人们熟知的现象,在昆虫中首次于灯蛾(*Diacrisia virginica*)的幼虫中发现。在刚开始,这些幼虫会在不同的植物种类及个体上进行搜寻和试食,像真正的杂食性昆虫一样。当为其提供在它们栖境中没有的矮牵牛(Petunia hybrida)叶片时,它们也会贪婪地取食。然而,在摄食24小时以后会产生不适(反刍、活动减弱、运动失调、轻度痉挛、胸部膨大)反应。当其恢复后,将其放置于一个有其他植物的空间内,和那些没有经验的幼虫相比,它们会避免取食矮牵牛(图8.12)[42]。这种学习行为对于生存的价值是非常显著的,然而在昆虫中这种厌食性学习的普遍性还不得而知。到目前为止,仅在一些鳞翅目幼虫以及几种蝗虫中观察到这种现象[12]。也许这种现象在杂食性昆虫中比专食性昆虫更为普遍[43]。在寡食性物种中,有一个典型的例子为东亚飞蝗(Locusta migratoria)成虫。对首次遇到的非寄主植物,昆虫会用触须对其进行诊断,并仅取食一小口,之后就会拒绝取食。以后再遇到这种植物时,紧靠触碰就能判断从而拒绝取食[21]。在这种情况下,当昆虫取食"不合口味"(具"惩罚反馈")的叶片时,位于触须的化学感受器会接受叶面的信息,学习行为通过联系这种接受信息而发生[20]。在另一种已经饥饿了两个小时以上的蝗虫(Schistocerca americana)中,这种学习反应仍未消失[29]。

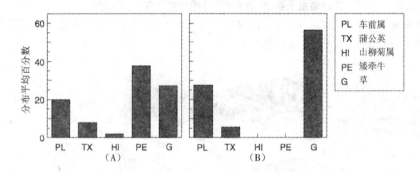

图8.12 两天的时间里,灯蛾(Diacrisia virginica)幼虫在野外环境下分布的平均百分数

(A)先前没有取食经历的幼虫;(B)仅取食矮牵牛(Petunia) 48小时,经恢复期后,幼虫的分布(德蒂尔,1988)[42]

2. 自选食谱(混合性膳食, mixed diet)

尽管多数植食性昆虫对寄主的选择都是高度特化的,而植物诱导偏好又对其寄主范围进行进一步的临时或永久的限制(见8.6.1b),而在一些物种中却观察到相反的行为,即对多种混合食物的渴望。这是通过昆虫在不同食物间频繁移动而表现的。当杂食性天幕毛虫(Malacosoma castrensis)幼虫被限制仅取食一种植物时,其生长缓慢并且死亡率很高。但当幼虫接触多种植物后,我们发现幼虫在多种植物间频繁转移,且生存率高于取食单一植物的幼虫。同样,豹灯蛾(Arctia caja)也表现出这种自动的取食转移行为,经转移取食后,它们生长得更好[86]。更换食物最为频繁的昆虫为蝗虫类,当仅以一种植物为食时,它们会死亡或表现出低的存活率和繁殖率[6,72]。在一项野外观察中指出,北美洲西南部巨蝗(Taeniopoda eques)的栖境中存在多种植物,而巨蝗(T. eques)一餐就要取食八种不同植物[18]。

一些学者认为,取食混合植物比单一植物的昆虫具两个生理优势。(1)以多种植物为食会使昆虫增加对各种营养物的摄入,从而达到更理想的营养配比;(2)防止对特定寄主植物有毒性的次生代谢物过量吸收。如果昆虫试图增加其必需的营养摄入,它会在混合植物的摄食条件下生长得更快。然而,如果昆虫想要通过混合摄食来稀释几种不同植物的毒素,也会摄入更多的食物[19]。

目前在蝗虫体内,已经发现引起混合摄食的三种机制,而每一种机制都可能与这种昆虫摄食的习惯和它们的栖境有关[15]。第一种机制为厌食学习。例如,当杂食性的蝗虫(*Schistocerca americana*)仅以菠菜喂养时,它会逐渐地减少食量,最终完全拒绝摄入菠菜。因为其化学感受器未发现任何变化,所以认为昆虫将取食菠菜产生的厌食结果(不适)与之前的味觉体验联系到一起[79]。而与此同时,供选择的其他食物就更具吸引力,由此暗示了对新奇事物(喜欢新奇)的偏好,会伴随已获得的厌食感一起发生。而对于一些简单栖境中的杂食性昆虫,因为几乎没有什么植物可供食用,因此这些昆虫也许会被迫长时间取食单一植物物种[15]。

"特定物质饥饿"会促使昆虫产生积极联系式学习。蝗虫之前取食的高蛋白食物中含有某种特殊的香味化合物,当它们被禁止摄食这种高蛋白食物一段时间后,蝗虫会被这种香味所吸引[123]。

第二种机制涉及化学感受器的调整。对氨基酸或糖类物质敏感的感受器,在应对血淋巴的营养成分改变时可能也发生改变。取食蛋白质含量较低的食物的昆虫,其血淋巴中氨基酸浓度较低,进而导致化学感受器对食物中氨基酸的敏感性增强。因此,昆虫就会转向更高含量蛋白质的新食物。对食物中碳水化合物也以相同的机制处理:如果昆虫的食物中碳水化合物的含量低,它的血淋巴中的糖水平下降,这会通过一系列未发现的途径来降低化学感受器对糖类敏感度的阈值[1]。例如杂食性灰翅夜蛾(*Spodoptera littoralis*)幼虫对蛋白质和碳水化合物的摄入,其化学感受器存在相似的变化[122]。而这种化学感受器调整的机制也存在于专食性物种中,例如仅取食稻科植物的烟草天蛾和东亚飞蝗。而分辨稻科植物要比双子叶植物困难得多,因为在释放的次生物质方面,双子叶植物的多样性更高[15]。有趣的是,在这些昆虫中,当它们面对不同的食物特征时,其上颚须的化学感受器数量会发生变化。这种与生物学相关的变化,反映了它们在取食行为上的变化[88]。

第三种机制为由新奇激发的刺激。例如,蝗虫(*Taeniopoda eques*)为"强迫性"的转变者,因为它们新发现的气味(即新的化学物质)能够产生一种诱导摄食的刺激信号。对于这种昆虫而言,多样性为生活的"兴奋剂"[18]。对于经常移动且栖境中植物种类丰富度高的蝗虫而言,这种机制尤为重要[15]。这个例子表明,对新奇食物的喜好不仅与厌食学习有关,如沙漠蝗(*Schistocerca americana*)所证明的,而且也与昆虫喜好食物种类的自动转变明显相关。

其他类群的一些昆虫也能从混合摄食中受益。例如,玻璃叶蝉(*Homalodisca coagulate*)在其漫长的一生中,为了增加存活率,会在不同的寄主植物间转换[27]。与之类似,两组以牧草为食的寡食性异翅亚目昆虫,取食四种寄主植物的昆虫,比仅取食一种寄主植物的昆虫有更高的存活率[44]。

3. 经验诱导产卵行为的改变

当某种雌性昆虫首次在一种特定的基质上完成产卵时,它们会从中获取经验。之后,这种经验会影响它们对产卵基质的选择。当苹果实蝇(*Rhagoletis pomonella*)在苹果上产卵后,它们会拒绝在提供的另一种寄主植物——山楂(*Crataegus mollis*)的果实上产卵。相反地,首次在山楂属上产卵的实蝇就会拒绝在苹果上产卵[95]。这种行为是联系式学习的一种形式,因为苹果实蝇将它们第一次遇到的果实的大小和表面化学物质与成功的产卵行为联系到一起。苹果实蝇不但可以区分不同的寄主植物种类,而且还可以区分苹果的不同栽培品种。这是因为它们从第一次产卵的经验中对特定的品种产生了偏好[94]。

一些蝴蝶可以利用"视觉线索"来进行产卵位置的选择。例如,美洲蓝凤蝶(*Battus philenor*)能够用视觉根据叶子的不同形状来区分马兜铃属(*Aristolochia*)的两种寄主植物[19]。

而这种对特定叶片形状的识别会通过对寄主植物的化学感受经历而得到强化[91]。釉蛱蝶属(*Heliconius*)的雌性个体第一次遇到一种植物时,会将叶片形状与它们的化学感受联系在一起,之后会寻找相同形状的叶片[56]。雌性会使用"搜寻印象"以叶片形状识别某种植物[98],而这种由神经存储的印象通常也包含挥发性化合物的信号[78]。

基于一些室内的研究,对于夜行性鳞翅目昆虫,菜粉蝶(*Pieris rapae*)[139]和棉铃虫(*Helicoverpa armigera*)的雌虫,同样表明先前的产卵经验可以造成寄主选择行为的改变[34]。总体而言,成虫的产卵学习行为是一种相当普遍的现象[92]。然而,如食物诱导偏好现象在一些昆虫中并不存在一样——不是每一种昆虫都会根据之前的经验对一种特定的植物产生产卵偏好。因此,当雌性虎凤蝶(*Papilio glaucus*,已知的食性最广的燕尾蝶)接触一种寄主植物两天后,在后来的多选测试中没有表现出任何的产卵偏好性[116]。

对于飞行昆虫而言,雌性对产卵地的寻找和识别要比对选择食物需要更多且更复杂的信息。因为在整个过程中,除常规飞行所需要的视觉和化学(嗅觉和味觉)刺激外,还需要在相距甚远的两地进行连续产卵,造成了在自然条件下,对它们的分析尤为困难。

8.7 幼虫和早期成虫的经验

在很早之前,一些昆虫学家就指出,幼虫在寄主植物上获得的经验,会对成虫的取食和产卵场所偏好产生影响。美国昆虫学家霍普金斯最先发现了与此现象相关的证据,并且认为这一现象与对寄主植物偏好的进化有关。"霍普金斯寄主选择原则"指出,某种能在多种植物上繁殖的昆虫会更倾向于在其已适应的植物上继续繁殖[33]。由于涉及进化过程,许多科学家已经开始寻找昆虫幼虫时期的摄食经验对成虫迁移行为的影响,但是都没有得到理想的结果[7,133]。但这并不能说明霍普金斯的推测是错误的。证明其假设的实验很难重复,因为他的实验所采用的昆虫(如一些天牛类)均很难处理且生命周期很长[33]。

之后,杰耐克提出了"新霍普金斯寄主选择原则"[65]。这个原则指出昆虫在成虫早期阶段与某一特定植物接触,同样会增强它们对该植物的偏好。而这种早期成虫的经验已经在果蝇(*Drosophila*)中得到证实[137]:将发育完全的幼虫和/或刚化蛹的幼虫彻底清洗后,去除之前饲养幼虫的食物,而成虫羽化后,之前的食物对未经清洗的个体比清洗过的表现出更强的吸引力。据这种思路,科比特提出了"化学遗留假说"[32]。他强调即使蛹被清洗了,幼虫时期的食物的化学痕迹也会存留在昆虫的血淋巴中或是存在于昆虫蛹的外部。这些痕迹会改变刚羽化的成虫的感受和行为。

与成虫早期经历会影响后续行为的推测类似,昆虫可能会选择幼虫期取食的植物部位羽化,例如一些在茎秆、水果和种子里发育的物种,尽管目前还未得到实验证实。

目前关于幼虫期的学习行为延续至成虫期的证据非常少。在一个有趣的实验中,以一种特定气味的(Pavlovian)调节训练果蝇(*Drosophila*)幼虫避开电击。在后来果蝇的变态过程中,尽管有大量神经系统的重建活动,而这种避开气味的行为一直被保留了下来[8,140]。这个实验证明,幼虫期的记忆保留到成虫期是可能的。在墨西哥豆瓢虫(*Epilachna varivestis*)中,在幼虫时期对摄食抑制剂习惯化后,这种行为似乎也能保留到成虫期[3]。还有其他几项以小菜蛾(*Plutella xylostella*)、粉纹夜蛾(*Trichoplusia ni*)[2]、灰翅夜蛾(*Spodoptera littoralis*)[4]和胡蜂[101]为实验材料的研究,都证明变态过程不会影响记忆的储存。在这些实例中的记忆保留现象虽然可以用科比特的"化学遗留假说"解释,但也不能排除中枢神经系统的信息储存作用。

8.8 经验诱导寄主植物偏好改变的适应意义

目前对于经验诱导寄主偏好改变的所有研究都是在实验室内进行的。因此我们不知道在自然界中这些结果是否和室内一致。但是假设上述的现象在自然条件下也存在,那么这样的改变适应性意义是什么呢?

在无选择的条件下,仅给昆虫供给一种单一化合物,且其浓度在不至完全抑制摄食时,就会发生对抑制物的习惯化。但是,习惯化并不会随抑制强度的增加或存在非寄主植物而产生。因此对摄食抑制习惯化的适应性意义可能在于昆虫在行为上克服如环境等因素引起的寄主植物抑制物浓度的轻微升高。而在缺乏理想的可供取食植物的情况下,昆虫还可以接受具有轻度抑制作用的植物。

厌食学习的适应性意义是不言而喻的:能够阻止昆虫摄入含有毒物质的植物和营养不适宜植物。杂食性的昆虫尤其容易摄入有毒植物,因此,自然选择促进了这类昆虫厌食学习能力的进化。专食性昆虫对寄主植物的选择非常挑剔,阻止了其取食引起生理不适的植物。因此,这类昆虫可能较缺乏厌食学习的能力[43],但我们对于这方面了解仍十分欠缺。

目前对诱导取食偏好的适应性意义仍缺乏证据。一些学者认为,昆虫频繁更换食物种类会导致其对食物的利用率下降[60,71,115]。如果昆虫被迫摄食一种新植物,一些昆虫(不是所有的)会遭遇生理高成本,可表现为成长速度减缓及其他健康参数下降(表8.3)。当昆虫取食一种寄主植物一段时间后,会在生理或生化水平上适应这种植物。诱导偏好的作用可能为在其生境中有很多寄主植物可供选择的情况下,尤其是在各种植物的树枝和树叶相互缠绕的情况下,防止昆虫在不同植物间转移。尽管这种假设很具说服力,然而,在一个陌生的生境中,即使存在适合某种昆虫取食的植物,昆虫却时常饥饿至死,对这种情况很难被理解[67]。而且,强的取食诱导偏好会迫使昆虫搜寻其熟悉的食物,即使在这个生境中有其他的适宜取食的寄主植物,这会使幼虫的成长速度减慢,并延长昆虫暴露于捕食者和寄生天敌的时间。当幼虫从寄主植物上掉落下来后,它们对这种植物气味的敏感性会提升,这能够帮助它们重新找到寄主。增强正常食物的摄入量可以通过加强刺激和减少摄食过程中的中断次数来实现。

表 8.3 纹黄豆粉蝶(*Colias philodice*)幼虫在其两种寄主植物——紫苜蓿(*Medicago sativa*)和草木樨(*Melilotus alba*)上转移后的各种性能参数(卡罗,1989)[71]

植物食物 喂养/测试	RGR	龄期持续时间 (h)	总食物消耗 (mg 干重)	蛹重量 (mg 干重)
紫苜蓿	0.26	104	125	29
紫苜蓿	0.08	150	81	14
草木樨	0.31	124	162	34
草木樨	0.11	141	89	16

注:供试幼虫为5龄。相对生长率(RGR)=增长干重(mg)/干体重(mg/天)。

在资源分布不均的区域,诱导产卵偏好被认为可以提高昆虫搜寻食物的效率,减少雌性昆虫在没有适宜食物的区域浪费时间徒劳寻找食物[92,96]。研究发现,专食性昆虫比广食性昆虫识别寄主植物需要的时间更少[66]，由此表明，对广食性昆虫而言，学习行为起到非常重要的作用。另外能在更多种类的寄主植物上产卵的雌虫，其后代有更高的存活率[99]。产卵雌虫的学习行为可能间接有助于幼虫的生长，使其幼虫需要更多的植物来完成发育过程。当雌虫集中在某种寄主植物上产卵时，其幼虫转移到其他植物上取食的可能性最低[35]。有时雌虫只固定在栖境中的一种寄主植物上产卵，却拒绝在其他与此植物同时存在的、适宜的寄主植物上产卵[70,99]，而对于这种情况，却无法用此理论进行解释[70,99]。

对植食性昆虫种群或个体而言，取食诱导偏好和产卵诱导偏好，均被认为是对昆虫固有的寄主特异性的一种暂时或永久性的限制。很有可能的是，并非所有的寄主诱导偏好行为都是具适应性的。像在"奢华宴会上饿死"的现象或拒绝在其他适合的基质上产卵的现象，也许表示昆虫神经系统将刚学会的一种行为转变为另一种固有行为。因此，诱导偏好可能为昆虫神经系统的惯性造成的一种"视野狭窄"现象[14,133]。

根据目前有限的资料显示，关于昆虫的习惯化和诱导偏好性，总体而言，这些现象在广食性物种中比专食性物种中更为明显[16]。至于这些现象是否（全部或部分）受昆虫神经系统的限制作用影响还亟待阐明。

总而言之，尽管联系式学习形式的适应性优势非常明显[12,52]，但对所有类型的经验诱导行为的适应性优势的完整答案，目前仍未给出[67]。

8.9 结论

由于存在寄主植物偏好现象，因此昆虫种内的取食和产卵行为可能存在很大的变化。

这种变化将有助于我们对资源利用和其他生态过程的理解，并且可以针对由于农业活动而形成的环境压力[40,73]。

由于植物与昆虫协同演化，因此单独就昆虫的行为变异或适应性进行分析是非常复杂的。昆虫的营养质量会随时间发生变化，基因组成也会随时间和地点发生改变。一种假设为，昆虫在个体发育期间，由于其生长、繁殖、迁移等不同阶段的营养需求发生改变，这种改变可能会影响昆虫的取食偏好。

在昆虫种内或种间发生取食转化的现象，可能是由于经验不同或者是母体影响，亦可能是由于基因构成的差异。在昆虫的一些主要类群中，均可观察到诱导取食偏好的现象，表明这是一种基本现象。但是也有一些物种似乎完全缺乏这种能力。

研究发现，有关先前经验的记忆并不仅储存于中枢神经系统中，也可能会（部分）储存于外周化学感受器神经元中，这为先前的经验会以某种方式调节所有细胞和器官的假设提供了证据。在神经元中所储存的记忆，是对其他类型的记忆模式的补充，如消化系统的解毒酶可以适应不同类型的食物（见5.3.5）。

本章讨论的要点为，昆虫与其寄主植物间的关系可能并非如最初认为的那样绝对和固定，这种关系具有很大的灵活性。这种灵活性在保证维持这种关系的同时，还是新的进化发展的先决条件。引用杰米的观点，诱导取食和产卵偏好，仅是在植食性昆虫种群或个体中是对取食和产卵场所挑剔行为的一种限制[67]。

总之，具备所有行为特征的昆虫是不存在的。每一个个体都是基于先天遗传和后天获得特性的独一无二的结合体。

8.10 参考文献

1. Abisgold, J. D. and Simpson, S. J. (1988). The effects of dietary protein levels and haemolymph composition on the sensitivity of the maxillary palp chemoreceptors of locusts. *Journal of Experimental Biolog*. 135, 215–29.
2. Akhtar, Y. and Isman, M. B. (2003). Larval exposure to oviposition deterrents alters subsequent oviposition behavior in generalist, *Trichoplusia ni* and specialist, *Plutella xylostella* moths. *Journal of Chemical Ecology*, 29, 1853–70.
3. Akhtar, Y. and Isman, M. B. (2004). Feeding responses of specialist herbivores to plant extracts and pure allelochemicals: etfects of prolonged exposure. *Entomologia Experimentalias et Applicata*, 111, 201–8.
4. Anderson, P, Hilker, M, and Löfqvist, J. (1995). Larval diet influence on oviposition behaviour in *Spodoptera littoralis*. *Entomologia Experimentalis et Applicata*, 74, 71–82.
5. Barbosa, P, Martinet, P, and Waldvogel, M. (1986). Development, fecundity and survival of the herbivore *Lymantria dispar* and the number of the plant species in its diet. *Ecological Entomology*, 11, 1–6.
6. Barbosa, O. L. (1965). Further tests of the effects of food plants on the migratory grasshopper. *Journal of Economic Entomology*, 58, 475–9.
7. Barron, A. B. (2001). The life and death of Hopkins host-selection principle. *Journal of Insect Behavior*, 14, 725–37.
8. Barron, A. B. and Corbet, S. A. (1999). Preinmaginal conditioning in Drosophila revisited. *Animal Behaviour*, 58, 621–8.
9. Bayalis, M. and Pierce, N. (1993) The efferts of ant mutualism on the foraging and diet of lycaenid caterpillars. In *Caterpillars. Ecological and evolutionary constraints on foraging* (ed. N. E. Stamp and T. M. Casey), pp. 404–21. Chapman & Hall, New York.
10. Beckerman, A. P. Uriarte, M., and Schmitz, O. J. (1997). Experimental evidence for a behavior-mediated trophic cascade in a terrestrial food chain. *Proceedings of the National Academy of Science of the USA*, 94, 10735–8.
11. Bennett, A. F. Interindividual variability: an underuailiaed resource. In *New direction in ecological physiology* (ed. M. E. Feder, A. F. Bennett, W. W. Burggren, and R. B. Huey), pp. 147–69. Cambridge University Press, Cambridge.
12. Bernays, E. A. (1995). Effects of experience in host-plant selection. In *Chemical ecology* (ed. R. Carde and W. Bell), pp. 47–64. Chapman & Hall, New York. an & Hall, New York.
13. Bernays, E. A. (1995). Effects of experience on feeding. In *Regulatory mechanisms in inset feeding* (ed. R. F. Chapman and G. de Bore), pp. 297–306. Chapman & Hall, New York.
14. Bernays, E. A. (2001). Neural limitations in phytophagous insects: implications for diet breath and evolution of host affiliation. *Annual Review of Entomology*, 46, 703–27.
15. Bernays, E. A. and Bright, K. L. (1993). Mechanisms of dietary mixing in grasshoppers: a review. *Comparative Physiology and Biochemistry A*, 104, 125–31.

16. Bernays, E. A. and Funk, D. J. (1999). Specialists make faster decisions than generalists: experiments with aphids. *Proceedings of the Royal Society, Series B*, 266, 151 – 6.
17. Bernays, E. A. and Weiss, M. R. (1996). Induced food in caterpillars: the need to identify mechanisms. *Entomologia Experimentalis et Applicata*, 78, 1 – 8.
18. Bernays, E. A. Bright, K, Howard, J. J, and Raubenheimer, D. (1992). Variety is the spice of life: frequent switching between foods in the polyphagous grasshopper *Taeniopoda eques* Burmeister (Orthoptera: Acrididae). *Animal Behaviour*, 44, 721 – 31.
19. Bernays, E. A., Bright, K. L., Gonzalez, N., and Angel, J. (1994). Dietary mixing in a generalist herbivore: tests of two hypotheses. *Ecology*, 75, 1997 – 2006.
20. Blaney, W. M. and Simmonds, M. S. J. (1985). Food selection by locusts: the role of learning in selection behaviour. *Entomologia Experimentalis et Applicata*, 39, 273 – 8.
21. Blaney, W. M. and Winstanley, C. (1982). Food selection behaviour in Locusta migratoria. In *Proceedings of the 5th International Symposium on Insect-Plant Relationships* (ed. J. H. Visser and A. K. Minks), pp. 365 – 6. Pudoc, Wageningen.
22. Bomford, M. K. and Isman, M. B. (1996). Desensitization of fifth instar *Spodoptera litura* (Lepidoptera: Noctuidae) to azadirachtin and neem *Entomologia Experimentalis et Applicata*, 81, 307 – 13.
23. Bongers, W. (1970). Aspects of host-piant relationships of the Colorado beetle. *Mededelingen Landbouwhogeschool Wageningen*, 70 – 10, 1 – 77.
24. Bossart, J. L. (2003). Covariance of preference and performance on normal and novel hosts in a locally monophagous and locally polyphagous butterfiy population. *Oecologia*, 135, 477 – 86.
25. Bossart, J. L. and Scriber, J. M. (1999). Preference variation in the polyphagous tiger swallowtail butterfly (Lepidoptera: papilionidae). *Environmental Entomology*, 28, 628 – 37.
26. Boys, H. A. (1978). Food selection by *Oedaleus senegalensis* (Acrididae: Orthoptera) in grassland and millet fields. *Entomologia Experimentalis et Applicata*, 24, 278 – 86.
27. Brodbeck, B. V., Andersen, P. C., and Mizell, R. F. (1995). Differential utilization of nutrients during development by the xylophagous leafhopper, *Homalodisca coagulata*. *Entomologia Experimentalis et Applicata*, 75, 279 – 89.
28. Carlsson, M. A., Anderson, P., Hartlieb, E., and Hansson, B. S. (1999). Experience-dependent modification of orientational response to olfactory cues in larvae of *Spodoptera Littoralias*. *Journal of Chemical Ecology*, 25, 2445 – 54.
29. Chapman, R. F. and Sword, S. (1993). The importance of palpation in food selection by a polyphagous grasshopper (Orthoptera: Acrididae). *Journal of Insect Behavior*, 6, 79 – 91.
30. Chapman, R. F., Bernays, E. A., Singer, M. S., and Hartmann, T. (2003). Experience influences gustatory responsiveness to pyrrolizidine alkaloids in the polyphagous caterpillar, *Estigmene acrea*. *Journal of Comparative Physiology A*. 189, 833 – 41.
31. Chyb, S., Eichenseer, H., Hollister, B., Mullin, C. A., and Frazier, J. L. (1995). Identification of sensilla involved in taste mediation in adult western corn rootworm (*Diabrotica virgifera* LeConte). *Journal of Chemical Ecology*, 21, 313 – 29.
32. Corbet, S. A. (1985). Insect chemosensory responses: a chemical legacy hypothesis. *Ecologi-*

cal Entomology, 10, 143 – 53.

33. Craighead, F. C. (1921). Hopkins-host selection principle as related to certain cerambycid beetles. *Journal of Agricultural Research*, 22, 189 – 220.
34. Cunningham, J. P., Mustapha, F. A., Jallow, M. F. A., Wright, D. J., and Zalucki, M. P. (1998). Learning in host selection in *Helicoverpa armigera* (Hübner) (Lepidoptera: Noctuidae). *Animal Behaviour*, 55, 227 – 34.
35. Cunningham, J. P., West, S. A., and Zalucki, M. P. (2001). Host selection phytophagous insects: a new explanation for learning in adults. *Oikos*, 95, 537 – 43.
36. Cottrell, C. B (1984). Aphytophagy in butterflies: its relationship to myrmecophily. *Zoological Journal of the Linnean Society*, 79, 1 – 57.
37. Darwin, C. (1875). *The variation of animals and plants under domestication* (2nd edn). Methuen, London.
38. De Boer, G. and Hanson, F. E. (1984). Food plant selection and induction of feeding preference among host and non-host plants in larvae of the tobacco hornworm *Manduca sexta*. *Entomologia Experimentalis et Applicatca*, 35, 177 – 93.
39. Del Campo, M. L., Miles, C. I., Schroeder, F. C., Mueller, C., Booer, R., and Renwick, J. A. A. (2001). Host recognition by the tobacco hornworm is mediated by a host plant compound. *Nature*, 411, 186 – 9.
40. Denno, R. F. and McClure, M. S. (1983). *Variable plants herbivores in natural and managed systems*. Academic Press, San Diego.
41. Dethier, V. G. (1980). Food-aversion learning in tow polyphagous caterpillars, *Diacrisia virginica* and *Estigmene congrua*. *Physiological Entomology*, 5, 321 – 5.
42. Dethier, V. G. (1988). Induction and aversion-learning in polyphagous arctiid larvae (Lepidoptera) in an ecological setting. *Canandian Entomologist*, 120, 125 – 31.
43. Dethier, V. G.. and Yost, MT. (1979). Oligophagy and absence of food-aversion learning in tobacco hornworms *Manduca sexta*. *Physiological Entomology*, 4, 125 – 30.
44. Di Giulio, M. and Edwards, P. J. (2003). The influence of host plant diversity and food quality on larval survival of plant feedig heteropteran bugs. *Ecological Entomology*, 28, 51 – 7.
45. Dicke, M. and Grostal, P. (2001). Chemical detection of natural enemies by arthropods: an ecological perspective. *Annual Review of Ecology and Systematics*, 32, 1 – 23.
46. Dickinson, J. L. (1992). Egg cannibalism by larvae and adults of the milkweed leaf beetle (*Labodomera civicollis*, Coleoptera: Chrysomelidae). *Ecological Entomology*, 17, 209 – 18.
47. Dixon, A. E. G. (1973). *Biology of aphids*. Edward Arnold, London.
48. Dixon, A. F. G.. (1985). *Aphid ecology*. Blackie, Glasgow.
49. Dres, M. and Mallet, J. (2002). Host races in plantfeeding insects and their importance in sympatric speciation. *Philosophical Transactions of the Royal Society of London.* B357, 471 – 92.
50. Drosopoulos, S. (1997). Biosystematic studies on the Muellerianella complex (Dephacidae, Homoptera. Auchenorrhyncha). *Mededelingen Landbouwhog school Wageningen*, 77 – 14, 1 – 133.

51. Du, Y. J, van Loon, J. J. A, and Renwick, J. A. A. (1995). Contact chemoreception of oviposition-stimulating glucosinolates and an oviposition-deterrent cardenolide in two subspecies of *Pieris napi*. *Physiological Entomology*, 20, 164 – 74.
52. Dukas, R. and Bernays, E. A. (2000). Learning promotes growth rate in grasshoppers. *Proceedings of the National Academy of Sciences of USA*, 97, 2637 – 40.
53. Fox, L. (1975). Cannibalism in natural poplations. *Annual Review of Ecology and Systematics*, 6, 87 – 106.
54. Fox, L. R. and Morrow, P. A. (1981). Specialization: species propeity or local phenomenon? *Science*, 211, 887 – 93.
55. Fritz, R. S. and Simms. E. L. (1992). *Plant resistance to herbivores and pathogens. Ecology, evolution and geneties*. University of Chicago Press, Chicago.
56. Gilbert, L. E. (1975). Ecological consequences of a coevolved mutualism between butterfilies and plants. In *Coevolution of animals and plants*. (ed L. E. Gilbert and P. E. Raven), pp. 210 – 40. University of Texas Press, Austin.
57. Glendinning, J. I. (2002). How do herbivorous insects cope with noxious secondary plant compounds in their diet? *Entomologia Expermentalis et Applicata*, 104, 15 – 25.
58. Glendinning, J. I. Domdom, S., and Long, E. (2001). Selective adaptation to noxious foods by a herbivorous insect. *Journal of Experimental Biology*, 204, 3355 – 67.
59. Gould, F. (1993). The spital scale of gnnetic variation in insect populations. In *Evolution of insect pests. Patterns of variation* (ed. K. C. Kim and B. A. Mcpheron), pp. 67 – 85. john Wiley, New York.
60. Grabstein, E. M. and Scriber, J. M. (1982). Host plant utilization by *Hyalophora cecropia* as affccted by prior feeding exoerience. *Entomologia Experimentalis et Applicata*, 32, 262 – 8.
61. Hare, J. D. and Kennedy, G. G. (1986). Genetic variation in plant-insect associations: survival of *Leptinotarsa decemlineata* populations on *Solanum carolinense*. *Evolution*, 40, 1031 – 43.
62. Howard, G. and Bernays, E. A. (1994). Cited in Ref. 130.
63. Hsiao, T. H. (1985). Ecophysiological and genetic aspects of geographic variations of the Colorado potato beetle. In *Proceedinds of the Symposium on the Colorado potato beetle, XVIIth International Congress of Entomology* (ed: D. N. Ferro and R. H. Voss), pp. 63 – 77. Massachusetts Agricultural Experiment Station, University of Massachusetts at Amherst.
64. Huang, X. P. and Renwick, J. A. A. (1995). Cross habituation to feeding deterrents and acceptance of a marginal host plant by *Pieris rapae* larvae. *Entomologia Experomentali et Applicata*, 76, 295 – 302.
65. Jaenike, J. (1983). Induction of host preference in *Drosophila melanogaster*. *Oecologia*, 58, 320 – 5.
66. Janz, N. (2003). The cost of polyphagy: oviposition decision time vs error rate in a butterfly. *Oikos*, 100, 493 – 6.
67. Jermy, T. (1987). The role of experience in the host selection of phytophagous insects. In *Perspectives in chemoreception and behavior* (ed: R. F. Chapman, E. A. Bernays, and J. G.

Stoffolano), pp. 142–57. Springer, New York.
68. Jermy, T., Hanson, F. E., and Dethier. V. G. (1968). Induction of specific food preference in lepidopterous larvae. *Entomologia Experimentalis et Applicata*, 11, 211–30.
69. Jermy, T., Bernays, E. A., and Szentesi, Á. (1982). The effect of repeated exposure to feeding deterrents on their acceptability to phytophagous insects. In *Proceedings of the 5th internationsl on insect-plant relationships* (ed. J. H. Visser and A. K. Minks), pp. 25–32. Pudoc, Wageningen.
70. Jones, R. e. and Ives, P. M. (1979.) The adaptiveness of searching and host seletion behaviour in *Pieris rapae*(*L.*). *Australian Journal of Ecology*, 4, 75–86.
71. Karowe, D. N. (1989). Facultative monophagy as a consequence of prior feeding experience: behavioral and physiological specialization in *Colias philodice* larvae. *Oecologia*, 78, 106–11.
72. Kaufman, T. (1965). Biological studies of some Bavarian Acridoedea (Orthoptera) with special reference to their feeding habits. *Annals of the Entomological Society of America*, 58, 791–800.
73. Kim, K. C. and McPheron, B. A. (1993). *Evolution of insect Pests. Patterns of variation*. John Wiley, New York.
74. Kirby, W. and Spence, W. (1863). *An introduction to entomology* (7th edn). Longman, Green, Longman, Roberts & Green, London.
75. Klos, R. (1901). Zur Lebensgeschichte von *Tephroclystia virgaureata*. *Verhandlungen der Kaiserlichen-Königlichen Zoologisch-Botanischen Gesellschaft in Wien*, 51, 785.
76. Kogan, M. (1977). The role of chemical factors in insect/plant relationships. *Proceedings of the 15th International Congress of Entomology*, Washington, DC, 211–27.
77. Kundu, R. and Dixon, A. F. G. (1995). Evolution of complex life cycles in aphids. *Journal of Animal Ecology*, 64, 245–55.
78. Landolt, P. J, and Molina, O. (1996). Host-finding by cabbage looper moths (Lepidoptera: Noctuidae): learning of host odor upon contact with host foliage. *Journal of Insect Behavior*, 9, 899–908.
79. Lee, J. C. and Bernays, E. A. (1988). Declining acceptability of food plat for a polyphagous grasshopper *Schistocerca americana*: the role of food aversion learning. *Physiological Entomology*, 13, 291–301.
80. Leibrecht, B. C. and Askew, H. R. (1980). Habituation from a comparative perspective. In *Comparative psychology: an evolutionarysis of animal behavior* (ed. M. R. Denny), pp. 208–229. John Wiley, New York
81. Llewellyn, M. (1982). The energy economy of fluid-feeding herbivorous insects. In *Proceedings of the 5th international symposium on insect-plant relationships* (ed. J. H. Visser and A. K. Minks), pp. 243–51. Pudoc, Wageningen.
82. Lu, W. and Logan, P. (1994). Genetic variation in oviposeteon between and within populations of *Leptinotarsa decemlineata* (Coleoptera: Chrysomelidae) *Annals of the Entomological Society of America*, 87, 634–40.
83. Ma, W. C. (1972). Dynamics of feeding responses in *Pieris brassicae* Linn. as a function of

chemosensory input a behavioural, ultrastructural and electro-physiological study. *Mededelingen Landbouwhogeschool Wageningen*, 72 – 11, 1 – 162.

84. Mackenzie, A. and Dixon, A. F. G (1990). Host alternation in aphids: constraint versus optimization. *American Naturalist*, 136, 132 – 4.

85. Maier, C. T. (1985). Rosaceous hosts of *Phyllonorycter* species (Lepidoptera: Gracillariidae) in New England. *Annals of the Entomological Society of America*, 78, 826 – 30.

86. Merz, E. (1959). Pflanzen und Raupen. Über einige Prinzipien der Futterwahl bei Grofschmetterlingsraupen. *Biologisches Zentralblatt*, 78, 152 – 88.

87. Moran, N. A. (1992). The evolution of aphid life cycles. *Annual Review of Entomology*, 37, 321 – 48.

88. Opstad, R, Rogers, S. M. , Behmer, S. T. , and Simpson, S. J. (2004). Behavioural correlates of phenotypic plasticity in mouthpart chemoreceptor numbers in locusts. *Journal of Insect Physiology*, 50, 725 – 36.

89. Panda, N. and Khush, G. S. (1995). *Host plant resistance to insects.* CAB International, Oxon.

90. Papaj, D. (1986a). An oviposition 'mistake' by *Battus phhilenor* L. (Papilionidae). *Journal of the Lepidopterists' Society*, 40, 348 – 9.

91. Papaj, D. (1986b). Conditioning of leaf shape discrimination by chemical cues in the butterfly *Battus philenor*. *Animal Behaviour*, 34, 1281 – 8.

92. Papaj, D. R. and Lewis, A. C. (1993). *Insect learning. Ecology and evolutionary perspectives.* Chapman & Hall, New York.

93. Pescador, A. R. (1993). The effects of a multispecies sequential diet on the growth and survival of tropical polyphagous caterpillars. *Entomologia Experimentaliset Applicata*, 67, 15 – 24.

94. Prokopy, R. J. and Papaj, D. R. (1988). Learning of apple fruit biotypes by apple maggot flies. *Journal of Insect Behavior*, 1, 67 – 74.

95. Prokopy, R. J. , Averill, A. L. , Cooley, S. S. , and Roitberg, C. A. (1982). Associative learning of apple fruit biotypes by apple maggot flies. *Science*, 218, 76 – 7.

96. Prokopy, R. J. , Papaj, D. R. , Cooley, S. S. , and Kallet, C. (1986). On the nature of learning in oviposition site acceptance by apple maggot flies. *Animal Behaviour*, 34, 98 – 107.

97. Provenza, F. D. and Cincotta, R. P. (1993). Foraging as a self-organizational learning process: accepting adaptability at the expense of predictability. In *Diet selection* (ed. R. N. Hughes), pp. 78 – 101. Blackwell, Oxford.

98. Rausher, M. D. (1978). Search image for leaf shape in a butterfly. *Science*, 200, 1071 – 3.

99. Rausher, M. D. (1980). Host abundance, juvenile survival, and oviposition preference in *Battus philenor*. *Evolution*, 34, 342 – 55.

100. Rausher, M. D. (1985). Variability for host preference in insect populations: mechanistic and evolutionary models. *Journal of Insect Physiology*, 31, 873 – 89.

101. Rayor, L. S. and Munson, S. (2002). Larval feeding experience influences adult predator acceptance of chemically defended prey. *Entomologia Experimentalis et Applicata*, 104, 193 – 201.

102. Reavy, D. and Lawton, J. H. (1991). Larval contribution to fitness in leaf-eating insects. In

Reproductive behaviour of insects: *individuals and populations* (ed. W. J. Bailey and J. Ridsdill-Smith), pp. 293 – 329. Chapman & Hall, London.

103. Renwick, J. A. A. (2001). Variable diets and changing taste in plant-insect relationships. *Journal of Chemical Ecology*, 27, 1063 – 76.
104. Renwick, J. A. A. and Huang, X. P. (1995). Rejection of host plant by larvae of cabbage butterfly: diet-dependent sensitivity to an antifeedant. *Journal of Chemical Ecology*, 21, 465 – 75.
105. Renwick, J. A. A. and Lopez, K. (1999). Experience-based food consumption by larvae of *Pieris rapae*: addiction to glucosinolates? *Entomologia Experimentalis et Applicata*, 91, 51 – 8.
106. Renwick, J. A. A., Zhang, W., Haribal, M., Attygalle, A. B., and Lopez, K. D. (2001). Dual chemical barriers protect a plant against larval stages of an insect. *Journal of Chemical Ecology*, 27, 1575 – 83.
107. Sandlin, E. A. and Willig, M. R. (1993). Effects of age, sex, prior experience, and intraspecific food variation on diet composition of a tropical folivore (Phasmatodea: Phasmatidae). *Environmental Entomology*, 22, 625 – 33.
108. Saxena, K. N. and Schoonhoven, L. M. (1978). Induction of orientation and feeding preferences in *Manduca sexta* larvae for an artificial diet containing citral. *Entomologia Experimentalis et Applicata*, 23, 72 – 8.
109. Schalk, J. M., Kindler, S. D., and Manglitz, G. R. (1969). Temperature and the preference of the spotted alfalfa aphid for resistant and susceptible alfalfa plants (*Therioaphis maculata*: Hem., Hom., Aphididae). *Journal of Economic Entomology*, 62, 1000 – 3.
110. Schoonhoven, L. M. (1967). Loss of host plant specificity by *Manduca sexta* after rearing on an artificial diet. *Entomologia Experimentalis et Applicata*, 10, 270 – 2.
111. Schoonhoven, L. M. (1969). Sensitivity changes in some insect chemoreceptors and their effect on food selection behaviour. *Proceedings of the Koninklijke Nederlandse Akademie van Wetenschappen, Amsterdam*, C72, 491 – 8.
112. Schoonhoven, L. M. (1977). On the individuality of insect feeding behaviour. *Proceedings of the Koninklijke Nederlandse Akademie van Wetenschappen, Amsterdam*, C80, 341 – 50.
113. Schoonhoven, L. M. and van Loon, J. J. A. (2002). An inventory of taste in caterpillars: each species its own key. *Acta Zoologica Academiae Scientiarum Hungaricae*, 48 (Suppl. 1), 215 – 63.
114. Schoonhoven, L. M., Blaney, W. M., and Simmonds, M. S. J. (1987). Inconstancies of chemoreceptor sensitivities. In *Insects-plants. Proceedings of the 6th international symposium on insect-plant relationships* (ed. V. Labeyrie, G. Fabres, and D. Lachaise), pp. 141 – 5. Junk, Dordrecht.
115. Scriber, J. M. (1982). The behavioural and nutritional physiology of southern armyworm larvae as a function of plant species consumed in earlier instars. *Entomologia Experimentalis et Applicata*, 31, 359 – 69.
116. Scriber, J. M. (1993). Absence of behavioral induction in multi-choice oviposition preference studies with a generalist butterfly species, *Papilio glaucus*. *Great Lakes Entomologist*, 28,

81 – 95.

117. Scriber, J. M. (2002). Evolution of insect-plant relationships: chemical constraints, coadaptation, and concordance of insect/plant traits. *Entomologia Experimentalis et Applicata*, 104, 217 – 35.

118. Scriber, J. M. and Lederhouse, R. C. (1992). The thermal environment as a resource dictating geographic patterns of feeding specialization of insect herbivores. In *Effects of resource distribution on animal-plant interactions* (ed. M. D. Hunter, T. Ohgushi, and P. Price), pp. 430 – 466. Academic Press, San Diego.

119. Scriber, J. M., Giebink, B. L., and Snider, D. (1991). Reciprocal latitudinal clines in oviposition behavior of *Papilio glaucus* and *P. canadensis* across the great Lakes hybrid zone: possible sex-linkage of oviposition preferences. *Oecologia*, 87, 360 – 8.

120. Sergel, R. (1987). On the occurrence and ecology of the *Rhododendron-leafhopper Graphocephala fennahi* Young 1977, in the western Palearctic region (Homoptera, Cicadellidae). *Anzeiger für Schädlungskunde, Pflanzenkrankheiten und Pflanzenschutz*, 60, 134 – 6.

121. Shaposhnikov, G. C. (1987). Evolution of aphids in relation to evolution of plants. In *Aphids. Their biology, natural enemies and control*, Vol. A (ed. A. K. Minks and P. Harrewijn), pp. 409 – 414. Elsevier, Amsterdam.

122. Simmonds, M. S. J., Simpson, S. J., and Blaney, W. M. (1992). Dietary selection behaviour in *Spodoptera littoralis*: the effects of conditioning diet and conditioning period on neural responsiveness and selection behaviour. *Journal of Experimental Biology*, 162, 73 – 90.

123. Simpson, S. J. and White, P. (1990). Associative learning and locust feeding: evidence for a 'learned hunger' for protein. *Animal Behaviour*, 40, 506 – 13.

124. Simpson, C. L., Chyb, S., and Simpson, S. J. (1990). Changes in chemoreceptor sensitivity in relation to dietary selection by adult *Locusta migratoria*. *Entomologia Experimentalis et Applicata*, 56, 259 – 68.

125. Singer, M. C. (1983a). Determinants of mutiple host use by a phytophagous insect population. *Evolution*, 37, 389 – 403.

126. Singer, M. C. (1983b). Heritability of oviposition preference and its relationship to offspring performance within a single insect population. *Evolution*, 37, 575 – 96.

127. Singer, M. C. and Lee, J. R. (2000). Discrimination within and between host species by a butterfly: implications for design of preference experiments. *Ecology Letters*, 3, 101 – 5.

128. Singer, M. C. and Parmesan, C. (1993). Sources of variation in patterns of plant-insect association. *Nature*, 361, 251 – 3.

129. Stockoff, B. A. (1993). Ontogenetic change in dietary selection for protein and lipid by gypsy moth larvae. *Journal of Insect Physiology*, 39, 677 – 86.

130. Sword, G. A. and Chapman, R. F. (1994). Monophagy in a polyphagous grasshopper, *Schistocerca shoshone*. *Entomologia Experimentalis et Applicata*, 73, 255 – 64.

131. Sword, G. A. and Dopman, E. B. (1999). Developmental specialization and geographic structure of host plant use in a polyphagous grasshopper, *Schistocerca emarginata* (= lineata) (Orthoptera, Acrididae). *Oecologia*, 120, 437 – 45.

132. Szentesi, Á. and Bernays, E. A. (1984). A study of behavioural habituation to a feeding deterrent in nymphs of *Schistocerca gregaria*. *Physiological Entomology*, 9, 329-40.

133. Szentesi, Á. and Jermy, T. (1990). The role of experience in host plant choice by phytophagous insects. In *Insect-plant interactions*, Vol. II (ed. E. A. Bernays), pp. 39-74. CRC Press, Boca Raton.

134. Szentesi, Á. and Jermy, T. (1993). A comparison of food-related behaviour between geographic populations of the Colorado potato beetle (Coleoptera, Chrysomelidae), on six solanaceous plants. *Entomologia Experimentalis et Applicata*, 66, 283-93.

135. Tarpley, M. D., Breden, F., and Chippendale, G. M. (1993). Genetic control of geographic variation for cannibalism in the southwestern corn borer, *Diatraea grandiosella*. *Entomologia Experimentalis et Applicata*, 66, 145-52.

136. Thompson, J. N. (1994). *The coevolutionary process*. University of Chicago Press, Chicago.

137. Thorpe, W. H. (1939). Further studies in preimaginal olfactory conditioning in insects. *Proceedings of the Royal Society*, B127, 424-33.

138. Tinbergen, L. (1960). The natural control of insects in pine woods. I. Factors influencing the intensity of predation by songbirds. *Archives Néerlandaises de Zoologie*, 13, 265-336.

139. Traynier, R. M. M. (1986). Visual learning in assays of sinigrin solution as an oviposition releaser for the cabbage butterfly, *Pieris rapae*. *Entomologia Experimentalis et Applicata*, 40, 25-33.

140. Tully, T., Cambiazo, V., and Kruse, L. (1994). Memory through metamorphosis in normal and mutant *Drosophila*. *Journal of Neuroscience*, 14, 68-74.

141. Ueckert, D. N. and Hansen, R. M. (1971). Dietary overlap of grasshoppers on sandhill rangeland in northeastern Colorado. *Oecologia*, 8, 276-95.

142. Van Loon, J. J. A. (1996). Chemosensory basis of feeding and oviposition behaviour in herbivorous insects: a glance at the periphery. *Entomologia Experimentalis et Applicata*, 80, 7-13.

143. Via, S. (1999). Cannibalism facilitates the use of a novel environment in the flour beetle, *Tribolium castaneum*. *Heredity*, 82, 267-75.

144. Zwölfer, H. (1970). Der Regionale Futterpflanzwechse', bei phytophagen Insekten als evolutionäres Problem. *Zeitschrift für Angewandte Entomologie*, 65, 233-7.

145. Zwölfer, H. (1988). Evolutionary and ecological relationships of the insect fauna of thistles. *Annual Review of Entomology*, 33, 103-22.

第9章 植物对昆虫内分泌系统的影响

9.1　发育
　　9.1.1　形态
　　9.1.2　滞育
9.2　繁殖
　　9.2.1　成熟
　　9.2.2　交尾
9.3　结论
9.4　参考文献

　　在地球上,任何一个地方都不会存在恒定不变的、永久适宜植物生长的气候条件。通常情况下,适宜的气候会与低温、干旱等气候条件交替出现。在气候恶劣的时期,植物会通过一些方式进行调节,如叶片脱落,亦或地上枝叶完全枯萎。对于一年生植物而言,其种子会在土壤中进行休眠,等待适宜的生长季节。昆虫可能会有停止发育或繁殖等一系列生理活动,进入滞育阶段,以保持相对静止的状态。而即使在气候适宜的季节,植物的营养成分(第5章)及体内次生代谢物质的数量也会改变(第4章)。当昆虫试图对生长周期短的植物,或植物的特定生长阶段实现最优利用时,保持与寄主植物生活周期的同步性,则体现出非常重要的价值。显然,对专食性昆虫而言,这种同步性比广食性昆虫更为重要。如果昆虫和植物对如光周期、温度和降雨量等一些气候因素的响应是同步的,那么便能很好地实现生活周期同步性。昆虫通过对这些气候因素的感知,能准确地预测未来环境的改变,并且做好应对的准备。如果昆虫能准确感知寄主植物不同发育阶段释放的信号,那么就更有可能与植物保持精确的物候同步性。有时,由于某种原因,寄主植物的生活周期可能会比依照临界日长等因素预测的周期提前或滞后。而事实上,许多昆虫都可以监测到寄主植物生理变化的早期征兆,从而激发其神经内分泌机制,在适当时机调整自身的生长发育和繁殖。
　　本章主要讨论的内容为:植食昆虫会利用一些植物的因素,尽可能地与寄主植物生活周期和潜在的生理机制保持紧密一致。

9.1　发育

　　昆虫发育的整个过程可能会受到寄主植物信号的调控,主要包括对昆虫的形态影响和诱导滞育。

9.1.1　形态

　　蚜虫具有一个非常显著的特点,即环境会影响蚜虫的形态特征。蚜虫通常会因季节变换和寄主更换而呈现不同形态[15]。

在众多种类中,夏季种群中既有有翅类型也有无翅类型(图9.1),这两种类型都能进行孤雌生殖。如果在植物被过度取食、营养成分开始降低的情况下,有翅类型具有较强的更换寄主的能力。因此营养状况似乎是影响翅形成的重要因素。一项以豌豆蚜(*Acyrthosiphon pisum*)为对象的研究表明,寄主植物的年龄也会影响翅的发育。而在种群密度过大的情况下,无翅雌虫能产生有翅后代。寄主植物的年龄和种群密度对于翅发育的影响可能会有累加效应(图9.2)。

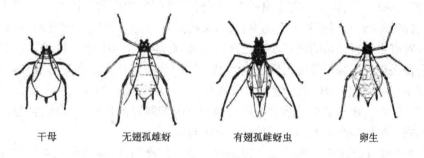

干母　　　无翅孤雌蚜　　　有翅孤雌蚜虫　　　卵生

图9.1　雌性野豌豆蚜(*Megoura viciae*)多态性(利斯,1961)[23]

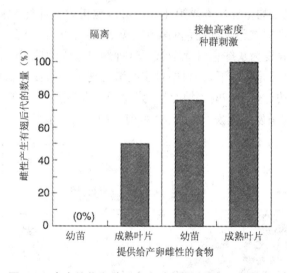

图9.2　寄主植物和种群密度对有翅豌豆蚜形成的作用

所有蚜虫都在蚕豆(*Vicia faba*)幼苗上隔离培育,成虫接触(或不接触)高密度种群刺激之后被放置在幼苗或完全展开的绿色成熟叶片上(萨瑟兰德,1967)[43]

继而产生的问题是,究竟是何种内在因素诱发了翅的发育? 这还要取决于蚜虫—植物组合体的类型,因为即使是相同生理条件也会有多种变化。氨基酸、糖类及(或)水分都是翅形成的关键营养因素。通常情况下,植物树液营养价值的降低必然会引起翅的形成,从而提高其迁移能力,并且提高找到更好寄主的可能性。需要注意的是,翅的形成并不必然意味着营养不足,而是指在食物质量改变前,对植物因素的一种反应。有翅类型在营养适宜的植物上可以正常生长和繁殖[15,18]。

以人工饲料(化学成分确定的流质食物)喂养蚜虫,能为确定食物成分对形态结构的影响提供一个直接的方法。以人工饲料喂养绿桃蚜(*Myzus persicae*)时,饲料标准配方中维生素C

的缺失会造成有翅后代的相对比例从 28% 提高到了 62%。然而当氨基酸浓度减半时,产生有翅后代的比率会降低[44]。显然翅的发育要受这些特定营养物质的影响。而在所有的类群中,这种影响并不绝对,可能是因为自然条件下的调控信号更为复杂。

在一些蚜虫类群中,在夏末和秋季,不同性别的形成,同样也受到寄主植物的影响。迄今为止,我们对老化的植物中产生怎样的变化刺激了性母的产生仍不甚了解[18]。有一个典型的例子,棉蚜(*Eriosoma pyricola*)寄居在梨树根部。在夏末和秋初这段时期内,有翅的性母发育完毕,它们破土而出后飞到第一寄主榆树上。梨树诱导产生性母的原因可能与根系的终止生长有关,后者由于受到光周期和温度的变化,对梨树的地上部分产生影响继而影响到根系,这些环境变量对蚜虫的直接影响通过实验已经排除,但对其有效的植物因素目前尚未确定[40]。

究竟是什么生理机制将环境诱因(如食物质量等)与决定成虫的翅是否形成联系起来呢?在这个过程中,保幼激素(JH)似乎起到了关键作用[13]。较低水平的保幼激素促使幼虫发育为有翅雌虫,反之,较高水平的保幼激素可通过调节基因抑制翅的形成。然而,食物成分能在多大程度上影响产生保幼激素的调控原件依然有待解决。

尽管大多数关于环境多态性的例子都是在蚜虫中发现的,但是这种现象同样存在于其他昆虫类群中。例如,尺蠖(*Nemoria arizonaria*)为北美的一种二化性蛾类幼虫,其不同世代的形态及行为均表现出了惊人的差异。春季孵化的幼虫以橡树柔荑花序为食,发育成高度隐蔽的模仿者,表皮为黄色且具褶皱还有红褐色的类似于雄蕊的圆点(图 9.3 中 A)。而夏季孵化的幼虫以橡树叶为食,其表皮为灰绿色且无褶皱。这种幼虫在受到惊扰时,会采取典型尺蠖式姿势(图 9.3 中 B)。人工饲养条件下,饲料中丹宁酸浓度高于自然食物水平时,幼虫就会发育成嫩枝型。相反,饲料中缺乏丹宁酸时,幼虫就会发育成柔荑花序型。在这个例子中,植物刺激(如丹宁酸水平)是如何引发不同发育反应的,目前我们依然不清楚[12]。

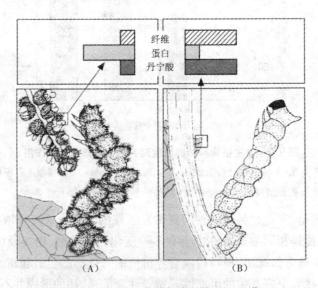

图 9.3　一种二化性蛾类幼虫——尺蠖

(A) 尺蠖(*Nemoria arizonaria*)的幼虫在春季世代以橡树的柔荑花序为食,并发育成柔荑花序型;(B) 在夏季世代,幼虫以橡树叶为食,并发育成嫩枝型。与叶片相比,柔荑花序含有较少的纤维和多酚(丹宁酸),但有更多的非结合蛋白(unbound protein)

昆虫的拟态除了在视觉特征方面模拟外,可能还进一步涉及化学方面。以尺蛾科(*Biston robustum*)为例,这种杂食性幼虫体表的质地、颜色和形状取决于它们所取食的寄主植物种类。此外,幼虫表皮的蜡质成分也取决于摄入的食物,其成分类似于幼虫栖息的嫩枝表面的化学物质。幼虫和嫩枝之间的化学相似性源自幼虫消化了寄主植物叶片,表明这是一种食物诱导适应性。而对于形态适应所涉及的机制仍有待阐明。将幼虫从其生长的寄主植物上转移到另一种它们还未适应的寄主植物上,可以说明这种"植物拟态"的作用。在第一寄主植物上,捕食性的蚂蚁完全忽略了这种幼虫,而在新的寄主上,它们会注意到并对其进行攻击[1]。

9.1.2 滞育

马铃薯甲虫经完全变态之后,会破土而出,并开始取食马铃薯或一些近缘植物的叶片。在长日照光周期条件下,马铃薯甲虫很快开始繁殖,但在短日照的光周期中,它们会再次返回土壤中,并进入滞育阶段。即便在长日照的情况下,相对于取食马铃薯新叶的甲虫,取食马铃薯老叶的甲虫也能表现出明显的滞育趋势(图9.4)。

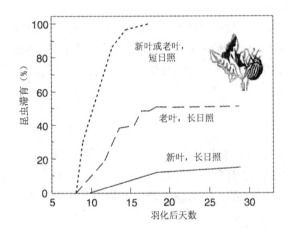

图9.4 食物的性质(老叶或新叶)和昼长(长日照与短日照)对马铃薯甲虫成虫滞育的影响(丹尼列夫斯基,1965)[4]

而这种反应不能简单地归咎于营养匮乏,因为将活跃昆虫的咽侧腺植入滞育个体中时,正常的活动和繁殖就会重新出现,甚至再取食老化叶片也不会改变[9]。在野外的观察表明,以欧白英(*Solanum dulcamara*,茄属)和马铃薯(*S. tuberosum*)为食的昆虫种群,滞育率存在着很大的差异,这说明了植物因素与诱导滞育相关。相对于寄生在马铃薯上的昆虫,寄生在欧白英植物上的昆虫较早开始滞育,这可能是因为,季末欧白英植物中糖苷生物碱的浓度较高,会导致昆虫较高的死亡率[14]。

在寄主植物中,由于其年龄变化会导致植物化学成分发生变化。一旦这种改变的成分进入植食性昆虫体内,甚至连带体内的寄生蜂都有可能发生滞育。寄生蜂的这种现象可能主要受其寄主昆虫调节,而非源自植物,但也不能排除这种直接影响的可能性[30]。

表 9.1 引起昆虫出现滞育的寄主植物的变化因素

昆虫	阶段	诱导滞育因素	参考文献
鳞翅目			
二化螟属(*Chilo* spp.)	幼虫	蛋白质和水含量降低	39
蛀杆虫(*Coniesta ignefusalis*)	幼虫	蛋白质和水含量降低	45
黑脉金斑蝶(*Danaus plexippus*)	成虫		11
鞘翅目			
马铃薯甲虫(*Leptinotarsa decemlineata*)	成虫	老龄的寄主植物	9,14
棉铃红虫(*Pectinophora gossypiella*)	幼虫	脂类,低的水分含量	2
四纹豆象(*Callosobruchus maculatus*)	成虫	寄主植物的花缺失	48
双翅目			
橄榄实蝇(*Bactrocera oleae*)	成虫	橄榄果实缺失	20
异翅亚目			
玫瑰菜蝽(*Eurydema rugosa*)	成虫	寄主植物的叶片缺少	27
膜翅目			
蚜虫寄生蜂		寄主食用的植物	30

在其他一些文献中也提到了食物引起几种植食昆虫滞育的例子,但在大多数研究中,其诱因本质还不太明确(表9.1)[5,41]。一些研究发现,水分含量和脂类水平的改变会诱导滞育[5],但是其他的一些因素,如植物生长调节剂,也可能作为调节神经内分泌过程的信号,从而调控滞育。

9.2 繁殖

植食昆虫经常在其取食的植物上或植物周围进行产卵。产卵雌虫会推迟其交尾行为和产卵时间,直至确认寄主植物正处于合适的发育期,这一策略具有显著的优势。事实上,许多昆虫的交尾行为和生殖腺的成熟,都会因为缺少寄主植物刺激剂而推迟。

9.2.1 成熟

在自然种群中,沙漠蝗虫(*Schistocerca gregaria*)成虫的生殖行为能推迟长达9个月的时间。然而,在一个种群内,个体间的发育过程却非常同步。一些环境因素会引发昆虫的性成熟,在雨季开始时,所有雌虫都准备产卵。刺激产卵的信号可能源自于昆虫的食物。当蝗虫成虫取食没药属(*Commiphora*)植物,如没药树(*C. myrrha*)的嫩芽时,所摄入的物质会引发性成熟。没药属植物嫩芽中含有大量精油,而精油的成分含萜类物质,例如松萜[46]、丁子香酚[22]和柠檬烯[34]等,在实验室条件下已被证实可以诱导昆虫性发育。但当刚羽化的成虫只取食老化的卷心菜叶时,其性成熟就会被抑制很长一段时期(图9.5)。当喂给它们正常的成熟叶片时,在成虫羽化三周后即开始产卵。用丁子香酚单独处理昆虫,或者将赤霉酸[25](嫩叶中含有的一种高浓度植物生长激素)添加到食物中,那么它们就能在营养不足的(老化的)植物上产卵(图9.5)[10]。显然,这些化合物作为激活神经内分泌系统的信号,因此可以刺激昆虫的性发育。

这种类似情况,在其他昆虫类群中也有发现。当寄主植物(卷心菜)不存在时,小菜蛾(*Plutella xylostella*)的生殖前期阶段就会变长。通过寄主植物特定挥发物对昆虫性发育的影响所进行的实验表明,寄主对昆虫卵巢发育的影响,能够被一种寄主挥发物——异硫氰酸烯丙酯模拟(表9.2)[17]。

第9章 植物对昆虫内分泌系统的影响

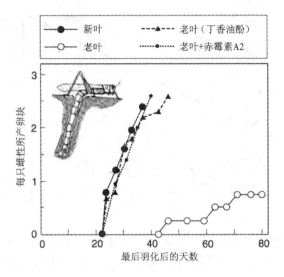

图 9.5 沙漠蝗（*Schistocerca gregaria*）的产卵行为受到食物的年龄和类型的影响

所有的昆虫都以卷心菜叶饲养，部分昆虫摄食老叶（老化的叶片），部分摄取新鲜叶片（生长完全的正常叶片），其中一组摄食老叶的昆虫每天还要额外摄入 1 mg 赤霉素 A2(gib)；另一组摄食老叶的昆虫在其羽化后七天，用 1 mg 丁子香酚进行外部处理（埃利斯等，1965）[10]

表 9.2 寄主植物存在 3 天后，对小菜蛾（*Plutella xylostella*）雌虫产卵的影响
（希利尔和索尔斯坦森，1969）[17]

刺激物	雌虫数目	每个卵巢管内卵的平均数目
对照（无植物）	47	2.4
卷心菜	38	5.7
异硫氰酸烯丙酯 0.1 ml	32	4.3
异硫酸烯丙脂 0.2 ml	31	5.1

在菜豆象（*Acanthoscelides obtectus*）中，寄主植物的气味并不起作用，只有菜豆象的下颚须接触到寄主植物——豆类的种子、叶片或豆荚后，才会刺激排卵。与豆类的广泛接触，会造成昆虫在几天内排卵量增加。然而，将豆类表面涂上涂层，或昆虫下颚须脱落时，昆虫的卵便不会成熟，这表明了在感受味觉刺激时，下颚须起到十分关键的作用。经漂洗过的豆类的种子，也能引发菜豆象排卵，但比未漂洗过的豆类种子的影响要微弱[31]。虽然这些化学因素对刺激性成熟起主要作用，然而，来自昆虫的嗅觉和视觉感受，对于其生理机制的影响也不能排除（图 9.6）。例如，当胡桃实蝇（*Rhagoletis juglandis*）的寄主胡桃果实存在时，相比胡桃果实不存在的情况而言，胡桃实蝇会提前进入排卵期，同时排卵进程加快。当将昆虫暴露于模拟其寄主的黄色球体中时，可以观察到产卵活动的加强。显然，来自寄主的视觉（也可能是触觉）刺激，使胡桃实蝇产生了生理变化，进而加速了产卵的进程[21]。

一些文献也记录了由寄主植物的利它素加速昆虫卵成熟的例子（表 9.3）[38]。所有例子中（有一个除外）均涉及寡食性昆虫，说明这种情况的出现绝非偶然。寄主植物对昆虫的影响，如之前对寡食性昆虫和杂食性沙漠蝗的讨论，可被昆虫用来预测其后代是否能获得充足食物。昆虫对于生殖发育的灵活调整策略，使其能够对寄主的突发性改变做出应对。

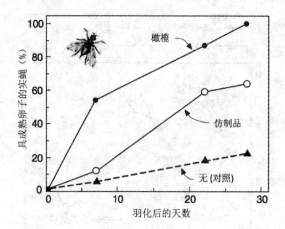

图9.6 存在或缺少橄榄果实或仿制品(橘色的蜡质球体)的情况下,雌性橄榄实蝇 (*Dacus oleae*) 卵的成熟情况(科威斯和萨纳卡基斯,1990)[19]

表9.3 昆虫卵的成熟受其寄主植物利它素的刺激

昆虫	取食的寄主范围	寄主植物	参考文献
直翅目			
蝗科			
沙漠蝗虫(*Schistocerca gregaria*)	杂食性	各种植物	10
同翅类			
杂斑木虱科			
木虱(*Euphyllura phillyreae*)	寡食性	木樨科	32
鳞翅目			
菜蛾科			
菜蛾(*Acrolepiopsis assectella*)	寡食性	韭菜	38
小菜蛾(*Plutella xylostella*)	寡食性	十字花科	17,29
螟蛾科			
斑螟(*Homoeosoma electellum*)	寡食性	向日葵	25
麦蛾科			
麦蛾(*Scrobipalpa ocellatella*)	寡食性	甜菜	38
麦蛾(*Phtorimaea operculella*)	寡食性	马铃薯	38
麦蛾(*Sitotroga cerealella*)	寡食性	谷物	38
卷蛾科(*Zeiraphera diniana*)	寡食性	落叶松	38
双翅目			
实蝇科			
实蝇(*Philophylla heraclei*)	寡食性	伞形科	38
实蝇(*Dacus oleae*)	寡食性	橄榄果	38,19
黄潜蝇科			
瑞典麦秆蝇(*Oscinella frit*)	寡食性	稻科	38
鞘翅目			
豆象科			
豆象(*Acanthoscelides obtectus*)	寡食性	豆类种子	38,31
豆象(*Careydon serratus*)	寡食性	花生(种子)	6

9.2.2 交尾

饲养蛾类的昆虫学爱好者们经观察了解到,通常寄主植物的存在会刺激交尾的发生。里迪福德和威廉斯姆首次报道了寄主植物对多音天蚕(*Antheraea polyphemus*)交尾行为的影响[37]。这种昆虫的交尾行为,只有在它们的寄主植物橡树叶的挥发物存在时才会发生。此后,也有一些例子证明,当雌蛾更接近寄主植物时,就会更早地开始了信息素的释放或"召唤"雄蛾的行为[24,36,47]。一种巢蛾(*Yponomeuta cagnagellus*)羽化一周后就能达到性成熟,寄主植物的挥发物会引发召唤行为。如果没有寄主存在,召唤行为就会被推迟,甚至被永久性抑制(图9.7)[16]。在这种情况下,寄主植物适合的挥发性物质,与雌虫开始召唤行为的时间二者共同作用,就能促使巢蛾属(*Yponomeuta*)昆虫的物种分化,因为选择不同的寄主植物作为其后代的食物会促使生殖隔离,形成姊妹群[26]。

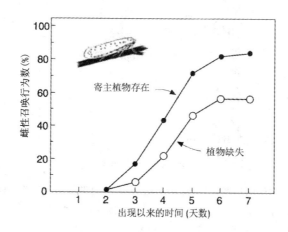

图9.7 在寄主植物存在或缺失时,巢蛾(*Yponomeuta evonymellus*)雌虫的召唤行为(萨德裹克斯和沃斯伯尼梅尔,1987)[16]

植物气味同样会对雄性昆虫产生影响。在一些夜行性物种中,当性引诱剂与植物绿叶挥发物混合时,雄性的反应会更为强烈[22]。绿叶挥发物和一些萜类物质能够与昆虫的性信息素相互作用,从而调节性信息素感受器的敏感性[28,42]。在野外的试验同样表明,当性信息素与寄主植物挥发物结合时,雄性的反应会增强。例如,雄性甜菜夜蛾(*Spodoptera exigua*)相比仅用性信息素来引诱,其更易被与植物挥发物和雌性性信息素混合的引诱吸引,说明某些寄主植物挥发物能够提高雄虫对在寄主植物上的雌虫的定位反应[7]。

有时寄主植物的某一特定部分会加速雌虫的召唤行为,如向日葵蛾(*Homoeosoma electellum*)会优先选择新开的花作为其产卵地。大多数雌蛾在羽化后的第一天,就开始召唤行为,然而当缺乏花粉时,召唤行为可能会被推迟多达两周的时间(图9.8)。向日葵蛾的初孵幼虫由于一些生物及非生物的因素,需要在短时间内获得花粉。雌蛾能迅速开始或推迟其召唤行为(卵巢的同时发育)以回应花粉的可获得性,因此可以满足后代对食物的需求[25]。

到目前为止,我们对于植物化合物(作为植物利他素)对召唤行为的影响仍知之甚少,而对于控制信息素和召唤行为产生的生理机制有了新的进展[34]。同其他物种一样,棉铃虫谷实夜蛾(*Helicoverpa zea*)的雌虫在交尾之前,先定居在寄主植物上。一旦它们为后代找到了适合的食物时,就会开始产生和释放性信息素。而召唤行为似乎是由玉米须的挥发物所引发的。

一些纯的化合物,如苯乙醛,能激发同样的反应。乙烯是一种参与水果催熟过程的植物激素,也具有同样的效果(图9.9)。

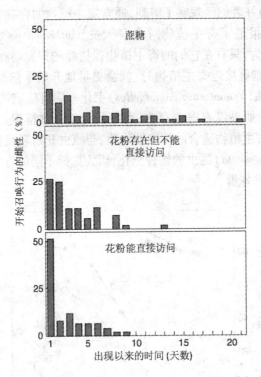

图 9.8　雌性向日葵蛾(*Homoeosoma electellum*)开始召唤行为的时间

昆虫被放置的实验环境为:只有蔗糖、蔗糖和不能直接访问的向日葵花粉、蔗糖与可直接访问的向日葵花粉(迪莱尔,1989b)[25]

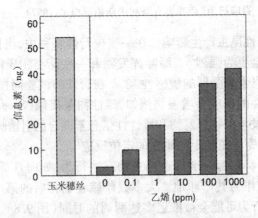

图 9.9　在玉米须(左边)或各种浓度乙烯(右边)存在的条件下,谷实夜蛾(*Helicoverpa zea*)信息素的产生情况,按每只雌性个体每 18 小时分泌的 ng 数计量(拉伊纳等,1992)[36]

谷实夜蛾(*H. zea*)幼虫取食于多种植物的果实,因此这些化合物可能起到通常的寄主所起到的诱导作用[36]。用硝酸银(银离子可以抑制植物中乙烯的反应)处理谷实夜蛾(*H. zea*)的触角,发现即使乙烯存在时,性信息素的释放也会被抑制,这表明雌虫的嗅觉系统也包括乙烯感受器。

当雌性棉铃虫感受到寄主植物的气味刺激时,其心侧体会释放出一种生物合成信息素激活神经肽(PBAN),以刺激腹部信息素腺体产生信息素(图 9.10)[33,36]。通过钙和环腺苷酸(cAMP)作为第二信使,PBAN 可直接作用于信息素腺体细胞[3]。

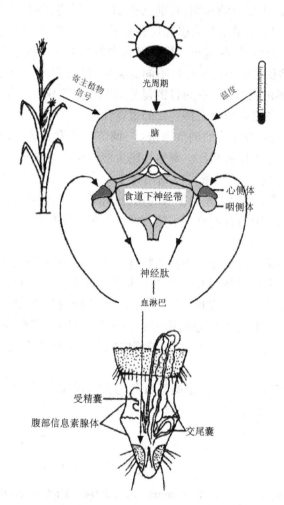

图 9.10 神经激素控制雌性谷实夜蛾(*Helicoverpa zea*)性信息素产生的示意图

在食道下神经节产生信息素生物合成激活神经肽,再转移到心侧体。一些外部因素如寄主植物气味、光周期和温度等产生的刺激会先传导至心侧体,再进入血淋巴,然后刺激腹部信息素腺体(PG)产生信息素。(拉伊纳,1988)[35]

并不是所有昆虫都利用信息素激活神经肽(PBAN)来调节信息素的生物合成。例如,小蠹虫在取食寄主植物后,通过从咽侧体释放的保幼激素来增加信息素的生物合成[46]。

以上所述例子充分表明,许多鳞翅目的处女雌虫,在它们开始吸引异性之前,首先会找到合适的寄主,经交尾后开始产卵。这种方式可以帮助生命周期较短的昆虫在有限的时间内,达到对资源的最优利用。

9.3 结论

在过去的50年中,寄主植物化学物质对于产卵雌虫的寄主选择过程,以及之后急待取食的幼虫对食物的识别过程中所起的作用,目前已得到了广泛论证(第6和7章)。相比之下,寄主植物对昆虫发育过程的调节却并未引起足够重视。本章探讨了昆虫通过调控自身的生命周期与寄主植物保持同步性的一些例子。尽管只有少数的例子明确了植物对昆虫的影响,但从分类学角度来看,这些例子中的昆虫分属不同的类别,说明在昆虫中,这可能是一种相对普遍的现象(如表9.3提供的数据)。不同目的昆虫,以及处于不同发育阶段的昆虫都会受影响,这进一步暗示了昆虫与寄主植物保持一致的重要性。

植食昆虫必须借助其自身独特的调节系统(与调节生长发育的系统完全不同)处理寄主生物发出的信号,这与寄生性昆虫相反。寄生性昆虫能利用寄主激素作为它们的信号分子,从而使其生活周期与寄主同步。兔跳蚤可以利用寄主激素,使它们的生殖周期与寄主相适应。植食昆虫则必须依赖与其生理物质差距很大的化合物信号。这些化合物信号表明了植物的存在,或其生理的细节状态。有时它们是寄主植物所特有的,有时,它们是更常见的化合物,如植物激素。

在很大程度上,相关的大部分化学物质的特性还有待阐明,同时昆虫对化学信号的探测和解读方式也需要进一步调查。同样的,关于昆虫体内不同过程间的链接反应问题还未解决,但至少有证据充分表明,所有生理反应和行为反应都是由神经内分泌系统调控的。

与对光周期的反应不同,植物信号的影响有一个显著特点:它们并非是引发有或无的反应。在多数情况下,即使持续缺乏寄主信号,种群中的部分个体也会完成发育,虽然可能有一些延迟。

可以预见,"昆虫为适应寄主植物而调整自身的生命周期"这一研究方向,将会为昆虫与植物关系的研究开拓更为丰富的领域。

9.4 参考文献

1. Akino, T., Nakamura, K. I., and Wakamura, S. (2004). Diet-induced chemical phytomimesis by twig-like caterpillars of *Biston robustum* Butler (Lepidoptera:Geometridae). *Chemoecology*, 14, 165 – 74.

2. Bull, D. L. and Adkisson, P. L. (1960). Certain factors influencing diapause in the pink bollworm *Pectinophora gossypiella*. *Journal of Economic Entomology*, 53, 793 – 8.

3. Choi, M. Y. and Jurenka, R. A. (2004). PBAN stimulation of pheromone biosynthesis by inducing calcium influx in pheromone glands of *Helicoverpa zea*. *Journal of Insect Physiology*, 50, 555 – 60.

4. Danilevski, A. S. (1965). *Photoperiodism and seasonal development of insects*. Oliver & Boyd, London. P242

5. Danks, H. V. (1987). *Insect dormancy: an ecological perspective*. Biological Survey of Canada, Ottawa.

6. Delobel, A. (1989). Influence des gousses d'arachide (*Arachis hypogea*) et de l'alimentation imaginale sur l'ovogénèse, l'accouplement et la ponte chez la bruche *Careydon serratus*. *Entomologia Experimentalis et Applicata*, 52, 281 – 9.

7. Deng, J. Y., Wei, H. Y., Huang, Y. P., and Du, J. W. (2004). Enhancement of attraction to sex pheromonesof *Spodoptera exigua* by volatile compounds producedby host plants. *Journal of Chemical Ecology*, 30, 2037 – 45.

8. Dettner, K. (1989). Chemische Ökologie. Ein interdisziplinäres Forschungsgebiet zwischen Biologie und Chemie. *Zeitschrift für Umweltchemie und Ökotoxikologie*, 4, 29 – 36.

9. De Wilde, J., Bongers, W., and Schooneveld, H. (1969). Effects of hostplant age on phytophagous insects. *Entomologia Experimentalis et Applicata*, 12, 714 – 20.

10. Ellis, P. E., Carlisle, D. B., and Osborne, D. J. (1965). Desert locusts: sexual maturation delayed by feedingon senescent vegetation. *Science*, 149, 546 – 7.

11. Goehring, L. and Oberhauser, K. S. (2002). Effects of photoperiod, temperature, and host plant age on induction of reproductive diapause and developmenttime in *Danaus plexippus*. *Ecological Entomology*, 27, 674 – 85.

12. Greene, E. (1989). A diet-induced developmental polymorphism in a caterpillar. *Science*, 243, 643 – 6.

13. Hardie, J. and Lees, A. D. (1985). Endocrine control of polymorphism and polyphenism. In *Comprehens iveinsect physiology, biochemistry and pharmacology*, Vol. 8 (ed. G. A. Kerkut and L. I. Gilbert), pp. 441 – 490. Pergamon Press, Oxford.

14. Hare, J. D. (1983). Seasonal variation in plant-insect associations: utilization of *Solanum dulcamara* by *Leptinotarsa decemlineata*. *Ecology*, 64, 345 – 61.

15. Harrewijn, P. (1978). The role of plant substances in polymorphism of the aphid *Myzus persicae*. *Entomologia Experimentalis et Applicata*, 24, 198 – 214.

16. Hendrikse, A. and Vos-Bünnemeyer, E. (1987). Role of the host-plant stimuli in sexual behaviour of the small ermine moths (*Yponomeuta*). *Ecological Entomology*, 12, 363 – 71.

17. Hillyer, R. J. and Thorsteinson, A. J. (1969). The influence of the host plant or males on ovarian development or oviposition in the diamondback moth *Plutella maculipennis* (Curt.). *Canadan Journal of Zoology*, 47, 805 – 16.

18. Kawada, K. (1987). Polymorphism and morph determination. In *Aphids, their biology, natural enemies, and control; world crop pests*, Vol. 2 A (ed. A. K. Minksand P. Harrewijn), pp. 255 – 268. Elsevier, Amsterdam.

19. Koveos, D. S. and Tzanakakis, M. E. (1990). Effect of the presence of olive fruit on ovarian maturationin the olive fruit fly, *Dacus oleae*, under laboratory conditions. *Entomologia Experimentalis et Applicata*, 55, 161 – 8.

20. Koveos, D. S. and Tzanakakis, M. E. (1993). Diapause aversion in the adult olive fruit-fly through effects ofthe host fruit, bacteria, and adult diet. *Annals of the Entomological Society of America*, 85, 668 – 73.

21. Lachmann, A. D. and Papaj, D. R. (2001). Effect of host stimuli on ovariole development in

the walnut fly, *Rhagoletis juglandis* (Diptera, Tephritidae). *Physiological Entomology*, 26, 38 – 48.

22. Landolt, P. J. and Phillips, T. W. (1997). Host plant influences on sex pheromone behavior of phytophagous insects. *Annual Review of Entomology*, 42, 371 – 91.

23. Lees, A. D. (1961). Clonal polymorphism in aphids. *Symposia of the Royal Entomological Society of London*, 1, 68 – 79.

24. McNeil, J. N. and Delisle, J. (1989a). Are host plants important in pheromone-mediated mating systems of Lepidoptera? *Experientia*, 45, 236 – 40.

25. McNeil, J. N. and Delisle, J. (1989b). Host plant pollen influences calling behavior and ovarian development of the sunflower moth, *Homoeosoma electellum*. *Oecologia*, 80, 201 – 5.

26. Menken, S. B. J., Herrebout, W. M., and Wiebes, J. T. (1992). Small ermine moths (*Yponomeuta*): their host relations and evolution. *Annual Review of Entomology*, 37, 41 – 66.

27. Numata, H. and Yamamoto, K. (1990). Feeding on seeds induces diapause in the cabbage bug, *Eurydema rugosa*. *Entomologia Experimentalis et Applicata*, 57, 281 – 4.

28. Ochieng, S. A., Park, K. C., and Baker, T. C. (2002). Host plant volatiles synergize responses of sexpheromone-specific olfactory receptor neurons in male *Helicoverpa zea*. *Journal of Comparative Physiology A*, 188, 325 – 33.

29. Pittendrigh, B. R. and Pivnick, K. A. (1993). Effects of a host plant, *Brassica juncea*, on calling behaviour and egg maturation in the diamondback moth, *Plutella xylostella*. *Entomologia Experimentalis et Applicata*, 68, 117 – 26.

30. Polgár, L. A. and Hardie, J. (2000). Diapause induction in aphid parasitoids. *Entomologia Experimentalis et Applicata*, 97, 21 – 7.

31. Pouzat, J. (1978). Host plant chemosensory influence on oogenesis in the bean weevil, *Acanthoscelides obtectus* (Coleoptera: Bruchidae). *Entomologia Experimentalis et Applicata*, 24, 601 – 8. P243

32. Prophetou-Athanasiadou, D. A. (1993). Diapause termination and phenology of the olive psyllid, *Euphyllura phillyreae* on two host plants in coastal northern Greece. *Entomologia Experimentalis et Applicata*, 67, 193 – 7.

33. Rafaeli, A. (2001). Neuroendocrine control of pheromone biosynthesis in moths. *International Review of Cytology*, 213, 49 – 92.

34. Rafaeli, A. (2005). Mechanisms involved in the control of pheromone production in female moths: recent developments. *Entomologia Experimentalis et Applicata*, 115, 7 – 15.

35. Raina, A. K. (1988). Selected factors influencing neurohormonal regulation of sex pheromone production in *Heliothis* species. *Journal of Chemical Ecology*, 14, 2063 – 9.

36. Raina, A. K., Kingan, T. G., and Mattoo, A. K. (1992). Chemical signals from host plant and sexual behavior in a moth. *Science*, 255, 592 – 4.

37. Riddiford, L. M. and Williams, C. M. (1967). Volatile principle from oak leaves: role in sex life of the polyphemus moth. *Science*, 155, 589 – 90.

38. Robert, P. C. (1986). Les relations plantes-insects phytophages chez les femelles pondeuses:

le rôle desstimulus chimiques et physiques. Une mise au point bibliographique. *Agronomie*, 6, 127 – 42.

39. Scheltes, P. (1978). The condition of the host plant during aestivation-diapause of the stalk borers *Chilopartellus* and *Chilo orichalcociliella* in Kenya. *Entomologia Experimentalis et Applicata*, 24, 479 – 88.
40. Sethi, S. L. and Swenson, K. G. (1967). Formation of sexuparae in the aphid *Eriosoma pyricola*, on pear roots. *Entomologia Experimentalis et Applicata*, 10, 97 – 102.
41. Steinberg, S., Podoler, H., and Applebaum, S. W. (1992). Diapause induction in the codling moth, *Cydia pomonella*: effect of larval diet. *Entomologia Experimentalis et Applicata*, 62, 269 – 75.
42. Stelinski, L. L., Miller, J. R., Ressa, N. E., and Gut, L. J. (2003). Increased EAG responses of tortricid moths after prolonged exposure to plant volatiles: evidence for octopamine-mediated sensitization. *Journal of Insect Physiology*, 49, 845 – 56, 1083.
43. Sutherland, O. R. W. (1967). Role of host plant in production of winged forms by a green strain of pea aphid *Acyrthosiphon pisum* Harris. *Nature*, 216, 387 – 8.
44. Sutherland, O. R. W. and Mittler, T. E. (1971). Influence of diet composition and crowding on wing production by the aphid *Myzus persicae* Journal of Insect Physiology, 17, 321 – 8.
45. Tanzubil, P. B., Mensah, G. W. K., and McCaffery, A. R. (2000). Diapause initiation and incidence in the millet stem borer, *Coniesta ignefusalis* (Lepidoptera: Pyralidae): the role of the host plant. *Bulletin of Entomological Research*, 90, 365 – 71.
46. Tillman, J. A., Seybold, S. J., Jurenka, R. A., and Blomquist, G. J. (1999). Insect pheromones—an overview of biosynthesis and endocrine regulation. *Insect Biochemistry and Molecular Biology*, 29, 481 – 514.
47. Yan, F., Bengtsson, M., and Witzgall, P. (1999). Behavioral response of female codling moths, *Cydia pomonella*, to apple volatiles. *Journal of Chemical Ecology*, 25, 1343 – 50.
48. Zannou, E. T., Glitho, I. A., Huignard, J., and Monge, J. P. (2003). Life history of flight morph females of *Callosobruchus maculatus* F.: evidence of a reproductive diapause. *Journal of Insect Physiology*, 49, 575 – 82.

第 10 章　生态学:共生却各自独立

10.1　植物对昆虫的影响
　10.1.1　植物物候学
　10.1.2　植物化学
　10.1.3　植物形态学
　10.1.4　替代食物
10.2　植食昆虫对植物的影响
10.3　地上部分与地下部分的昆虫—植物交互作用
10.4　微生物和昆虫—植物的交互作用
10.5　脊椎动物和昆虫—植物的交互作用
10.6　群体内间接的物种交互作用
　10.6.1　资源利用性竞争
　10.6.2　似然竞争
　10.6.3　营养级联
10.7　物种交互与表型可塑性
10.8　自上而下和自下而上的驱动力
10.9　食物网和化学信息网
　10.9.1　食物网
　10.9.2　化学信息网
10.10　群落
　10.10.1　为什么会有如此多的稀有植食昆虫存在?
　10.10.2　群集性
　10.10.3　群落发展
10.11　分子生态学
10.12　结论
10.13　参考文献

　　生态学探讨的主要内容是:生物体与生物或非生物环境如何相互作用? 以及这种相互作用如何解释群落的组成与动态[117,141]? 可以对生物群落以不同的级别进行研究,但是将这些级别联系起来,对生态学家却是非常严峻的挑战。例如,群落生态学家们主要研究在种群动态的环境下,群落物种的组成如何波动。而行为生态学家主要关注昆虫个体如何对生物或非生物环境做出反应,以及这些反应是否(从成本和收益角度考虑)为其繁殖后代考虑。生物个体的特征可以影响与其他生物体的关系。由个体、细胞、亚细胞层面决定了种群动态及群落组成。基因表达如何影响个体的表型及种群内个体间的关系,最终影响群落进程,这将成为研究生态和环境基因组学领域的新课题,同时也将成为 21 世纪生物学的一个主题。

从第2章中我们得知,每一种植物都可能成为某类植食昆虫的食物。另外,肉食性节肢动物可能寄生在植物内部,或利用植物产生某种物质。此外,植物上的昆虫群落还要面对许多其他的进攻者,如细菌、真菌、线虫、软体动物和螨虫,还包括脊椎动物,如鸟类和哺乳动物(见图10.1)。同样的,这些生物都有其对应的天敌。最后,植物与植物竞争来争夺资源,与根瘤菌等共生体之间也存在交互作用。因此,昆虫—植物群落所处的环境相当复杂。

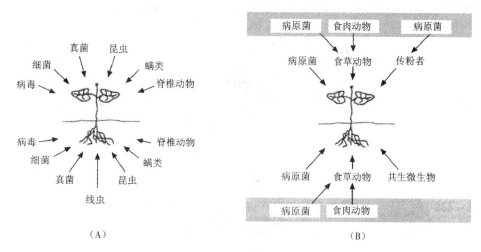

图10.1 植物生活在复杂的环境中

(A)植物受到大量的不同种群的生物攻击;(B)每一个进攻者都有自己的天敌,且所有进攻者之间也存在联系。此外,植物与它们的共生生物,如传粉者和根瘤菌,也存在着交互作用

植物与昆虫的交互作用既发生在地表以上,也发生在地表以下,但是任一环境中,植物与昆虫的交互作用都是非常复杂的。本章主要探究植物和昆虫如何交互,以及其他生物体如何影响这些交互作用。最后我们将探讨处于食物网和群落中的植物与昆虫的多级营养交互是如何进行的,以及昆虫—植物间的交互作用同群落动态之间的关系。

10.1 植物对昆虫的影响

之前的章节中提到,植物能以多种不同的方式影响植食昆虫。对昆虫而言,植物不仅是食物,植物的生理结构、三维构造和化学信息等因素都能对植食昆虫产生影响。此外,以同样的方式,植物还可能与传粉者和肉食性昆虫产生交互作用,这种作用又会影响到植食昆虫[36,148]。肉食性动物在植物上的分布程度会受到植物特征的影响。换而言之,植物可能为植食昆虫提供一个无天敌或有天敌的空间。植物的防御具有直接或间接的特性。直接防御由植物的特征进行调节,能够直接影响植食昆虫的生理和行为反应,如毒素和刺棘。与之相反,间接防御通过增强肉食性昆虫的性能,例如提供庇护所、可选择的食物或者化学信息素(图10.2)等[35]。将植物转变成一个植食昆虫天敌的高密度空间,这是间接防御的基本特征之一。

传粉者在花期访问植物,植物通过视觉引诱、气味和食物回报等行为刺激传粉(第12章)。此外,对于传粉者来说,花朵也可能是一种无天敌(或有天敌)空间,这可能都会影响授粉成功率[43]。

在群落内,每一种关系都有可能会影响其他关系,因此,对一个群落的全面了解不能仅仅考虑一种交互作用,也不能将所有交互作用简单叠加。在这章中,我们将探讨植物个体特性对肉食性昆虫和传粉者的影响。植物的特性可能会直接对肉食性昆虫产生影响,如植物的毛状体对食肉昆虫行为的影响(第3章)。另外植物还可以通过影响植食昆虫,间接地影响肉食性昆虫。例如,植物通过影响植食昆虫的体型的同时可能会影响寄生性天敌的产卵选择。因而植物对植食昆虫体型的影响,可能会影响植食昆虫与寄生性天敌间的交互作用,从而对寄生性天敌的群落产生重要的影响[93]。这两种影响将在下面章节中进行阐述。

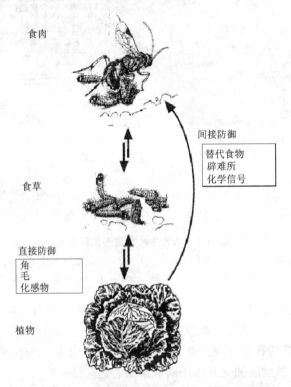

图10.2 植物能通过直接防御影响植食昆虫,来保护自己免受植食昆虫侵害,
或通过间接防御增强肉食性天敌对植食昆虫的影响

10.1.1 植物物候学

植物物候可能对植食昆虫产生重要影响。气候变化会干扰昆虫和植物物候现象的同步性,可能严重影响昆虫的种群动态。多数以叶片为食的鳞翅目昆虫,如尺蛾(*Operophthera brumata*)幼虫,仅在橡树发芽到第一批树叶舒展的时间内取食橡树(*Quercus robur*)。图10.3表明,如果萌芽先于卵的孵化时间(B)或者卵的孵化先于发芽的时间(A),大部分幼虫会处于饥饿状态。因此,这两个事件发生的一致度(C)越高,幼虫的性能就越好(图10.4),而橡树的脱叶程度就越严重[45,46]。显然,提前或延后发芽时间的树木,可能通过"物候逃避"避免受到昆虫的侵袭。随着近年来气候的变化,植物物候与昆虫的关系尤为重要。在过去的25年里,春季的温度已有所升高,而不同的物种可能以不同的方式做出应对。气候的变化已经导致荷兰尺蛾(*Operophthera brumata*)卵孵化和橡树发芽的同步性日趋降低,可能是由于影响橡树物候和蛾类的非生物变量间关系的改变[179]。

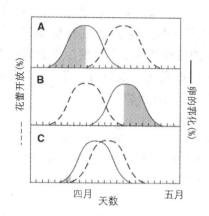

图 10.3　尺蛾(*Operophthera brumata*)卵孵化的时间和橡树发芽时间

注：阴影区域表示幼虫处于饥饿状态（菲尼，1976）[46]。

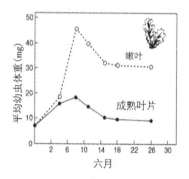

图 10.4　饲养在橡树的新叶与成熟叶片上的尺蛾(*Operophthera brumata*)幼虫的平均鲜重

最初，在四龄和五龄的取食阶段，幼虫的体重会增加，之后，从预蛹阶段开始，直到蛹化完成，体重一直在下降。随着橡树叶龄的增长，蛋白质含量逐渐降低而丹宁酸含量逐渐升高，导致了取食两种叶片幼虫的性状差异（菲尼，1976）[46]

10.1.2　植物化学

关于植物的次生代谢物对植食性昆虫的影响，在第 5 章中已进行了全面的阐述。这些化学物质可能会延缓植食昆虫的发育，能够使它们中毒，甚至杀死它们，或者可以被植食昆虫分解。这些不同的结果是由特定的次生代谢物与植食性昆虫决定的。当次生代谢物延缓植食昆虫发育时，可能会造成昆虫受天敌攻击的时间延长。例如，寄生在营养状态不良的柳树上的叶甲(*Galerucella*)，其死亡率较高且甲虫的发育速度较慢[57]。当两种昆虫同时出现且具有相同的寄生性天敌时，其中发育速度慢的昆虫会增加寄生性天敌攻击的时间，并会提高死亡率，最终会局部灭绝[19]。

一些寡食性植食昆虫，能利用有毒的植物化合物进行防御。寡食性植食昆虫通常具有较高的抗毒性，因此常可以分解植物的化学防御物质。因此，它们能够更好地抵御天敌，如病原体、捕食者和寄生性天敌。在烟草、专食性昆虫烟草天蛾、广食性昆虫粉纹夜蛾(*Trichoplusia ni*)和病原菌(*Bacillus looper*)间的互作关系中，尼古丁的效果为一个具体的实例。尼古丁对广食性昆虫有不利的影响，但对于寡食性昆虫的影响极小。当专食性昆虫被暴露在病原菌(*B.*

looper)中时,植物中高浓度的尼古丁甚至可以发挥有益的作用。相对于摄入尼古丁含量低的毛虫,摄入尼古丁含量高的毛虫受病苏云金芽孢杆菌(*B. thuringiensis*)感染的致死率更低(图10.5)。同样,尼古丁可能会保护寡食性昆虫免受寄生性天敌或者捕食者的侵害。处于捕食者与寄生性天敌之中的寡食性昆虫,能更好地利用来自寄主的防御化合物。而寡食性昆虫更善于防御一般的自然天敌而非其专化性天敌[7]。

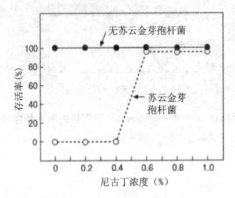

图 10.5 以含有六个不同浓度的尼古丁、含有(或不含)苏云金芽孢杆菌(*Bacillus thuringiensis*)的人工饲料喂养烟草天蛾(*Manduca sexta*)幼虫的化蛹率(克里斯克等,1988)[85]

植食昆虫能够引起植物化学物质的重大改变,例如毒素和消化还原剂(第4章)。例如,对烟草而言,机械损伤或植食昆虫的侵害会造成尼古丁浓度的大幅度提高,以保护植物抵御不适应尼古丁的植食昆虫的侵害[5]。然而如前所述,寡食性昆虫可能会利用植物的化学防御来免受其天敌的危害。一个有趣现象是,植物不会无视自己的化学防御物质被滥用。如果野生烟草受到寡食性昆虫烟草天蛾(*M. sexta*)的攻击时,其释放尼古丁的浓度会减弱。这是由植食昆虫反刍物中的诱导因子间接调控的植物减缓释放尼古丁的反应[78]。因此,植物可以根据侵袭它们的昆虫类别来调控其次生代谢物。

植食性昆虫的取食也可能会促进植物次生代谢物的产生,这些次生代谢物能够增强对植食昆虫天敌的吸引,因此植物可以从中获益。例如,植食行为会使植物挥发物的性质和质量均发生显著变化,这些挥发物可以吸引植食昆虫的天敌,如寄生性或捕食性天敌[35,163]。甚至植食昆虫的产卵行为也能够引发植物挥发物释放,以吸引天敌[70]。对于天敌而言,植物挥发物解决了它们搜寻食物时所面临的一个重要的问题:可靠性—可检测性问题(图10.6)[175]。显然,植食性昆虫释放的化学物质是天敌定位植食昆虫最可靠的线索。然而,植食昆虫在自然选择的压力下,它们会尽量减少释放能被天敌利用的线索。相比之下,由于植物的生物量较大,来自植物的线索要丰富得多,尽管除了植食昆虫诱导产生的挥发物之外,其他定位植食昆虫的线索的可靠性都较低。不同的植物经植食昆虫诱导产生的挥发物是特异性的,而不同的昆虫甚至不同的龄期的昆虫诱导产生的挥发物也是不同的。因此,植食昆虫诱导的植物挥发物是能够揭示植食昆虫的类别、状态和密度的可靠线索[175]。另一方面,经观察表明,植食昆虫诱导的植物挥发物可能会吸引其他的植食昆虫,因为受损植物比完整植株的挥发物释放率更高。例如,粉纹夜蛾(*Trichoplusia ni*)会遭受同种幼虫侵害的卷心菜释放的挥发性物质吸引,但接近气味源后,它就会在附近无损的卷心菜上产卵[87]。

总之,植物经诱导产生的化学变化对昆虫群落的构成具有显著影响[153,154]。

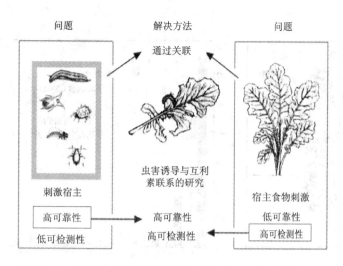

图10.6 食肉性节肢动物都要面临可靠性—可检测性的问题

这与第一和第二营养级生物的可利用化学信息有关,将二者结合是解决这个问题的关键(沃特等,1991)[174]

10.1.3 植物形态学

植物的形态能够对植食昆虫和肉食昆虫的行为产生影响(第3章)。因此,植物形态会影响寄生于植物上的植食和肉食昆虫群落,对植食昆虫—肉食昆虫间的交互作用和种群动态产生重要影响。例如,豌豆植株形态结构的一个简单突变,会对其上的昆虫种群动态产生非常重要的影响。相对寄生在豌豆无叶突变体上的蚜虫,寄生于野生豌豆植株上的蚜虫能够更好地逃脱于瓢虫的捕食。因为瓢虫从野生植株上跌落的频率大约为突变体植株的两倍。因此,无论在实验室还是在野外条件下,野生植株上蚜虫数量的增长都比在突变体植株上高很多[80]。

10.1.4 替代食物

为了吸引传粉者,植物赋予花能够产生花蜜的功能。然而,花蜜和花粉也可能被其他植食或肉食昆虫消耗。蚂蚁等食肉性昆虫经常到花中搜集花蜜(图10.7A)。在一项研究中,雅诺证明了与去除花朵的十字花科蔊菜(*Rorippa indica*)相比,蚂蚁对开花植株的来访量大且持久[190]。这些蚂蚁在植株上巡逻,一旦遇到植食昆虫,就会将对方杀死并带回蚁巢。菜粉蝶(*Pieris rapae*)是使蔊菜(*R. indica*)受损最严重的植食昆虫,蚂蚁的来访大大降低了菜粉蝶的危害(图10.7B)。而传粉昆虫不会因蚂蚁的访问影响其传粉行为,无论植物是否有蚂蚁到访,其种子的产量都基本相同。因此,花蜜的产生可以间接地保护植物免受植食昆虫的侵害。

花粉也能作为一些捕食性昆虫的食物,如草蛉、瓢虫、蝽象和植绥螨。这些捕食性昆虫可以在花上收集花粉,但它们还可以取食花下面的叶片上掉落的花粉。另外,花粉还是一些传粉者(如蜜蜂和熊蜂等)或一些植食昆虫(如灰蝶、象甲、叶甲)和许多缨翅目昆虫的食物来源。

因此,花粉能调控一系列的交互作用,那么对于个体植株而言,这些交互作用的净结果是什么呢?由雄株黄瓜、西花蓟马(*Frankliniella occidentalis*)和捕食性螨(*Iphiseius degenerans*)组成了一个植物—植食昆虫—捕食性动物系统[172]。

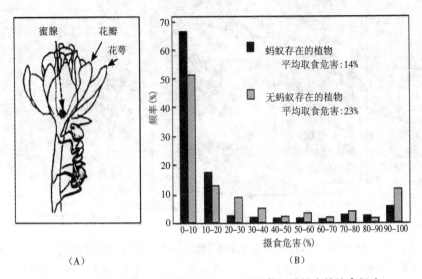

图 10.7 蚂蚁会从葶菜花中收集花蜜且减少菜粉蝶幼虫的植食行为

(A) 蚂蚁 (*Lasius niger*) 从十字花科葶菜 (*Rorippa indica*) 花中收集花蜜;(B) 蚂蚁 (*L. niger*) 的存在会减少菜粉蝶 (*Pieris rapae*) 幼虫在葶菜上的植食行为 (雅诺,1994)[190]

蓟马和螨虫都能摄食花粉并且在花粉上繁殖。在温室盆栽的实验条件下,在每一植株的一片成熟叶片上添加花粉,结果导致捕食者种群数量的增加以及蓟马种群数量的下降。观察发现,与植食昆虫相反,捕食者都聚集在添加花粉的叶片上。因此,尽管植物生产花粉是以生殖为目的,但是它也可能具有其他重要作用,即间接防御植食昆虫。另外,如果捕食性节肢动物频繁地访花,那么花朵就会成为天敌密集的空间,而致使传粉昆虫努力学习躲避花朵。例如,蜜蜂会学习避开有死去同种蜜蜂或蜘蛛的花朵的行为[43]。

植物可以为肉食性动物提供花外蜜,这些花外蜜是由叶片或叶柄上的蜜腺分泌的,由昆虫取食或机械损伤诱导产生(图 10.8)[83]。例如,血桐(*Macaranga tanarius*)[66]和野生利马豆(*Phaseolus lunatus*)[67]植株以产生花外蜜来应对损伤或茉莉酸。相对其他诱发性植物反应(第 4 章),花外蜜的产生更类似一个局部现象而非整体现象。例如,在棉花植株中,诱导花外蜜的产生主要取决于植株的损伤程度,同时仅对新叶产生轻微的影响[184]。肉食性节肢动物,如蚂蚁、寄生蜂、捕食螨和蜘蛛都是常见的花外蜜消耗者[10,83],通过捕食植食昆虫,这些"警卫"守卫着植物的健康[131]。

10.2 植食昆虫对植物的影响

植食昆虫通过取食植物组织来对植株造成影响。植食昆虫的取食行为可导致植物在幼苗阶段死亡,但是成熟植物却很少因昆虫取食致死。植食昆虫可对植株产生一系列的影响,但一般情况下,不会使植株死亡[27]。而植物特征的改变会导致其与植食昆虫之间,及与肉食性昆虫之间的交互作用发生改变,进而影响植物、植食昆虫和食肉性动物的种群动态。昆虫的取食行为可能会导致植物出芽、生根、开花和结种的过程都发生改变,取食行为也会改变植物的许多特征,如化学和形态学特征,或者花蜜的产量(见 10.1.4)。植物形态改变的极端形式是由一些膜翅目或鞘翅目昆虫诱导产生虫瘿(第3章)。植食昆虫取食植物顶端的分生组织以改变植物的形态。例如,植食昆虫取食了蓟(*Cirsium pitcheri*)的顶端分生组织,导致植株补偿生长行为[142]。植物在受到植食昆虫侵害时,其补偿再生的潜力极大[166]。当损害发生在生长初期时,再生长的能力通常会更强[160]。

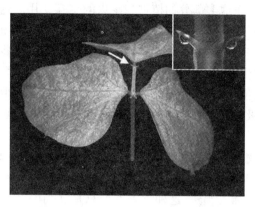

图10.8 受红叶螨(*Tetranychus urticae*)侵害的利马豆的花外蜜腺的分泌物

植食昆虫能以不同的方式影响植物的生长繁殖。取食种子或花的昆虫,如双翅目、鳞翅目、鞘翅目和膜翅目的很多昆虫对植株有显著的影响。以叶、根、茎为食的昆虫会减少植株种子的产量,而吸食韧皮部或木质部汁液的昆虫所产生的影响最为显著[27]。昆虫的取食对植物的花和营养体造成影响后,也将影响植物的传粉昆虫的访花率。例如,以黑心金光菊(*Rudbeckia hirta*)为食的沫蝉,会造成黑心金光菊(*R. hirta*)开花数量的减少和传粉昆虫访问的数量减少[61]。采集花粉或花蜜的传粉昆虫在花间传粉,而取食种子的类群可能会传播植物种子。例如,许多热带植物的种子富含油脂(即油质体)以刺激蚂蚁将其带回巢穴,蚂蚁将有营养的油脂层剥离后,将种子遗弃[21]。

在很长一段时间里,人们一直认为植食昆虫对植物种群动态及分布的影响很小[25]。但在过去的10年,越来越多的证据表明这种影响比以往所认为的要显著得多。例如,将蓟(*Cirsium canescens*)植株上取食花和种子的昆虫移除,会导致种子产量的增加,植株周围幼苗数量也会增加,同时开花植株的数量也会增加[91]。此外,植食性昆虫也可能影响植物分布。十字花科碎米荠(*Cardamine cordifolia*)生性喜阴,分布于落基山脉中部区域。由于昆虫的取食行为在阳光下比在阴凉处更为激烈,这是造成植物分布限制在阴面的主要因素[92]。基于一个模型表明,昆虫的取食对植物种群动态的影响似乎比预期更为广泛[95],然而仍需更多的研究证实。

植食昆虫也可以通过影响肉食性昆虫而间接影响植物。蚂蚁为了收集蚜虫的蜜露,需要保护蚜虫,除此之外,还要巡查植物并攻击植食昆虫。植物因为蚜虫的存在而受到蚂蚁的保护,这对植物有很大益处。在一个有趣的实验中,韦尔林和他的同事发现,由于一种蚜虫(*Aphis jacobaeae*)的侵袭会吸引蚜虫的保护者——蚂蚁的到来,从而保护了千里光(*Senecio jacobaea*)植株免受寡食性辰砂飞蛾(*Tyria jacobaeae*)的危害[182]。因为植株的直接防御对辰砂飞蛾不起作用。而与辰砂飞蛾相反,千里光的化学防御物质(吡咯里西啶生物碱,PAs)会对蚜虫(*A. jacobaeae*)产生负面影响。蚜虫的存在,对已经适应植物化学防御的寡食性昆虫可以起到抵御作用。千里光植物种群因PAs的含量差异而存在多样性,而且这种多样性依赖于歧化选择。在辰砂飞蛾数量较多的年份中,PAs含量低的植株达到最佳生长状态,而在辰砂飞蛾数量较少的年份中,PAs含量高的植株则达到最佳生长状态[182]。

10.3 地上部分与地下部分的昆虫—植物交互作用

目前为止,我们只关注于地上部分的植物与昆虫的交互作用。这反映了当前生态学中对昆虫与植物交互作用这一领域的研究重点。而关于地下部分昆虫与植物的交互作用的研究相对较少,但近年来其数量在迅速增加。基于目前的数据表明,已知的地上部分的植物和昆虫间的交互作用同样也可能发生在地下。植物的次生代谢物也会影响地下的植食昆虫,而地下的植食昆虫也可诱发植物次生代谢物的变化。一篇发表的文章表明,取食植物地下部分的植食昆虫,也能诱导植物产生吸引食肉性昆虫的化合物。黑葡萄象甲幼虫的摄食活动会诱导植株释放一些化学物质,能吸引远处的天敌——线虫[173]。此外,地表之上和之下的植物组织与昆虫的交互作用,可能通过植物特性产生系统性的改变而实现关联[17,104]。因此,地上和地下的植食昆虫可能会相互竞争,例如取食根部的水稻象甲(*Lissorhoptrus oryzophilus*)和取食叶片的草地夜蛾(*Spodoptera frugiperda*)[159]。当然,这种竞争不会总是互利的[16]。此外,食根昆虫可能会改变植物的特性,从而间接地影响食叶昆虫的肉食性天敌,以防御食叶昆虫侵袭。例如,食根的金针虫(*Agriotes lineatus*)幼虫能诱导寄主植物产生10倍于原水平的花外蜜,从而增加食肉性昆虫的访问量[183]。经研究表明,地下昆虫能影响植被演替的速率和方向,以及植物的物种丰富度[30]。有些现象似乎只存在于地表,而事实上,其地下部分可能也发生了相应的改变。食叶昆虫能诱导烟草叶片的尼古丁含量增加,也引发烟草根部产生尼古丁[4]。

取食植物地上部分的昆虫,也能够对植物地下部分的共生真菌产生影响。车前子(*Plantago lanceolata*)依靠菌根获得营养,当其遭受植食昆虫侵害时,会通过丛枝菌根减少植物的定植分布,因此降低了叶片受损的程度,这样,就在植食昆虫和共生菌间建立了一个平衡的交互作用关系[48]。

诱导植物反应可以调控地上和地下部分植食昆虫—肉食性动物间的交互作用,昆虫—植物间的交互作用可能会涉及多重营养的交互作用,这比人们预想得更为复杂[167]。事实上,对于地上部分和地下部分的交互作用的划分,主观人为的因素很大,其依据并非完全科学,而建立二者间的联系将成为未来几年中主要的且非常有价值的研究热点。

10.4 微生物和昆虫—植物的交互作用

微生物在生态系统中广泛存在,但通常较为隐蔽。昆虫—植物的交互作用通常要涉及微生物(见5.5),近年来,微生物对昆虫—植物关系所起的作用逐渐得到关注[8,34,76]。

植物病原体微生物与植食昆虫一样,都需要利用植物相同的资源,且二者之间交互作用的许多机制也早为人所知[112]。为了对抗二者的危害,植物演化出诱导防御机制。经研究发现,一些植物病原体可以促进植食昆虫取食,而有些病原体却能干扰昆虫取食。植食昆虫通常以十八碳酸途径(第4章)产生抗性,病原体以水杨酸途径诱导植物产生抗性。这些途径之间会相互影响。在番茄中,十八碳酸途径诱发了其对植食昆虫的防御,但对植物病原体的抗性降低,而水杨酸途径可能会引起相反的效果[152]。而昆虫取食行为也可能意外激活水杨酸途径,同时微生物也可能激活十八碳酸途径[47,107,113,186]。因此,各种途径间会存在交互作用,这已被研究证实,并被描述为植物对病原体与植食昆虫的交叉反应。植食昆虫激活的水杨酸途径,可能会降低植物对植食昆虫的防御力,而增强对竞争者——病原微生物的防御力[47,107]。

除了植物病原体,植物内还藏匿着多种微生物共生体,如能影响植物的营养价值或毒素水平的菌根真菌和根瘤菌。微生物也可在细胞内存在,如在半翅目或鞘翅目昆虫肠腔中发现的细胞内共生菌[41,42](第5章)。这些微生物能为植食昆虫提供如氨基酸等必需的营养物质。最近,在蚜虫[23]和粉虱[29]体内还发现了次级共生菌的存在。昆虫肠腔内的微生物也许可以消化食物或诱发植物防御,随昆虫粪便排出体外的代谢物,可被寄生性天敌用作利他素。例如,葱谷蛾(*Acrolepiopsis assectella*)的粪便中的克雷伯氏菌,会产生含硫的代谢产物,可被寄生蜂(*Diadromus pulchellus*)作为利他素[155]。

在昆虫与植物的交互作用中,对于昆虫,微生物也能发挥病原体的作用。目前,已经从应用角度(如控制虫害方面)对植食昆虫的微生物病原体进行了较深入的研究,研究重点为苏云金芽孢杆菌(*Bacillus thuringiensis*)(第13章)。植物可以对植食昆虫和病原体间的交互作用产生影响。例如,杂食性的尺蛾(*Operophtora brumata*)取食橡树叶片后,相对于阿拉斯加云衫(*Sitka spruce*)或石南属植物植物,它很快就被核型多角体病毒 NPA 杀死[121]。此外,相对于正常的白菜叶片,取食白菜黑斑菌(*Alternaria brassicae*)感染的大白菜叶片后,辣根猿叶甲(*Phaedon cochleariae*)的生长速度会减慢许多,导致其更易受到金龟子绿僵菌(*Metarhizium anisopliae*)的侵染[129]。

微生物也可能对肉食性昆虫产生重要的影响,但影响得非常隐蔽。到目前为止,对捕食性昆虫或寄生蜂的病原体的关注仍然很少[18]。然而,这类昆虫没有理由不受病原体感染,且有证据表明,微生物可能会干扰植物—肉食性昆虫间的互利共生作用。例如,杆状病毒通过影响寄生蜂(*Microplitis croceipes*)的飞行能力来影响其搜寻植食性寄主的能力[62]。感染杆状病毒会导致活力下降和翅变形。智利捕植螨(*Phytoseiulus persimilis*)引感染病原体后,会干扰其对植食昆虫诱发的植物挥发物的感知力[135]。感染病原体的捕食螨不能区分受损植株和无损植株散发的挥发物,因此在捕食时便不会受到受损植物挥发物的吸引。微生物共生体似乎广泛存在于寄生蜂中,也有可能存在于捕食性节肢动物中。在很多寄生蜂[147]和一些捕食性螨类[20]中都发现了沃尔巴克氏体(Wolbachia),沃尔巴克氏体种群数量的增长能影响其寄主的生殖方式,因此也能影响捕食者(寄生性天敌)—猎物间的种群动态变化。

令人欣慰的是,目前对于微生物—昆虫—植物间的交互作用的关注越来越多,希望在不久的将来,可以获得更多令人兴奋的发现。

10.5 脊椎动物和昆虫—植物的交互作用

对昆虫—植物关系的研究,通常不涉及脊椎动物。然而,植食性脊椎动物也可能对植食昆虫产生重要的直接或间接的影响。例如,脊椎动物对植物种群动态的影响比对植食昆虫的影响更显著。脊椎动物对植物群落成分和植食昆虫的食物成分都会产生重要的影响[125],进而对昆虫群落成分造成潜在影响。此外,植食性脊椎动物也可能偶尔捕食植食性昆虫。有很多例子表明植食性脊椎动物会对植食性昆虫产生负面影响。例如,脊椎动物可以在不影响昆虫物种丰富度的情况下,大量减少昆虫数量[54,120],或者通过改变植物的结构间接减少昆虫数量[49]。这些影响均可能迫使植食昆虫选择最不易受植食性脊椎动物侵袭的植物寄居[187,191]。

一些脊椎动物,如食虫的鸟类,能通过捕食植食昆虫促进植物生长[96]。鸟类能从远处根据植物提供的线索区分受昆虫侵害的植株和无损的植株,而非仅依靠对植物的侵害程度的视觉判断[94]。然而有趣的是,一些研究表明,昆虫也可能对植食性脊椎动物产生较强的负面影响。例如,在金合欢(*Acacia drepanolobium*)荆棘里筑巢的蚂蚁,能完全阻止哺乳动物取食金合欢(图10.9),这能显著提高其寄主对植食性哺乳动物的抵御能力[143]。尽管植物具有荆棘,但是对植食性脊椎性动物不会造成强烈的影响,主要原因是蚂蚁增强了植株抵御脊椎动物的能力。

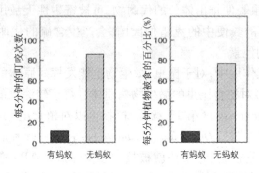

图10.9 在蚂蚁存在或不存在的情况下,金合欢被山羊咬食的平均次数及百分比(斯特普利,1988)[143]

植食性脊椎动物可以通过营养循环影响植物。植食行为可以加快或减慢营养循环的速度。由于体积庞大,植食性脊椎动物通常被认为是影响营养循环的最重要因素。然而昆虫也可以加快营养循环和植物产量[9]。目前,对脊椎动物和无脊椎动物影响营养循环的交互作用仍然未知。或许昆虫只是在一个小的空间范围内,对土壤结构产生很小的影响,相对而言,大型的、运动能力强的哺乳类食草动物可以将植物连根拔起、严重践踏并且紧实土壤,因而对土壤结构的影响更为强烈。

脊椎动物与植物—昆虫间的交互作用是一个仍有待探索的研究领域,其潜在的重要作用值得我们展开进一步研究。

10.6 群落内间接的物种交互作用

在生物群落内部不同的生物体间,除了存在直接的交互作用外,还存在着间接的交互作用。间接交互作用为通过一种物种调节其他两种物种间的交互作用。例如,一种植食性昆虫可以取食两种寄主植物,如果一种植物对该昆虫耐受性弱,而对另一种耐受性强,在耐受性强的植物上建立的昆虫种群,会对耐受性弱的植物产生负面影响。因此昆虫对两种植物产生了间接的交互作用。由一种植食昆虫影响两种植物竞争的例子有很多。例如在野外实验中,在绒毛草(*Holcus mollis*)和猪殃殃属(*Galium saxatile*)的植物群落中喷洒杀虫剂,会消灭食草蚜虫(*Hocaphis holci*),但也会增强植物间竞争,最终导致猪殃殃(*G. saxatile*)的数量下降[26]。

直接交互作用一直被认为是生态系统中最重要的交互作用。毕竟这种交互作用所导致的干扰、食物消耗、死亡以及生殖等过程,都对种群动态有很明显的直接影响。然而,近年来间接相互作用的重要性也逐渐被人们了解。例如,溪流中的蜉蝣稚虫通过化学信息素来感知天敌的存在时,它们会控制藻类的摄入,并且花费相对较多的时间隐蔽,避免为天敌捕食。而这对藻类生物量的增加有积极的作用[100]。一项研究潜叶蛾与寄生蜂之间的交互作用的结果表明，二者之间存在许多潜在的间接交互作用。其中，最强烈的间接交互作用存在于一些取食相同寄主植物的潜叶蛾之间，因为它们会吸引几乎相同的寄生天敌[130]。

通常认为，间接交互作用与直接交互作用相比，其影响更弱。如果一个物种的添加或者移除并没有引起目标物种数量的显著变化，便可认为其影响"较弱"[13]。尽管间接交互作用的平均影响较弱，但是"较弱"的交互作用之间的强度变化可能会很大——往往比强交互作用的强度变化还要大。这就意味着，在某些情况下，弱交互作用会比强交互作用的效果更强，因此，不能因为某些交互作用的平均强度较弱就忽略它们[13]。

间接交互作用主要以两种途径产生：(1)一个物种种群密度的改变，会通过一个中间物种的变化再直接作用于第二个物种(图 10.10 A)。(2)一个中间物种可以调控其他两个物种之间的交互作用(图 10.10B)。间接交互作用可以受许多因素调节，如形态、化学和行为。一些间接交互作用的类型已经明确(图 10.10 中 C – E)[188]。具体的实例将在下面章节中进行讨论。

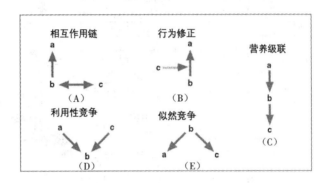

图 10.10　不同类型的间接交互作用

(A)物种 c 通过一系列直接交互作用影响物种 a；(B)物种 c 通过改变寄主 a 与 b 的交互作用，从而对 a 产生间接影响；(C—E)间接影响的三种常见类型：(C)营养级联；(D)资源利用性竞争；(E)似然竞争(伍顿，1994)[188]

10.6.1　资源利用性竞争

当两种生物在不发生身体接触的情况下竞争同一资源时，就出现了资源利用性竞争(Exploitative competition)(图 10.10D)。长期以来，这个问题一直是群落生态学研究的核心问题。以往，人们一直认为植食昆虫的种间竞争不太重要[149]。一种论点认为，植物通常会向植食昆虫提供开放生态位。通过对某种植食昆虫在不同地理区域的同一种植物上的出现率进行比较，便可得知空缺生态位(vacant niches)的数量。图 10.11 所示在 6 个不同地域内，欧洲蕨(*Pteridium aquilinum*)被利用的程度，表明：(1)通过对不同区域如北美与英国的生境进行比较，得知欧洲蕨存在着许多的空缺生态位；(2)相同的资源可以被多种昆虫利用，例如在英格兰有 10 种昆虫取食欧洲蕨的羽片叶。经劳顿等人推断，在进化过程中不同的植食昆虫在欧洲蕨上定居的程度是一个随机过程，而非受到植食昆虫种间作用的限制[89]。但是，更多的最新证据表明，种间竞争对植食昆虫的重要性很可能超出人们之前对此的预想。基于两项独立的研究显示，种间竞争的发生非常频繁，尽管在不同系统间，竞争的强度和频率有很大差异[28,33]。在一项对 104 个植物—植食昆虫关系的比较研究中，德诺等人讨论了 193 对存在潜在竞争的物种数据[33]。在这个讨论中,作者也将两种植食昆虫之间存在的竞争或互惠关系包括了进来。数据显示,其中 3/4 的交互作用为种间竞争。在吸食植物汁液的昆虫中的种间竞争比咬食叶片的独立生活的昆虫高得多,并且亲缘关系较近的类群间的竞争比无关联类群间的竞争要高得多。吸食植物汁液的昆虫、蛀干昆虫及以水果和种子为食的昆虫,这些类群的种间竞争比例最高(表 10.1)。由于较多的对摄食种子和果实的昆虫的研究都是在实验室内进行的(在 21 项室内研究中,有 8 项以摄食种子和果实的昆虫为研究对象),因此对其种间竞争比例的研究结果可能会误差较大。但总体而言,在 193 项研究中的 166 项都是基于野外环境的研究。这些研究表明,种间竞争发生的频率是非常频繁的,尽管在研究中受测的昆虫均为低密度物种,可能缺少代表性。由于植物经昆虫取食后,其诱导产生的抗性可能会起长期作用,因此暗示了种间竞争也可能会在之前没有考虑过的时间和空间尺度上发生。例如,食根蚜和食叶蚜存在种间竞争,尽管它们之间没有直接的身体接触[104]。

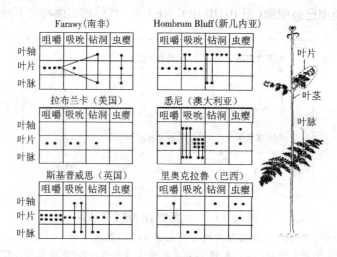

图 10.11 植食性昆虫在 6 个不同地区取食欧洲蕨(*Pteridium aquilinum*)的取食位点与取食方式

注:每个圆点代表一种昆虫;昆虫取食超过一种的植株组织以连接直线表示。

表 10.1 不同取食方式的植食昆虫种间竞争的频率(德诺等,1995)[33]

取食方式	种间竞争的比例%	交互作用总的数量
吸食性	88	48
蛀干类	93	14
蛀木类	93	15
取食种子和水果	100	21
咬食叶片	57	53

10.6.2 似然竞争

两个生物体通过共同的捕食者来产生的竞争为似然竞争(图 10.10E)。其原理为,某物种的密度会通过影响中间捕食者的密度随后对另一物种的密度产生影响[72]。例如,通过对蚜虫的寄主植物施肥,可以促使蚜虫种群密度增长,从而导致其捕食者瓢虫数量的增加,最后造成临近区域的另一蚜虫种群密度的降低[105]。长期以来,似然竞争一直被认为是非常少见的竞争类型,而近年来,相关文献的数量也在迅速增加[22,105]。有一种特殊的似然竞争是由于替代性食物的出现而产生的,如花粉可以被肉食性节肢动物取食。例如,生长在蒲公英附近的苜蓿上的蚜虫,被捕食者瓢虫(*Coleomegilla maculate*)捕食的几率更高。由于蒲公英的花粉提高了瓢虫(*C. maculate*)的种群密度,并由此导致苜蓿上的蚜虫密度降低(图10.12)[63]。与此相似,似然竞争可以发生在植物之间,在这些植物上拥有相同的植食昆虫[157]。

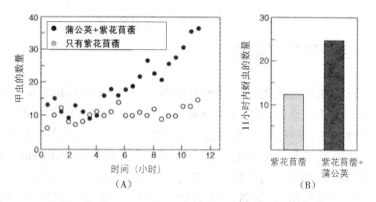

图10.12　蒲公英花粉提高了瓢虫种群密度并导致苜蓿上的蚜虫密度降低

(A)在养虫笼中一侧是受到了蚜虫侵害的苜蓿植株,另一侧除苜蓿(有蚜虫)外,还混有完整无损的蒲公英植株(无蚜虫),比较两侧的瓢虫数目;(B)两侧被取食的蚜虫数目(汉姆等,2000)[63]

10.6.3　营养级联

营养级联(图10.10C)是食物网动态研究的主要议题,同时也是研究间接作用的一个典型范例。在营养级联中,一条食物链上的一个营养级会影响下一个营养级,肉食性动物丰富度的变化会改变植物的分布和丰富度[110,115,132]。关于"地球为绿色"的一个假想的原因是由于捕食者对植食昆虫的捕食,会使大量的植物幸存下来[58]。营养级联理论为这个推测提供了理论依据。然而,一个营养级联的出现与肉食性—植食昆虫的交互作用,以及植食昆虫—植物间的交互作用的相对强度,还有食物网中联系的数量有关。食物网越是线性的,肉食动物的消失对植物丰富度的影响就越大。随着网状关系的增加,肉食动物的消失可通过调整食物网中其他营养关系而得到补偿。在一些生态学文献中,关于营养级联产生的论题展开了激烈的讨论。已经充分证明了营养级联对水生生态系统的重要性,但对于陆地生态系统而言,其重要性显著降低[114,150]。就此可能的原因为:同肉食动物对植食昆虫的影响相比,由于陆生植物对植食昆虫的多重防御机制,植食昆虫对植物的影响相对减弱。此外,如果在一个群落中起主导作用的物种(捕食者)的数量有限,依据营养级联的理论,这个群落的物种多样性则较低。由于陆地生态系统中高度复杂食物网,由此某一物种可能与网中的不同链接联系而产生不同的作用[150]。然而,一篇综述表明营养级联可能比人们之前认为的更为普遍。施米茨等人进行了41项陆地生态系统的营养级联分析,并报道了60项独立测试,其中所有的植食动物都为节肢动物[132]。基于这些研究结果,营养级联是普遍存在的:在60项测试中有45项当移除肉食性动物后,植物的一些可变因素,如损伤、生物量以及繁殖力等因素都发生了变化。而相较于植物的生物量和繁殖,移除肉食性动物对植株损伤的影响更为强烈。而对植株损伤的影响,往往不会继续对植株的繁殖产生影响,因为植株可以补偿损伤并且/或能耐受植食昆虫的损伤。

在某些情况下，也能观察到营养级联影响的减弱，如肉食动物对植食昆虫的直接影响，可能比肉食动物对植物的间接影响更为强烈。其中的一些情况，可以用植物对植食昆虫的防御机制来解释，而施米茨等人的分析表明，复杂的食物网可能会抑制营养级联的影响[132]。总之，基于目前的相关信息表明，营养级联的现象可能普遍存在，但是其影响的重要性仍然有待验证，尤其是在野外条件下的研究[73]。这也涉及对群落进程的一些次要影响因素的重要性的争论（见10.7）[98]。

10.7　物种交互与表型可塑性

通常认为种间关系的固定模式为：一个种群的所有个体享有共同的特征，以相同的方式与其他生物体产生交互作用。然而，生物体却具有表型可塑性，即它们可以根据生物或非生物环境而展现不同的表型[2]。表型可塑性可能在任何营养级上发生。例如，植食昆虫的取食行为可以引起植物发生大量的化学和物理变化，而这些表型变化是由造成损伤的植食昆虫的种类决定的（第4章）。诱导产生的植物反应能影响植物与其他植食昆虫、植物与肉食性动物以及存在竞争的植物之间的交互作用[36]。例如，由于天蛾（*Manduca*）幼虫取食野生烟草而诱发了防御机制，这会导致该植株与邻近的非诱导同种植株进行资源竞争时，付出更大的代价。启动诱导防御机制的植株与邻近的非诱导同种植株相比，生长更慢，产种量更低。有趣的是，存在竞争关系的诱导植株，其植株的生长率和产种量与存在竞争的非诱导植株的情况大体相同（图10.13）[165]。其中的原理仍需阐明，但是数据显示，当未受损烟草植物位于众多诱导植株周围时，会有机会受益。

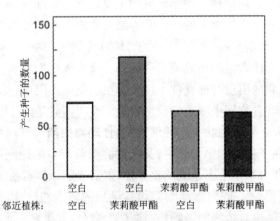

图10.13　烟草植株（*Nicotiana attenuata*）和临近植株经不同处理后烟草植株的产种量
植株经过茉莉酸甲酯（MeJA）处理能诱发植株产生防御机制，结果显示将对照（未处理）植株置于经MeJA处理的植株周围时，存在明显的繁殖优势（范达姆和鲍尔温，1998）[165]

植食昆虫表型的变化可由植物引起。植食昆虫的搜寻行为会受到植物群落组成的影响。对于寄主植物，植食昆虫可以依据之前的觅食经历形成特定的搜寻图像，进而在不同环境下引发不同的寄主选择行为（第8章）[111]。例如，果蝇对标记信息素的反应，也取决于水果大小和可选择的食物的密度[119]。

肉食昆虫也可以诱导植食昆虫表型的改变。肉食昆虫的存在就可以引起植食昆虫的行为发生变化,例如逃避行为、改变觅食与藏匿的时间分配,以及昆虫的形态变化[38]。例如,当黄瓜甲虫(*Diabrotica undecimpunctata howardi*)感受到狼蛛(*Hogna helluo*)释放的化学信息素时,会减少进食,而当感受到其他三种危险性较小的捕食蜘蛛的化学信息素时,则不会出现这一反应[139]。肉食性昆虫在面对可利用资源[176]和天敌的情况发生变化时,可以表现出不同的应对策略[71]。肉食昆虫之前的经验也可以影响其在不同阶段的觅食行为[176]。例如,寄生蜂可通过联系式学习对植食昆虫诱导产生的植物挥发物做出反应,因此暂时限定在某一特定植物上进行觅食[162]。

由于表型可塑性的存在,生态系统中的交互作用是相互关联的,并且一个生态系统中出现的现象并不一定在其他生态系统中出现。而关联性意味着交互作用不仅受昆虫基因型和生理状态的影响,还要受到资源和天敌状态的影响[119]。

10.8 自上而下和自下而上的驱动力

关于群落的形成是自上而下(消费者驱动)还是自下而上(生产者驱动)驱动的,这一问题引发了长期的激烈讨论[74,185]。这一问题也引起了对植物多样性影响植食昆虫行为的讨论,特别是对于农业生产中单种栽培与混合作物栽培的讨论。在单作物栽培中,寄主植物上特定的植食昆虫数量往往比混合栽培时,同种植株上的同种植食昆虫的数量高得多(图10.14)。一种资源集中假说(The resourceconcentration hypothesis)强调自下而上驱动力,并提出植食昆虫更容易找到并长期停留在生长密集或物种单一环境的寄主植物上。在简单的环境下,专化性强的物种的相对密度通常较高[127]。而天敌假说(The enemy hypothesis)则强调自上而下的驱动力,提出肉食性昆虫的丰富度和多样化越高,植物群落的多样性就越高,因为肉食性昆虫导致了植食昆虫数量的减少[127]。

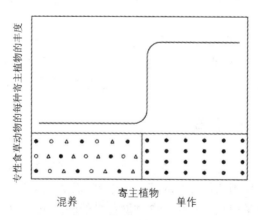

图 10.14 混合栽培与单作物栽培条件下,寄主植株对于特定植食昆虫丰富度的影响(斯特朗等,1984)[149]

海尔斯顿和同事在一篇很有影响的论文中提到,尽管存在种类繁多且数量庞大的植食昆虫,但"我们的世界依然是绿色的",是因为肉食性天敌对植食昆虫种群的数量进行了控制[58]。然而,也有一些其他的观点支持自下而上的驱动。例如,据研究报道,诱导性植物反应是造成秋白尺蛾（*Epirrita autumnata*）种群周期性变化的主要原因[64]，同时植物诱导抗性的模式（第4章，防御和抵抗之间的差异）显示出自下而上的驱动，能够在没有其他与密度相关的因素下调节和驱动植食昆虫种群数量持续波动[164]。直到最近十年我们的认识才逐渐明晰：自上而下和自下而上的驱动往往是紧密联系在一起的，而并非是相互排斥的。自上而下的驱动会影响自下而上的驱动，反之亦然[1,72,188]，这一认识与以往对间接交互作用的认识类似。如果植物的特性可以影响昆虫群落的构成，进而也会对自上而下的驱动力产生影响，这种是自上而下的驱动。此外，植物还有许多特性可以影响肉食动物的效率[35,116]。例如虫菌穴或虫害诱导挥发物等植物特性，可以增强肉食动物对植食昆虫的影响。因此，对"群落的形成主要为两种驱动中的哪一种"的研究开始转向对这两种驱动对于群落的相对贡献的研究。另外，自上而下和自下而上的驱动都可能会发生变化，这在衡量它们对植食昆虫的影响方面很重要[55,79,126]。例如，植食昆虫决定是否接受当前遇到的寄主植物作为其产卵地，或继续搜寻更适合的，考虑的因素一方面取决于寄主植物的分布、密度及其质量变化，另一方面取决于肉食动物的分布及它们引起的死亡风险。例如，对植食昆虫来说，如果灾难的风险很大，如寄主植物被植食性的哺乳动物完全消耗掉，在更多的寄主植株上繁殖后代会付出更大的代价[126]。

有一个关于自上而下和自下而上驱动的争论，是围绕为什么大多数的植食昆虫都是专食性的这一问题所产生的[14,75,134]。在长期以来，这个问题被认为可以从植物化学角度进行解释。植物次生代谢物种类繁多，昆虫只有分别对应相关寄主植物的特定次生代谢物，才能应对这种不寻常的复杂性。在一篇开创性的论文中，伯尼斯和格雷厄姆证实了肉食动物也同样重要[14]。专食性昆虫能迅速决定取食场所，并更易逃避天敌。快速选择寄主植物极短的取食时间，对于躲避捕食者和寄生蜂是非常重要的[15]。在昆虫的觅食行为中，（避免）与天敌交互作用的重要性已经得到了越来越多的支持。例如，寄生蜂是造成粉蝶（*Pieris*）死亡的主要原因，因此粉蝶会选择质量次优但天敌相对较少的植物作为其寄主，而不会选择营养价值高但寄生蜂频繁出现的植物[108]。在叶甲（*Oreia elongate*）中也发现了类似的现象[6]。令人兴奋的是这些研究都是基于野外研究而得出的实验。

显然，植食昆虫对植物食物的选择，可能与食物质量和天敌攻击的风险有关（见11.7）。因此，自上而下和自下而上的驱动不应该是非此即彼的，而是相辅相成的。研究不同生态系统中两种驱动的相对贡献，比研究哪种驱动发挥作用更能加深我们对这一领域的了解。

10.9　食物网和化学信息网

在生态系统中，通过资源与消费者的联系，通过信息化学物质的介导，物种间建立了密切的联系。食物网为具交互作用的物种间的营养关系网，而通过信息化学物质进行调节的关系称为化学信息网。一个化学信息网可能比食物网包含的联系更多，因为在有机体之间，即使不存在营养关系，也有可能通过信息化学物质产生联系。

10.9.1　食物网

想要全面地对一个完整的食物网进行分析是非常困难的，因为所涉及的物种数量太多。因此对于大多数的食物网而言，仅分析包含物种间交互作用比较强烈的子集。食物网的三种主要类型为：(1)连接性食物网（connectancewebs），提供了不同的营养链（trophic links）之间连接的相对频率。(2)半定量食物网(semiquantitative webs),能提供营养链相对丰富度的信息。(3)定量食物网(quantitative webs),所有营养链都以相同的能量单位衡量[130]。

为了说明定量食物网的复杂性,现以细蛾(*Phyllonorycter*,鳞翅目:细蛾科)和其四种对应植物上的寄生蜂组成的食物网为例进行说明(图10.15)[130]。在两年的时间里,从4种树木上培养出了12种细蛾,共吸引到27种寄生蜂。寄主植物对食物网的结构产生很大的影响。我们必须认识到,食物网是一个反映"群落动态的快照",会随时间发生变化,如种群动态的非同步性及每个物种的世代数的变化等。因此,食物网会随时间改变而呈动态变化。在两年的时间里,在约1000 m^2 的研究区域内,细蛾(*Phyllonorycter*)的数量约为7 500万只。而在寄生蜂与寄主可能存在的324(12×27)种交互作用中,观测到的有132种(41%),在植食昆虫与植物可能存在的48(12×4)种交互作用中,观察到的仅有12种(25%)[130]。我们也应该认识到,尽管以上实验对象是形成食物网的主要成员,但事实上它们仅为这个区域群落的一小部分,还有一些成员没有被包括进来,例如:(1)细蛾(*Phyllonorycter*)的捕食者和病原体;(2)其他植食昆虫和对应的寄生蜂、捕食者以及病原体;(3)其他植物和对应的植食昆虫以及寄生蜂、捕食者和病原体。一些食物网还包括了第四营养级——重寄生蜂[106]。迄今为止,对植物与传粉昆虫的食物网已进行了较深入的研究[101],对植食或肉食昆虫与传粉昆虫间的直接或间接交互作用也进行了研究[43,148]。

在一个基于食物的连接性食物网中,描述了植食昆虫、寄生蜂、肉食动物和病原体与英国南部的金雀花之间的联系,共记录了154种类群的370条营养链。这个连接性食物网没有提供营养链频率的详细信息,但提供了一系列植食昆虫与它们不同类型的天敌的数据,显示了植食昆虫与捕食者之间的联系(实际观测值除以可能的最大值)与寄生蜂的联系更加紧密。因为捕食者往往比寄生蜂消耗更多的昆虫。

分析定量食物网可以获取群落物种间潜在的直接或间接的交互作用信息,并且也可以进行比较分析。例如,欧麦克等人比较了由感染或未感染内生菌的种子萌发的黑麦草(*Italian ryegrass*),与其上的蚜虫和寄生蜂形成的食物网[109]。结果表明,内生菌的存在会影响蚜虫的丰富度,并随之影响食物网的复杂度。未感染内生菌植株形成的食物网复杂性更强,这是由于与每个物种相关的营养交互作用数量增加,以及与同一寄生蜂形成间接联系的数量增加。

10.9.2 化学信息网

食物网中每个成员产生的信息化学物质都能直接影响交互作用。消费者利用生产者产生的信息化合物,同时生产者也可利用消费者产生的信息化合物。例如,植食昆虫利用植物挥发物定位植物[36,178](第6章),利用天敌的信息化合物来避免被捕食[38]。此外,释放到环境中的信息化合物可能被群落中任何生物体利用来满足其需要。信息化合物还可以调节间接交互作用。例如,植食昆虫诱导植物产生的挥发物可排斥或吸引植食昆虫,但这些挥发物也能通过吸引肉食动物,间接地影响植食昆虫—肉食动物的交互作用。由于不同肉食动物引发应对挥发物的不同反应,可能会调节其争夺相同资源或种群内部捕食作用的程度。肉食节肢动物在植食昆虫或远处的小生境上的定位,在很大程度上依赖于植食昆虫诱导的植物挥发物。相对而言,植食昆虫产生的信息化合物只在近距离起作用。此外,植食昆虫诱导的植物挥发物可以通过影响邻近植株的表型,继而影响邻近植株的植物—植食昆虫—肉食动物间的交互作用[37]。

事实上,信息化合物可以影响群落中各种类型的行为,而一种行为可以被一系列不同来源

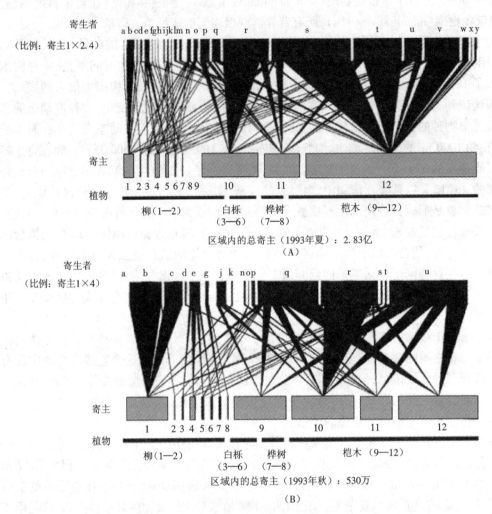

图 10.15 英国南部一个区域的两个世代的细蛾(*Phyllonorycter*)与其寄生蜂及植物之间的植物——寄主——寄生蜂定量食物网

12 种细蛾(*Phyllonorycter*,编号 1 至 12)被饲养在四种植物上。细蛾受 27 寄生蜂(a—y)的寄生。括号内的数字代表植物上寄生蜂种类的编号。条形的宽度代表植物、植食昆虫及寄生蜂占这个区域内预计总数的百分比,由此可见在 1993 年夏季,寄生蜂 s、细蛾 12 和桤木(*Alnus*)是丰富度最高的物种(罗特和戈费雷,2000)[130]

的信息化合物影响。例如,一只待产卵的植食昆虫可利用的信息化合物有:植物类别和质量[169],是否存在竞争者[133],是否存在其他有益(能提供保护)的植食昆虫[137]以及是否有天敌的存在[38]。最佳摄食理论(Optimal foraging theory)假定动物是无所不知的,能适时做出决定,实现其"最大化适应"[145]。尽管昆虫绝非"无所不知"且昆虫个体所能获得的信息通常是有限的,但越来越多的证据表明,在获取周围环境情况的信息方面,昆虫仍具有非凡的才能[84,177]。当昆虫似乎未做出最佳决定时,并不能仅以"失误"进行解释。有时,我们可能忽略了一些其他的信息,而这些信息却是昆虫做出决定的基本依据。例如,植食昆虫选择寄主植物,主要考虑是否能长期适应,而非考虑其后代的健康,这说明了植食昆虫可能并非是"称职的母亲"。然而这个理论仍需结合一个新的假说,关于寄主植物对昆虫性能的影响[97]。

总之,食物网被一个更为复杂的化学信息网所覆盖[35,136],化学信息网与食物网相互影响,如植食昆虫诱导植物挥发物的例子所示(图10.16)。

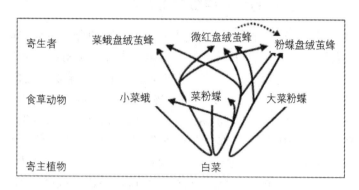

图 10.16　化学信息网

卷心菜受不同植食昆虫幼虫的侵袭,导致吸引了寄生蜂,并影响植食昆虫的产卵偏好(盐尼等,2001,2002;[136,137]海尔弗利特,1994,1998[51,52])

一个生物体释放的信息化合物不仅能改变其表型,如生物的外表从不显眼到显眼,还可以对群落成员产生多方面影响,继而改变食物网的交互作用和群落的构成[181]。目前,对信息化合物调控的交互作用方面的研究,已专门用于研究食物网内独立的交互作用(isolated interactions)。

10.10　群落

群落是有(或可能有)交互作用的物种群体。如前所述,物种间的交互作用不仅局限于直接作用。事实上,当考虑昆虫与植物群落时,不能仅关注主要类群。植物与昆虫还要与许多其他生物(从微生物到哺乳动物)发生交互作用。橡树子、舞毒蛾、鼠、鹿、蜱和莱姆病原体之间的联系便是一个有趣的例子[77]。每2到5年,橡树会产生大量的橡树子,而橡树子是美国东部的一种白爪老鼠的主要食物,这种老鼠也是舞毒蛾(*Lymantria dispar*)蛹的重要捕食者。舞毒蛾可以造成橡树突发性的大面积脱叶,这会对橡树产生相当大的影响。大量的橡树子引诱白尾鹿进入橡树林区,而白尾鹿和白爪鼠又是黑腿蜱虫(*Ixodes scapularis*)的主要寄主。这种蜱携带能引起人类莱姆病的螺旋菌,蜱从鹿身上掉落并繁殖,同时幼虫受老鼠身上细菌的传染。最终橡树子的密度能影响鹿和鼠,结果导致舞毒蛾的爆发和由蜱传播的疾病。

由于英国南部兔子种群的食草行为,使得当地草皮变得很短,然而当兔子感染了兔黏液瘤病时,其种群数量大幅度下降,草地则生长得又密又高。如果土地中草仅长高了几厘米而土壤温度下降的话,则会引起大蓝小灰蝶(*Maculinea arion*)幼虫的寄主红衣红蚁(*Myrmica sabuleti*)的数量迅速减少。在20世纪50年代,这导致了这种漂亮的珍稀蝴蝶消失在英国[158]。以上例子均说明,群落中不同类别的成员之间,存在直接或间接的交互作用。

尽管一些研究成果非常激动人心,但多数昆虫—植物群落的研究仍仅限于昆虫或植物。群落生态学的主要议题是了解群落构成和动态。群落的构成和动态是由物种的集群、灭绝和交互作用而共同决定的。因此,研究昆虫与植物的交互作用如何对群落中其他成员产生影响,将是今后研究的一个挑战。研究植物与植食昆虫的交互作用,要考虑不同的营养级。第三营养级的成员通常包括鸟类或其他脊椎动物以及肉食昆虫和寄生蜂等。

10.10.1 为什么会有如此多的稀有植食昆虫存在?

昆虫分类学专家们都知道,在昆虫类群中,几乎所有较高的分类阶元(属,科),甚至大多数种类都属"稀有"昆虫,换而言之,根据这些昆虫的生活史和寄主植物等情况,可以推断它们本应该出现的时间和地点,但实际上在以上时间和地点却往往很难发现它们的踪迹[50]。以下假说也许可以解释这种现象:植食昆虫很少受到食物的限制,但是经常受到天敌的约束[58]。这一假说得到了鲁特和卡普奇诺等人的支持,在为期 6 年的对植食昆虫与野生秋麒麟草(*Solidago altissima*)种群的关系研究中发现,调查的 138 种植食昆虫中,只有 7 种数量较多,而这 7 种的密度也很低,低至不会对其他类群的密度产生影响[128]。也许植食昆虫的寄主植物的营养价值并非最佳(第 5 章),可以对此进行解释。拉腾认为,这种理论可以解释为什么植食昆虫的密度比可利用食物资源能承载的密度要低得多[189]。此外,极端的天气条件也可能是造成许多物种"稀少"和在本地暂时性灭绝的原因,例如大蓝蝶(*Euphydrias*)等物种[44]。即使我们假设物种"稀少"是上述因素综合作用的结果,但仍有一个有趣的问题需要我们解答:为什么有些昆虫种类丰富,而其他物种(通常为亲缘关系密切的物种)"稀少"呢? 在对昆虫—昆虫、昆虫—植物交互作用的生态和进化过程进行概括之前,许多植食昆虫物种的"稀少性"是应该被充分考虑的事实[50]。

10.10.2 群集性

在一个地域范围内,每株植物上昆虫种类的数量会受植物覆盖区域的影响。通常情况下,与分布狭隘的植物相比,分布广泛的植物上的昆虫种类更多(图 10.17)。在小的地域范围内,植物的群集情况要受自身体积大小的影响。此外,与虫源的隔离情况和虫源的群体规模也是植物集群时的重要决定因素。例如,根据岛屿生物地理学中的平衡理论模型预测,岛屿上的物种的数量由岛屿面积和隔离程度共同决定。这种理论对真正的岛屿以及被非生境包围的生境斑块(habitat patches)都适用[53]。在物种数量较少的小生境中,会缺失如分解、传粉、寄生和捕食等生态系统功能[144]。而且,外食性(ectophagous)和内食性(endophagous)更专化的昆虫会受到不同的影响。

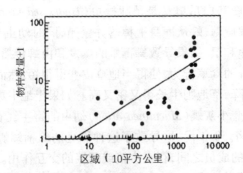

图 10.17 以英国作为一种多年生草本植物的地理分布范围时,草本植物对
植食昆虫物种数量的影响(劳顿和施罗德,1977)[88]

在对野豌豆(Vicia sepium)的实地考察中发现,在草甸上,野豌豆的分布范围是决定豆荚中内食性昆虫物种多样性和种群丰富度的重要因素[86]。在一项单独的试验中,将盆栽野豌豆放置于远离草甸的区域,随着隔离距离(即与草甸的距离)的增加,内食性昆虫定居的成功率将大幅降低。而对于寄生蜂的影响要比对植食昆虫的影响更为强烈(图10.18)。因此,随着寄主植物覆盖区域面积的减少和隔离距离的增加,植食昆虫的寄生率将显著降低[86]。在更小的区域内(只有12米),研究花粉甲虫在油菜上的寄生率也呈现出类似的结果[156]。

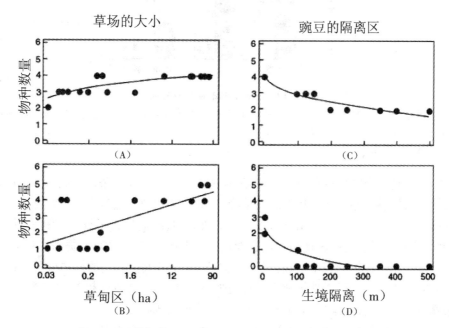

图 10.18　草甸的面积对物种丰富度的影响以及物种丰富度
与小斑块蚕豆(vicai)生境隔离距离的关系

(A)植食昆虫物种数;(B)寄生类物种数,对比植食动物与寄生动物的回归线显示,其斜率和截距都有显著差异;(C)植食昆虫物种数;(D)寄生类物种数,对比植食动物与寄生动物的回归线显示,其斜率无显著差异,而截距有显著差异(克吕士和克恩克,2000)[86]

总之,景观结构对当地的物种构成和群落中种间交互作用起决定性的作用。

10.10.3　群落发展

以下两种类型的实验可以用于研究群落的发生:(1)使用杀虫剂去除共栖物(defaunation);(2)将植物引入一个新的区域。目前已经对第一种其他类型的实验投入了大量的精力,并已由森博洛夫和威尔逊实现[138]。而很多人也在无意之中(以其他实验目的)开展了第二类实验:将许多植物引入新的区域,它们置于新的昆虫种群集聚地。从而开启了新寄主植物的群落发生研究(图10.19)。如以金雀花为例,调查本地和外来生境的群落构成情况[102]。在外来生境中,杂食性植食昆虫为优势种群,而在本地生境中,寡食性植食昆虫为优势种群。在本地和外来的生境中,每株植物上的杂食性植食昆虫的平均丰度并没有差异。在外来的生境中,发现了空生态位(empty niches),例如缺失取食花和种子的植食昆虫。这表明,最先定居新引入植物的昆虫为杂食性植食昆虫,而寡食性动物则相对需要更长的时间。随着植食昆虫生物量的增加,天敌(捕食者和寄生蜂)也随之增加,表明植物首先吸引植食昆虫聚集,随后是肉食动物。许多其他引入的植物也具有同样的群落发生模式[149]。

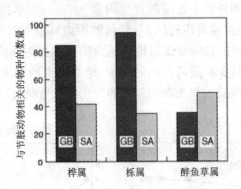

图 10.19　三种植物的群落发生研究

三种生长于英国(GB)和南非(SA)的植物：欧洲白桦(*Betula pendula*)、英国栎(*Quercus robur*)和醉鱼草(*Buddleia* spp.)。欧洲白桦和英国栎是英国的本土植物，后来被引入南非，而醉鱼草则相反。与引入植物相比，本土植物上植食性节肢动物的物种丰富度更高

群落的发生是一个随机过程，依赖于本地物种库的情况和环境条件。此外，现有的不同群落可能会给昆虫提供生态机会来拓殖新的植物，例如新寄主植物通过靠近正常寄主，从而使昆虫实现移居。在一项关于叶螨定植于黄瓜植株的研究中，对此进行了证明。利马豆作为寄主植物适合度很高且大量存在，因此植食昆虫能在迅速进行适应性调整后，提高在利马豆附近的原本不适合作为寄主植物的黄瓜上建群率[56]。此外，受外来植物入侵的植物群落，可以通过植食昆虫来影响外来物种。而本地植物多样性越强，植食昆虫生态位就越多，外来植物入侵本地植物群落就越困难。此外，本地植物群落中特定的物种构成相当重要，因为这会共享一些植食昆虫[118]。因此在不同条件下，不同地区的群落组成和发生可能会不尽相同[161]。

外来植食昆虫能否在本地建群，在一定程度上取决于种间竞争的程度。在之前这被认为是相对不重要的因素，而最近的研究表明，事实上，种间竞争比通常认为的更为重要，尤其是对吸食性植食昆虫来说[28,33]。在一项研究入侵节肢动物的种间竞争——竞争替代(competitive displacement)中，发现一种极端而有趣的情况：在 48 例替代案例中，有 37 例是由入侵物种造成的[122]。除了竞争以外，食性的专化程度和天敌的影响都会对入侵昆虫群落的建立产生强烈的影响。一般情况下，入侵昆虫普遍为杂食性并且缺乏天敌。

10.11　分子生态学

近年来，生态学家开始采用分子方法来解决传统方法很难或根本不可能解决的问题。分子技术开辟了昆虫—植物关系领域许多激动人心的新方向。多年来，生化学家和分子生物学家以及生态学家都在各自的领域独立地研究植物—昆虫间的交互作用。而现在，这种隔阂正在消失，并且正在不断产生一些综合研究方法。例如，对昆虫诱导植物防御的分子机制方面的了解越来越多。参与植物防御的一系列基因已被描述，同时新的 cDNA 芯片技术可以监测昆虫诱导下基因的表达[69,123,124]。据一些资料显示，烟草由于受到烟草天蛾(*Manduca sexta*)的取食，而造成一些基因(包括对压力、创伤和病原体反应的基因，以及参与防御过程分配碳和氮的基因)表达上调，这很符合植物权衡其发育和防御的生态理论[68]。通过比较机械损伤与昆虫

取食的影响，可以发现哪些基因上调或下调来回应取食行为，而这也可能表明植食昆虫是否操纵植物的防御，或植物是否以植食昆虫作为诱因启动其特定的防御机制。例如，经菜粉蝶（*Pieris rapae*）取食拟南芥（*Arabidopsis thaliana*）后，会导致参与植物防御的一系列基因表达上调，而经机械损伤后，则不会诱发这种上调行为（图10.20）[123]。这表明，植食昆虫可以调控植物防御行为，这在经烟草天蛾（*Manduca sexta*）取食的野生烟草中也有记录[78]。此外，cDNA芯片技术结合特定突变及转基因技术，研究单一位点突变或基因改变对总基因组的表达谱的影响[123]。基因表达是植物应对变化环境而改变的第一步。因此，通过诺瑟杂交（Northern blot）或基因芯片得到的基因表达，已用于研究植物对一些因素的反应，如邻近植物挥发物[3]、寡食性和杂食性昆虫的植食行为[124,180]、不同的营养水平[90]、不同类型的进攻者，如不同的植食昆虫[65]、植食昆虫对病原微生物、机械损伤对虫害诱导因子[59]。这对于植物是否（及怎样）对一定的处理做出反应，及是否对不同的处理有不同的反应，提供了相关的原始信息。而这将引出基因表达模式对于表型表达的促进作用，以及与群落成员间的交互作用等方面的研究。例如，烟草天蛾（*Manduca sexta*）幼虫摄取食野生烟草植物，会调控茉莉酸信号通路。使得这个通路的基因表达上调。通过使茉莉酸信号通路中的三个基因（即脂氧化酶、氧化物裂解酶和丙二烯氧合酶）沉默，可以在实验室内[60]和野外条件下研究这些基因在烟草植物诱导防御中的重要性[82]。结果表明，脂氧化酶基因在植物诱导防御机制中尤为重要，当沉默此基因后，植物受到昆虫的侵害加剧，不论是烟草天蛾这类适合此植物的昆虫，还是原本不取食野生烟草的叶蝉（*Empoasca*）[82]。

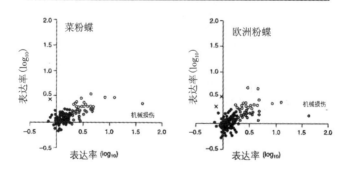

图10.20　对比机械损伤与菜粉蝶（*Pieris rapae*）或欧洲粉蝶（*Pieris brassicae*）取食损伤，
拟南芥（*Arabidopsis thaliana*）中150个基因的表达水平

目前对于昆虫与十字花科植物的交互作用，尤其是与芸苔属（*Brassica*）植物相关昆虫的生态学知识已经相对深入[24,127,136]。拟南芥（*Arabidopsis thaliana*）的全基因组已经测序完成，已成为植物学研究的模式植物，也可能成为生态学研究的模式物种[103]。而令人惊讶的是，迄今为止，对昆虫与拟南芥交互作用的研究仍十分有限。然而，对十字花科植物中的初步研究结果对生态学的发展做了很大的贡献。例如，拟南芥与其他十字花科植物以相同的方式回应昆虫取食[171]：由昆虫取食诱导的直接[99,146]或间接[168,170]防御机制，并且拟南芥中参与防御的信号转导通路与其他十字花科植物也相类似[39]。为拟南芥植物个体特征的研究提供了极好的机会，因为可以获得一个或有限位点的突变体或转基因植株,这使得一种比较生态学的研究方法从梦想变成现实。这种方法也被用于研究一些其他植物种,如野生烟草(*Nicotiana attenuata*)等[5]。基因沉默方法的发展,帮助我们获得了大量原本不可获得的突变体,这在理解特定基因在生物体的生态作用方面是一个突破[40,82]。而将精心设计的实验在野外条件下进行实地研究将是这种方法的重要组成部分[82]。

植物次生代谢物在昆虫—植物关系中发挥着重要作用(第 5 章),不同的细胞色素 P450 单氧酶(P450 s)既参与植物的生物合成过程,也参与昆虫的解毒过程[11]。据报告,拟南芥(*A. thaliana*)基因组中存在 280 种的 P450 s 基因,而果蝇(*Drosophila melanogaster*)的基因组中有 90 种。许多 P450 s 参与植物中呋喃香豆素的合成。而黑凤蝶(*Papilio polyxenes*)有 2 种以上的 P450 s 解毒呋喃香豆素,将黑凤蝶幼虫置于呋喃香豆素的环境中时,P450 s 基因会被激活。这类基因在特定的组织中表达,限于中肠内,或很小一部分在脂肪体。诱导的过程受基因中香豆素应答元件的调节[12]。值得注意的是,此元件在加拿大凤蝶(*P. canadensis*)中也存在,而对食物中含有的呋喃香豆素几乎不做应答。一个可能的解释为,加拿大凤蝶可能用 P450 s 为其他植物次生代谢物解毒,并且这些基因可以由不同的化合物诱导[12]。

分子生物学技术可以用于分析植食昆虫种群的遗传特征,由此提供了昆虫应对植物防御的遗传基础。这已被应用于跳甲(*Phyllotreta nemorum*)和十字花科欧洲山芥(*Barbarea vulgaris*)的研究,用以评估种群的亚结构。在大约 100 米到 1 公里的植物斑块间,或相距约 44 公里的聚居地之间,植物种群出现了遗传分化现象。结果表明遗传分化与地理距离关系密切[31]。

使用分子生态学方法,可以构建新的基因型,以研究一些植物性状,如次生代谢物的作用等。例如,将高粱(*Sorghum bicolor*)中的蜀黍氰甙(dhurrin)以三基因途径(three-gene pathway)合成后转移至拟南芥。这会导致跳甲(*P. nemorum*)对植物取食的损伤降低,幼虫死亡率提高,表明蜀黍氰甙能赋予尚未进化出阻抗机制的植物以抵抗植食昆虫的能力[151]。

全基因组测序实现了在不同情况下对基因组表达变化的监测。这种基因组研究法已经超出了生物化学和生理学方面的基因表达和基因功能分析的范围。由于分子生物学技术的迅速发展,可以帮助我们实现对基因生态功能的分析[40,82,123]。这是生态或环境基因组学研究领域中新兴的研究方向。一些设计精细的实验有助于人们了解与基因生态功能相关的知识。这一领域的迅速发展对我们了解昆虫—植物的生态关系有巨大的推动作用[40,81]。

10.12 结论

昆虫—植物交互作用的生态学研究,已从植物—植食昆虫的交互作用向多重营养级交互作用的方向发展,包括与其他群落成员之间,从微生物到哺乳动物,地上和地下动物的交互作用等。此外,对昆虫—植物交互作用的生态研究,已迅速包括了从亚细胞到个体水平的范围,以致发明了不同的设计巧妙的工具来研究物种的特性对群落成员交互作用乃至对群落进程的影响。

物种间的间接交互作用、影响较小和非致命性的影响,能对群落的发展进程和构成有显著的影响。为了解昆虫与植物的生态学关系,在许多研究中显示(或暗示)存在两种驱动力即自上而下和自下而上的作用。在很长一段时间内主要讨论哪种作用最为重要。然而,这两种驱动力通常交互作用,因此在今后关于植物—昆虫的生态关系研究中,应主要关注如何把两种作用结合起来。

分子生物学为生态学家提供了进行昆虫—植物生态关系研究的理想工具。既可监测在不同条件下个体的反应，又可以研究个体特性的改变对群落生态的影响。将机理研究与生态功能研究相结合，会使对植物与昆虫生态关系的研究受益无穷。

10.13 参考文献

1. Abrams, P. A., Menge, B. A, and Mittelbach, G. G. (1996). The role of indirect effects in food webs. In *Food webs: integration of patterns and dynamics* (ed. G. A. Polis and K. O. Winemiller), pp. 371-95. Chapman & Hall, New York.
2. Agrawal, A. A. (2001). Phenotypic plasticity in the interactions and evolution of spicies. *Science*, 294, 321-6.
3. Arimura, G., Ozawa, R., Shimoda, T., Nishioka, T., Boland, W., and Takabayashi, J. (2000). Herbivory-induced volatiles elicit defence genes in lima bean leaves. *Nature*, 406, 512-15.
4. Baldwin, I. T. (1998). Jasmonate-induced responses are costly but benefit plants under attack in native populations. *Proceedings of the National Academy of Sciences of the USA*, 95, 8113-18.
5. Baldwin, I. T., Halitschke, R., Kessler, A., and Schittko, U. (2001). Merging molecular and ecological approaches in plant-insect interactions. *Current Opinion in Plant Biology*, 4, 351-8.
6. Ballabeni, P., Wlodarczyk, M., and Rahier, M. (2001). Does enemy-free space for eggs contribute to a leaf beetle's oviposition preference for a nutritionally inferior host plant? *Functional Ecology*, 15, 318-24.
7. Barbosa, P. (1988). Natural enemies and herbivore-plant interactions: influence of plant allelochemicals and host specificity. In *Novel aspects of insect-plant interactions* (ed. P. Barbosa and D. K. Letourneau), pp. 201-29. John Wiley, New York.
8. Barbosa, P., Krischik, V. A., and Jones, C. G. (1991). *Microbial mediation of plant-herbivore interactions*. John Wiley, New York.
9. Belovsky, G. E. and Slade, J. B. (2000). Insect herbivory accelerates nutrient cycling and increases plant production. *Proceedings of the National Academy of Sciences of the USA*, 97, 14412-17.
10. Bentley, B. L. (1977). Extrafloral nectaries and protec-tion by pugnacious bodyguards. *Annual Review of Ecology and Systematics*, 8, 407-27.
11. Berenbaum, M. R. (1995). The chemistry of defense: theory and practice. *Proceedings of the National Academy of Sciences of the USA*, 92, 2-8.
12. Berenbaum, M. R. (2002). Postgenomic chemical ecology: from genetic code to ecological interactions. *Journal of Chemical Ecology*, 28, 873-96.
13. Berlow, E. L. (1999). Strong effects of weak inter-actions in ecological communities. *Nature*, 398, 330-4.

14. Bernays, E. and Graham, M. (1988). On the evolution of host specificity in phytophagous arthropod. *Ecology*, 69, 886-92.
15. Bernays, E. A. (2001). Neural limitations in phytophagous insects: implications for diet breadth and evolution of host affiliation. *Annual Review of Entomology*, 46, 703-27.
16. Bezemer, T. M., Wagenaar, R., Van Dam, N. M., and Wäcker, F. L. (2003). Interactions between above and below-ground insect herbivores as mediated by the plant defense system. *Oikos*, 101, 555-62.
17. Bezemer, T. M., Wagenaar, R., Van Dam, N. M., Van Der Putten, W. H., and Wäcker, F. L. (2004). Above- and below-ground terpenoid aldehyde induction in cotton, *Gossypium herbaceum*, following root and leaf injury. *Journal of Chemical Ecology*, 30,53-67.
18. Bjørnson, S. and Schütte, C. (2003). Pathogens of mass-produced natural enemies and pollinators. In *Quality control and production of biological control agents—theory and testing procedures* (ed. J. C. Van Lenteren), pp. 133-65. CABI, Oxon.
19. Bonsall, M. B. and Hassell, M. P. (1997). Apparent competition structures ecological assemblages. *Nature*, 388, 371-3.
20. Breeuwer, J. A. J. and Jacobs, G. (1996). *Wolbachia*: intracellular manipulators of mite reproduction. *Experimental and Applied Acarology*, 20, 421-34.
21. Brew, C. R., O'Dowd, D. J., and Rae, I. D. (1989). Seed dispersal by ants: behaviour-releasing compounds in elaiosomes. *Oecologia*, 80, 490-7.
22. Chaneton, E. J. and Bonsall, M. B. (2000). Enemy-mediated apparent competition: empirical patterns and the evidence. *Oikos*, 88, 380-94.
23. Chen, D. Q., Montllor, C. B., and Purcell, A. H. (2000). Fitness effects of two facultative endosymbiotic bacteria on the pea aphid, *Acyrthosiphon pisum*, and the blue alfalfa aphid, *A. kondoi*. *Entomologia Experimentalis et Applicata*, 95, 315-23.
24. Chew, F. S. and Renwick, J. A. A. (1995). Host plant choice in *Pieris* butterflies. In *Chemical ecology of insects* (ed. R. T. Carde' and W. J. Bell), pp. 214-38. Chapman & Hall, New York.
25. Crawley, M. J. (1989). Insect herbivores and plant population dynamics. *Annual Review of Entomology*, 34, 531-64.
26. Crawley, M. J. (1989). The relative importance of vertebrate and invertebrate herbivores in plant population dynamics. In *Insect-plant interactions*, Vol. 1 (ed. E. A. Bernays), pp. 45-71. CRC Press, Boca Raton.
27. Crawley, M. J. (1997). Plant-herbivore dynamics. In *Plant ecology* (ed. M. J. Crawley), pp. 401-74. Blackwell Science, Oxford.
28. Damman, H. (1993). Patterns of interaction among herbivore species. In *Caterpillars. Ecological and evolutionary constraints on foraging* (ed. N. E. Stamp and T. M. Casey), pp. 132-69. Chapman & Hall, New York.
29. Darby, A. C., Birkle, L. M., Turner, S. L., and Douglas, A. E. (2001). An phid-borne bacterium allied to the secondary symbionts of whitefly. *FEMS Microbiology Ecology*, 36, 43-

50.

30. De Deyn, G. B. Raaijmakers, C. E., Zoomer, H. R., Berg, M. P., De Ruiter, P. C., Verhoef, H. A., et al. (2003). Soil invertebrate fauna enhances grassland succession and diversity. *Nature*, 422, 711 – 13.

31. De Jong, P. W., De Vos, K., and Nielsen, J. K. (2001). Demic structure and its relation with the distrution of an adaptive trait in Danish flea beetles. *Molecular Ecology*, 10, 1323 – 32.

32. Deleted.

33. Denno, R. F., McClure, M. S., and Ott, J. R. (1995). Interspecific interaction in phytophagous insects: competition reexamined and resurrected. *Annual Review of Entomology*, 40, 297 – 331.

34. Dicke, M. (1996). The role of microorganisms in tritrophic interactions in systems consisting of plants, herbivores, and carnivores. In *Microbial diversity in time and space* (ed. R. R. Colwell, U. Simidu, and K. Ohwada), pp. 71 – 84. Plenum, New York.

35. Dicke, M. and Vet, L. E. M. (1999). Plant-carnivore interactions: evolutionary and ecological consequences for plant, herbivore and carnivore. In *Herbivores: between plants and predators* (ed. H. Olff, V. K. Brown, and R. H. Drent), pp. 483 – 520. Blackwell Science, Oxford.

36. Dicke, M. and Van Loon, J. J. A. (2000). Multitrophic effects of herbivore-induced plant volatiles in an evolutionary context. *Entomologia Experimentalis et Applicata*, 97, 237 – 49.

37. Dicke, M. and Bruin, J. (2001). Chemical information transfer between plants: back to the future. *Biochemical Systematics and Ecology*, 29, 981 – 94.

38. Dicke, M. and Grostal, P. (2001). Chemical detection of natural enemies by arthropods: an ecological perspective. *Annual Review of Ecology and Systematics*, 32, 1 – 23.

39. Dicke, M. and Van Poecke, R. P. M. (2002). Signalling in plant-insect interactions: signal transduction in direct and indirect plant defence. In *Plant signal transduction* (ed. D. Scheel and C. Wasternack), pp. 289 – 316. Oxford University Press, Oxford.

40. Dicke, M., Van Loon, J. J. A., and De Jong, P. W. (2004). Ecogenomics benefits community ecology. *Science*, 35, 618 – 19.

41. Dillon, R. J. (2000). Re-assessment of the role of the insect gut microbiota. *Abstracts Book XXXIst International Congress of Entomology*, 1, 51 – 5.

42. Douglas, A. E. (1998). Nureitional interactions in insect-microbial sumbioses: aphids and their symbiotic bacteria *Buchnera*. *Annual Review of Entomology*, 43, 17 – 37.

43. Dukas, R. (2001). Effects of preceived danger on flower choice by bees. *Ecology Letters*, 4, 327 – 33.

44. Ehrlich, C. B., White, R. R., and Brown, I. L. (1980). Extinction, reduction, stability and increase; the responses of checkerspot butterfly (*Euphydryas*) populations to the California drought. *Oecologia*, 46, 101 – 5.

45. Feeny, P. (1970). Seasonal changes in oak leaf tannins and nutrients as a cause of spring feeding by winter moth caterpillars. *Ecology*, 51, 565 – 81.

46. Feeny, P. (1976). Plant apparency and chemical defense. *Recent Advances in Phytochemistry*, 10, 1-40.
47. Felton, G. W. and Eichenseer, H. (1999). Herbivore saliva and its effect on plant defense against herbivores and pathogens. In *Induced plant defenses against pathogens and herbivores. Biochemistry, ecology and agriculture* (ed. A. A. Agrawal, S. Tuzun, and E. Bent), pp. 19-36. APS Press, St Paul.
48. Gange, A. C., Bower, E., and Brown, V. K. (2002). Differential effects of insect herbivory on arbuscular mycorrhizal colonization. *Oecologia*, 131, 103-12.
49. Gardner, S. M., Hartley, S. E., Davies, A., and Palmer, S. C. F. (1997). Carabid communities on heather moorlands in Northeast Scotland: the consequences of grazing pressure for communituy diversity. *Biological Conservation*, 81, 175-286.
50. Gaston, K. J. (1994). *Rarity*. Chapman & Hall, London.
51. Geervliet, J. B. F., Vet, L. E. M., and Dicke, M. (1994). Volatiles from damaged plants as major cues in long-range host-searching by the specialist parasitoid *Cotesia rubecula*. *Entomologia Experimentalis et Applicata*, 73, 289-97.
52. Geervliet, J. B. F., Vreugdenhil, A. I., Vet, L. E. M., and Dicke, M. (1998). Learning to discriminate between infochemicals from different plant-host complexes by the parasitoids *Cotesia glomerata and C. rubecula* (Hymenoptera: Braconidae). *Entomologia Experimentalis et Applicata*, 86, 241-52.
53. Gillespie, R. G. and Roderick, G. K. (2002). Arthropods on islands: colonization, speciation, and conservation. *Annual Review of Entomology*, 47, 595-632.
54. Gomez, J. M. and Gonzalez-Megias, A. (20002). Asymmetrical interactions between ungulates and phytophagous insect: being diggerent matters. *Ecology*, 83, 203-11.
55. Gouinguene, S., Degen, T., and Turlings, T. C. J. (2001). Variability in berbivore-induced odour emissions among maize cultivars and their wild ancestors (teosinte). *Chemoecology*. 11, 9-16.
56. Gould, F. (1979). Rapid host range evolution in a population of the phytophagous mite *Tetranychus urticae* Koch. *Evolution*, 33, 791-802.
57. Haggstrom, H. and Larsson, S. (1995). Slow larval growth on a suboptimal willow results in high predation moetality in the leaf beetle *Galerucella lineola*. *Oecologia*, 104, 308-15.
58. Hairston, N. G., Smith, F. E., and Slobodkin, L. B. (1960). Community structure, popilation control, and comprtition. *American Naturalist*, 94, 421-5.
59. Halitschke, R., Gase, K., Hui, D. Q., Schmidt, D. D., and Baldwin, I. T. (2003). Molecular interactions between the specialist herbivore *Manduca sexta* (Lepidoptera, Sphingidae) and its natural host *Nicotiana attenuata*. VI. Microarray analysis reveals that most herbivore-specific transcriptional changes are mediated by fatty acid-amino acid conjugates. *Plant Physiology*, 131, 1894-902.
60. Halitschke, R., Ziegler, J., Keinanen, M., and Baldwin, I. T. (2004). Silencing of hydroperoxide lyase and allene oxide synthase reveals substrate and defense signaling crosstalk in

Nicotiana attenuata. *Plant Journal*, 40, 35–46.
61. Hamback, P. A. (2001). Direct and indirect effects of herbivory: feeding by spittlebugs affects pollinator visitation rates and seedset of *Rudbeckia hirta*. *Ecoscience*, 8, 45–50.
62. Hamm, J. J., Styer, E. L., and Lewis, W. J. (1988). A baculovirus pathogenic to the parasitoid *Microplitis croceipes* (Hymenoptera: Braconidae). *Journal of Invertebrate Pathology*, 52, 189–91.
63. Harmon, J. P., Ives, A. R., Losey, J. E., Olson, A. C., and Rauwald, K. S. (2000). *Coleomegilla maculata* (Coleoptera: Coccinellidae) predation on pea aphids promoted by proximity to dandelions. *Oecologia*, 125, 543–8.
64. Haukioja, E. and Hakala, T. (1975). Herbivore cycles and periodic outbreaks. Formulation of a general hypothesis. *Reports from the Kevo Subarctic Research Station*, 12, 1–9.
65. Heidel, A. J. and Baldwin, I. T. (2004). Microarray analysis of salicylic acid – and jasmonic acid-signalling in responses of *Nicotiana attenuata* to attack by insects from multiple feeding guilds. *Plant Cell and Environment*, 27, 1362–73.
66. Heil, M., Koch, T., Hilpert, A., Fiala, B., Boland, W., and Linsenmair, K. E. (2001). Extrafloral nectar production of the ant-associated plant, *Macaranga tanarius*, is an induced, indirect, defensive response elicited by jasmonic acid. *Proceedings of the National Academy of Sciences of the USA*, 98, 1083–8.
67. Heil, M. (2004). Induction of two indirect defences benefits Lima bean (*Phaseolus lunatus*, Fabaceae) in nature. *Journal of Ecology*, 92, 527–36.
68. Herms, D. A. and Mattson, W. J. (1992). The dilemma of plants: to grow or to defend. *Quarterly Review of Biology*, 67, 283–335.
69. Hermsmeier, D., Schittko, U., and Baldwin, I. T. (2001). Molecular interactions between the specialist herbivore *Manduca sexta* (Lepidoptera, Sphingidae) and its natural host *Nicotiana attenuata*. I. Largescale changes in the accumulation of growth- and defense-related plant mRNAs. *Plant Physiology*, 125, 683–700.
70. Hilker, M. and Meiners, T. (2002). Induction of plant responses to oviposition and feeding by herbivorous arthropods: a comparison. *Entomologia Experimentalis et Applicata*, 104, 181–92.
71. Höller, C., Micha, S. G., Schulz, S., Francke, W., and Pickett, J. A. (1994). Enemy-induced dispersal in a parasitic wasp. *Experientia*, 50, 182–5.
72. Holt, R. D. and Lawton, J. H. (1994). The ecological consequences of shared natural enemies. *Annual Review of Ecology and Systematics*, 25, 495–520.
73. Holt, R. D. (2000). Trophic cascades in terrestrial ecosystems. Reflections on Polis *et al*. *Trends in Ecology and Evolution*, 15, 444–5.
74. Hunter, M. D. and Price, P. W. (1992). Playing chutes and ladders: heterogeneity and the relative roles of bottom-up and top-down forces in natural communities. *Ecology*, 73, 724–32.
75. Jermy, T. (1988). Can predation lead to narrow food specialization in phytophagous insects? *Ecology*, 69, 902–4.

76. Jones, C. G. (1984). Microorganisms as mediators of plant resource exploitation by insect herbivores. In *A new ecology. Novel approaches to interactive systems* (ed. P. W. Price, C. N. Slobodchikoff, and W. S. Gaud), pp. 53-99. Wiley, New York.
77. Jones, C. G., Ostfeld, R. S., Richard, M. P., Schauber, E. M., and Wolff, J. O. (1998). Chain reactions linking acorns to gypsy moth outbreaks and Lyme disease risk. *Science*, 279, 1023-6.
78. Kahl, J., Siemens, D. H., Aerts, R. J., Gäbler, R., Kühnemann, F., Preston, C. A., *et al*. (2000). Herbivoreinduced ethylene suppresses a direct defense but not a putative indirect defense against an adapted herbivore. *Planta*, 210, 336-42.
79. Karban, R., Agrawal, A. A., and Mangel, M. (1997). The benefits of induced defenses against herbivores. *Ecology*, 78, 1351-5.
80. Kareiva, P. and Sahakian, R. (1990). Tritrophic effects of a simple architectural mutation in pea plants. *Nature*, 345, 433-4.
81. Kessler, A. and Baldwin, I. T. (2002). Plant responses to insect herbivory: the emerging molecular analysis. *Annual Review of Plant Biology*, 53, 299-328.
82. Kessler, A., Halitschke, R., and Baldwin, I. T. (2004). Silencing the jasmonate cascade: induced plant defenses and insect populations. *Science*, 305, 665-8.
83. Koptur, S. (1992). Extrafloral nectary-mediated interactions between insects and plants. In *Insect-plant interactions*, Vol. 4 (ed. E. A. Bernays), pp. 81-129. CRC Press, Boca Raton.
84. Krebs, J. R. and Kacelnik, A. (1991). Decision making. In *Behavioural ecology: an evolutionary approach* (ed. J. R. Krebs and N. B. Davies), pp. 105-36. Blackwell Scientific, Oxford.
85. Krischik, V. A., Barbosa, P., and Reichelderfer, C. F. (1988). Three trophic level interactions: allelochemicals, *Manduca sexta* (L.), and *Bacillus thuringiensis* var. kurstaki Berliner. *Environmental Entomology*, 17, 476-82.
86. Kruess, A. and Tscharntke, T. (2000). Species richness and parasitism in a fragmented landscape: experiments and field studies with insects on *Vicia sepium*. *Oecologia*, 122, 129-37.
87. Landolt, P. J. (1993). Effects of host plant leaf damage on cabbage looper moth attraction and oviposition. *Entomologia Experimentalis et Applicata*, 67, 79-85.
88. Lawton, J. H. and Schröder, D. (1977). Effect of plant type, size of geographical range and taxonomic isolation on number of insect species associated with British plants. *Nature*, 265, 137-40.
89. Lawton, J. H., Lewinsohn, T. M., and Compton, S. G. (1993). Patterns of diversity for insect herbivores on bracken. In *Species diversity in ecological communities* (ed. R. E. Ricklefs and D. Schluter), pp. 178-84. University of Chicago Press, Chicago.
90. Lou, Y. G. and Baldwin, I. T. (2004). Nitrogen supply influences herbivore-induced direct and indirect defenses and transcriptional responses to *Nicotiana attenuata*. *Plant Physiology*, 135, 496-506.

91. Louda, S. M. and Potvin, M. A. (1995). Effect of inflorescence feeding insects on the demography and lifetime fitness of a native plant. *Ecology*, 76, 229–45.
92. Louda, S. M. and Rodman, J. E. (1996). Insect herbivory as a major factor in the shade distribution of a native crucifer (*Cardamine cordifolia* A. Gray, bittercress). *Journal of Ecology*, 84, 229–37.
93. Luck, R. F. and Podoler, H. (1985). Competitive exclusion of *Aphytis lingnanensis* by A. *melinus*: potential role of host size. *Ecology*, 66, 904–13.
94. Mantyla, E., Klemola, T., and Haukioja, E. (2004). Attraction of willow warblers to sawfly-damaged mountain birches: novel function of inducible plant defences? *Ecology Letters*, 7, 915–18.
95. Maron, J. L. and Gardner, S. N. (2000). Consumer pressure, seed versus safe-site limitation, and plant population dynamics. *Oecologia*, 124, 260–9.
96. Marquis, R. J. and Whelan, C. J. (1994). Insectivorous birds increase growth of white oak through consumption of leaf-chewing insects. *Ecology*, 75, 2007–14.
97. Mayhew, P. J. (2001). Herbivore host choice and optimal bad motherhood. *Trends in Ecology and Evolution*, 16, 165–7.
98. McCann, K. S., Hastings, A., and Huxel, G. R. (1998). Weak trophic interactions and the balance of nature. *Nature*, 395, 794–8.
99. McConn, M., Creelman, R. A., Bell, E., Mullet, J. E., and Browse, J. (1997). Jasmonate is essential for insect defense in *Arabidopsis*. *Proceedings of the National Academy of Sciences of the USA*, 94, 5473–7.
100. McIntosh, A. R. and Townsend, C. R. (1996). Interactions between fish, grazing invertebrates and algae in a New Zealand stream: a trophic cascade mediated by fish induced changes to grazer behaviour? *Oecologia*, 108, 174–81.
101. Memmott, J. (1999). The structure of a plant-pollinator food web. *Ecology Letters*, 2, 276–80.
102. Memmott, J., Fowler, S. V., Paynter, Q., Sheppard, A. W., and Syrett, P. (2000). The invertebrate fauna on broom, *Cytisus scoparius*, in two native and two exotic habitats. *Acta Oecologca*, 21, 213–22.
103. Mitchell-Olds, T. (2001). *Arabidopsis thaliana* and its wild relatives: a model system for ecology and evolution. *Trends in Ecology and Evolution*, 16, 693–700.
104. Moran, N. A. and Whitham, T. G. (1990). Interspecific competition between root-feeding and leaf-galling aphids mediated by host-plant resistance. *Ecology*, 7, 1050–8.
105. Muller, C. B. and Godfray, H. C. J. (1997). Apparent competition between two aphid species. *Journal of Animal Ecology*, 66, 57–64.
106. Muller, C. B., Adriaanse, I. C. T., Belshaw, R., and Godfray, H. C. J. (1999). The structure of an aphidparasitoid community. *Journal of Animal Ecology*, 68, 346–70.
107. Musser, R. O., Hum-Musser, S. M., Eichenseer, H., Peiffer, M., Ervin, G., Murphy, J. B., *et al.* (2002). Caterpillar saliva beats plant defences—a new weapon emerges in the ev-

olutionary arms race between plants and herbivores. *Nature*, 416, 599-600.
108. Ohsaki, N. and Sato, Y. (1994). Food plant choice of *Pieris* butterflies as a trade-off between parasitoid avoidance and quality of plants. *Ecology*, 75, 59-68.
109. Omacini, M., Chaneton, E. J., Ghersa, C. M., and Muller, C. B. (2001). Symbiotic fungal endophytes control insect host-parasite interaction webs. *Nature*, 409, 78-81.
110. Pace, M. L., Cole, J. J, Carpenter, S. R., and Kitchell, J. F. (1999). Trophic cascades revealed in diverse cosystems. *Trends in Ecology and Evolution*, 14, 483-8.
111. Papaj, D. R. (1986). Shifts in foraging behavior by a *Battus philenor* population: field evidence for switching by individual butterflies. *Behavioral Ecology and Sociobiology*, 19, 31-9.
112. Paul, N. D., Hatcher, P. E., and Taylor, J. E. (2000). Coping with multiple enemies: an integration of molecular and ecological perspectives. *Trends in Plant Science*, 5, 220-5.
113. Pieterse, C. M. J. and Van Loon, L. C. (1999). Salicylic acid-independent plant defence pathways. *Trends in Plant Science*, 4, 52-8.
114. Polis, G. A. (1999). Why are parts of the world green? Multiple factors control productivity and the distribution of biomass. *Oikos*, 86, 3-15.
115. Polis, G. A., Sears, A. L. W., Huxel, G. R., Strong, D. R., and Maron, J. (2000). When is a trophic cascade a trophic cascade? *Trends in Ecology and Evolution*, 15, 473-5.
116. Price, P. W., Bouton, C. E., Gross, P., McPheron, B. A., Thompson, J. N., and Weis, A. E. (1980). Interactions among three trophic levels: influence of plant on interactions between insect herbivores and natural enemies. *Annual Review of Ecology and Systematics*, 11, 41-65.
117. Price, P. W. (1997). *Insect ecology*. Academic Press, New York.
118. Prieur-Richard, A. H., Lavorel, S., Linhart, Y. B., and Dos Santos, A. (2002). Plant diversity, herbivory and resistance of a plant community to invasion in Mediterranean annual communities. *Oecologia*, 130, 96-104.
119. Prokopy, R. J. and Roitberg, B. D. (2001). Joining and avoidance behavior in nonsocial insects. *Annual Review of Entomology*, 46, 631-65.
120. Rambo, J. L. and Faeth, S. H. (1999). Effect of vertebrate grazing on plant and insect community structure. *Conservtion Biology*, 13, 1047-54.
121. Raymond, B., Vanbergen, A., Pearce, I., Hartley, S. E., Cory, J. S., and Hails, R. S. (2002). Host plant species can influence the fitness of herbivore pathogens: the winter moth and its nucleopolyhedrovirus. *Oecologia*, 131, 533-41.
122. Reitz, S. R. and Trumble, J. T. (2002). Competitive displacement among insects and arachnids. *Annual Review of Entomology*, 47, 435-65.
123. Reymond, P., Weber, H., Damond, M., and Farmer, E. E. (2000). Differential gene expression in response to mechanical wounding and insect feeding in *Arabidopsis*. *Plant Cell*, 12, 707-19.
124. Reymond, P., Bodenhausen, N., Van Poecke, R. M. P., Krishnamurthy, V., Dicke, M.,

and Farmer, E. E. (2004). A conserved transcript pattern in response to a specialist and a generalist herbivore. *Plant Cell*, 16, 3132–47.

125. Ritchie, M. E. and Olff, H. (1999) Herbivore diversity and plant dynamics: compensatory and additive effects. In *Herbivores: between plants and predators* (ed. H. Olff, V. K. Brown, and R. H. Drent), pp. 175–204. Blackwell Science, Oxford.

126. Roitberg, B. D., Robertson, I. C., and Tyerman, J. G. A. (1999). Vive la variance: a functional oviposition theory for insect herbivores. *Entomologia Experimentalis et Applicata*, 91, 187–94.

127. Root, R. B. (1973). Organization of a plant-arthropod association in simple and diverse habitats: the fauna of collards (*Brassica oleracea*). *Ecological Monographs*, 43, 95–124.

128. Root, R. B. and Cappuccino, N. (1992). Patterns in population change and the organization of the insect community associated with goldenrod. *Ecological Monographs*, 62, 393–420.

129. Rostas, M. and Hilker, M. (2003). Indirect interactions between a phytopathogenic and an entomopathogenic fungus. *Naturwissenschaften*, 90, 63–7.

130. Rott, A. S. and Godfray, H. C. J. (2000). The structure of a leafminer-parasitoid community. *Journal of Animal Ecology*, 69, 274–89.

131. Ruhren, S. and Handel, S. N. (1999). Jumping spiders (Salticidae) enhance the seed production of a plant with extrafloral nectaries. *Oecologia*, 119, 227–30.

132. Schmitz, O. J., Hamback, P. A., and Beckerman, A. P. (2000). Trophic cascades in terrestrial systems: a review of the effects of carnivore removals on plants. *American Naturalist*, 155, 141–53.

133. Schoonhoven, L. M. (1990). Host-marking pheromones in Lepidoptera, with special reference to two Pieris spp. *Journal of Chemical Ecology*, 16, 3043–52.

134. Schultz, J. C. (1988). Many factors influence the evolution of herbivore diets, but plant chemistry is central. *Ecology*, 69, 896–7.

135. Schütte, C., Van Baarlen, P., Dijkman, H., and Dicke, M. (1998). Change in foraging behaviour of the predatory mite *Phytoseiulus persimilis* after exposure to dead conspecifics and their products. *Entomologia Experimentalis et Applicata*, 88, 295–300.

136. Shiojiri, K., Takabayashi, J., Yano, S., and Takafuji, A. (2001). Infochemically mediated tritrophic interaction webs on cabbage plants. *Population Ecology*, 43, 23–9.

137. Shiojiri, K., Takabayashi, J., Yano, S., and Takafuji, A. (2002). Oviposition preferences of herbivores are affected by tritrophic interaction webs. *Ecology Letters*, 5, 186–92.

138. Simberloff, D. S. and Wilson, E. O. (1969). Experimental zoogeography of islands: the colonization of empty islands. *Ecology*, 50, 278–96.

139. Snyder, W. E. and Wise, D. H. (2000). Antipredator behavior of spotted cucumber beetles (Coleoptera: hrysomelidae) in response to predators that pose varying risks. *Environmental Entomology*, 29, 35–42.

140. Southwood, T. R. E., Moran, V. C., and Kennedy, C. E. J. (1982). The richness, abundance and biomass of the arthropod communities on trees. *Journal of Animal Ecology*, 51,

635-6.
141. Speight, M. R., Hunter, M. D., and Watt, A. D. (1999). *Ecology of insects. Concepts and applications*. Blackwell Science, Oxford.
142. Stanforth, L. M., Louda, S. M., and Bevill, R. L. (1997). Insect herbivory on juveniles of a threatened plant, *Cirsium pitcheri*, in relation to plant size, density and distribution. *Ecoscience*, 4, 57-66.
143. Stapley, L. (1998). The interaction of thorns and symbiotic ants as an effective defence mechanism of swollen-thorn acacias. *Oecologia*, 115, 401-5.
144. Steffan-Dewenter, I. and Tscharntke, T. (1999). Effects of habitat isolation on pollinator communities and seed set. *Oecologia*, 121, 432-40.
145. Stephens, D. W. and Krebs, J. R. (1986). *Foraging theory*. Princeton University Press, Princeton.
146. Stotz, H. U., Pittgendrigh, B. R., Kroymann, J., Weniger, K., Fritsche, J., Bauke, A., et al. (2000). Induced plant defense responses against chewing insects. Ethylene signaling reduces resistance of Arabidopsis against Egyptian cotton worm but not diamondback moth. *Plant Physiology*, 124, 1007-17.
147. Stouthamer, R., Breeuwer, J. A. J., and Hurst, G. D. D. (1999). *Wolbachia pipientis*: microbial manipulator of arthropod reproduction. *Annual Review of Microbiology*, 53, 71-102.
148. Strauss, S. Y. (1997). Floral characters link herbivores, pollinators, and plant fitness. *Ecology*, 78, 1640-5.
149. Strong, D. R., Lawton, J. H., and Southwood, T. R. E. (1984). *Insects on plants: community patterns and mechanisms*. Harvard University Press, Cambridge, MA.
150. Strong, D. R. (1992). Are trophic cascades all wet? Differentiation and donor-control in speciose ecosystems. *Ecology*, 73, 747-54.
151. Tattersall, D. B., Bak, S., Jones, P. R., Olsen, C. E., Nielsen, J. K., Hansen, M. L., et al. (2001). Resistance to an herbivore through engineered cyanogenic glucoside synthesis. *Science*, 293, 1826-8.
152. Thaler, J. S., Fidantsef, A. L., Duffey, S. S., and Bostock, R. M. (1999). Trade-offs in plant defense against pathogens and herbivores: a field demonstration of chemical elicitors of induced resistance. *Journal of Chemical Ecology*, 25, 1597-609.
153. Thaler, J. S. (2002). Effect of jasmonate-induced plant responses on the natural enemies of herbivores. *Journal of Animal Ecology*, 71, 141-1.
154. Thaler, J. S., Stout, M. J., Karban, R., and Duffey, S. S. (2001). Jasmonate-mediated induced plant resistance affects a community of herbivores. *Ecological Entomology*, 26, 312-24.
155. Thibout, E., Guillot, J. F., Ferary, S., Limouzin, P., and Auger, J. (1995). Origin and identification of bacteria which produce kairomones in the frass of *Acrolepiopsis assectella* (Lep, Hyponomeutoidea). *Experientia*, 51, 1073-5.
156. Thies, C. and Tscharntke, T. (1999). Landscape structure and biological control in agroeco-

systems. *Science*, 285, 893 – 5.

157. Thomas, C. D. (1986). Butterfly larvae reduce host plant survival in vicinity of alternative host species. *Oecologia*, 70, 113 – 17.

158. Thomas, J. and Lewington, R. (1991). *The butterflies of Britain and Ireland*. Dorling Kindersly, London.

159. Tindall, K. V. and Stout, M. J. (2001). Plant-mediated interactions between the rice water weevil and fall armyworm in rice. *Entomologia Experimentalis et Applicata*, 101, 9 – 17.

160. Trumble, J. T., Kolodny-Hirsch, D. M., and Ting, I. P. (1993). Plant compensation for arthropod herbivory. *Annual Review of Entomology*, 38, 93 – 119.

161. Tscharntke, T., Vidal, S., and Hawkins, B. A. (2001). Parasitoids of grass-feeding chalcid wasps: a comparison of German and British communities. *Oecologia*, 129, 445 – 51.

162. Turlings, T. C. J., Wäckers, F. L., Vet, L. E. M., Lewis, W. J., and Tumlinson, J. H. (1993). Learning of host-finding cues by hymenopterous parasitoids. In *Insect learning: ecological and evolutionary perspectives* (ed. D. R. Papaj and A. C. Lewis), pp. 51 – 78. Chapman & Hall, New York.

163. Turlings, T. C. J., Loughrin, J. H., McCall, P. J., Rose, U. S. R., Lewis, W. J., and Tumlinson, J. H. (1995). How caterpillar-damaged plants protect themselves by attracting parasitic wasps. *Proceedings of the National Academy of Sciences of the USA*, 92, 4169 – 74.

164. Underwood, N. (1999). The influence of plant and herbivore characteristics on the interaction between induced resistance and herbivore population dynamics. *American Naturalist*, 153, 282 – 94.

165. Van Dam, N. M. and Baldwin, I. T. (1998). Costs of jasmonate-induced responses in plants competing for limited resources. *Ecology Letters*, 1, 30 – 3.

166. Van der Meijden, E., Wijn, M., and Verkaar, H. J. (1988). Defence and regrowth, alternative plant strategies in the struggle against herbivores. *Oikos*, 51, 355 – 63.

167. Van der Putten, W. H., Vet, L. E. M., Harvey, J. H., and Wäckers, F. L. (2001). Linking above- and below-ground multitrophic interactions of plants, herbivores, pathogens, and their antagonists. *Trends in Ecology and Evolution*, 16, 547 – 54.

168. Van Loon, J. J. A., De Boer, J. G., and Dicke, M. (2000). Parasitoid-plant mutualism: parasitoid attack of herbivore increases plant reproduction. *Entomologia Experimentalis et Applicata*, 97, 219 – 27.

169. Van Loon, J. J. A. and Dicke, M. (2000). Sensory ecology of arthropods utilizing plant infochemicals. In *Sensory ecology* (ed. F. G. Barth and A. Schmid), pp. 253 – 70. Springer, Heidelberg.

170. Van Poecke, R. M. P., Posthumus, M. A., and Dicke, M. (2001). Herbivore-induced volatile production by *Arabidopsis thaliana* leads to attraction of the parasitoid *Cotesia rubecula*: chemical, behavioral, and gene-expression analysis. *Journal of Chemical Ecology*, 27, 1911 – 28.

171. Van Poecke, R. M. P. and Dicke, M. (2004). Indirect defence of plants against herbivores:

using *Arabidopsis thaliana* as a model plant. *Plant Biology*, 6, 387 – 401.
172. Van Rijn, P. C. J., Van Houten, Y. M., and Sabelis, M. W. (2002). How plants benefit from providing food to redators even when it is also edible to herbivores. *Ecology*, 83, 2664 – 79.
173. Van Tol, R. W. H. M., Van der Sommen, A. T. C., Boff, M. I. C., Van Bezooijen, J., Sabelis, M. W., and Smits, P. H. (2001). Plants protect their roots by alerting the enemies of grubs. *Ecology Letters*, 4, 292 – 4.
174. Vet, L. E. M., Wäckers, F. L., and Dicke, M. (1991). How to hunt for hiding hosts: the reliability-detectabilityproblem in foraging parasitoids. *Netherlands Journal of Zoology*, 41, 202 – 13.
175. Vet, L. E. M. and Dicke, M. (1992). Ecology of infochemical use by natural enemies in a tritrophic context. *Annual Review of Entomology*, 37, 141 – 72.
176. Vet, L. E. M., Lewis, W. J., and Cardé, R. T. (1995). Parasitoid foraging and learning. In *Chemical ecology of insects* (ed. R. T. Cardé and W. J. Bell), pp. 65 – 101. Chapman & Hall, New York.
177. Vet, L. E. M. (1996). Parasitoid foraging: the importance of variation in individual behaviour for population dynamics. In *Frontiers of population ecology* (ed. R. B. Floyd and A. W. Sheppard), pp. 245 – 56. CSIRO Publishing, Melbourne.
178. Visser, J. H. (1986). Host odor perception in phytophagous insects. *Annual Review of Entomology*, 31, 121 – 44.
179. Visser, M. E. and Holleman, L. J. M. (2001). Warmer springs disrupt the synchrony of oak and winter moth phenology. *Proceedings of the Royal Society of London*, *Series B*, *Biological Sciences*, 268, 289 – 94.
180. Voelckel, C. and Baldwin, I. T. (2004). Generalist and specialist lepidopteran larvae elicit different transcriptional responses in *Nicotiana attenuata*, which correlate with larval FAC profiles. *Ecology Letters*, 7,770 – 5.
181. Vos, M., Moreno Berrocal, S., Karamaouna, F., Hemerik, L., and Vet, L. E. M. (2001). Plant-mediated indirect effects and the persistence of parasitoid-herbivore communities. *Ecology Letters*, 4, 38 – 45.
182. Vrieling, K., Smit, W., and Van der Meijden, E. (1991). Tritrophic interactions between aphids (*Aphis jacobaeae* Schrank), ant species, *Tyria jacobaeae* L. and *Senecio jacobaea* L. lead to maintenance of genetic variation in pyrrolizidine alkaloid concentration. *Oecologia*, 86, 177 – 82.
183. Wäckers, F. L. and Bezemer, T. M. (2003). Root herbivory induces an above-ground indirect defence. *Ecology Letters*, 6, 9 – 12.
184. Wäckers, F. L., Zuber, D., Wunderlin, R., and Keller, F. (2001). The effect of herbivory on temporal and spatial dynamics of foliar nectar production in cotton and castor. *Annals of Botany*, 87, 365 – 70.
185. Walker, M. and Jones, T. H. (2001). Relative roles of top-down and bottom-up forces in

terrestrial tritrophic plant-insect herbivore-natural enemy systems. *Oikos*, 93, 177 – 87.
186. Walling, L. L. (2000). The myriad plant responses to herbivores. *Journal of Plant Growth Regulation*, 19, 195 – 216.
187. Williams, I. S., Jones, T. H., and Hartley, S. E. (2001) The role of resources and natural enemies in determining the distribution of an insect herbivore population. *Ecological Entomology*, 26, 204 – 11.
188. Wootton, J. T. (1994). The nature and consequences of indirect effects in ecological communities. *Annual Review of Ecology and Systematics*, 25, 443 – 66.
189. Wratten, S. (1992). Population regulation in insect herbivores—top-down or bottom-up? *New Zealand Journal of Ecology*, 16, 145 – 7.
190. Yano, S. (1994). Flower nectar of an autogamous perennial *Rorippa indica* as an indirect defense mechanism against herbivorous insects. *Researches on Population Ecology*, 36, 63 – 71.
191. Zamora, R. and Gomez, J. M. (1993). Vertebrate herbivores as predators of insect herbivores: an asymmetrical interaction mediated by size differences. *Oikos*, 6, 223 – 8.

第11章 进化:昆虫与植物,永恒的斗争

11.1 昆虫与植物相互作用的化石记录
11.2 物种形成
   11.2.1 生殖隔离
   11.2.2 物种形成的速率
   11.2.3 共生物种的形成
11.3 昆虫的基因变异
   11.3.1 种间变异
   11.3.2 种内变异
   11.3.3 偏好与表型之间的关系
   11.3.4 基因突变和寄主植物的适应性
11.4 植物抗性基因的突变
11.5 选择与适应
11.6 昆虫多样性的演变
11.7 专食性寄生昆虫的进化
   11.7.1 应对植物次生代谢产物
   11.7.2 竞争
   11.7.3 降低由天敌引发的死亡率
   11.7.4 系统发育关系
11.8 植食性昆虫与寄主间的进化
   11.8.1 对协同进化理论的批判
   11.8.2 对协同进化理论的支持
11.9 结论
11.10 参考文献

昆虫有惊人的能力以适应不断变化的环境条件。例如,自20世纪40年代引进化学杀虫剂后,昆虫迅速地进化,从而形成大量对杀虫剂的抗性[32],为此,寻找新的杀虫剂正面临着越来越严重的问题(图11.1)。这也为植食昆虫对植物次生代谢物的适应提供了一个很好的例证[27]。通过研究一些害虫(如麦瘿蝇(*Hessian fly*)或褐飞虱)对新繁育的有剧毒的抗性作物生物型的适应速度,可以发现植食昆虫具有惊人的适应能力[103]。同样值得注意的是,在人为选择下(如通过植物育种学家们),植物也能产生很大的变化。化石记录了几亿年来,植物和昆虫新物种的进化过程。通过自然选择,昆虫与植物的相互作用正在不断地发生改变。然而,这种植物与昆虫间的相互作用机制却不易被发现,并且这一直是有争议的话题。施塔尔在1888年写道:"对于植物而言,周围的动物不仅会影响它们的形态,还会影响它们的化学结构"[52]。埃尔利希和雷文于1964年提出昆虫和植物协同进化的理论[43]。对此引发了激烈的讨论,并由此

开展了许多新的调查研究,因此人们对昆虫与植物相互关系的理解也处于不断的发展中。在本章中,我们将着重论述关于专化(specialization)、物种形成(speciation)以及昆虫与植物关系的进化这些仍存在争议的问题。

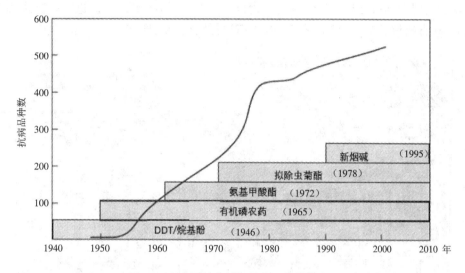

图 11.1　不同杀虫剂与抗病品种数的关系

1939年,合成杀虫剂DDT开始生产并投入使用,而首次有记录的对DDT出现抗性的案例是在1946年。在随后的60年里,随着杀虫剂的发展和使用,产生抗药性的节肢动物种类也在极速增加。长条表示特定类型的杀虫剂被使用的时期,括号内的时间表示抗药性首次被证明的年份

11.1　昆虫与植物相互作用的化石记录

亿万年来,地球表面的植物发生了巨大的改变(图11.2)。最古老的昆虫和植物化石可以追溯到约4亿年前。在中生代时期(大约4亿年前至1亿4千万年前),昆虫与植物的种类开始多样化。植食性昆虫开始进化出多种方式来开发新的植物作为食物。在与昆虫相关的植物化石中,发现了大量被节肢动物作为食物而毁坏的痕迹(图11.3)[82]。

化石记录表明,最早的植食性节肢动物出现于泥盆纪(大约4亿年前)早期,以刺吸式口器获取孢粉为食[80]。据报道从外界获取食物和磨损食物的行为分别出现在宾夕法尼亚纪(3亿1千万年前至—2亿9千万年前)中后期。通过凿洞取食的方法获取叶子出现在二叠纪(大约2亿9千万年前)[80]。不同的取食体系中,最有可能的演变顺序是从吮吸到咀嚼,继而是挖掘[121]。

植物比昆虫出现得稍微早些,但是当最大的植物群(被子植物)出现在白垩纪时期时,大量的昆虫已经在当时出现了[81]。到白垩纪结束时,主要植食性昆虫群体的多样化和被子植物的迅速多样化之间的关系一直是许多讨论的主题,而这种关系也许能帮助人们确定植物是否促进昆虫的进化或者昆虫是否促进植物的进化。

几个现存的植食性昆虫目,尤其是鳞翅类、鞘翅类(叶甲科、象甲科)、双翅目(潜蝇科、瘿蚊科)和膜翅目(瘿蜂目),在被子植物出现和传播之后广泛蔓延[121]。这表明,开花植物出现和进化加速了这些群体的演变。然而其他的古生物学资料显示,一些现代的昆虫家族性的辐射

形进化开始于2亿4千5百万年前,比1亿4千4百万年前至6千6百万年前被子植物的出现和迅速发展早了1亿年[81]。这些也许能够支持昆虫的多样性影响被子植物多样性的假说。图11.4的主轴图表明鞘翅类、鳞翅类、双翅目以及晚中生代出现的膜翅目的家族辐射形进化,与开花植物的出现和其占优势地位的时间相吻合,因此开花植物也许能促进很多昆虫辐射形进化。然而,在直翅目、同翅目和半翅目中却没有这样明显的表现。

图11.2 石炭纪晚期的植物

蕨类植物在温暖的沼泽中达到它生长的顶峰。超过50 m高的蕨类植物和浓密的林下叶层以及叶状的蕨类植物以及杉叶藻生长在一起。种子植物,例如科达属针叶树(右上角),能长到将近30 m高,树干直径1 m。这种植物有最多1 m长15 cm宽的表带状树叶(马德夫,1959年)[87]

11.2 物种形成

昆虫是物种最丰富的多细胞生物,因此构成了生物多样性的重要组成部分。自从达尔文出版了《物种起源》[30]后,物种的起源(即生殖隔离形成的物种)问题一直受到人们的广泛关注[94]。同样,对昆虫的物种起源和它们的专化性的问题也在20世纪进行了深入讨论,尤其是在昆虫与植物相互作用的相互演化方面[43,74,125]。总体上,斯苏仁已经详细地总结了物种起源理论的历史,尤其是昆虫物种的起源理论的历史[113]

大多数昆虫是专食性的[51]。经典学说认为物种的形成是在不同地区发生的,因为由于种群所处地理位置孤立不能交换基因。然而,经过长时间的科学争论,现在已经清楚物种的形成也可以发生在同一区域[11],植食性昆虫是被证明的自例。正如布什和他的同事们对苹果实蝇的研究所示[18],苹果实蝇最初在山楂上被发现,但苹果后来才被认为是它们的寄主。偏好在苹果上产卵的苹果实蝇也会在苹果上交配,然而偏好在山楂上产卵的苹果实蝇也会在山楂上产

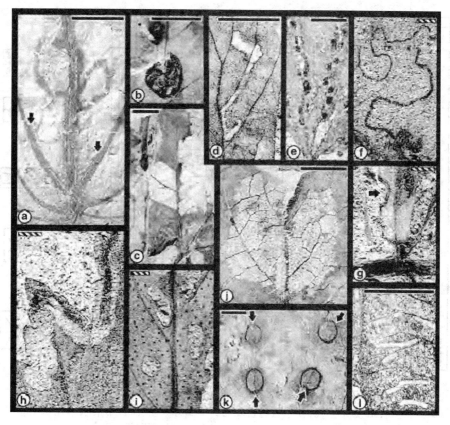

图 11.3　来自北达科他州西南部的威利斯顿盆地的一系列植物昆虫的关系

从左上角最早的古新世延续到右下角最晚的白垩纪早期。所有的材料来自丹佛自然科学博物馆或是耶鲁皮博迪植物博物馆。(a) 双子叶植物 (*Paranymphaea crassifolia*) 上,两个线性箭头显示的是昆虫的产卵位点,接下来是次级叶脉,之后是主要叶脉,最后是一个很大的化蛹室;(b) 在铁线莲主要叶脉上的虫瘿;(c) 昆虫取食对悬铃木 (*Platanus raynoldsi*) 的毁坏;(d) 叶片的骨架;(e) 一种植物 (*Trochodendroides nebrascensis*) 上许多的虫瘿;(f) 在月桂树 (*Marmarthia pearsonii*,樟科) 叶片上挖掘的初始阶段;(g) 对拟蓝果树水杉 (*Metasequoia sp.*) 的取食 (箭头所示);(h) 微蛾科 (鳞翅目) 叶片上的浅叶痕迹;(i) 荨麻目未知的科植物叶片上的取食;(j) 悬铃木 (*Erlingdorfia montana*,悬铃木科) 的一般骨架;(k) 集中在悬铃木 (*E. montana.*) 的主要叶脉上的大规模的昆虫集群;(l) 悬铃木科的未知属上的取食痕迹 (拉万什比等,2002)[82]

卵,这种现象被称为选型交配。此外,二者生活史的特点也体现出差别(见 11.2.1)。因此,在原来的种群中,因为对山楂的偏好导致生殖隔离形成亚群,最终导致同域物种的形成。植食性昆虫中还有许多其他例子[11],如蚱蜢[109]、甲虫[104]、蚜虫[21]、小巢蛾[95],也可以证明同域物种形成的存在。因此,同域物种形成的概念现在也被广泛接受。

尽管异域性物种的形成发生在较大的地理范围内,同域物种的形成导致与亲本种群距离非常接近的新物种镶嵌式的进化。这也对我们如何看待生物多样性的动态变化和空间范围有一定影响。

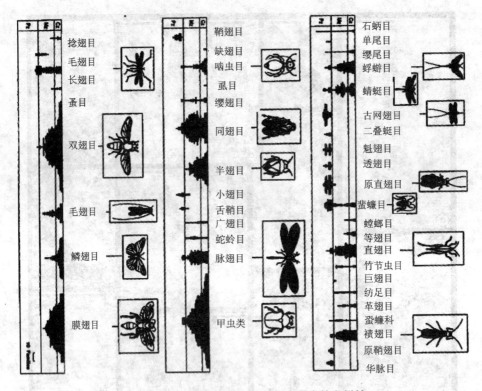

图 11.4　在显生宙的不同阶段化石昆虫的多样性

Pz 为古生代，Mz 为中生代，被子植物出现在约 2/3 的中生代（拉万代拉和塞科斯基，1993，1991 年美国科学进步联合会，已获许可）

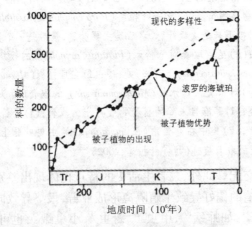

图 11.5　从三叠纪到现在昆虫的多样性

虚线说明多样性指数的变化可能开始于三叠纪并持续到白垩纪早期。Tr 代表三叠纪；J 代表侏罗纪；K 代表白垩纪；T 代表第三纪（拉万代拉和塞科斯基，1993）[81]

11.2.1　生殖隔离

当一些障碍阻止或限制两个种群之间的基因流动时，新的物种才会出现。下面我们将讨论在昆虫中出现的几类阻碍。有的阻碍会导致交配前的生殖隔离(空间和行为阻碍,生活史不同步)或交配后的隔离(精卵不相容)。

1. 空间障碍

对基因流的最明显的障碍是山脉、大海、河流、沙漠等地理上的种群隔绝。这些阻碍的有效性在很大程度上取决于昆虫的扩散能力。例如,金花虫科的金花虫(*Oreina cacaliae*)和金花虫(*O. globosa*)是仅以蜂斗叶(*Petasites senecio*)和菊科(*Adenostyles*)植物为食的寡食性植物,却表现出相当大的遗传分化,这是用电泳法对距离只有40~250公里远的瑞士和德国之间的种群进行分析的结果[40]。令人惊讶的是,10米甚至几百米的距离就可能足够导致空间隔离了。丹麦的跳甲(*Phyllotreta nemorum*)的R基因在种群中也表现出了强大的遗传分化。具有R基因的甲虫具备了一种特殊能力,可以取食十字花科山芥属植物,而本身缺乏R基因的甲虫,在取食这种植物时会死亡[31,98]。

2. 行为障碍

植食性昆虫取食和/或产卵偏好的差异性,可以导致有效的隔离,因此这很有可能给同域物种的形成提供机会。例如,苹果蛆(*Rhagoletis pomonella*)成虫被苹果中特定的化学物质所吸引,而其近缘种蓝莓蛆(*R. mendax*)则被蓝莓果实中不同的化学物质所吸引[53]。虽然这两个品种可以很容易地在实验室进行杂交,遗传分析表明,在自然条件下它们之间没有基因流动,因为交配只发生在各自寄主植物上的个体之间[20,47,48]。同样地,蚜虫(*Aulacorthum solani* s. str.)是杂食性的但是不吃肺草(*Pulmonaria officinalis*),在自然条件下不能与专门以地肺草(*P. officinalis*)为食的蚜虫(*A. solani langei*)杂交。两个亚种由于它们各自喜好的寄主植物不同而完全分开,即使它们的寄主植物混合生长,甚至彼此紧挨着。然而另一个亚种蚜虫(*A. s. aegopodi*)以羊角芹(*Aegopodium podagraria*)为食,可能偶尔会与以肺草为食的蚜虫(*A. solani* s. str)杂交(图11.6)[100]。

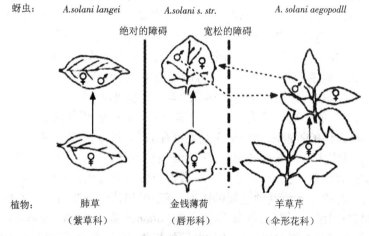

图11.6 蚜虫(*Aulacorthum*)交配前的隔离,由于对寄主植物的偏好不同而导致隔离强度的不同(木勒,1985)[100]

其他类型的行为障碍是由性信息素的差异造成的,例如,在北美[76]和法国[105]两个欧洲玉米钻心虫(*Ostrinia nubilalis*)种群之间的性信息素成分的差异,是在飞虱(同翅目)[26]中性信息素成分的差异导致的。

3. 生活史不同步

不同步的生活史可能出现在昆虫的生活史各个阶段中,特别是在交配期间,这可能会导致生殖隔离。例如,北美角蝉复合体(*Enchenopa binotata*),包含9个同域生存的个体,以共同的寄主植物为食。它们不同步的生活史是保持生殖隔离的主要因素[141]。如图11.7所示,3个苹果蛆种群在生活史中只在很短的时间内出现了重叠模式,从而大大减少种群之间的交配[19]。然而可能出现的问题是,这种不同步的生活史是否会引起新的种化事件。

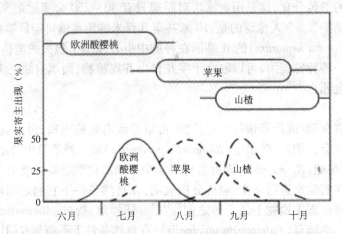

图11.7 威斯康星州的三个浆果实蝇(*Rhagoletis pomonella*)种群成虫羽化的时间(布什,1975)[19]

4. 杂交不相容

同一物种的不同种群可能会有不同程度的遗传分化,因此杂交种的受精卵有可能是不育的。例如,樱桃实蝇(*Rhagoletis cerasi*)的欧洲种群至少分为两个地理亚群,图11.8这显示了不同亚群的不相容性:南部的雄性(圆圈)和东北部的雌性(三角形)进行杂交,产生的卵的孵化率很低,可能是由于遗传因素导致的。另一个原因可能与微生物共生体的存在相关[17]。根据汤普森的理论[123],共生体与环境因子在物种形成过程中有着重要的作用。在一个环境中,一个特定的共生体和其宿主之间的相互作用可能是对立的,但在另一个环境下可能是共生的。这样,共生体与宿主的相互作用在不同的环境中产生的微小差异可能放大为不同昆虫种群间的差异,从而导致物种的形成。含有立克次氏体(*Rickettsia*)的紫苜蓿象鼻虫(*Hypera postica*)种群和不含立克次氏体的种群之间的杂交是不可育的[66]。沃尔巴克氏共生体(*Wolbachia*)分布相当广泛,可以对其寄主产生深远的影响,包括细胞质不相容,这种共生体通常被认为在昆虫物种的形成中发挥着作用。然而,最近的一个数据分析认为得出这样的结论还为时过早,在昆虫物种形成中可能忽视了一些其他重要的因素,核基因和其他一些共生体的相互作用,或者与沃尔巴克氏共同起作用[137]。例如,在蟋蟀复合体(*Allonemobius fasciatussocius*)中,分子数据表明,三种蟋蟀均起源于存在沃尔巴克氏体的共同的祖先,因此表明沃尔巴克氏体可能在物种形成的爆发期起着微小的作用[90]。

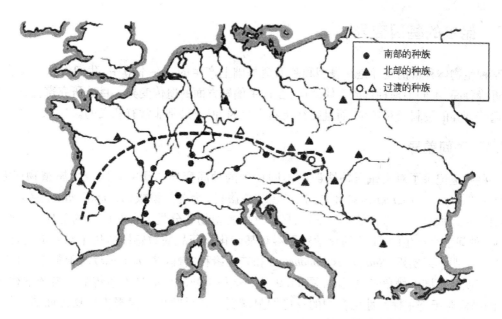

图11.8　樱桃绕实蝇(*Rhagoletis cerasi*)的不同地理种群显示了不相容性(博勒等,1976)[17]

11.2.2　物种形成的速率

单位时间内出现的新物种(子代)的数量,主要取决于亲本中的遗传变异,其次取决于自然选择和遗传漂变。对相对年轻的海洋岛屿的研究,如对40万年前在太平洋上出现的夏威夷群岛中的夏威夷大岛(Big Island)的研究,为遗传变异产生的作用提供了证据。如上所述,随着原始的祖先迁入岛屿后,不同物种的分化速率表现出非常大的差异。这可以解释为不同的基因组分化和重组能力存在很大的差异[22]。为什么有些基因组非常稳定而一些其他基因组却演变迅速仍然是一个待解决的问题[93]。

物种形成的速率也取决于一个世代所需的时间,大多数突变是在细胞减数分裂过程中出现的。植食性昆虫在这方面表现出极大的变异性,例如,北美的十七年蝉的一个世代周期将持续17年[139],而在热带区域的小菜蛾可以达到每年28代[64]。显然,假设的突变率都是相同的,世代时间短的物种(如大多数昆虫)比那些世代周期长的物种(如多年生植物)可能更容易迅速形成新物种。因此,假设一个物种施加于另一个物种同样强大的选择压力,一个植食性昆虫适应其寄主植物变化的速度,可能远远高于植物对昆虫攻击产生的反应速度。

11.2.3　共生物种的形成

作为生物体之间相互作用的结果,互利共生物种的形成被称为协同进化[124]。埃尔利希和雷文猜想植食性昆虫和植物之间的相互作用可能会导致双方的物种形成[43]。在大量有关于协同进化观点的文献中,汤普森解释了不同的生态、遗传和进化的条件如何影响种化和协同进化的过程[125]。我们将在第11.8节中分析这个问题。

11.3 昆虫的基因变异

某种生物体种群的变异是在表型和基因型基础上产生的。在本节中,我们将重点关注昆虫的近缘种间,不同地理种群或个体之间,在寄主偏好方面的遗传变异。目前研究寄主植物偏好的遗传变异的资料相对较少,而且往往是不完整的,但成果是非常有意义的。

11.3.1 种间差异

一些报道记录了对巢蛾属姊妹群的不同物种的种间变异的研究。以卫矛属植物(卫矛科)为食的巢蛾(*Y. cagnagellus*)和专门以苹果(蔷薇科)为食的巢蛾(*Y. malinellus*)杂交产生的 F1 代,可以取食亲代喜食的两种植物,尽管这两种植物在分类上存在巨大的差异。一些数据表明,种间杂交产生的 F1 代的化学感受器对某些化学药品的敏感性遗传了双亲的特性[131]。弗雷等人证明苹果实蝇(*Rhagoletis pomonella*)和其近缘种实蝇(*R. mendax*)对于苹果、山楂的味道有着明显不同的触角点位反应(图 11.9)[55]。这种不同的反应是可遗传的。与亲本相比,F1 代的触角对于寄主挥发性化合物的反应明显减弱。这反映出寻找寄主植物方面能力的减弱,很可能是其基因流缺失的原因[54]。

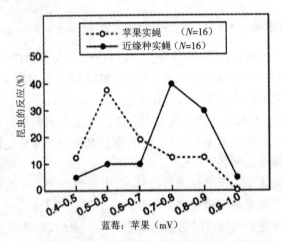

图 11.9 苹果实蝇(*Rhagoletis pomonella*)和其近缘种实蝇(*R. mendax*)
的触角电位反应图(弗雷和布什,1990)[53]

北美洲西部的雌性凤蝶(*Papilio zeliacon*)可以在两种伞形植物厚翅芹(*Lomatium grayi*)和其近缘种厚翅芹(*Cymopeterus terebinthus*)上产卵,但是其近缘种凤蝶(*P. oregonius*)只能在(菊科,*Artemisia dracunculus*)上产卵。实验室中,两物种的杂交后代对产卵地点的选择偏好与常染色体和性染色体都有一定的相关性(表 11.1)。杂交后代与亲本相比,通常有更大的寄主选择范围。

表 11.1 种间和种内不同种群杂交，对寄主植物所产生的偏好性

种类	食物—植物偏好 产卵偏好	表现	A 或 X	标志	参考文献
种间杂交					
烟叶蛾 v. ×s.	优势地位		A		5
烟叶蛾 v. ×s.	优势地位		A	性能由多个基因决定	117
巢蛾（Yponomeuta）c. ×m.	亲本双方		A	在杂交，化学感受器感受信息是敏感的	131
蛱蝶（Lymenitis）a. ×a.	亲本双方		A	在杂交中，诱导偏好的信息	63
凤蝶 o. ×z.		中间	A	影响表现的基因与影响产卵选择的基因不同	128
凤蝶 o. ×z.	优势地位		X(A)	X 染色体具有最大的效应，通过常染色体基因修饰效应	122
凤蝶	优势地位			一个亲本种是单食性的，另一个是多食性的	120
凤蝶 g. ×c.			X	产卵地选择可能基于相对较少的基因位点	116
实蝇（Procecidochares）a. ×A	优势地位			产卵地的选择基于单基因、双等位基因系统	67
种内杂交 a. ×A (不同种类或品质)					
乳草蟓	优势地位		A	产卵地偏好是多基因遗传的	84
三裂果蝇	优势地位		A	产卵地偏好是多基因遗传的	69
三裂果蝇	优势地位		A	食物偏好与产卵地选择无关	68
绿烟夜蛾	优势地位		X	产卵地选择的偏好性与父本显性遗传有关	135

注：A，包含常染色体；X，包含性染色体；P，部分与大部分其他动物相反，即鳞翅目（包括鸟类）有性别决定系统，雌性是异配性别（heterogametic sex）。

11.3.2 种内变异

目前在多种昆虫中均发现，在同一物种的不同种群中，对寄主植物的偏好是可遗传的。例如，实验室培养的寡食性凤蝶（Papilio zeliacon）和单食性凤蝶（P. oregonius）均表现出种内变异，即在其他的寄主植物上产卵的倾向[124]。来自密西西比州的夜蛾（Helicoverpa virescens）比来自美国维尔京群岛的种群更倾向于在棉花上产卵。

一些种群的家蚕（Bombyx mori）可以很容易地接受除桑叶外的几种其他植物，因为它们具有能够显著降低对这些植物次级化合物敏感性的受体，而这些化合物对于普通蚕具有很强的抑制进食作用[114]。蚜虫（Uroleucon ambrosiae）在寄主植物偏好方面会表现出很强的地理差异[4]。在北美东部，蚜虫专门选择巨豚草（Ambrosia trifida）作为寄主，而在美国西南部，许多种紫菀科的其他植物也可以作为其寄主[58]。

植食性昆虫种群表现出对寄主植物的不同偏好，可能受当地的实际情况，如寄主植物种类的相对丰富度的影响，通过自然选择，一些基因型可能更受昆虫的喜爱。

11.3.3 偏好与表型之间的关系

目前的研究,关于种群中不同产卵偏好与幼虫表现的基因差异的研究产生了不一致的结果[127,133]。可能是因为多数研究是以种群平均数的测量为基础的,而基因差异的进化是个体之间可遗传差异水平上的变化。或者说,选择压力在成体与幼体上的作用不同。例如,成体可能表现出优化觅食成功率的选择,而这与它们后代所受到的寻找最优寄主植物的选择不完全吻合[111]。在一些案例中的差异被认为是存在于个体水平上的,可以检测到显著的差异[23,101,112],但跳甲(*Phyllotreta nemorum*)除外[102]。

北美粉蝶属的一些物种可以在十字花科植物(*Thlaspi arvense*)上产卵,虽然这些植物对它们的幼虫是有毒的。这些植物在 19 世纪末被引入,可能是时间太短,幼虫尚未进化出适应性。

11.3.4 基因突变和寄主植物的适应性

自 20 世纪 90 年代以来,在昆虫与植物相互作用的进化过程中,对种群结构的进化过程和作用方面的研究引起了人们极大的兴趣[98,125]。种群结构中的局部适应,如集合种群(meta-populations)可能会导致紧密相连的局部适应基因复合体。如果寄主植物的基因型不同可以给植食性动物种群提供更多的机遇,就可能会因为局部选择更倾向于能够适应当地寄主植物基因型的特定基因而产生基因型不同的植食性动物亚种群。不同的基因会产生联系,从而形成共同适应的基因复合体,而这些基因复合体会在和其他亚种的个体杂交期间分解[31]。种群结构对于我们理解适应性进化中的种内基因突变的作用有很大的帮助。以在 G 型欧洲山芥(*Barbarea vulgaris* spp.)种群上收集的跳甲(*Phyllotreta nemorum*)为例,它们都是 R 基因的纯合子,R 基因用来抑制植物的抵御作用。而在其他寄主植物上取样的跳甲种群中,发现的带有 R 基因的跳甲数量要少很多(图 11.10)。

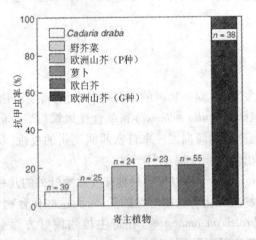

图 11.10 在不同的寄主植物上收集的具有 R 基因的跳甲的比例

分子工具对于阐明共同适应的基因复合体中的基因机制具有很大的帮助。这方面的研究涉及影响迁徙、基因流动和适应的种群结构与选择机制[39,99]。需要注意的是,分子基因工具最初只用于模式生物,现在也越来越多地应用于非模式生物。因此,这个领域有望在未来几年取得重大进展。

11.4 植物抗性基因的突变

植物可能会受到很多敌害的攻击,植物有很多选择来抵御植食性昆虫(第3、第4、第10章),而不同的防御机制可能会在彼此之间产生负面作用。因此,在植物抗虫方面发现基因的多态性也就不足为奇了。例如,狗舌草(*Senecio jacobaea*)对不同浓度的生物碱呈现出多态性,这种高浓度的次级代谢物能够支持植物有效防御很多植食性物种(如蚜虫),但是不能防御专食性物种朱砂蛾(*Tyria jacobaeae*)。相反,这种低浓度的次级代谢物能够支持植物通过蚂蚁来间接防御朱砂蛾(*Tyria jacobaeae*),蚂蚁到达植株并收集蚜虫蜜露,同时蚂蚁也会捕食朱砂蛾(*Tyria jacobaeae*)的毛毛虫。因此,在带有高浓度烷生物碱的植物个体上朱砂蛾(*Tyria jacobaeae*)数量极少,而在带有低浓度烷生物碱的植物个体上(*Tyria jacobaeae*)的数量则很多[134]。来自专食性昆虫和广食性昆虫选择压力的变化可以解释植物种群的基因变异。除了生态成本,如狗舌草的防御涉及的,也有直接成本,如与植物防御相关的适应性降低[6,65]。

一个种群中的不同个体的植物所受的损害是不同的,可能是环境的不同(如光线和之前的损害),也可能是基因的不同。另外,环境因素的不同也可能是由基因导致的,例如基因不同可以影响对之前的损害所产生的诱导抗性。抗虫的基因变异已经在大量植物中(包含农作物和天然物种)得到记录。然而,大量关于农作物抗性的基因机制的信息表明,很多基因机制会介导抗虫性[77]。例如,苜蓿对豌豆蚜(*Acyrthosiphon pisum*)的抗性作用就是通过一个或几个基因介导的,而对有斑纹的苜蓿蚜(*Therioaphis maculata*)的抗性作用则是通过很多基因介导的。另外,甜瓜(*Cucumis melo*)对棉蚜(*Aphis gossypii*)的抗性是单基因作用的,其他的基因起到的作用微乎其微[77]。对美国两个洲的野生烟草物种进行比较,发现通过胰蛋白酶抑制剂产生了不同的基本和诱导性的直接防御,以及通过植食性物种诱导的植物挥发物产生了**诱导性的间接防御**,而这种防御程度与植食性动物的损害程度有关。胰蛋白酶抑制剂的增加导致植物的适应性降低[61]。防风草织蛾(*Depressaria pastinacella*)对野生的欧洲防风草(*Pastinaca sativa*)的损害在种群内部也会有变化,有时会导致适应性的完全丧失。基因研究表明,伞状花序种子是可遗传的,并且与个体差异的可遗传变异有关[9]。

总之,植物抗虫的基因不同已经被广泛证实,因此,植物种群和植食性昆虫种群组成了很多的基因型复合体,这些基因型在抗性和反抗性方面的成本和收益不同。这种基因变异为自然选择提供了素材,能在最大程度上适应的基因型得以遗传。

11.5 选择与适应

大量的实验数据表明,植食性的节肢动物能够快速适应其寄主植物[76,133]。例如,杂食性二斑叶螨(*Tetranychus urticae*)能够很快适应一种新的寄主植物。在10代以内,在西红柿和西兰花上的叶螨比在青豆上的叶螨表现出更强的适应性和更低的死亡率[57]。大量的植物次级代谢物质是植食性昆虫必须克服的障碍(第4、5章)。这种选择也是为了植食性昆虫的适应。例如,对大量不适应的植食性物种而言,芥子油苷或尼古丁都是有毒的,而一部分专食性物种已经适应了这些毒素。例如,烟草的专食性昆虫——烟草天蛾幼虫能有效地排出消化的尼古丁,这样就不会中毒。然而,这种适应并非没有代价,因为毛毛虫在抑制尼古丁生物合成从而不产生尼古丁的烟草植株上生长得更好。[119]

除了基本防御,植物也具有诱导性防御,这种防御一般被视为能够降低防御代价的机制。尽管降低防御代价是诱导性防御的一个重要方面,另一个优势是诱导性防御可以使植物表现出不同的表现型,这能够降低植食性昆虫对植物防御的适应性[2,60]。例如,来自一个森林不同地点的蒜芥(*Alliaria petiolata*)在防御化合物的水平方面会有一些变化,而当这些来自不同地方的植物在温室里生长时,这些水平是相似的[25]。这些野外的变化不能反应在基因水平上。植物防御的表型变异会影响植食性昆虫并减缓植食性昆虫的适应性。另外,植食性昆虫在搜寻食植和选择产卵地的过程中,也会形成对某种特定植物物种的偏好[111]。

与昆虫适应其寄主植物的大量实验证据形成鲜明对比,植物适应植食性昆虫的实验证据似乎很少(第12章)。仅仅存在很少的例子,例如植食性动物的压力决定了十字花科植物的分布,这种植物主要生长在其他植物物种的树荫下,因此这种植物很可能适合生长在隐蔽的条件下[86]。

在拟南芥(*Arabidopsis thaliana*)中,通过对植食性动物和病原体的消除实验表明,这些生物体会选择浓度高的硫代葡萄糖苷和高密度的毛状体。然而,由于病原体和昆虫在实验中被消除,仍然不清楚每一个去除的有机体对选择压力的相对影响[91]。此外,众所周知,植物的特征可以通过植物育种家的人工选择[103]或人工选择进行修改[96]。过去人们认为,昆虫对植物不会有强大的选择性,因为昆虫不会对植物造成很大的伤害。不过,也有昆虫可能会摧毁当地的所有寄主,如朱砂蛾可以摧毁其寄主植物狗舌草[134]或烟草天蛾幼虫可以摧毁所有的寄主植物野生烟草[106]。植食性昆虫可以显著缩短其寄主植物自然条件下的寿命[85]。此外,植食性动物损害和对植物适应性产生的效果的关系不一定是线性的;较少的损坏可能会使植物产生较大的适应性。例如,低密度的瘿蜂(*Andricus quercus-calicis*)对夏栎(*Quercus robur*)植物造成的种子产量减少程度比不受损害的夏栎(*Quercus robur*)的种子产量减少程度轻[28]。因此,有充分的理由认为昆虫对植物的影响比通常假定的要大。最近进行的研究为此提供了实验支持。研究表明,昆虫对植物的生长和生存[49]有很强的影响,昆虫通过栖息地的专化性影响热带森林的多样性[89]。生长在营养贫瘠土壤中的植物和营养丰富土壤中的植物对昆虫的影响是不同的。营养贫瘠土壤中的植物可通过丹宁酸与蛋白质的高比例来进行自我防护,而营养丰富土壤中的植物自我防护较差,但增长速度较快[49]。这项调查不仅表明昆虫可以对植物施加强大的压力,而且在进行分析时,不同的因素,如生物和非生物压力,都应被考虑。有趣的是,这项研究为詹森在30年前提出的一个假说提供了实验支持,即植物的防护是植物在营养贫乏的土壤上专一进化的重要因素,而不是适应低营养条件的结果[72]。

最后,即使在这些情况下,昆虫对植物的健康只有轻微的影响,在进化中主要因素是对植物健康的影响是否在不同植物个体之间存在差异以及这种差异是否有遗传基础。

11.6 昆虫多样性的演变

昆虫是物种最丰富的类群。几种已被验证的假设解释了昆虫,尤其是植食性昆虫惊人的进化。

根据生态饱和假说(ecological saturation hypothesis),昆虫物种或较高的昆虫分类阶元占用的生态位的数量是大体不变的[92]。只有当另一个昆虫类群被竞争排除在外并灭绝以后,一个新的昆虫分类阶元才能建立。

扩大资源假说(expanding resource hypothesis)的支持者认为,植物为昆虫提供的资源是可以增加的[138],复杂的植物会庇护更多的植食性昆虫(见10.6)。这一假设还意味着,在一般情况下,植食性昆虫的多样性主要由植物决定。

正如第2章提到的,植食性昆虫占所有昆虫种类的45%左右。系统分析表明,植食性类群的特点比与其紧密关联的非植食性种类显示出更多的多样性[125]。据估计,在现有的类群中,植食性昆虫的数量至少增加50倍[97]。在13个姐妹群中有11个类群,它们的植食性种群的数量几乎是植食性种群的两倍[97]。白垩纪末期发生了昆虫的大灭绝。对美国北达科他州西南部的跨越白垩纪和古新纪的化石分析表明,在此之前种类多样、数量丰富的专食性动物在大灭绝事件后损失惨重且不易恢复,而广食性动物却可以迅速恢复,并且年代不是很久远的化石的数量丰富[82]。这个例子关注到与昆虫和植物相互作用的进化有关的几个重要方面:提出了植食性昆虫专食性生活方式的优缺点、选择压力和物种形成以及植食性昆虫的进化。

11.7 专食性寄生昆虫的进化

记录植食性昆虫中的专食性现象比广食性现象更为容易,但解释就要难很多。专食性往往被认为是从广食性中演化而来的。不过,也有一些例子不支持这种观点[125]。例如,蛱蝶族(Nymphalini)就没有产生越来越专化的倾向。祖先很可能专食荨麻科植物,而杂食性是一种衍生的现象[71]。正如我们在11.3和11.5中所看到的,在一个种群中的昆虫个体在寄主植物的选择和偏好方面存在差异,对新的寄主植物的适应也可能会出现偏好。各种关于植食性昆虫食性的演变因素已经被提出。主要因素是:(1)应对植物次生代谢产物;(2)避免竞争;(3)降低天敌造成的死亡率。

11.7.1 应对植物次生代谢产物

植食性昆虫会接触到许多的植物次生代谢产物,其中包括剧毒化合物,仅举几个例子如生物碱、硫代葡萄糖苷和呋喃豆香素类物质等(第4章)。植食性昆虫有避免或解毒这些植物化合物的机制。能够解毒一类植物化合物的昆虫通常不能解毒一个完全不同类的次级代谢产物。食用相同的植物类群的专食性昆虫进化出不同的排毒或排泄机制以避免同一种植物次生代谢的影响(第5章)[107,140]。解毒或排泄次生植物代谢物,常常被认为能避免这些植物化学物质对植食性造成的负面影响。然而最近的几项研究表明这是不正确的[3,119]。举例来说,十字花科特定的次生代谢产物能降低专食十字花科的植食性昆虫菜青虫的适应性[3]。因此,尽管专食性的植食性动物能够以某些富含次生代谢产物的植物为食,但这并不意味着植物毒素不会影响植食性动物的生理机能和适应性。然而,能解毒植物次生代谢物或避免接触这些代谢物,使得植食性动物取食的植物不适合成为其他植食性动物的食物。

有调查已经表明,一个单一的植物基因可以决定该植物是否可以供植食性动物食用。脂氧化酶基因是调控茉莉酸信号转导通路的一个关键基因,通过沉默脂氧化酶基因,野生烟草植物会受到叶蝉的侵袭,而自然条件下叶蝉不会取食野生烟草植物。

然而,野生烟草植物却容易受到如烟草天蛾幼虫的攻击[78]。因此,在那些能对植食性昆虫做出反应而诱导产生大量尼古丁和其他化学物质来进行防御的野生植株上,适应性昆虫如烟草天蛾幼虫,可以取食这些植株,但存在于同一环境中的它们潜在的竞争对手却不能取食。

很多情况下,抑制物并非毒素,可能在物种形成和专化方面发挥作用。例如,豌豆蚜(*Acyrthosiphon pisum*)的寄主专化主要决定因素是植物的接受行为而不是毒素的影响。对抑制剂的行为反应介导寄主植物的选择和选择性交配,因为蚜虫在它们的寄主植物上交配[21]。抑制剂在寄主植物的选择和在感官对化学品刺激的反应演变中的作用的广泛概述请参见7.9。因此,来自它们食用的植物中的抑制剂在这些以韧皮部为食的植食性动物专食化和物种形成方面发挥作用。

11.7.2 竞争

在一篇有关昆虫与植物相互作用进化的里程碑式的论文中,矣尔利希和雷文提到从植食性昆虫的种间竞争中逃离可能是导致寄主植物专化的一个因素[43]。很长一段时间内,竞争作为寄主植物专化中的一个重要问题被丢弃,因为植食性昆虫之间的竞争被认为是不重要的。然而,最近越来越多的证据表明这种观点并不正确(第10章)[29,33]。植食性昆虫以各种方式争夺食物,包括个体之间的竞争在同一时间攻击食用植物的不同组织[130],暂时分离的攻击(例如通过诱导植物反应),以及在空间上分开的攻击(例如通过营养流的转移或者诱导系统抵抗力)影响同一植株个体其他部分的植食性动物。

越来越多的有关植食性昆虫之间竞争的证据表明,竞争在植食性昆虫专食化的生活方式演变过程中的作用需要重新被评估。竞争的作用有可能比当前认为的更为重要。

11.7.3 降低由天敌引发的死亡率

植食性昆虫有许多敌人,如节肢动物的捕食者和寄生蜂或脊椎动物捕食者。各种各样的例子说明,植物是如何提高植食性昆虫的天敌效力的,如昆虫天敌、寄生蜂[37]甚至猛禽[88]。然而,植食性昆虫可能会利用其寄主植物的防御系统来满足自己的利益,也就是说它们用来防御敌害,如病原体或寄生蜂(第10章)。例如,植物次生代谢物可能会被植食性昆虫分离和利用来抵御其天敌。要做到这一点,它们必须适应植物的防御机制,以避免对自己产生消极影响。专食性的植食性昆虫可能能够更好地逃脱它们的天敌,如广食性捕食者[16]。例如,在一系列的温室实验中,黄蜂捕食者以更多的广食性毛毛虫为食,而不以专食性的毛毛虫为食[12]。专食性的植食性动物也需要较少的时间来接受寄主植物[15,42],拥有适应性的转化性口器能更有效地取食,这可能会提高逃避天敌的概率[14]。毕竟,取食寄主植物时,成为牺牲品的概率比休息时大很多(高达100倍)[13]。例如,在对毛虫(*Uresiphita reversalis*)进行观察期间,发现大多数毛虫在被花蟹杀害时正处于取食阶段,即使取食花费的时间比休息花费的时间要少得多[13]。

有观点认为,对专食性昆虫而言,天敌引发的死亡率较低是寄主植物专化性的结果,而不是造成的原因[75]。然而,这表明没有单一因素能造成植食性动物专食化。事实上迄今为止,没有一个单一的因素已被确定为寄主植物专一化的影响因素,而几个因素共同作用的证据很多[16,108,115]。不同因素的相对贡献可能取决于作用的系统。

11.7.4 系统发育关系

对寄主植物的选择可以进行实验研究,并取得了植食性昆虫对寄主植物适应的有趣结果(见11.5)。但是,寄主植物选择的进化史的演变过程却很难估计。可以通过一个支序图或进化树来进行解释,这说明不同的现存物种来自一个共同的祖先。到目前为止,产生了一些植食性昆虫的进化树[45,46],但有限的数据显示,相关的昆虫物种以相关的寄主植物为食[59](图11.11):在调查的大多数昆虫中,由于寄主植物的变化影响昆虫物种形成的发生率小于17%[46]。

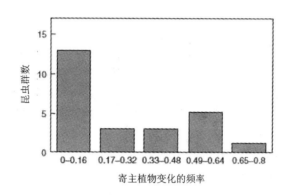

图 11.11 在 25 个植食性昆虫中,由于寄主发生变化导致昆虫新物种形成的概率分布(法雷尔等,1992)

另一种方法是评估昆虫和寄主植物的系统发育之间的对应关系,已经区分出这种关系的四种类型(图 11.12)[74]。在 A 类中,亲缘关系很近的昆虫物种杂食性或单食性地生活在亲缘关系较远的植物物种上(不一致的系统发育)。例如,欧洲巢蛾(Yponomeuta)仅以四个寄主植物群落为食,这四个群落属于三个不同的植物种类[95]。B 类代表亲缘关系很近的昆虫,杂食性(或者部分单食性)地生活在亲缘关系很近的寄主植物上(部分与进化树相同)的情况。典型的例子是菜青虫,暗脉菜粉蝶和日本纹白蝶,大致上喜欢相同的十字花科物种[24]。在 C 型中,亲缘关系亲密的昆虫物种杂食性地生活在亲缘关系很近的寄主植物上,并且与进化树完全符合。这样的一个例子体现在叶甲属金花虫科和其寄主植物关系中,见图 11.13。在 D 型中,杂食性物种以不同科的植物物种为食。B 型和 C 型表明了系统发育的保守性:在植食性昆虫中,物种的形成往往伴随着在亲缘关系密切的植物类群之间转化。关于昆虫与植物系统发育匹配程度上的调查非常稀少。对 14 种,至少部分系统发育是可行的组合分析表明,超过半数的组合的系统发育一般超过 0.5 在 0~1 之间(图 11.14)[46]。不过,这种匹配只有在 25% 的情况下是有意义的[45]。可以认为,在 B 型和 C 型的亲缘关系匹配(图 11.11)中,有远超过 0.5 的对应关系来自相类似的系统发育,而 A 和 D 预计将小于 0.5(图 11.14)。最近,天牛和马利筋属植物之间的关系是有关一致发育的新例子[45]。因此,虽然信息是有限的,接近一致的系统发育是罕见的,部分一致的系统发育仍然占一个相当大的比例。然而,研究系统发育匹配的数量是非常少的。此外,有趣的是,有关一致的系统发育最好的例子与甲虫有关,甲虫的成虫和幼虫都以寄主植物为食,因此它们与寄主植物有亲密依赖的关系,这与鳞翅目的情况相反。

11.8 植食性昆虫与寄主间的进化

1964 年,埃尔利希和雷文提出植食性昆虫和寄主植物可以通过相互进化或协同进化进行竞争[43]。在接下来的几十年里,学术界已经多次进行了关于协同进化的讨论[73,74,108,126]。协同进

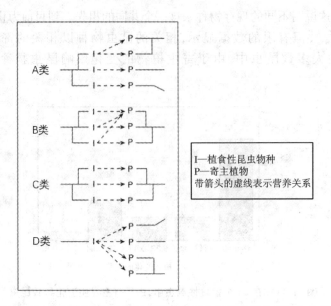

图 11.12　亲缘关系很近的昆虫物种或一个单一的昆虫物种与它们的寄主植物之间的进化类型

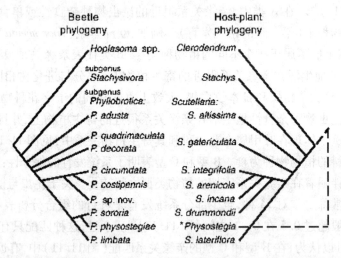

图 11.13　萤叶甲(*Phyllobrotica*)以及其近缘种与它们寄主植物的进化图

化被定义为,一个种群中一个个体特征的演变是对另一个种群中某一个体的特征做出的反应,接着第二个种群也对发生演变的第一个种群特征做出反应[73]。在近几十年中,协同进化理论已经产生了很多研究。1989年,英国生态协会表明:生态学家把动植物间的协同进化列入生态学中最重要的50个概念之一[125]。

目前已经发现了不同类型的协同进化:(a) 经典协同进化,即在两个物种中发生了相互进化[43];(b) 扩散协同进化,即认为协同进化发生在一个群落中,并非两物种间的相互作用[50];(c) 协同进化的地理镶嵌理论,考虑到种群内部发生的空间变化,以至有一种或多种物种间地理空间的不停变化而导致的协同进化[125]。在下面的段落中,我们首先提出协同进化这个概念,再来解释目前昆虫和植物的关系,接着提供一些证据证明协同进化。

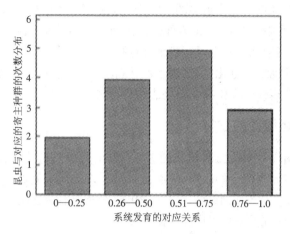

图 11.14 14 个独立的昆虫种群以及它们系统发育对应的寄主种群的次数分布

注：采用克莱斯一致指数表示，从 0（没有对应关系）到 1（完全对应）。

11.8.1 对协同进化理论的批判

已有的一种观点认为，植食性昆虫和它们寄主植物间的协同进化是不会发生的。主要的论据是，如植食性昆虫和它们寄主植物这两个合作伙伴是不平等的，是非对称的交互作用。在交互作用中，植物影响植食性昆虫的选择，但是昆虫不会影响植物的选择，因此不会发生相互选择作用[74,121]。因此，昆虫没有影响植物的进化。例如，在某种昆虫的寄主植物大量存在的地方，昆虫并不总是大量出现，这意味着昆虫的选择压力对同一物种的寄主植物中的个体而言是不一样的，且它们的力量太弱了，还不能导致抗性能力的提高。埃尔利希和雷文的经典协同进化理论确实描述了协同进化，特别是在物种水平上[43]。然而，进化机制没有发生在物种水平上，而发生在个体水平上。经典协同进化理论的批判者们是正确的，因为支持这个理论的证据非常稀少。尽管这可能表明，这个协同进化理论不是普遍适用的，但这并不意味着协同进化理论不能在植食性昆虫与它的寄主植物间发生作用（见 11.8.2）。

一些不赞同协同进化，支持植物与昆虫间不对称的相互作用的学者提出了顺序进化理论，也就是说，植食性昆虫的进化会遵循所食植物的进化过程，但是反过来却并不会发生[74]。这一理论是由杰米提出[74]，他认为昆虫利用了现有的植物所提供的大量有效生态位。对寄主植物的选择可能是主要由昆虫的化学感受系统控制的行为过程。因此，建立新的昆虫与植物寄生的关系，大部分源于昆虫的化学感应系统的进化。对新寄主植物的营养质量的适应是一个随之产生的过程[74]。然而，这种理论并不包括一些众所周知的现象，即通过相关学习昆虫可以修正它们的行为（第 8 章）。植食性昆虫会通过学习使用某些线索来避免有毒或营养物质不适合的寄主植物。因此，植物体内的毒素和营养物质通过反作用机制可以影响昆虫对寄主植物的选择行为。一种昆虫的行为是对由昆虫的化学感应系统传来的信息进行解释的结果。一系列的有关联的研究表明，由于之前不同的经验，昆虫可以用不同的方式应对相同的化学信息。因此，只要昆虫可以察觉到植物提供的线索，判断接受或者拒绝，会因为一些其他因素改变选择，比如其毒性或营养成分。昆虫开始食用一种新的植物而没有中毒，并能够识别该植物中的化学信息，随后可以用这些信息来找到合适的食物来源。

然而植物比植食性昆虫要面临更多的自然选择。病原微生物、植食性哺乳动物和非生物逆境,都将对植物产生影响。顺序选择的理论并没有表明什么是植物进化的主要选择因素。目前的假设是,由几个因素结合起来对植物施加选择。

植物能影响植食性昆虫的进化,反之则不然,这似乎与古生物的数据相冲突,古生物的数据表明被子植物的种群多样化发生在主要昆虫种群多样化之后[81]。

11.8.2 对协同进化理论的支持

虽然埃尔利希和雷文针对昆虫与植物的关系[43]提出的协同进化理论遭到了严厉地批评,而且似乎很少有人支持原先制定的版本,因而也激励了许多研究者在昆虫与植物关系的选择压力方面进行大量的研究。近几十年来,协同进化理论处于不断进化中,也包含了在亚种水平上的变异。下面,我们提供了一些支持性证据。

首先是在物种水平上的支持。例如,叶甲(*Phyllobrotica*)和其寄主植物匹配的进化树:亲缘关系很近的昆虫仅以亲缘关系很近的植物为食(图11.13)[44]。然而,必须认识到,目前对于这种完全匹配的进化树的报道很少[44,45],平行进化树研究的总量是非常有限的[46]。此外,对甲虫(*Blepharida*)和其橄榄科的寄主植物基于分子钟的谱系分析比较表明,在过去大约1亿多年中,植物的防御和昆虫的反防御取食特征在同步演变[7]。

其次是种群水平上的支持,对不同种群中的野生欧洲防风草(*Pastinaca sativa*)和对它专食性的昆虫防风草织蛾(*Depressaria pastinacella*)的分析表明,由于香豆素含量的变化,植物的表型会产生变化,植物种群中出现了多态性。植食性昆虫种群也出现了多态性研究发现,它们代谢不同类型的香豆素的能力不同,可以由此将它们进行分类。因此,植物和植食性昆虫都存在共同进化。此外,当考虑到植物和植食性昆虫表型匹配时,在3/4的种群中相匹配的频率非常大(图11.15)[8]。

通过对植物区系分布以及植物化学物质对植食性昆虫的毒性进行分析,这为埃尔利希和雷文所预言的"逃生和辐射"理论提供支持[43]:一个分布范围小的、较新演变的次生代谢产物比分布较广的、很久前演变的次生代谢产物具有更大的毒性[27]。通过大量分析植物化学文献为协同进化理论提供了强有力的支持。

调查进化动力时,重点应该是在个体和种群水平上的过程,因为选择是以个体为单位,而不是以物种为单位。这是汤普森提出的地理镶嵌式协同进化理论[125]。这一理论指出,协同进化理论比对个别种群的研究或在进化树中寻找特征的研究明显更具活力。毕竟,植食性物种及其寄主植物的分布没有明显的重叠,而且寄主植物可以用不同的方式逃离植食性昆虫,包括定植在还不太适合植食性昆虫的栖息地。能确定的是,植物和昆虫种群能形成原种群(即当地亚种群与适应性种群的组合)[31,62]。这会影响迁移、选择和对当地的适应。因此,在物种水平上开展的相互适应分析可能无法代表适应实际发生的水平。另外除了处理个体,而不是种群或物种,个体的表型是具有内在可塑性的。有相互作用的个体之间可能会调整其表型去回应它们各自的合作伙伴。这可以反映出个体对遇到的变异的反应的进化[1]。虽然表型可塑性如诱导的植物防御已经被认为能"简单反映出植物在压力条件下体内代谢的变化,导致产生一些昆虫不能忍受的植物化学成分的改变"[74]。

此外数学建模已经表明,协同进化整个周期内的竞争是有可能的。已建成的模型基于一些关键的假设,如植物的抵抗力的成本和效益,以及植食性动物在毒性或解毒能力上这些性状

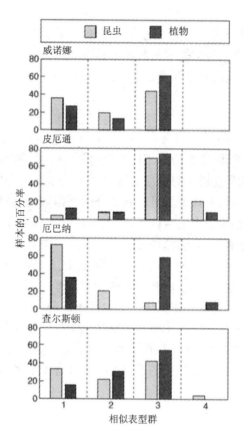

**图 11.15　四个种群中,昆虫防风草织蛾(*Depressaria pastinacella*)
和植物欧洲防风草(*Pastinaca sativa*)的表型匹配情况**

注:采集的植物标本与收集的蛹的位置临近。*代表植物与昆虫之间严重不匹配。

的水平变化的成本如何[10]。然而,建模和实验研究都没有得到协调,这也许是为什么这个模型没有实验数据支持的原因之一。

总之,支持经典协同进化的数据似乎很少。然而,这可能从协同进化在整体水平上被研究调查的频率远远高于自然选择下的基本元素——在个体水平这个事实中可以得到部分解释。

传统上,人们认为协同进化发生在一个双营养范围内,涉及直接的植物防御。第二种植物防御的主要形式——间接防御,是植物和植食性动物的天敌之间的共生[35,37,110]。互利共生一直被认为比对立的相互作用更容易进行协同进化演变[125]。不过,即使植物和它的天敌都从虫害诱导植物挥发物的间接防御中受益,那也不见得这样的挥发物从一开始就作为间接防御发展而来[35,110]。它们最初的功能很有可能是不同的,随后通过食肉节肢动物的选择塑造了诱导植物挥发物产生的特征。这种诱导融合可以特定地用于寄生植物的植食性昆虫,肉食性的昆虫也可以用这种融合来区分引诱它们的植食性昆虫[38]。植物的这种特定反应支持了"植物的防御受植食性昆虫选择的影响"这一理论。食肉节肢动物的反应在成功繁殖方面体现了选择性优势[56,132]。然而,在博弈理论框架中,据推测,防御特性由于周边竞争对手的不断变化而随之发生变化[110]。毕竟,防御不仅保护具有防御特性的有机体,也可能保护它们的邻居[36,110]。这可能导致防御表现的变化,也可以解释为什么有些植物在直接防御和间接防御方面几乎不用投资[110]。

总之,协同进化的理论引出了对影响植物—昆虫相互作用的基础进化过程的激烈争论。争论仍在继续批判地分析进入科学领域的新证据,并可能刺激进一步研究植物和植食性昆虫在化学武器、联盟和避难之间的竞争演变。

11.9 结论

植物是固着生长的有机体,需要应对不同的生物的、非生物的压力。它们能够广泛地生活在我们的星球上,这表明它们能够很好地应对这些不同的压力。在对植物进化至关重要的选择力的探索过程中产生了热烈的讨论。虽然有充分的证据表明是植食性昆虫适应植物,植物适应植食性昆虫的证据却很少。人们常认为,昆虫与植物的相互关系具有不对称选择压力的特点。然而,有充分的证据表明,我们低估了昆虫在寄主植物上的选择压力。特别是在个体水平和在微进化的时间过程中已经发现了相互选择压力,这方面的假说在指导实验性的研究方面非常重要。关于协同进化的争论和由此产生的相互选择的实验研究方法在接下来的10年将会继续发展,这是很值得期待的。协同进化的地理镶嵌理论(强调进化过程中的区域差异)有可能发挥重要作用。分子生物学技术将在识别区域变量过程方面发挥很大的作用。以昆虫与植物同步演化为基础的机制,包括一些在陆地生态系统中最富有的聚集物,将仍然是一个重要的研究主题。对昆虫与植物相互作用的进化的了解将有助于我们了解生物多样性的起源和整个生物学。

11.10 参考文献

1. Agrawal, A. A. (2001). Phenotypic plasticity in the interactions and evolution of species. *Science*, 294, 321–6.
2. Agrawal, A. A. and Karban, R. (1999). Why induced defenses may be favored over constitutive strategies in plants. In *The ecology and evolution of inducible defenses* (ed. R. Tollrian and C. D. Harvell), pp. 45–61. Princeton University Press, Princeton.
3. Agrawal, A. A. and Kurashige, N. S. (2003). A role for isothiocyanates in plant resistance against the specialist herbivore *Pieris rapae*. *Journal of Chemical Ecology*, 29, 1403–15.
4. Asaoka, K. (2000). Deficiency of gustatory sensitivity to some deterrent compounds in 'polyphagous' mutant strains of the silkworm, *Bombyx mori*. *Journal of Comparative Physiology A*, 186, 1011–18.
5. Avison, T. I. (1988). Cited in Ref. 117.
6. Baldwin, I. T. (1998). Jasmonate-induced responses are costly but benefit plants under attack in native populations. *Proceedings of the National Academy of Sciences of the USA*, 95, 8113–18.
7. Becerra, J. X. (2003). Synchronous coadaptation in an ancient case of herbivory. *Proceedings of the National Academy of Sciences of the USA*, 100, 12804–7.
8. Berenbaum, M. R. and Zangerl, A. R. (1998). Chemical phenotype matching between a plant

and its insect herbivore. *Proceedings of the National Academy of Sciences of the USA*, 95, 13743–8.

9. Berenbaum, M. R., Zangerl, A. R., and Nitao, J. K. (1986). Constraints on chemical coevolution: wild parsnips and the parsnib webworm. *Evolution*, 40, 1215–28.

10. Bergelson, J., Dwyer, G., and Emerson, J. J. (2001). Models and data on plant-enemy coevolution. *Annual Review of Genetics*, 35, 469–99.

11. Berlocher, S. H. and Feder, J. L. (2002). Sympatric speciation in phytophagous insects: moving beyond controversy? *Annual Review of Entomology*, 47, 773–815.

12. Bernays, E. A. (1989). Host range in phytophagous insects: the potential role of generalist predators. *Evolutionary Ecology*, 3, 299–311.

13. Bernays, E. A. (1997). Feeding by lepidopteran larvae is dangerous. *Ecological Entomology*, 22, 121–3. 14.

14. Bernays, E. A. (1998). Evolution of feeding behavior in insect herbivores—success seen as different ways to eat without being eaten. *BioScience*, 48, 35–44.

15. Bernays, E. A. (2001). Neural limitations in phytophagous insects: implications for diet breadth and evolution of host affiliation. *Annual Review of Entomology*, 46, 703–27.

16. Bernays, E. and Graham, M. (1988). On the evolution of host specificity in phytophagous arthropods. *Ecology*, 69, 886–92.

17. Boller, E. F., Russ, K., Vallo, V., and Bush, G. L. (1976). Incompatible races of european cherry fruit fly, *Rhagoletis cerasi* (Diptera-Tephritidae), their origin and potential use in biological control. *Entomologia Experimentalis et Applicata*, 20, 237–47.

18. Bush, G. L. (1969). Sympatric host race formation and speciation in frugivorous flies of the genus *Rhagoletis* (Diptera, Tephritidae). *Evolution*, 23, 237–51.

19. Bush, G. L. (1975). Sympatric speciation in phytophagous parasitic insects. In *Evolutionary strategies of parasitic insects and mites* (ed. P. W. Price), pp. 187–206. Plenum, New York.

20. Bush, G. L. (1987). Evolutionary behaviour genetics. In *Evolutionary genetics of invertebrate behavior* (ed. M. D. Huettel), pp. 1–5. Plenum, New York.

21. Caillaud, M. C. and Via, S. (2000). Specialized feeding behavior influences both ecological specialization and assortative mating in sympatric host races of pea aphids. *American Naturalist*, 156, 606–21.

22. Carson, H. L. (1987). Colonization and speciation. In *Colonization, succession and stability* (ed. A. J. Gray, M. J. Crawley, and P. J. Edwards), pp. 187–206. Blackwell Scientific, Oxford.

23. Chew, F. S. (1977). Coevolution of pierid butterflies and their cruciferous foodplants. 2. Distribution of eggs on potential foodplants. *Evolution*, 31, 568–79.

24. Chew, F. S. and Renwick, J. A. A. (1995). Host plant choice in Pieris butterflies. In *Chemical ecology of insects* (ed. R. T. Cardé and W. J. Bell), pp. 214–38. Chapman & Hall, New York.

25. Cipollini, D. (2002). Variation in the expression of chemical defenses in *Alliaria petiolata* (Brassicaceae) in the field and common garden. *American Journal of Botany*, 89, 1422–30.
26. Claridge, M. F. (1995). Species concepts and speciation in insect herbivores. Planthopper case-studies. *Bollettino di Zoologia*, 62, 53–8.
27. Cornell, H. V. and Hawkins, B. A. (2003). Herbivore responses to plant secondary compounds: a test of phytochemical coevolution theory. *American Naturalist*, 161, 507–22.
28. Crawley, M. J. (1985). Reduction of oak fecundity by low-density herbivore populations. *Nature*, 314, 163–4.
29. Damman, H. (1993). Patterns of interaction among herbivore species. In Caterpillars. Ecological and evolutionary constraints on foraging (ed. N. E. Stamp and T. M. Casey), pp. 132–69. Chapman & Hall, New York.
30. Darwin, C., (1859). *On the origin of species by means of natural selection, or the preservation of favoured races in the struggle for life*. Murray, London.
31. De Jong, P. W. and Nielsen, J. K. (2002). Host plant use of *Phyllotreta nemorum*: do co-adapted gene complexes play a role? *Entomologia Experimentalis et Applicata*, 104, 207–15.
32. Denholm, I., Devine, G. J., and Williamson, M. S. (2002). Evolutionary genetics. Insecticide resistance on the move. *Science*, 297, 2222–3.
33. Denno, R. F., McClure, M. S., and Ott, J. R. (1995). Interspecific interaction in phytophagous insects: competition reexamined and resurrected. *Annual Review of Entomology*, 40, 297–331.
34. Denno, R. F., Peterson, M. A., Gratton, C., Cheng, J. A., Langellotto, G. A., Huberty, A. F., et al. (2000). Feedinginduced changes in plant quality mediate interspecific competition between sap-feeding herbivores. *Ecology*, 34. 81, 1814–27.
35. Dicke, M. and Sabelis, M. W. (1989). Does it pay plants to advertize for bodyguards? Towards a costbenefit analysis of induced synomone production. In *Causes and consequences of variation in growth rate and productivity of higher plants* (ed. H. Lambers, et al.), pp. 341–358. SPB Publishing, The Hague.
36. Dicke, M. and Vet, L. E. M. (1999). Plant-carnivore interactions: evolutionary and ecological consequences for plant, herbivore and carnivore. In *Herbivores: between plants and predators* (ed. H. Olff, V. K. Brown, and R. H. Drent), pp. 483–520. Blackwell Science, Oxford.
37. Dicke, M. and Van Loon, J. J. A. (2000). Multitrophic effects of herbivore-induced plant volatiles in an evolutionary context. *Entomologia Experimentalis et Applicata*, 97, 237–49.
38. Dicke, M., Van Poecke, R. M. P., and De Boer, J. G. (2003). Inducible indirect defence of plants: from mechanisms to ecological functions. *Basic and Applied Ecology*, 4, 27–42.
39. Dicke, M., Van Loon, J. J. A., and De Jong, P. W. (2004). Ecogenomics benefits community ecology. *Science*, 305, 618–19.
40. Dobler, S., Mardulyn, P., Pasteels, J. M., and Rowell-Rahier, M. (1996). Host-plant switches and the evolution of chemical defense and life history in the leaf beetle genus *Oreina*. *Evolution*, 50, 2373–86.

41. Drukker, B., Bruin, J., and Sabelis, M. W. (2000). Anthocorid predators learn to associate herbivoreinduced plant volatiles with presence or absence of prey. *Physiological Entomology*, 25, 260-5.

42. Dukas, R. (2004). Causes and consequences of limited attention. *Brain Behavior and Evolution*, 63, 197-210.

43. Ehrlich, P. R. and Raven, P. H. (1964). Butterflies and plants: a study in coevolution. *Evolution*, 18, 586-608.

44. Farrell, B. and Mitter, C. (1990). Phylogenesis of insect/plant interactions: have *Phyllobrotica* leaf beetles (Chrysomelidae) and the lamiales diversified in parallel? *Evolution*, 44, 1389-403.

45. Farrell, B. D. and Mitter, C. (1998). The timing of insect/plant diversification: might *Tetraopes* (Coleoptera: Cerambycidae) and *Asclepias* (Asclepiadaceae) have co-evolved? *Biological Journal of the Linnean Society*, 63, 553-77.

46. Farrell, B. D., Mitter, C., and Futuyma, D. J. (1992). Diversification at the insect-plant interface. *BioScience*, 42, 34-42.

47. Feder, J. L., Chilcote, C. A., and Bush, G. L. (1990a). Regional, local and microgeographic allele frequency variation between apple and hawthorn populations of *Rhagoletis pomonella* in western Michigan. *Evolution*, 44, 595-608.

48. Feder, J. L., Chilcote, C. A., and Bush, G. L. (1990b). The geographic pattern of genetic differentiation between host associated populations of *Rhagoletis pomonella* (Diptera: Tephritidae) in the eastern United States and Canada. *Evolution*, 44, 570-94.

49. Fine, P. V. A., Mesones, I., and Coley, P. D. (2004). Herbivores promote habitat specialization by trees in Amazonian forests. *Science*, 305, 663-5.

50. Fox, L. R. (1988). Diffuse coevolution within complex communities. *Ecology*, 69, 906-7.

51. Fox, L. R. and Morrow, P. A. (1981). Specialization: species property or local phenomenon? *Science*, 211, 887-93.

52. Fraenkel, G. S. (1959). The raison d'être of secondary plant substances. *Science*, 129, 1466-70.

53. Frey, J. E. and Bush, G. L. (1990). *Rhagoletis* sibling species and host races differ in host odor recognition. *Entomologia Experimentalis et Applicata*, 57, 123-31.

54. Frey, J. E. and Bush, G. L. (1996). Impaired host odor perception in hybrids between the sibling species *Rhagoletis pomonella* and *R. mendax*. *Entomologia Experimentalis et Applicata*, 80, 163-5.

55. Frey, J. E., Bierbaum, T. J., and Bush, G. L. (1992). Differences among sibling species *Rhagoletis mendax* and *R. pomonella* (Diptera, Tephritidae) in their antennal sensitivity to host fruit compounds. *Journal of Chemical Ecology*, 18, 2011-24.

56. Fritzsche-Hoballah, M. E. and Turlings, T. C. J. (2001). Experimental evidence that plants under caterpillar attack may benefit from attracting parasitoids. *Evolutionary Ecology Research*, 3, 553-65.

57. Fry, J. D. (1989). Evolutionary adaption to host plants in a laboratory population of the phytophagous mite *Tetranychus urticae* Koch. *Oecologia*, 81, 559 – 65.
58. Funk, D. J. and Bernays, E. A. (2001). Geographic variation in host specificity reveals host range evolution in *Uroleucon ambrosiae* aphids. *Ecology*, 82, 726 – 39.
59. Futuyma, D. J. and Mitter, C. (1996). Insect-plant interactions: the evolution of component communities. *Philosophical Transactions of the Royal Society of London*, Series B, 351, 1361 – 6.
60. Gardner, S. N. and Agrawal, A. A. (2002). Induced plant defence and the evolution of counter-defences in erbivores. *Evolutionary Ecology Research*, 4, 1131 – 51.
61. Glawe, G. A., Zavala, J. A., Kessler, A., Van Dam, N. M., and Baldwin, I. T. (2003). Ecological costs and benefits correlated with trypsin protease inhibitor production in *Nicotiana attenuata*. *Ecology*, 84, 79 – 90.
62. Hanski, I. (1999). *Metapopulation ecology*. Oxford University Press, Oxford.
63. Hanson, F. E. (1976). Comparative studies on induction of food preferences in lepidopterous larvae. *Symposia Biologica Hungarica*, 16, 71 – 7.
64. Ho, T. H. (1965). The life history and control of the diamondback moth in Malaya. *Ministry of Agriculture Co-operation Bulletin*, 118, 1 – 26.
65. Hoballah, M. E., Kollner, T. G., Degenhardt, J., and Turlings, T. C. J. (2004). Costs of induced volatile production in maize. *Oikos*, 105, 168 – 80.
66. Hsiao, T. H. and Hsiao, C. (1985). Hybridization and cytoplasmic incompatibility among alfalfa weevil strains. *Entomologia Experimentalis et Applicata*, 37, 155 – 9.
67. Huettel, M. D. and Bush, G. L. (1972). The genetics of host selection and its bearing on sympatric speciation in *Procecidochares* (Diptera: Tephritidae). *Entomologia Experimentalis et Applicata*, 15, 465 – 80.
68. Jaenike, J. (1985). Genetic and environmental determinants of food preference in *Drosophila tripunctata*. *Evolution*, 39, 362 – 9.
69. Jaenike, J. (1987). Genetics of oviposition preference in *Drosophila tripunctata*. *Heredity*, 59, 363 – 9.
70. Janz, N. (1998). Sex-linked inheritance of host-plant specialization in a polyphagous butterfly. *Proceedings of the Royal Society of London. Series B, Biological Sciences*, 265, 1675 – 8.
71. Janz, N., Nyblom, K., and Nylin, S. (2001). Evolutionary dynamics of host-plant specialization: a case study of the tribe Nymphalini. *Evolution*, 55, 783 – 96.
72. Janzen, D. H. (1974). Tropical blackwater rivers, animals, and mast fruiting by the Dipterocarpaceae. *Biotropica*, 6, 69 – 103.
73. Janzen, D. H. (1980). When is it coevolution? *Evolution*, 34, 611 – 12.
74. Jermy, T. (1984). Evolution of insect host plant relationships. *American Naturalist*, 124, 609 – 30.
75. Jermy, T. (1988). Can predation lead to narrow food specialization in phytophagous insects? *Ecology*, 69, 902 – 4.
76. Karban, R. and Agrawal, A. A. (2002). Herbivore offense. *Annual Review of Ecology and*

Systematics, 33, 641-64.

77. Kennedy, G. G. and Barbour, J. D. (1992). Resistance variation in natural and managed systems. In *Plant resistance to herbivores and pathogens. Ecology, evolution and genetics* (ed. R. S. Fritz and E. L. Simms), pp. 13-41. University of Chicago Press, Chicago.

78. Kessler, A., Halitschke, R., and Baldwin, I. T. (2004). Silencing the jasmonate cascade: induced plant defenses and insect populations. *Science*, 305, 665-8.

79. Klun, J. A. and Maini, S. (1979). Genetic basis of an insect chemical communication system: the European corn borer (Lepidoptera, Pyralidae). *Environmental Entomology*, 8, 423-6.

80. Labandeira, C. C. (1998). Early history of arthropod and vascular plant associations. *Annual Review of Earth and Planetary Sciences*, 26, 329-77.

81. Labandeira, C. C. and Sepkoski, J. J. (1993). Insect diversity in the fossil record. *Science*, 261, 310-15.

82. Labandeira, C. C., Johnson, K. R., and Wilf, P. (2002). Impact of the terminal Cretaceous event on plant-insect associations. *Proceedings of the National Academy of Sciences of the USA*, 99, 2061-6.

83. Lawton, J. H. and McNeill, S. (1979) Between the devil and the deep blue sea: on the problem of being a herbivore. In *Population dynamics* (ed. R. M. Anderson, B. D. Turner, and L. R. Taylor), pp. 223-44. Blackwell, Oxford.

84. Leslie, J. F. and Dingle, H. (1983). A genetic basis of oviposition preference in the large milkweed bug, *Oncopeltus fasciatus. Entomologia Experimentalis et Applicata*, 34, 215-20.

85. Louda, S. M. and Potvin, M. A. (1995). Effect of inflorescence feeding insects on the demography and lifetime fitness of a native plant. *Ecology*, 76, 229-45.

86. Louda, S. M. and Rodman, J. E. (1996). Insect herbivory as a major factor in the shade distribution of a native crucifer (*Cardamine cordifolia* A. Gray, bittercress). *Journal of Ecology*, 84, 229-37.

87. Mägdefrau, K. (1959). *Vegetationsbilder der Vorzeit* (3rd edn). G. Fischer, Jena.

88. Mantyla, E., Klemola, T., and Haukioja, E. (2004). Attraction of willow warblers to sawfly-damaged mountain birches: novel function of inducible plant defences? *Ecology Letters*, 7, 915-18.

89. Marquis, R. J. (2004). Herbivores rule. *Science*, 305, 619-21.

90. Marshall, J. L. (2004). The *Allonemobius-Wolbachia* host-endosymbiont system: evidence for rapid speciation and against reproductive isolation driven by cytoplasmic incompatibility. *Evolution*, 58, 2409-25.

91. Mauricio, R. and Rausher, M. D. (1997). Experimental manipulation of putative selective agents provides evidence for the role of natural enemies in the evolution of plant defense. *Evolution*, 5, 1435-44.

92. May, R. M. (1981). *Theoretical ecology: principles and applications*. Blackwell, Oxford.

93. Mayr, E. (1954). Change of genetic environment and evolution. In *Evolution as a process*

(ed. J. Huxley, A. C. Hardy, and E. B. Ford), pp. 157 – 80. Allen & Unwin, London.

94. Mayr, E. (2004). 80 years of watching the evolutionary scenery. *Science*, 305, 46 – 7.

95. Menken, S. B. J., Herrebout, W. M., and Wiebes, J. T. (1992). Small ermine moths (*Yponomeuta*): their host relations and evolution. *Annual Review of Entomology*, 37, 41 – 88.

96. Michel, A., Arias, R. S., Scheffler, B. E., Duke, S. O., Netherland, M., and Dayan, F. E. (2004). Somatic mutation-mediated evolution of herbicide resistance in the nonindigenous invasive plant hydrilla (*Hydrilla verticillata*). *Molecular Ecology*, 13, 3229 – 37.

97. Mitter, C., Farrell, B., and Wiegmann, B. (1988). The phylogenetic study of adaptive zones—has phytophagy promoted insect diversification? *American Naturalist*, 132, 107 – 28.

98. Mopper, S. (1996). Adaptive genetic structure in phytophagous insect populations. *Trends in Ecology and Evolution*, 11, 235 – 8.

99. Morin, P. A., Luikart, G., and Wayne, R. K. (2004). SNPs in ecology, evolution and conservation. *Trends in Ecology and Evolution*, 19, 208 – 16.

100. Müller, F. P. (1985). Genetic and evolutionary aspects of host choice in phytophagous insects, especially 99. aphids. *Biologisches Zentralblatt*, 104, 225 – 37.

101. Ng, D. (1988). A novel level of interactions in plant-insect systems. *Nature*, 334, 611 – 13.

102. Nielsen, J. K. and De Jong, P. W. (2005). Temporal and host-related variation in frequencies of genes that enable *Phyllotreta nemorum* to utilize a novel host plant, *Barbarea vulgaris*. *Entomologia Experimentalis et Applicata*, 115, 265 – 70.

103. Panda, N. and Khush, G. S. (1995). *Host plant resistance to insects*. CABI, Wallingford.

104. Pappers, S. M., Van der Velde, G., Ouborg, N. J., and Van Groenendael, J. M. (2002). Genetically based polymorphisms in morphology and life history associated with putative host races of the water lily leaf beetle, *Galerucella nymphaeae*. *Evolution*, 56, 1610 – 21.

105. Pelozuelo, L., Malosse, C., Genestier, G., Guenego, H., and Frerot, B. (2004). Host-plant specialization in pheromone strains of the European corn borer *Ostrinia nubilalis* in France. *Journal of Chemical Ecology*, 30, 335 – 52.

106. Preston, C. A., Laue, G., and Baldwin, I. T. (2001). Methyl jasmonate is blowing in the wind, but can it act as a plant-plant airborne signal? *Biochemical Systematics and Ecology*, 29, 1007 – 23.

107. Ratzka, A., Vogel, H., Kliebenstein, D. J., Mitchell-Olds, T., and Kroymann, J. (2002). Disarming the mustard oil bomb. *Proceedings of the National Academy of Sciences of the USA*, 99, 11223 – 8.

108. Rausher, M. D. (1988). Is coevolution dead? *Ecology*, 69, 898 – 901.

109. Rodriguez, R. L., Sullivan, L. E., and Cocroft, R. B. (2004). Vibrational communication and reproductive isolation in the *Enchenopa binotata* species complex of treehoppers (Hemiptera: Membracidae). *Evolution*, 58, 571 – 8.

110. Sabelis, M. W., Van Baalen, M., Bakker, F. M., Bruin, J., Drukker, B., Egas, M., *et al.* (1999). The evolution of direct and indirect plant defence against herbivorous arthropods. In *Herbivores: between plants and predators* (ed. H. Olff, V. K. Brown, and R. H. Drent),

pp. 109-66. Blackwell, Oxford.

111. Scheirs, J. and De Bruyn, L. (2002). Integrating optimal foraging and optimal oviposition theory in plant-insect research. *Oikos*, 96, 187-91.

112. Scheirs, J., De Bruyn, L., and Verhagen, R. (2000). Optimization of adult performance determines host choice in a grass miner. *Proceedings of the Royal Society of London. Series B, Biological Sciences*, 267, 2065-9.

113. Schilthuizen, M. (2001). *Frogs, flies and dandelions: the making of species*. Oxford University Press. Oxford.

114. Schneider, J. C. and Roush, R. T. (1986). Genetic differences in oviposition preference between two populations of *Heliothis virescens*. In *Evolutionary genetics of invertebrate behavior* (ed. M. D. Huettel), pp. 163-71. Plenum, New York.

115. Schultz, J. C. (1988). Many factors influence the evolution of herbivore diets, but plant chemistry is central. *Ecology*, 69, 896-7.

116. Scriber, J. M., Giebink, B. L., and Snider, D. (1991). Reciprocal latitudinal clines in oviposition behavior of *Papilio glaucus* and *P. canadensis* across the Great Lakes hybrid zone: possible sex-linkage of oviposition preferences. *Oecologia*, 87, 360-8.

117. Sheck, A. L. and Gould, F. (1993). The genetic basis of host range in *Heliothis virescens*: larval survival and growth. *Entomologia Experimentalis et Applicata*, 69, 157-72.

118. Simms, E. L. and Rausher, M. D. (1989). The evolution of resistance to herbivory in *Ipomoea purpurea*. II. Natural selection by insects and costs of resistance. *Evolution*, 43, 573-85.

119. Steppuhn, A., Gase, K., Krock, B., Halitschke, R., and Baldwin, I. T. (2004). Nicotine's defensive function in nature. *PLoS Biology*, 2, e217.

120. Stride, G. O. and Straatman, R. (1962). The host plant relationship of an Australian swallowtail *Papilio aegeus*, and its significance in the evolution of host plant selection. *Proceedings of the Linnean Society of New South Wales*, 87, 69-78.

121. Strong, D. R., Lawton, J. H., and Southwood, T. R. E. (1984). *Insects on plants: community patterns and mechanisms*. Harvard University Press, Cambridge, MA.

122. Tabashnik, B. E. (1986). Evolution of host utilization in *Colias* butterflies. In *Evolutionary genetics of invertebrate behavior* (ed. M. D. Huettel), pp. 173-84. Plenum Press, New York.

123. Thompson, J. N. (1987). Symbiont-induced speciation. *Biological Journal of the Linnean Society*, 32, 385-93.

124. Thompson, J. N. (1988). Evolutionary genetics of oviposition preference in swallowtail butterflies. *Evolution*, 42, 1223-34.

125. Thompson, J. N. (1994). The *coevolutionary process*. University of Chicago Press, Chicago.

126. Thompson, J. N. (1999). Specific hypotheses on the geographic mosaic of coevolution. *American Naturalist*, 153, S1-S14.

127. Thompson, J. N. and Pellmyr, O. (1991). Evolution of oviposition behavior and host prefer-

ence in Lepidoptera. *Annual Review of Entomology*, 36, 65–89.

128. Thompson, J. N., Wehling, W., and Podolsky, R. (1990). Evolutionary genetics of host use in swallowtail butterflies. *Nature*, 344, 148–50.

129. Tooker, J. F., Koenig, W. A., and Hanks, L. M. (2002). Altered host plant volatiles are proxies for sex pheromones in the gall wasp *Antistrophus rufus*. *Proceedings of the National Academy of Sciences of the USA*, 99, 15486–91.

130. Van der Putten, W. H., Vet, L. E. M., Harvey, J. A., and Wäckers, F. L. (2001). Linking above-and below-ground multitrophic interactions of plants, herbivores, pathogens, and their antagonists. *Trends in Ecology and Evolution*, 16, 547–54.

131. Van Drongelen, W. and Van Loon, J. J. A. (1980). Inheritance of gustatory sensitivity in F1 progeny of crosses between *Yponomeuta cagnagellus* and *Yponomeuta malinellus* (Lepidoptera). *Entomologia Experimentalis et Applicata*, 28, 199–203.

132. Van Loon, J. J. A., De Boer, J. G., and Dicke, M. (2000). Parasitoid-plant mutualism: parasitoid attack of herbivore increases plant reproduction. *Entomologia Experimentalis et Applicata*, 97, 219–27.

133. Via, S. (1990). Ecological genetics and host adaptation in herbivorous insects: the experimental study of evolution in natural and agricultural systems. *Annual Review of Entomology*, 35, 421–46.

134. Vrieling, K., Smit, W., and Van der Meijden, E. (1991). Tritrophic interactions between aphids (*Aphis jacobaeae* Schrank), ant species, *Tyria jacobaeae* L. and *Senecio jacobaea* L. lead to maintenance of genetic variation in pyrrolizidine alkaloid concentration. *Oecologia*, 86, 177–82.

135. Waldvogel, M. and Gould, F. (1990). Variation in oviposition preference of *Heliothis virescens* in relation to macroevolutionary patterns of heliothine host range. *Evolution*, 44, 1326–37.

136. Wasserman, S. S. and Futuyma, D. J. (1981). Evolution of host plant utilization in laboratory populations of the southern cowpea weevil *Callosobruchus maculatus* F. (Coleoptera: Bruchidae). *Evolution*, 35, 605–17.

137. Weeks, A. R., Reynolds, K. T., Hoffmann, A. A., and Mann, H. (2002). *Wolbachia* dynamics and host effects: what has (and has not) been demonstrated? *Trends in Ecology and Evolution*, 17, 257–62.

138. Whittaker, R. H. (1977). Evolution of species diversity in land communities. *Evolutionary Biology*, 10, 1–67.

139. Williams, K. S. and Simon, C. (1995). The ecology, behavior, and evolution of periodical cicadas. *Annual Review of Entomology*, 40, 269–95.

140. Wittstock, U., Agerbirk, N., Stauber, E. J., Olsen, C. E., Hippler, M., Mitchell-Olds, T., et al. (2004). Successful herbivore attack due to metabolic diversion of a plant chemical defense. *Proceedings of the National Academy of Sciences of the USA*, 101, 4859–64.

141. Wood, T. K. (1993). Diversity in the New World Membracidae. *Annual Review of Entomology*, 38, 409–35.

第 12 章　昆虫与植物：卓越的互利共生关系

12.1　互利共生
12.2　花的专一性
　12.2.1　昆虫对花的识别
　12.2.2　昆虫对花的采集
12.3　传粉的动力
　12.3.1　距离
　12.3.2　可接近性
　12.3.3　温度
　12.3.4　评价食物源
　12.3.5　奖励策略
　12.3.6　花蜜信号的状态
12.4　对多花序花的授粉运动
12.5　竞争
12.6　进化
12.7　自然保护
12.8　经济
12.9　结论
12.10　参考文献

当施普伦格尔在德国的一所路德教会学校当校长的时候，他经常为自己的职务苦恼，于是他的医生建议他学习自然科学，并教他一些植物学知识以缓解苦恼。施普伦格尔随后对花的形态、花蜜的分泌等花的多种功能进行了深入的研究，并在 1793 年发表了一篇题为《植物繁殖器官——花之迷》的文章[116]。

在这本具有里程碑意义的书中，通过对 500 多种植物的观察发现，大多数的植物需要由昆虫授粉来获得种子，虽然其中大部分是被子植物，得出了"so scheint die Natur es nicht haben zu wollen, dasz irgend eine Blume durch ihren eigenen Staub befruchtet werden solle"的结论。这一观点与当时植物是自体受精并结果的观念相冲突。由于当时的植物学家们认为施普伦格尔是非专业的人士，因此他的工作成果不被认可且被遗忘，直到施普伦格尔去世 60 年后才被达尔文再次注意到[128]。鉴于施普伦格尔没有提出为什么自然不允许自体受精这个问题，达尔文对异体受精的生物学意义进行了深刻的思考。通过施普伦格尔的研究，达尔文认识到异花受精会增加变异的几率，从而为他的自然选择理论打下了基础。达尔文著的《物种起源》是一本具有深远意义的书，在书中达尔文表示他认可施普伦格尔的结论并且讲述了他另外一个实验可以证明昆虫在授粉中所起到的作用[22]。达尔文在出版了《物种起源》后提到，"可怜的老施普伦格尔的观点的价值不只局限于现在，而在他死后的许多年后已经被充分认识到了"。达尔文用谦虚的措辞肯定了施普伦格尔的观点。

图 12.1　施普伦格尔描述昆虫授粉的图书封面

与动物世界截然相反,绝大多数的被子植物是雌雄同体的(大约80%)。它们的花朵既有雄蕊又有雌蕊。全世界只有大约10%的植物雄性花和雌性花开在分开的植株上(雌雄异株)。雌雄同株植物的这种天然性能很容易发生自体受精并且因此近亲繁殖,但是正如达尔文和施普伦格发现的,植物是避免自体受精的。正因为在植物进化中这个十分重要的特征,各种各样的机制的发展就是为了避免自花传粉。它们分两种机制:第一种是在同一时间或同一空间把雌雄分开(例如雄蕊和子房的成熟时间不同),第二种是不能自花传粉,这种方式是基于植物本身可以区分自身的花粉粒和外来植物花粉粒的能力,并且只允许来自不同植株的花粉萌发花粉管和发育胚珠。

大约有2/3的有花植物是通过昆虫传粉的,这种"服务"不是"免费"的。作为传送花粉的

"回报",植物提供花蜜和花粉给传粉者作为食物。花粉和花蜜是昆虫的理想营养物质:花蜜大概包含50%的糖分而花粉含有15%~60%的蛋白质和一些其他的必要元素[96,106]。由于双方是否能够生存下来几乎不能或者根本不能没有彼此的相互作用,所以说这是一个非常好的互利共生的例子。两个搭档在利益上存在合作关系的例子是非常普遍的,但是在生物界被子植物和昆虫授粉者之间也许是最壮观、规模最大的互利共生的例子。

虽然昆虫和花之间形成了一个以互利共生为原则的典型例子,但是周围不同物种之间互利共生的程度,以及花和传粉者相互依赖的程度还是分布广泛的。在这些互利共生的物种中有些比较极端的合作者已经高度的特化并且影响到了彼此的生存问题。例如,无花果(Figs)只能被专门的榕小蜂(Fig wasps)传粉受精——这个特化现象已经达到了最大化,因为每一种类的无花果只能被自己特有的一种黄蜂传粉受精。雌黄蜂在花中受精产卵,幼虫在种子里长大,从蛹成长到成熟体都一直在植株的果实中,直到交配。雌黄蜂飞离而去在大约距离10千米或更远的无花果上产卵。这个移居的雌蜂就成为无花果传送花粉的交通工具。产卵之后,她会死在无花果里[14,85]。它们之间的相互依赖是毋庸置疑的。

另一个极端关系是两者之间相敌对的关系[123]。例如,蜂兰花(Ophrys orchids)可以被它们的传粉者看待成"性寄生虫",这样在没有花的情况下也一样可以做得很好。其他的昆虫和花之间的关系存在于这两个极端范围之中的某处。

12.1 互利共生

许多植物种类是通过有色彩斑斓的颜色和浓郁香气的花来吸引昆虫传粉,尽可能使种子的生产达到最优(表12.1)[10]。例如,百脉根(Lotus corniculatus)在没有传粉者的时候几乎不产生种子。不过只要有一个蜜蜂来访就会使每一朵花里产生若干个种子,但是要想达到最大化的传粉受精,需要来访者数量达到12~25只[84]。虎耳草(Saxifraga hirculus)的花也必须被来访许多次才能确保最佳的结实率。在大约200个传粉者来访之后,在此期间大约有350粒花粉粒被置于柱头上,在这些花中,平均每一朵花可以产生30颗种子[91]。因此,植物为了能最优化自己的结实率,必须被一次以上的昆虫来访。不同的来访者可能带来来自不同父本的花粉或者是不相匹配的花粉。农业和园艺作物的受精和结实的情况经常不是很理想,原因主要在于没有足够的天然传粉者。如果将蜜蜂群体移入作物产区,那么作物的产量一定会有显著的提高(图12.2)[30]。

表12.1 在沼泽生态系统中,驱逐了传粉昆虫后,对4种杜鹃科植物种子产量的影响(瑞德,1975)[100]

植物	种子产量(%)	
	无昆虫	有昆虫
野迷迭香(马醉木)	0.7	33.6
沼泽的月桂树狭叶花石楠	0	55.6
拉布拉多茶(加茶杜香)	1.0	96.2
大果越橘(蔓越莓)	4.0	55.7

花粉和花蜜构成了昆虫重要的食物资源。蜂类(包括蜜蜂和黄蜂)甚至会同时从花中得

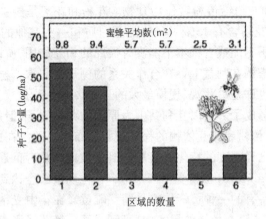

图 12.2 红三叶草(*Trifolium pratense*)的种子产量与蜜蜂聚集地距离的关系(布劳恩等, 1953)[9]

到这两种产物中的大部分营养,并且它们也有能力去收集有亲缘关系的大部分花粉。蜜蜂被密集的羽状毛所覆盖,当昆虫接触花的花药时,通过身上的小钩子有效地抓住并且带走花粉粒。花粉的黏性有利于黏附在昆虫身上,方便昆虫的授粉(图 12.3)。风媒传粉的花因缺乏油性的"花粉鞘"而没有黏性。除了利于花粉传播之外,昆虫的毛还起着隔离层的作用,可以在低温条件下保持自己的体温[47]。在飞行中蜜蜂会用自己的腿把收集到的花粉从腿上的绒毛梳理到它们位于两只后腿胫骨外侧的花粉篮里(图 12.4)[114]。

图 12.3 意蜂体表与头部的电镜照片

(A) 意蜂(*Apis mellifera*)体表的羽状毛;(B) 意蜂头部的毛上粘有的花粉粒(巴恩, 1985)

用这种手段,工蜂也能将 10~20 毫克的花粉运送到蜂巢。一个蜜蜂群落包括大约 10 000~50 000 只昆虫,每年消耗大约 20 千克的花粉和 60 千克的花蜜。花粉为蜜蜂提供生长和繁殖必需的蛋白质。为了养育一只蜜蜂需要消耗大约 125 毫克,这个重量相当于一只成年蜂的体重[113]。

在一些类似于韧皮汁液的花蜜中,10%~70% 的重量是糖分。其他的组分多为混合物,例如游离氨基酸、脂类、矿物质和次生化合物,这些物质占花蜜组分较少。对于传粉者,糖分是最

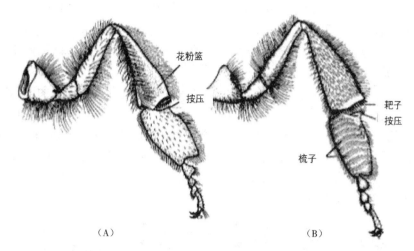

图 12.4 工蜂（A. mellifera）的后腿

（A）花粉篮（corbicula）的外表面，由一个空的凹面和边缘的硬毛组成，通过按压使得花粉进入到花粉筐内；（B）梳子（combs）和耙子（rakes）的内表面（斯诺德格拉斯，1956）[114]

有价值的回报，因为糖分能满足传粉者能量的需求，花蜜里的氨基酸对那些别无选择的传粉者也是很有吸引力的[2]。

令人奇怪的是，有些品种的植物会产生毒素或防护剂来阻碍采蜜者。现在已经有若干假设被提出并用来解释这些植物的反常现象，但是由于还没有获得确凿的试验性证据，所以还没有令人满意的解释[1]。

尽管植物和传粉者非常依靠彼此，但是任何互利共生体系中的参与者都对彼此有抵触，因为每一方都希望更多地利用对方。植物需要在它们的柱头上接受尽可能多的同种花粉粒并且传送自己的花粉粒到另外同种个体的柱头。理想的传粉媒介会在每一次采花时很好地接触花药和柱头，在相同的植物间迅速地移动并且为它们传粉，即使当时周围有很多其他种类的花也能很好地传粉。为了使它的传粉者采更多的花，植物就需要分泌足够的花蜜来吸引传粉者不断来访，但是很多的传粉者在每次旅途中仅仅需要访问一小部分的花，采集了足够的花蜜就飞回蜂巢。从植物的角度来讲，饥饿忙碌的昆虫是它们最理想的传粉者。从另一方面讲昆虫根据觅食理论，会尽量消耗最少的能量和时间去收集尽可能多的食物。这就意味着有丰富花蜜的花将会作为首选，并且在一次觅食期间去采不同品种的花也许会变得更有效。这种植物和传粉昆虫之间的利益冲突已经成为现在植物传粉机制演变的主要原因。

12.2　花的专一性

蜜蜂常常在觅食之旅中访问单一品种的花，而忽略许多其他合适的和有意义的替代植物物种的花朵。这种倾向被称为花的专一性。这种现象对于花的授粉生态学和进化学具有非常重要的作用，因此值得特别注意。

授粉的专一性不仅可以提高觅食效率，而且有利于维持植物物种之间的生殖隔离，从而维持物种差异。这种类型授粉的专业化涉及如何从远处识别花和获取花粉的技能，以及收集来自不同的体系结构花朵的花粉和花蜜。花的专一性优势主要归因于昆虫对同时处理不同类型花的有限的能力[132]。因此如果它的感觉系统和行为是暂时固定的一种特定方式,那么它们的觅食效率就会更有效。为了实现这个目标,昆虫似乎用一个感性的机制,类似于"搜索形象"找到鲜花,就像食草动物对植物的选择以及肉食动物对猎物的捕食模式。然而实际上对于这种假设的支持仍然很少[34,38]。

通过对花粉的检查发现花粉壁具有很显著的特征,花粉壁有纹路特别的、具刻点的、交错的、有刺的或者其他可辨别的特征(图12.5)[83]。花的专一性通常是作为觅食过程结束之后表示的个人百分比[21,68]。通过表12.2中所分析的蜜蜂对花粉的负荷可以看出许多蜂种对花都有很高的专一性。因为一个纯粹负荷的定义是严格的,所以它们甚至比数据更严格。"混合"花粉往往包含少量一个或多个其他植物物种的花粉。例如,表12.2中19%的蜜蜂的"混合负荷",每一个的纯度都是95%~99%[43]。

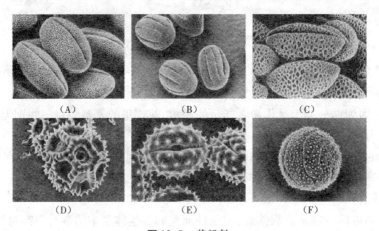

图12.5 花粉粒

(A)黑夏至草(*Ballota nigra*),三条纵沟;(B)拉拉滕(*Galium palustre*),多条纵沟;(C)蒲公英(*Leontodon saxatilis*);(D)狗舌草(*Senecio jacobaea*);(E)花蔺(*Butomus umbellatus*);(F)谷精草(*Eriocaulon aquaticum*)(普罗克等,1996)[96]

表12.2 蜜蜂总科的访花携粉数(Flower constancy 表示为从
某种植物中能携带花粉的个体的比率)

属	花粉量
蜜蜂(*Apis*)	81
切叶蜂(*Megachile*)	65
雄蜂(*Bombus*)	55
隧蜂(*Halictus*)	81
地花蜂(*Andrena*)	54
条蜂(*Anthophora*)	20

花的专一性的持续时间可能也有很大的差别。通常蜜蜂在一次觅食过程中保持一个花种。其他个体可能对于花的专一性较长,它们会对一种花保持几小时或者几天的专一性。蜜蜂群体中不同的个体会出现对不同品种花的专一性,不同的蜜蜂群体对花的专一性也不同。但是绝对的专一性会适得其反,使昆虫很难发现更有价值的资源。因此,独居蜂通过不断地检查不同的花卉品种来评估是否有更有价值的品种可供使用,因此它的专一性较蜜蜂低。在后

一种情况下,通过"侦察兵"的不断探索使得效率得以提高,"侦察兵"通过不断探索发现最好的食物源和信息,通过高度发达的信息系统传达给其他成员去食物源采集。

从花粉负荷分析和在野外的直接观察来看,并非所有的物种传粉媒介对一个花种是专一的。虽然社会性昆虫——蜜蜂要优于其他群体,但是连续访问同一种花的现象也在蜜蜂总科以外的昆虫中观察到了,比如一些蝴蝶、食蚜虻(Syrphidae)也出现较小程度的这种现象[41,74,134]。

12.2.1 昆虫对花的识别

花的高度专一性不仅仅需要快速学习并认识一个有价值的花卉品种,还需要迅速地识别出同种花所具有的特点以减少发生错误的机会。许多传粉者能迅速地识别出花的特征,如形状、颜色和气味。

在视觉上,色彩是一个非常重要的信号,另一方面可以进行定位,一方面可以与其他花相区别。对特征的记忆,例如颜色和食物源的奖励对觅食效率的提高很有帮助。蜜蜂在大约6次访问之后就可以可靠(即90%的准确性)地记住一种色彩(图12.6)。大菜粉蝶(*Pieris rapae*)可以做得更好:它们一次就可以识别一种色彩,准确率在85%以上[75]。与另一种凤蝶科的蝴蝶(*Battus philenor*)的实验表明蝴蝶的颜色学习能力还有很多种,其中两种能力让人印象深刻,即蝴蝶两种不同的刺激行为:觅食行为和产卵行为。这种双重的调节可以让雌性蝴蝶有效搜寻食物并在活动的区域产卵[136]。

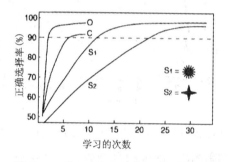

图12.6 学习次数与正确选择率的关系

蜜蜂对气味(O)、颜色(C)、形状(S)的学习曲线表明,对于气味的反应比颜色更快,精确性也更高;对于图形的认知比气味和颜色更难,对于复杂图形(S_1)比简单图形(S_2)的认知更快(门海泽尔等,1974[80];雪纳,1972[112])

花的彩色视觉系统的光谱选择性反射和传粉者会随着时间的推移相互作用并发展。有一个长期的误解是昆虫无法识别红色的花朵,其实红色是包含在它们的彩色世界里的[13]。

要利用蜜蜂最佳的稳定性和防止其传粉的"失误",一个植物应该有尽可能不同于同域异种鲜花花卉的特异性颜色。花的颜色显示不同被子植物之间强烈的差异,它们光的波长是昆虫感受到最敏感的波长范围[12]。对特定的栖息地进行整个植物系的检查表明,花的颜色更多样化,比昆虫和人类更离散。这些发现表明,由蜜蜂传粉媒介的感官系统随着花的颜色进化,或者说是一个共同进化的结果[79]。

形状和图案似乎比较难学,10~30次实验所达到的水平约相当于对色彩的精确记忆(图12.6)。有趣的是,更复杂的形状可以比简单的学习得快,因为先天具有对复杂图形的辨识力,即对比较大的长的图形的认识能力。吸引力的视觉线索可进一步增加花蜜指引的存在,再次增加花的视觉复杂性。

花的性状特别是对称的花,无论是径向还是双侧对称,蜜蜂(和其他昆虫)对它有着较高的辨识能力,它们能够发现并检测花的对称性。完美对称的花朵似乎比其他花朵能产生更多的花蜜,传粉者因为选择压力选择对称的花朵[36]。

对花的高度专一性要求不仅能够迅速学习有益的花卉品种,也要有能力在不够具体特点的基础上辨识同种花卉,以尽量减少错误的机会。花的专一性还体现在花可以产生出丰富的挥发性物质。相应的,传粉者的嗅觉系统同样要高度发达,并且它们的学习能力往往是令人非常深刻的。蜜蜂在接触花之后只需要一次就能记住花的气味,准确率达到93%~100%,而要记住一种色彩就需要4~6次的访问(图12.6)。有趣的是,单化合物的气味往往比复合气味更难记忆(图12.7)。混合的气味和颜色更容易让传粉者与食物奖励联系在一起。

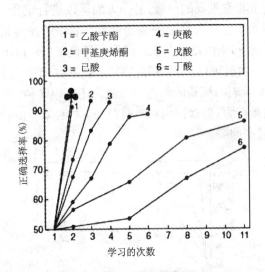

图12.7 蜜蜂对于花香和各种标准化合物气味的学习过程(克里斯顿,1971)[67]

花的气味印记是在特定理论配比浓度的复合挥发性化学物质[98]。通常情况下,这些化合物是单萜类和倍半萜类、含苯环的、苯丙酯类和脂肪酸衍生物组成的。尽管它们有极大的多样性,但是这些化合物生物合成的代谢途径是相对较少的(通常是重叠)[126]。

大多数花的香味中所含的挥发性物质至少有几个,但通常由很多成分组成。虽然混合往往由一个或几个主要成分占主导地位[24,56,64],但这并不一定意味着这些成分是昆虫最重要的信号。事实上,色谱分析表明,向日葵花香味中有多达144种化合物的成分。其中至少28种化合物是构成蜜蜂认为的"向日葵气味",这表明昆虫有一个高度适应的嗅觉系统[94]。一个带有气味的混合物引起了蜜蜂的行为反应,但是只有少数几种关键的化合物在起作用[69]。

鉴于大量的花的气味成分以及蜜蜂对气味的辨别能力,因此蜜蜂需要记住独特的气味来辨识不同的花。实验证据表明,蜜蜂的确可以学习并区别至少700种不同的花香。花卉的气味不仅仅是作为标识来帮助蜜蜂识别花卉的品种,气味还可以帮助蜜蜂更有效地采集花蜜。鲜花出于这种原因可能会在空中通过挥发性气味排成一定的路线,来引导传粉者采集花蜜(图12.8)。这种气味的线索包括触觉和味觉的刺激,可以帮助经验丰富的昆虫快速地找到花粉和蜜腺[24,25]。花卉还含有异味的化合物,对访问者做出有益或者有害的提示。某些情况下,花蜜含有与其他花卉不同的挥发物气味。可以想象,这种差异可以使传粉者进行快速检测得出结果从而减少花的处理时间[97]。

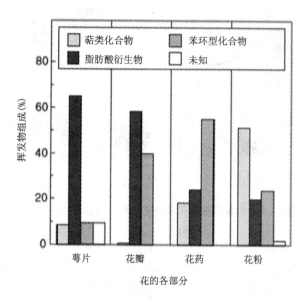

图 12.8 玫瑰(*Rosa rugose*)花的不同部位产生挥发物的不同组分(多布森等,1990)[25]

观察花粉产生挥发物的特点,整体的花香味往往是不同的,这被看作是花粉故意提供给昆虫传粉者奖励的证据。蜜蜂能区分不同植物的花粉气味支持这个假设。

花蜜的整体味道可能是由氨基酸的含量决定的。不同的花蜜可能给传粉昆虫相关的气味中的氨基酸含量存在巨大差异。成年蝴蝶的觅食选择仅仅是受到氮源的氨基酸浓度影响,蜜蜂对氨基酸变化的响应在花蜜模拟解决方案里。因此,不同的氨基酸出现在不同的花蜜上,很有可能是帮助传粉者寻找花,对花的专一性起到一定的作用[33,81]。

花除了颜色、性状和气味外还有花瓣的显微组织为特征。其表面覆盖着多种微观的单向皱纹、丘疹和盘可以作为授粉者的识别线索。蜜蜂可以通过这些纹理特征来区分花瓣的表面、背面以及花的前后。显然接触是另外一种能力,对那些比较复杂的花会通过接触来进行进一步的认识[60]。

12.2.2 昆虫对花的采集

当蜜蜂落在花朵上后会显现出它们与生俱来的探索响应,但是它们必须要学会如何更有效地从复杂的花中得到它们所需的东西。学习需要时间和精力的投入,一旦一种花被成功探索,就有利于蜜蜂继续探索取食。学会怎样熟练地从复杂的花种中采蜜不是一项简单的工作[70]。不同种的花的蜜腺通常隐藏在各自非常特定的位置,因此需要不同的处理技术,这就要求蜜蜂探索学习这些事,例如在花的哪里准确降落,以及怎样尽快地找到花的蜜腺的具体位置(图 12.9)。

伞形科平坦置顶的花序对蜜蜂来说找食物是相对容易的。当蜜蜂在四周快速移动时,它们只需要通过按压自己的身体至花的表面来从花中采集花粉。而对于相对复杂的花种,蜜蜂则需要更先进的技术,例如窄叶蛇头草(*Chelone glabra*)、玄参科,其花瓣必须被撬开、分离后蜜蜂才能得到花蜜;又比如乌头(乌头属,*Aconitum*)的花,蜜蜂必须从底部钻进去,沿着花药来找到被隐藏得很深的花蜜(图 12.10)。

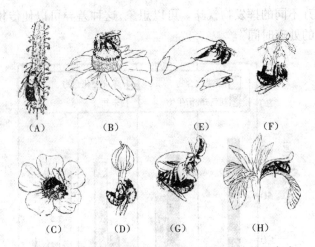

图 12.9 大黄蜂采集花蜜或花粉对不同的花的处理方法

(A)在梯牧草(*Phleum* sp.)的花序上行走,收集花粉;(B)从头状花序上收集花粉或花蜜;(C)抓住并震动蔷薇(*Rosa* sp.);(D)用腿和上颚固定住欧白英(*Solanum dulcamara*)的花;(E)进入蛇头草(*Chelone*)的花中;(F)从越橘(*Vaccinium*)的花中搜集花蜜;(G)在凤仙花属(*Impatiens* sp.)的花中抢夺花蜜;(H)从莺尾花中采蜜(海因里希,1976)[45]

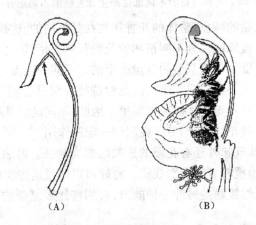

图 12.10 乌头花有两个垂直的花蜜瓣部分形状为筒状,在顶端有蜜腺

(A)阴毛乌头的花蜜瓣(箭头所指为筒入);(B)杂色乌头花,工蜂将舌深入到花蜜瓣中。蜜蜂在进入其底部后必须经过花药才能探索出两个花蜜瓣的顶端。乌头属植物只出现在大黄蜂出没之地(诺尔,1956[63],莱弗和普洛赖特,1988[71])

　　由于蜜蜂有限的神经,所以它从一种植物上学到的采花技术很可能会妨碍到它从其他植物上采集花蜜。熊蜂没有采集过形态复杂的花,比如水仙花(凤仙花属植物,*Impatiens biflora*)并且只有有限的次数接近花蜜,通常找不到丰富的花蜜。在它们充分掌握技能来提取花蜜之前需要60~100次的练习(图12.11)[46]。坚定不移地采一种花的策略一定会增加觅食效率,因为昆虫一旦知道一种特定的花的花蜜的具体位置以及怎样用最少的努力来得到,就会为它们节约能量和时间。

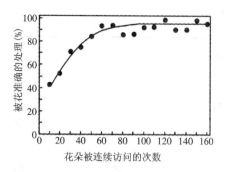

图 12.11　连续接触二花凤仙 (*Impatiens biflora*) 花 1~160 次后，大黄蜂觅食效率的提高 (海因里希，1979a)[46]

12.3　传粉的动力

互利共生的关系对双方来说都是有益的，这点也体现在两者越来越好的适应性上。根据最优化理论，生物体在每一个活动中通过平衡付出与收益来尝试最大化地增加它们的生存概率和生殖存活率。付出—收益分析对昆虫—花的应用已经证明在理解相互依靠的程度上相当有用。最佳觅食理论认为搜寻策略包括最大化提高食物的摄入量和净利用率，即每单位时间内的净热量增益。

积极授粉可以通过两种方法学习。第一种是基于模型的概念分析，促进觅食的行为可以通过生理机制来解释。第二种方法是通过进行田间实验测试模型得到能量平衡上的变化。影响昆虫觅食成本效益的因素是多方面的，但是至少包括食物源的距离、采集的难易、食物的数量和质量以及环境的温度等。这四个基本因素将在下面讨论。

12.3.1　距离

蜜蜂的觅食范围通常遍布蜂巢周围广阔的地域。一个关于落叶森林里一个群体的详细调查表明，最普遍的搜寻距离是 600~800 m，但是许多个体可以飞出蜂巢数公里。因为一个群体 95% 的觅食活动发生在半径 6 km 内，所以这个群体的食物源区域大概为 100 km²[113]。根据觅食环境和其他环境条件，搜寻范围有时会更大。同样，大黄蜂经常在距离蜂巢几百米甚至数公里的距离觅食[131]。

采集食物需要消耗很多的能量。一只重 500 mg 的大黄蜂在搜寻时每小时消耗多达 600 J 的能量，相当于 40 mg 葡萄糖的能量。到目前为止飞行所消耗的能量占了很大的比重。大型昆虫提取花蜜时通常会在花上盘旋飞行，这样的行为消耗的能量也是相当多的。为了节约能量，蜜蜂只会在有利可图的情况下才会飞行到距离较远的食物源去觅食。在一个实验点里工作人员通过放置高浓度的糖分使得较远的蜜蜂前来觅食。去远处觅食不仅需要考虑飞行所耗的能量，还需要考虑飞行所耗费的时间。因此距离与最小食物浓度的关系不是一个线性的关系而是一个指数的关系（图 12.12）。离蜂巢 3 km 的食物源要比蜂巢附近的食物源至少要多提供 3.4 倍的花蜜才能使它们更具吸引力。尽管会有很多能量的消耗，但是蜜蜂还是高效的飞行家[138]。蜜蜂飞行距离蜂巢 4.5 km 的食物源一来回所消耗的总能量仅为所采集花蜜的

10%的能量。

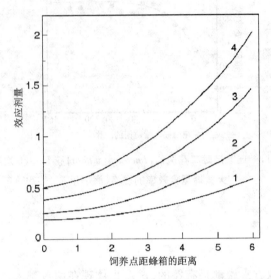

图12.12 饲养点距蜂箱的距离与效应剂量的关系

蜜蜂找到含有糖分的溶液,如果糖分的浓度高于一定阈值,才会招募蜂巢中的其他成员,阈值会随着距离的增加而增高,也会随着环境因素而改变,例如存在其他可选择的食物或者气候的因素(博世,1956)[8]

在自然界中,蜜蜂会遇到花中花蜜含量的变化。它们注意到这一点了吗?如果注意到的话,这会对它们有什么影响吗?在瑞尔和他的同事所做的一个简单实验中,昆虫在两种颜色不同的人造花上觅食。所有蓝颜色的花含有等量的糖水,而黄色的花含有的糖水量不同,但是其平均糖量与蓝色花相同。大黄蜂和胡蜂都更喜欢蓝色花,因为蓝色花的奖励差别不大。然而,当糖含量差异很大的花提高其平均值,昆虫就更喜欢去这些花上觅食。显然蜜蜂的觅食策略包含一定程度的风险规避,它们通过增加收益来补偿损失[101]。因此,距离是决定觅食的最基本原因,第二个则是可预测性,而额外的花蜜回报是第三个原因。

12.3.2 可接近性

花的形态对昆虫采集花粉或花蜜的时间有很大的影响。花如果开得浅的话只用较少的时间就能处理完,因为从花上的任何部位都能采到花蜜。像乌头(乌头属,*Aconitum*,图12.10)这样复杂的花,需要它们的访问者用更复杂的处理方法,因为确定花蜜的位置是很难的,因此也会花费很多时间。花通过提供丰富的花蜜来酬谢传粉者的投入和努力。

缺乏经验的大黄蜂学习采花所尝试的次数和需要的时间随着花的复杂程度而增长(见表格12.3)[70]。丰富的花蜜和授粉者对复杂的花形态的不断学习提升了花被采的概率。

表12.3 大黄蜂对于9种植物的处理时间与觅食成功率(莱弗,1994)[70]

花形与植物种类	植物种类处理时间(s)		觅食成功(%)
	1	55	1
碗形花			
罗布麻苍耳	5.5	0.4	100

花形与植物种类	植物种类处理时间(s)		觅食成功(%)
	1	55	1
茶叶花	14.1	0.3	100
开管花			
夏枯草	13.9	0.1	100
广布野豌豆	18.1	0.2	100
凤仙花鹗	20.4	1.7	70
封闭管花			
关龙胆	44.4	6.5	45
窄叶蛇头草	196.6	8.1	40
舟形乌头花			
川鄂乌头	134.7	13.6	35
欧乌头	153.5	3	29

12.3.3 温度

当温度接近 0 ℃ 或在北极低于 0 ℃ 时,依然能看到大黄蜂出来采集食物。蜜蜂在 10~16 ℃ 的季节环境里变得活跃。蜜蜂可以在低温下采蜜是因为它是恒温动物,并且在飞行时胸腔的最低温度可以保持在 30 ℃。它们通过飞行时新陈代谢产生的热来维持身体的高温,当不飞行时,则通过翅膀颤抖飞行肌来产热[47,130]。然而,在低温下寻找食物是非常消耗能量的。在 5 ℃ 下采集食物的大黄蜂比在 26 ℃ 时要多耗 2~3 倍的能量来维持它们 30 ℃ 的胸腔温度(图 12.13)。较高的温度是飞行肌正常运动的先决条件。为了维持高体温,蜜蜂拥有一个相当高活性的酶,即果糖 1,6 - 二磷酸酶,它能通过水解 ATP 来产生热量。一些大黄蜂体内这种酶的活性是蜜蜂的 40 倍,这就使它们可以在更低的温度下觅食。

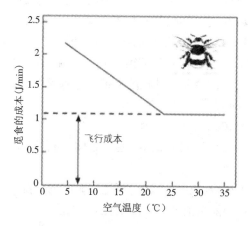

图 12.13 大黄蜂对于温度和访花消耗的热量,虚线之上的部分都用于体温调节(海因里希,1979b)[47]

在寒冷的天气条件下,额外的能量消耗会通过高能量的摄取来补偿。因此可以看到在寒冷的天气里,大黄蜂在可以生产丰富花蜜的杜鹃花上觅食而忽略樱桃(李属甜樱桃,*Prunus avium*)的花和狭叶山月桂(*Kalmia angustifolia*)的花,因为这些花在寒冷环境下产的花蜜太少以至于不能满足昆虫对能量的需求。

一旦知道了在不同温度下探索者的能量需求,还有单位时间内飞行所需消耗的能量,每朵花可预测的花蜜产量以及花与花之间的平均距离,那么预测就能做出来了,比如在不同的温度下哪些花会被采、哪些不会被采。海因里希在其著名的著作《大黄蜂经济学》一书中以实验为依据揭示了大黄蜂的能量策略,此能量策略把多种变化考虑在内,从而保证了确定的能量平衡得以维持[47]。

一大早开花的植物会被那些可以调节自己体温的大型昆虫造访。随着温度逐渐上升,小的传粉昆虫开始变得活跃。在大早晨,一种熊果(*Arctostaphylos otayensis*,杜鹃花科)的花每朵中含有 6.3 J 的糖分。当在 2 ℃ 甚至接近 0 ℃ 觅食时,大黄蜂需要每分钟从花中得到 3.4 J 的能量。中午每朵花只含有 1.3 J 的糖分,此时大型昆虫已经对花失去兴趣,花的主要觅食者是一些小型昆虫[48]。

为了使寒冷气候条件下的花能被更有效地采集,花蜜的产量必须提高或者此时的花间距应该比高温环境下更靠近,这样它们才能被快速且成功地采集。春天的花长成簇状可能是一种成功的传粉策略。事实上,长在北部的植物比同种低纬度的植物分泌更多的花蜜,在海拔梯度变化上也有相似的趋势表明花和它们的传粉者有很大的适应性。

尽管关于花蜜供应与传粉者能量需求的关系我们了解的还远远不够,但是有充分的证据表明它们之间的供需关系有着很好的协调性。

12.3.4 评价食物源

蜜蜂在找到食物源之后会返回蜂巢,通过著名的"蜜蜂舞"来向其他蜜蜂交流食物源的位置数量等细节。仔细分析它们形式化的行为和它们对于各种各样食物资源产生的微妙反应,使弗里希和他的同事可以确定蜜蜂是通过哪些因素的计算分析出食物的盈利能力[32]。他们发现在蜜蜂的舞蹈中包含了食物源的很多特征,比如说距离、花蜜的质量(即花蜜中糖的含量)、花蜜的黏度、采集花蜜的难易程度、风向、气味以及当时的天气状况等。除了直接的消耗(即飞行和觅食消耗的能量和时间),决定去采集的蜜蜂还可能会"估计"间接消耗,例如觅食时潜在的危险以及身体的磨损等[50]。它们对时间的衡量尚不明确,例如处理的时间,但是一些证据显示蜜蜂在测量觅食收益时,它们会在某种程度上把时间预算整合到能量预算中,因此蜜蜂可能通过对能量的预算来估计时间[129]。

因此,传粉动力学的研究包括以下 3 个因素:

(1)传粉者的觅食行为(觅食区域的距离、植物之间飞行距离、觅食决策、运动模式和速度)。

(2)觅食的种类和数量(花粉、花蜜的成分,热量和时间的分配)。

(3)传粉的消耗(传粉者的能量消耗和时间消耗)。

12.3.5 奖励策略

为了促进异性杂交,植物需要传粉者,例如昆虫和脊椎动物,但要为它们的服务提供奖励,花必须提供一定量的花蜜去吸引传粉者,但是它们必须限制这种奖励,使传粉者继续访问同属的其他植物[62]。每朵花或者每株植物分泌的花蜜会得到最充分的利用。而花蜜在一天中产生的时刻(图 12.14)、产蜜的季节,这些和这株植物打算"聘请"的传粉者有关。一只 100 mg 的大黄蜂落在一朵花上,每分钟需要消耗大约0.3 J的能量,而一只3 g的天蛾要汲取花蜜,盘旋在花上消耗的能量是大黄蜂的140倍[48]。

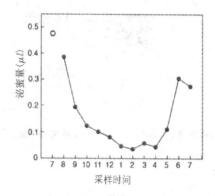

图 12.14　在一天中,没有昆虫访问时每小时分泌花蜜的量(吉伯等,1991)[35]

花蜜的产量明显受植物碳需求量上的流失的影响,正因为如此,很难去量化生产成本在生长或者繁殖方面的消耗。我们只能说,目前分泌花蜜所消耗的能量在不同物种植物之间的差异是很大的[23]。无论能量投入是什么水平,似乎可以合理地认为,植物尽可能地减少花蜜流动过程中损失的能量,并直接将能量用于生产种子。这种现象导致人们产生了这样的推测:有些植物个体可能会欺骗它们的同种个体,为了储存能量,只分泌一点花蜜或者根本不产生花蜜。一些适应一类花的蜜蜂,将会继续取食这些花,即使没有奖励。这种欺骗是自动拟态的表现[20]。蜜蜂之间的互动,产生花蜜的植物和以自动拟态欺骗的植物之间的"博弈"可以理解为很正常的现象,正如一项研究已经证明的,一株个体蜡花(紫草科)花蜜的流动,这些花中大约有25%可以产生大量的花蜜,但其余的花朵只分泌少量的花蜜。可以看出,高分泌腺和低分泌腺的比率与花显示的价值、产生的花蜜量、花之间的区别和传粉者的处理时间相适应[35]。

其他关于植物花蜜产量的不同也显示了植物可以采用不同的策略,欺骗的比率可以随时改变。一项研究表明:28 种植物中有 24 种属于 23 个属,其中 16 种产花蜜少,并且在这些种类中,一株植物上产花蜜少的花平均占 24%。这种现象说明,一些产花蜜少的花存得多,会有利于植物的健康。产蜜少的花不需要花费大量能量去产蜜,传粉者又无法区分产蜜多和产蜜少的花,这种事实就支持了一种假设,植物可达到一个稳定的比例使传粉者无法分辨花之间的区别[121]。

在原始被子植物的花蜜中发现了少量氨基酸。在其他植物类群中含有大量氨基酸,它们在草本植物中的浓度是 0.4～4.7 毫克。发现这种传粉者与花蜜上包含的氨基酸的关联使人们推测出这种关系的存在证明了传粉者没有适应氮资源,例如蝴蝶和蛾。因此管状的花一般被鳞翅类的昆虫传粉比以苍蝇为传粉的花的概率高[3]。在蝴蝶身上的实验显示,雌性菜粉蝶比照只含糖的花蜜更倾向于包含氨基酸的花蜜[2]。

一些植物的花有两种雄蕊:一种是正常产生花粉的雄蕊,另一种是"奖励雄蕊"(图12.15)。不育的雄蕊通常更显而易见,它们有着能吸引传粉者的鲜艳颜色,并且产生具有高营养价值的不育花粉。不育雄蕊可明显模拟可育雄蕊,通过欺骗的方式去吸引传粉者。当不育雄蕊演变时,这些昆虫自动地将它们当作可育雄蕊。我们可以假设认为,不育雄蕊的发展就植物的产量是有益的,但这还未经证明。

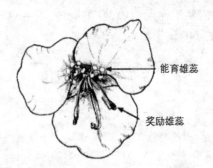

图 12.15　鸭跖草蓇葖(*Commelina tuberosa*)有两种雄蕊(汉斯，1990)[49]

12.3.6　花蜜信号的状态

蜜蜂要想有效提高觅食效率，一方面是要避免不产蜜的花朵，而不是随意地在花丛中飞行；另一方面，需要辨识出有充足食物资源的花。在高等昆虫学中能看到许多行为，比如说空气中残留的"足迹"气味以及其他访问者留下的挥发性信息素。收集花蜜的蜜蜂和熊蜂通过离开时的气味来标记采集过的花，这种气味持续的时间短。在这种气味的作用下，可以尽量避免采集这些花，从而提高觅食效率[110,118]。

熊蜂的觅食是更精确的，它们可以根据自己的能力调整觅食策略，比如鉴别花的种类和避免已经采访过的花等。研究显示，在抑制时期不同种植物之间似乎与花蜜的分泌率存在反比。例如，有些熊蜂属的熊蜂对聚合草属的豆瓣菜在3～10分钟内即显示出抵抗性，这种豆瓣菜有比较高的花蜜分泌率，然而对花蜜分泌率低的草木犀属的豆瓣菜和莲花根的抵抗性却能持续至少2～24小时[119]。

蜜蜂和熊蜂都能发出一种信息，表明这里的食物资源是值得采集的。这种大黄蜂分泌的信息素是由烷类和烯烃类构成的复合化合物[111]。通过识别自己或者其他同伴留下的信息素，可以使它们更有效、更快速地找到资源最丰富的花朵。同样的无刺蜂(Meliponini)觅食时常常对已经被采食过的花朵用气味进行标记[89]。显然，利用这种信号的标记，蜜蜂可以通过减少访问已被采集过的花和减少对花的搜寻时间来提高它们的觅食效率。

在植物方面，我们也可以获得花的发育情况。如果植物显示出它们曾经被传粉者访问过的特性，那么它们的传粉效率就会增加。许多植物正是通过改变花的颜色、气味以及外形结构来准确做出警示[117]。因此，莲花金雀花(Papilionaceae)的橙色花朵在授粉后会变成黄色，而一些其他种类的花在授粉后会因紫外线的反射发生改变[55]。同样的，羽扇豆(Papilionaceae)在授粉后也会发生颜色的改变。该物种的龙骨花瓣在授粉后开始变为粉红色，之后慢慢变化为带有白色横幅点的紫色。这种颜色的变化可能是由花粉管的生长导致的[90]。

角堇菜花授粉之后会发生由白色变化到紫色的壮观现象。这种变化是由三个花青素基因的表达使得花青素不断产生而导致的。想必是与授粉有关的激素，如乙烯和赤霉素促使这些基因激活随后控制合成了花色素[27]。

花的颜色也随着时间而改变。花的颜色至少有456种，在78种不同的被子植物的花上发生变化，当花朵盛开或者凋零的时候会产生颜色的变化[135]。性生存能力和花蜜的分泌在花变化之后会变低，因为它们缺乏花粉和出现不容性。那么为什么植物在花失去生殖能力后还要维持它呢？这很可能是要保留旧的花用以吸引远处的传粉者。在到达近处之后，蜜蜂通过学习可以很容易地分辨出不同阶段花的颜色,以避免已经授粉了的花[133,134]。因此,因为授粉和老化而改变颜色的花仍然在为植物担当吸引传粉者的任务,直到花朵凋谢。

对授粉之后花产生的香味变化的研究非常少[117],但是一些文献报道涉及了一些家族的变化情况[107,124]。例如,兰属(*Catasetum maculatum*)的花可以在授粉后几分钟就完全停止气味的产生,但是其他物种的花却需要几小时(如野生烟草,*Nicotiana attenuata*)甚至几天(如细距舌唇兰,*Platanthera bifolia*)的时间。通过这种方法植物可以节约资源并且可以为传粉者更好的指出还没有授粉的花[98]。

12.4 对多花序花的授粉运动

为了促进昆虫对花卉的认识和传粉,许多花的花序都聚集在一起,从而比单一的花朵更容易被采集。垂直生长的花序,如毛地黄(图12.16)、乌头、柳兰和羽扇豆具有特殊的益处,这种空间安排可以使雌雄同株(雌雄同体)的植物更好地进行授粉。

图12.16 毛地黄(*Digitalis lutea*)的直立花絮

蜜蜂和苍蝇等昆虫在垂直花序上觅食时往往遵循一定的觅食路线,通常是从底部开始往上飞行。较低的鲜花比上层的鲜花提供更多的花蜜,和最佳觅食理论一致,昆虫发现上部的花是最大的蜜源而花较小,下部花多而且含蜜数量是可以被接受的。此外,高处的糖浓度往往比下部的高一些。昆虫授粉运动的方向是向上的,这对植物也是有好处的,因为这些品种的花是雄蕊先熟的,也就是花药比柱头先成熟几天。每天都会有一个新的花在花序的顶端开放,取代底部已经衰老的花。下部的花在功能上是雌性,有成熟的柱头;上部的花在功能上是雄性,有成熟的花粉但是柱头还未成熟。从底部开始访问的方式,只对花序上部的花的花粉进行觅食的行为,很明显是促进了花的异花授粉以及减少了花的自花授粉的机会。植物的开花策略似乎很好地适应了昆虫的传粉行为[47,77]。

12.5 竞争

昆虫与花的关系涉及一些影响双方生存的基本条件：食物（昆虫方面）和生殖（植物方面）。因此，两者可能发生对可用资源的竞争。植物要想成功繁殖依赖于昆虫的授粉活动，因此会有花粉的投入；而昆虫在选择压力下，会比同类物种更有效地利用食物资源。

植物可以因为争夺授粉者产生更多的花。同时这就会增加同株授粉（同一植株之间的授粉）的风险，因此会限制花粉的产生。有几种解决这种两难境地的方法[62,115]。植物物种可以避免竞争，例如利用不同物种授粉器官的不同，花卉形态上的差别和开花时间上的差异。适应不同的传粉者，例如在清晨提供高回报利益以吸引大黄蜂，生长较长的花冠管使只有长舌的昆虫采到蜜腺[99]，这些无疑是一个非常有效的解决方案。开花时间由遗传控制，绽放的时间会避免与其他的种类发生冲突。显然这种资源的分割更有利于植物和传粉者。时间策略已经在一些简单的植物群落中被发现。因此，在落基山草地群落上的不同植物物种的开花时间是不同的，从而减少了对熊蜂授粉的竞争[95]。同样，大多数由昆虫授粉的沼泽植物会在不同的时间盛开以减少对大黄蜂的竞争。更详细的分析显示，物种对大黄蜂的依赖很大程度上显示在盛开时期，而在种之间的分离程度较少依赖于大黄蜂[47]。

在早春只有很少的植物开花，但是会有大量的传粉者竞争食物。因此，除了种子有较长的生长时间的优势外，提前开花的植物可能更有利于提高传粉的成功率。相反的，植物同样存在着被冻死的危险。

在春末夏初是植物大量开花的季节。这很好地体现了花蜜中的糖浓度以及蜜蜂舞蹈的季节性变化。在初夏，只有花蜜中的高蔗糖浓度才会引起传粉者的注意。在盛夏，天然的食物源不再丰富而且会有其他的昆虫参与竞争。这时食物传粉者之间的竞争是最激烈的。因此，吸引传粉者的蔗糖浓度是春末时的 1/16 倍（图 12.17）。显然蜜蜂在这种情况下调整了它们的觅食策略，采集质量比较一般的食物。

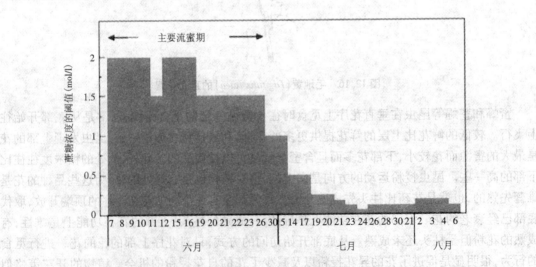

图 12.17　在夏季早期和中期，能够引起蜜蜂抢夺行为的糖浓度阈值（林道尔，1948）[76]

在春季，一些植物物种如杨柳（柳属）和蒲公英（*Taraxacum officinalis*），产生的花蜜和花粉的供应几乎是无限的。正如许多昆虫有丰富的食物来源,这已经表明这些植物使用了"倾销"的策略用以吸引更多的昆虫。作为竞争的结果,植物的结实率下降,食物的供给得到增强。因此,蒲公英能在苹果园内吸引传粉者远离苹果树。但是比较讽刺的是,蒲公英在很大程度上是单性生殖的。尽管可以产生如此丰富的花蜜,但是还必须有另一种促进传粉的方法。但是缺乏确凿的证据。

一个关于竞争的有趣例子,涉及两种大黄蜂在两个不同种花卉上的觅食。每种蜜蜂都对一种类型的花有明显的偏好。然而当大部分的蜂种被迁移之后,其他的蜂种除了访问自己偏爱的花,它们开始更频繁地访问那些空余的花的品种[51]。

总之,强有力的证据表明,传粉者和植物之间的竞争是资源分区(图12.18)[99]和性状替换等因素共同进化的结果。植物可以减少授粉媒介竞争的时间,在此期间是受精的最佳时期,不断变化的结构和生理特点减少了传粉者的范围,但确保它们有足够的食物资源[59,87]。

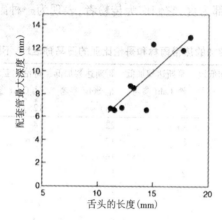

图12.18　9种大黄蜂喙(舌)的长度与花冠管的最大深度(兰塔和伦德伯格,1980)[99]

12.6　进化

被子植物是迄今为止最大的陆地植物群落。它们的特点是在花的大小、形状和颜色上有着眼花缭乱的多样性。林奈在他的《自然体系》(1735)一书中基于植物的性器官对开花植物进行了分类。

被子植物的进化成功,无疑是因为其生殖器官适应昆虫授粉的结果。根据古植物学和分类学,传粉机制是一种古老的机制。首先被子植物和一些其他的分支已经是虫媒授粉,虽然在早期可能是风媒和虫媒共存的[15,18]。

被子植物的快速辐射早在白垩纪(90万—130万年前)就已经结束了,它替代了古老的蕨类,楔叶类植物群落,在晚白垩纪和早第三纪多样化的花粉和花蜜替代了昆虫对裸子植物的采集(图12.19)。许多复杂的传粉系统具有被子植物起源时的特点[16]。

与风媒传粉相比,昆虫传粉的方式更为多样。然而,也存在不足之处,并且昆虫传粉是由风媒传粉(风媒传粉作用)反复进化而来的[15,19];在许多由昆虫授粉的植物中就有一少部分变成了风媒传粉,比如说白蜡(木犀科)、唐松草(毛茛科)还有安布罗西亚(菊科)。与风媒传粉相比较,昆虫传粉不需要浪费大量的花粉,并且对小花粉粒起作用,而不用像风媒传粉那样随风散布大量的有效花粉。昆虫传粉也可以在低植物密度和广泛分布在多种植物之间的个体进行精确的花粉传播。一些植物群落,比如在潮湿的热带雨林中几乎不存在风媒传粉,因为这里缺乏足够的空气流(表12.4)。

相反地,风媒传粉的植物类型分布在有风的地区,比如在北纬以及物种多样性较低的地区。因此风媒植物的比例随着纬度和海拔的升高而稳定提高,北极乔木中这一比例达到80%~100%(图12.20)[102]。尽管如此,风媒传粉和虫媒传粉是否涉及更高的能源成本仍需要进一步确定。

如今膜翅类在花粉传播中扮演着中心地位,但是在以前是其他类群占优势。在开始时,一系列非专化昆虫作为花粉携带者为传粉服务,其中以甲虫最为显著。鞘翅类和双翅目仍然是主要的被子植物传粉者。侏罗纪后期这些昆虫的种族就建立了,现如今可以追溯到一些为木兰、洋蜡梅这样的原始木本植物传粉的甲虫传粉者群体的进化起源[122]。在一些完整的种乃至科中的甲虫仍专门以花的局部为食,它们作为传粉者,表现为一种副作用。

表12.4 在哥斯达黎加的热带雨林和哥伦比亚的亚马孙区域,不同传粉系统的比率

花粉传播	哥斯达黎加低地 Can + Sub(参考.58)	哥斯达黎加低地 Can(参考.6)	哥斯达黎加低地 Sub(参考.66)	哥伦比亚水淹区 Can(参考.125)	哥伦比亚高地 Can(参考.125)	平均
中蜂或大蜂	29	44	22	24	17	27
小蜂	19	8	17	17	35	19
甲虫	9	—	16	5	4	7
蝴蝶	6	2	5	15	12	8
飞蛾	12	14	7	5	4	8
黄蜂	2	4	2	7	3	4
小昆虫	19	23	7	14	17	16
蝙蝠	4	4	4	—	4	4
蜂鸟	6	2	18	7	5	8
风	2	—	3	3	0	2
总数	110% ($n=145$)	100% ($n=52$)	100% ($n=220$)	100% ($n=74$)	100% ($n=68$)	

被子植物花的多样性演变通常被解释为其与传粉昆虫协同进化的结果[26]。一些昆虫群体已经比其他更具影响力。然而,多数学者用"协同进化"这一词语表示在有相互作用的种群之间互利进化的变化关系[123]。由于认识到这种释义,虫媒植物和传粉昆虫的进化关系在绝大多数情况下是不均匀的:花粉传播者对包括大多数辐射对称植物类群在内的开花植物进化起决定性作用,而植物几乎不对传粉昆虫的进化(如物种形成)产生影响。当意识到授粉昆虫可以聚合为具有选择压力的"功能组"(例如长舌的苍蝇或小工蜂)时,这种进化关系的不均衡性开始轻微地倾斜,由此刺激个别花种的显著特点的发展(例如细长的花冠管或花粉的样子)。传粉者中有效群体运用选择压力的方式也许是形成花种多样化的重要因素[29]。出于同样的目的,协同进化已引起那些群体紧密联系,就像无花果和榕小蜂以及丝兰和丝兰蛾之间的关系那样[123]。

其他昆虫种群在它们作用于开花植物的选择压力上存在很大差异。蜜蜂对花的进化起着非常重要的作用。因为蜜蜂在幼体以及成体阶段的生存都依赖于花的资源,它们又对花有很

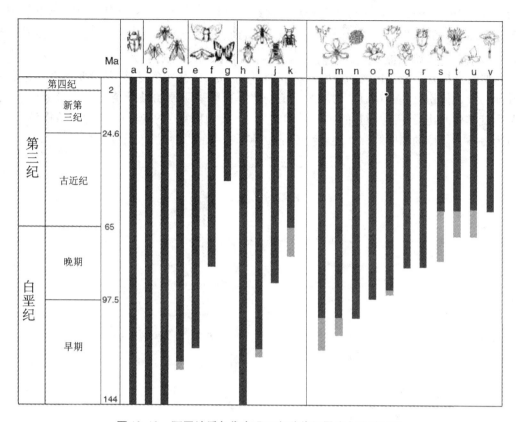

图 12.19 不同地质年代出现了多种传粉昆虫与有花植物

（a—k）为有传粉昆虫出现的地质年代，(l—v)为有花植物出现的地质年代，黑条是基于化石证据，灰条是源自化石证据的推论；Ma，距今百万年前；昆虫：(a) 鞘翅目；(b—d) 双翅目：(b) 大蚊科，(c) Mycethophilidae，(d) 舞虻科；(e—g) 鳞翅目：(e) 小翅蛾科，(f) 夜蛾科，(g) 凤蝶科；(h—k) 膜翅目：(h) 广腰亚目，(i) 泥蜂科，(j) 胡蜂科，(k) 蜜蜂科。植物：(l) 小而简单的花序；(m) 无环或半轮生花；(n) 小的单被花；(o) 环形，异形花被，放射状花；(p) 子房上位，异形花被的花；(q) 合瓣花；(r) 子房下位，单被花；(s) 两侧对称花；(t) 刷妆花；(u) 蝶状花；(v) 深漏斗形花序（克雷佩和弗里斯，1987[17]；弗里斯和克雷佩，1987[31]；格里马尔迪，1996[44]）

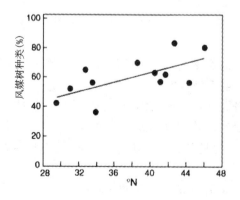

图 12.20 北美东部风媒植物的纬度函数（瑞格尔，1982）[102]

多适应性。它们的消化系统可以从花粉粒中提取营养物质[127]。很少有其他的昆虫种群能利用这种蛋白质丰富的植物产品。很好的学习能力结合它们飞行和导航的优势能力,使得蜜蜂可以让花稳定并广泛地分布[87]。

这些特点提升了开花植物的特殊性,同时它们也具有进化了的结构,比如那些花管束和一些其他与蜂媒传粉有关的花冠结构。开花植物化石遗骸表明原始被子植物有大量雄蕊、雌蕊以及螺旋排列的花瓣,就像现在的木兰和白色水百合(白睡莲)的花。随着时间的推移,这就发展为一种有规律的辐射对称并向着特定昆虫种群适应的方向发展(图12.21)。

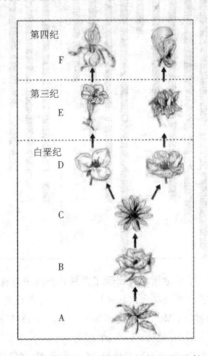

图12.21　千万年来花形状的进化趋势

以现生花为例,(A)为最早的花,没有可辨认的形状和对称性;(B)呈现半球形但是仍没有明显的对称性(例如木兰);(C)典型的辐射对称开放花序,例如阿多斯(Adonis sp.),随后分化为单子叶植物(左)和双子叶植物(右);(D)有了少而固定的花器,如紫鸭跖草(Tradescantia sp.)(左)、毛茛(Ranunculus sp.)(右);(E)两侧对称和隐藏的蜜腺,如小苍兰莱皮克(Freesia sp.)(左)和耧斗菜(右);(F)复杂的、严格两侧对称的花形,如枸兰(左)和乌头(右)(莱皮克,1971)[73]

这些措施包括萼片和花瓣数量的减少,花冠管状及带刺花冠的形成,以致它们只能供长舌昆虫觅食。到白垩纪后期,两侧对称的开花植物完成了进化。花的各部联合,例如凤蝶科就发生在第三纪早期而且高等开花植物的增殖影响了膜翅目昆虫与被子植物之间引人入胜的进化合作的开端。蜜蜂发现自身有两处难以训练:后背的中心处和头部下方。一些两侧对称的花,比如豆科植物,利用蜜蜂的灵敏性和它们对花粉的负载,使花粉装在难以达到的位置,进而防止花粉移入花粉篮。

后白垩纪到第三纪前期是新型被子植物和蜂媒传粉者大量涌现的时期,从某种程度上表明了其间的关联性[18,44,82]。像兰花及舟形乌头那些最终过渡到三维花型的植物存在两大优势。第一,得益于它们显眼的外形,传粉者可以从远处轻而易举地将其辨认出。第二,由于传粉者掌握熟练有效地获得回馈的方法,雄蕊雌蕊的位置可以适应植物的特定传粉者。由此,大黄蜂具有嗜好双边运动的本性是很有意思的,所以在它们第一次偶遇中蜜蜂更青睐于双向形花[104]。由此可以明显地看出协同进化为两侧花系统和先天优势打下了基础。

据观察数据,由膜翅类传粉的植物比其他种群传粉植物更为多样化。也就是说,膜翅目在昆虫与开花植物进化系统之间具有高级的地位。因此在加利福尼亚北侧,靠蜜蜂传粉的植物每属平均就有5.9种,但每属中只有3.4种为杂交昆虫传粉类植物。这一显著差异显示了在蜜蜂传粉类植物中增加的物种形成率[42]。另一个蜜蜂传粉的优势在它们多毛的皮层。这就让它们能一次性输送大量花粉粒,相应地,蜂媒传粉类植物的花粉量很高,进而导致了单株植物种子数量很高。

兰花属在适应昆虫传粉者改良多样性方面达到了极致。和25 000多种植物比较来看,单子叶兰花更能代表近来大多数高度进化的管状植物类群。在兰花植物中,花粉粒黏合在一起形成棒状的小包,叫作花粉包,通常每花两个。每个花粉块都有黏性可以黏贴在昆虫的头部或者昆虫身上。接着它被运到另一株花上,这取决于花的特殊形态,昆虫落在花粉包恰好所在的地方,这就取决于花的特殊形态。一个花粉包足够与单个传粉者身上的胚珠受精,提高后代数量。兰花花朵结构的多样化表现了对不同种类授粉者的适应性,约60%的兰花是蜂媒授粉。尤其像北半球大黄蜂和新发现的独居长舌花蜂这样的没有集群性的蜂类在大量的分离植物类群中很有效,通过花粉传递的精确性确保了杂交的进行。当传粉蜜蜂接触不同种兰花时,由于不同的花种将其花粉包咬合在昆虫的不同部位,通常会维持生殖隔离。多达13个可以放置花粉的地方已被记录。

兰花在花形以及花香上的多样性都非常明显。开花植物以花粉作为酬劳,这样昆虫可以促进在识别花形多样性和气味变化的本领上的学习[57]。植物也有很多欺骗行为,比如兰花通过模仿雌性信息素的方法吸引来访者。大约1/3的兰花种植物不会提供花蜜花粉作为回馈,它们通过模拟异性气味来吸引传粉者。例如兰属花,它释放的挥发物和传粉者对应的雌性制造的化学信息素非常相似(图12.22)[93]。通过兰属植物的研究,植物在授粉后增加了一种模仿雌性授粉者交配后产生的信号物质的产量。这种化合物是法尼基己酸,可抑制雄性与已受精的雌性交配,同样也减少了蜜蜂对已授粉花朵的探视[108]。

许多兰属花,除了会散发诱惑的香味,还会模仿与授粉者同种的雌性昆虫,进化出一种视觉和触觉上的刺激(图12.23)。巡逻的雄性昆虫会因此而产生性兴奋,并尝试去与那些花朵交配。这种"伪交配"很少会导致精子的释放,但可使"配偶"昆虫触碰到花粉,并黏附在它身上。当这只昆虫被另一朵花吸引时,授粉就可能随之发生。这种策略的优点是在访问过一朵花后,这只昆虫的性欲望并不会随之熄灭,下一朵花依然对它具有相同的吸引力。部分植物的这种情况也可看作一种寄生行为,因为这时昆虫是被无偿利用的。这种战术并不是特例。它在兰花和它们的拜访者之间独立演变了至少3次,偶尔也会在其他的植物类群中出现,甚至会涉及各种昆虫群体[120]。

12.7 自然保护

由于虫媒传粉在维持世界生态环境的植物多样性方面有重要作用,自然界任何传粉昆虫的大量减少都会对世界上的植物造成重要影响。事实上,蜜蜂在其中更是有着极为重要的地

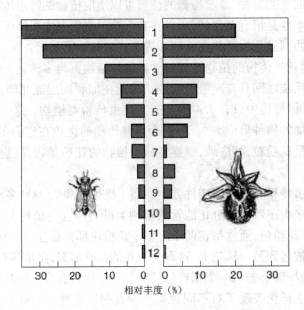

图 12.22 花的气味模拟昆虫的性信息素的例子

土蜂性信息素的组成与蜜蜂兰的挥发性气味进行比较,处女雌性的表皮提取物(左)包含了 12 个直链饱和和不饱和的烃类化合物,在蜜蜂兰(*O. sphegodes*,右)花的提取物中,也发现了相同的成分(席斯特尔等,1999)[109]

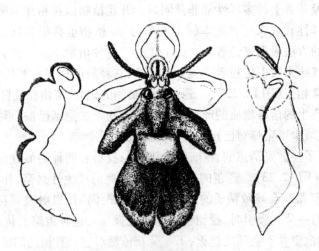

图 12.23 蜜蜂兰和土蜂

(A)蜜蜂兰的传粉者——土蜂的轮廓;(B)蜜蜂兰的毛状覆盖物;(C)花的轮廓

位。这些都已经被内夫和辛普森进行了适宜的陈述[87]:"在各种各样的植物群落中,毫无疑问蜜蜂是极其重要甚至是最重要的传粉群。事实上,我们很难想象出一个没有蜜蜂的世界。其他昆虫在一定程度上或许能担当蜜蜂的角色,但在许多群体中,大多数的植物依然专性地依赖蜜蜂进行传粉。如果蜜蜂突然被它们的生存系统淘汰,这些植物中有许多种都将会消失灭绝。"

尽管有这种观念,但在授粉力变化对植物群落的影响方面,生态保护学者却仅有很少的量化数据。两个偶然间进行的大规模"实验"也许可以表明,在自然栖息地或农业系统里,当本地授粉媒介数量减少,一些植物种的种苗也会随之减少。20世纪70年代初,加拿大新不伦瑞克省为控制大片森林地区的蚜虫(*Choristoneura fumifcrana*)暴发,喷洒了一种对蜜蜂有剧毒性的农药。这严重影响了蓝莓的授粉成功率。当停止使用这种农药后,就能发现这个影响会稳定的恢复[59,61]。1973年美国西北部为控制苜蓿田的蚜虫而滥用二嗪农农药,导致大量的黑彩带蜂死亡。两年后黑彩带蜂也仅恢复到它最初数量的25%[54]。同样地,会切断传粉者可选择食物来源的除草剂的大范围使用,也可能会因为对野生授粉昆虫的负面影响给自然植被带来深远的影响。

栖息地的破坏,包括边际土地和灌木篱墙的消移,导致饲料作物多样性下降和自然传粉者巢穴的减少,因此这是本土蜜蜂多样性和数量惊人下降的主要原因[11]。反过来,这样的下降也可能会反映到当地的植物群系上。

还有一种生境变化的形式是由人类活动造成的栖息地破碎化。植物种群的生存能力可能会由于授粉昆虫的缩减而受到当地栖息地分割化的影响。现已发现,这种影响会大规模出现在龙胆科(假龙胆属、芸薹种)植物种群中,在当地小的草原栖息地区域也有发现。在这种情况下,由于异花授粉水平的降低,发现灭绝率随着当地栖息地分裂的增加而增加。在一些大的地方栖息地碎片,花被大黄蜂访问的频率似乎比那些小栖息地碎片高4~6倍。因此,结实率及种子质量在较小的栖息地显著下降会造成植物种群生存力的显著性差异[72]。

引进外地的蜜蜂或黄蜂,对本地的授粉者来说无疑会承担一定的风险。例如,在塔斯马尼亚,迄今为止几乎在每个地方最丰富的访花昆虫都是被引进的蜜蜂,数量往往是所有其他访花昆虫的10倍甚至更多。这些花的大部分资源都被这些蜜蜂收集,它们通常在早晨其他本地蜜蜂到达之前就已经开始活跃起来。很可能本地的授粉者只能得到一小部分可用的植物资源。这必定会对种群数量甚至当地授粉物种的生存产生重大影响,反过来可能也影响当地植物的生存。因此,蜜蜂在一个地区的引入若超出了它们的活动范围,就可能对本地生物的多样性产生重大影响[39]。

现在我们认识到了正是杀虫剂的滥用、栖息地的破坏以及寄生虫和病原体的传播导致了野生授粉物种以及蜜蜂的减少。因此,一个全球性的授粉危机似乎迫在眉睫,不仅对自然生态系统,而且还对农业生产有严重影响[61]。

12.8 经济

一个不显著却很关键的贡献是昆虫授粉的作物是人类的食物来源。我们的食物约有30%都来自蜜蜂授粉的植物,蜜蜂的授粉效率高得令人惊奇。例如,100只蜜蜂可以在5小时内完成1公顷苹果的授粉。相比之下,蜜蜂作为蜂蜜生产者的角色就显得微不足道了。在全球农作物中,经蜜蜂授粉的农作物的价值超过每年蜂蜜产量价值的50倍。若蜜蜂全部消失,我们将很难确定农作物损失的确切数字,但据估计,美国每年仅63片由蜜蜂授粉的农作物产地所产出的价值就相当于5亿~14亿美元[65]。欧盟177片由蜜蜂授粉的农作物产地所产出的经济价值就相当于约40亿欧元[137]。低下的授粉水平不仅会降低作物产量,同样会降低诸如苹果、瓜类作物和其他水果的质量。

在西方约80%的作物通常由蜜蜂授粉，但这个数据可能会被过高地估计，蜜蜂的贡献并没有那么高。反复的研究已经表明，作为一个传粉者，本土蜂也如同蜜蜂一样是许多作物的重要助手，甚至它们比蜜蜂更优秀[65,92]。因此，对未来农业生产来说，这些已在欧洲和北美一些地区得到较完整记录的衰落的野生蜂种是需要我们认真关注的[61]。维持一些未开垦的土地作为避难的栖息场所(占25%[4])可以阻止那些原生蜜蜂的进一步减少，同时也为昆虫的天敌提供了避难场所，这样也有益于对有害物种的控制，最近关于这种措施的一些例子及其积极成果都令人倍受鼓舞[65,92]。

甚至是自花授粉的植物物种，在授粉良好栖息地附近长大也可能会有相当高的产量。例如在本地蜜蜂或引进蜜蜂种群授粉稳定的地区，咖啡灌木就会高产50%[105]。由于咖啡多种植在世界生物多样性最丰富或生物多样性受威胁的地区，所以以传粉者授粉为主的森林会使咖啡产量大大增加的发现，能十分明确地阐明保护森林在农业景观方面的潜在经济价值[105]。

为弥补当地自然授粉的不足，往往会租赁大量蜂群，并且有时会移动很远的距离[30,53]。额外的授粉能力可以通过饲养其他蜂种获得。为提高授粉的成功率，这些方法现在已经开始应用并发展到商业模式上，无论是在露天饲养(切叶蜜蜂)还是在温室(大黄蜂)。

12.9 结论

存在于昆虫与花朵之间的伙伴关系的迷人全景提供了展示生物史上最长关系之一的窗口。同时它又显示出在同一时间范围内昆虫与植物之间的复杂关系是其他任何类型的交互作用都难以超越的。这种复杂关系是由两个动力系统相互影响而引起的。植物与传粉者相互依靠、互利共生，同样，这两个合作伙伴也形成了有竞争力的相互作用系统：(1)植物为授粉者提供食物；(2)授粉者为花朵传粉[59]。这种复杂的相互作用的结果往往是难以预料的，但却是现今地球的生物群组成的核心。众所周知，如果没有昆虫授粉，地球陆地生态系统将永远都不会像现在这么丰富。花与授粉者之间的关系已经成为许多书籍共同探讨的话题。有一本非常成文、语言非常流畅并配有插图的书应该被提到，它就是由F. G. 巴斯撰写的《昆虫与花》[5]。引言是由古尔德和古尔德[37]，海因希里[47]以及普洛克等人写出来的[96]，并且近年来古尔德也写出了优美的书籍[40]。达夫尼著的手册提供了一个有用的方法和步骤用于授粉研究，它强调生态研究[21]。汤普森提出了演化和互利共生的原则[123]。

12.10 参考文献

1. Adler, L. S. (2000). The ecological significance of toxic nectar. *Oikos*, 91, 409–20.
2. Alm, J., Ohnmeiss, T. E., Lanza, J., and Vriesenga, L. (1990). Preference of cabbage white butterflies and honey bees for nectar that contains amino acids. *Oecologia*, 84, 53–7.
3. Baker, H. G. and Baker, I. (1986). The occurrence and significance of amino acids in floral nectars. *Plant Systematics and Evolution*, 151, 175–86.
4. Banaszak, J. (2000). Effect of habitat heterogeneity on the diversity and density of pollinating insects. In *Interchanges of insects between agricultural and surrounding landscapes* (ed. B. Ekbom, M. E. Irwin, and Y. Robert), pp. 123–40. Kluwer, Dordrecht.

5. Barth, F. G. (1985). *Insects and flowers. The biology of a partnership.* Princeton University Press, Princeton, NJ.
6. Bawa, K. S., Bullock, S. H., Perry, D. R., Coville, R. E., and Grayum, M. H. (1985). Reproductive biology of tropical lowland rain forest trees. II. Pollination systems. *American Journal of Botany*, 72, 346–56.
7. Beekman, M. and Ratnieks, F. L. W. (2000). Long-range foraging by the honey-bee, *Apis mellifera* L. *Functional Ecology*, 14, 490–6.
8. Boch, R. (1956). Die Tänze der Biene bei nahen und fernen Trachtquellen. *Zeitschrift für Vergleichende Physiologie*, 38, 136–67.
9. Braun, E., MacVicar, R. M., Gibson, D. A., Pankiw, P., and Guppy, J. (1953). Studies in red clover seed production. *Canadian Journal of Agricultural Science*, 3, 48–53.
10. Burd, M. (1994). Bateman's principle and plant recognition: the role of pollen limitation in fruit and seed set. *Botanical Review*, 60, 83–139.
11. Cane, J. H. (2001). Habitat fragmentation and native bees: a premature verdict? *Conservation Ecology*, 5, article no. 3.
12. Chittka, L. (1997). Bee color vision is optimal for coding flower color, but flower colors are not optimal for being coded—why? *Israel Journal of Plant Sciences*, 45, 115–27.
13. Chittka, L. and Waser, N. M. (1997). Why red flowers are not invisible to bees. *Israel Journal of Plant Sciences*, 45, 169–83.
14. Cook, J. M. and Rasplus, J. Y. (2003). Mutualists with attitude: coevolving fig wasps and figs. *Trends in Ecology and Evolution*, 18, 241–8.
15. Cox, P. A. (1991). Abiotic pollination: an evolutionary escapeforanimal-pollinatedangiosperms. *Philosophical Transactions of the Royal Society B*, 333, 217–24.
16. Crane, P. R., Friis, E. M., and Pedersen, K. R. (1995). The origin and early diversification of angiosperms. *Nature*, 374, 27–33.
17. Crepet, W. L. and Friis, E. M. (1987). The evolution of insect pollination in angiosperms. In *The origins of Angiosperms and their biological consequences* (ed. E. M. Friis, W. G. Chalconer, and P. R. Crane), pp. 181–201. Cambridge University Press, Cambridge.
18. Crepet, W. L., Friis, E. M, and Nixon, K. C. (1991). Fossil evidence for the evolution of biotic pollination. *Philosophical Transactions of the Royal Society B*, 333, 187–95.
19. Culley, T. M., Weller, S. G., and Sakai, A. K. (2002). The evolution of wind pollination in angiosperms. *Trends in Ecology and Evolution*, 17, 361–9.
20. Dafni, A. (1984). Mimicry and deception in pollination. *Annual Review of Ecology and Systematics*, 15, 259–78.
21. Dafni, A. (1992). *Pollination ecology. A practical approach.* IRL Press, Oxford.
22. Darwin, C. (1859). *The origin of species.* John Murray, London.
23. De La Barrera, E. and Nobel, P. S. (2004). Nectar: properties, floral aspects, and speculations on origin. *Trends in Plant Science*, 9, 65–9.
24. Dobson, H. E. M. (1994). Floral volatiles in insect biology. In *Insect-plantinteractions*, Vol. 5

(ed. E. A. Bernays), pp. 47-81. CRC Press, Boca Raton, FL.
25. Dobson, H. E. M., Bergström, G., and Groth, I. (1990). Differences in fragrance chemistry between flower parts of *Rosa rugosa* Thunb. (Rosaceae). *Israel Journal of Botany*, 39, 143-56.
26. Dodd, M. E., Silvertown, J., and Chase, M. W. (1999). Phylogenetic analysis of trait evolution and species diversity variation among angiosperm families. *Evolution*, 53, 732-44.
27. Farzad, M., Griesbach, R., Hammond, J., Weiss, M. R., and Elmendorf, H. G. (2003). Differential expression of three key anthocyanin biosynthetic genes in a color-changing flower, *Viola cornuta* cv. Yesterday, Today and Tomorrow. *Plant Science*, 165, 1333-42.
28. Feinsinger, P. (1983). Coevolution and pollination. In *Coevolution* (ed. D. J. Futuyma and M. Slatkin), pp. 282-310. Sinauer, Sunderland, MA.
29. Fenster, C. B., Armbruster, W. S., Wilson, P., Dudash, M. R., and Thomson, J. D. (2004). Pollination syndromes and floral specialization. *Annual Review of Ecology, Evolution and Systematics*, 35, 375-403.
30. Free, J. B. (1993). *Insect pollination of crops* (2nd edn). Academic Press, London.
31. Friis, E. M. and Crepet, W. L. (1987). Time of appearance of floral features. In *The origins of angiosperms and their biological consequences* (ed. E. M. Friis, W. G. Chalconer, and P. R. Crane), pp. 145-79. Cambridge University Press, Cambridge.
32. Frisch, K. von (1967). *The dance language and orientation of bees*. Harvard University Press, Cambridge.
33. Gardener, M. C. and Gillman, M. P. (2002). The taste of nectar—a neglected area of pollination ecology. *Oikos*, 98, 552-7.
34. Gegear, R. J. and Laverty, T. M. (1995). Effect of flower complexity on relearning flower-handling skills in bumble bees. *Canadian Journal of Zoology*, 73, 2052-8.
35. Gilbert, F. S., Haines, H., and Dickson, K. (1991). Empty flowers. *Functional Ecology*, 5, 29-39.
36. Giurfa, M., Dafni, A., and Neal, P. R. (1999). Floral symmetry and its role in plant-pollinator systems. *International Journal of Plant Sciences*, 160, S41-S50.
37. Gould, J. L. and Gould, C. G. (1988). *The honey bee*. W. H. Freeman, New York.
38. Goulson, D. (2000). Are insects flower constant because they use search images to find flowers? *Oikos*, 88, 547-52.
39. Goulson, D. (2003). Effects of introduced bees on native ecosystems. *Annual Review of Ecology, Evolution and Systematics*, 34, 1-26.
40. Goulson, D. (2003a). *Bumblebees. Their behaviour and ecology*. Oxford University Press, Oxford.
41. Goulson, D., Stout, J. C., and Hawson, S. A. (1997). Can flower constancy in nectaring butterflies be explained by Darwin's interference hypothesis? *Oecologia*, 112, 225-31.
42. Grant, V. (1949). Pollination systems as isolating mechanisms in angiosperms. *Evolution*, 3, 82-97.

43. Grant, V. (1950). The flower constancy of bees. *Botanical Review*, 16, 379–98.
44. Grimaldi, D. A. (1996). Captured in amber. *Scientific American*, 274(4), 71–7.
45. Heinrich, B. (1976). The foraging specializations of individual bumblebees. *Ecological Monographs*, 46, 105–28.
46. Heinrich, B. (1979a). 'Majoring' and 'minoring' by foraging bumblebees, Bombus vagans: an experimental analysis. *Ecology*, 60, 245–55.
47. Heinrich, B. (1979b). *Bumblebee economics*. Harvard University Press, Cambridge, MA.
48. Heinrich, B. and Raven, P. H. (1972). Energetics and pollination ecology. *Science*, 176, 597–602.
49. Hess, D. (1990). *Die Blüte* (2nd edn). Ulmer, Stuttgart.
50. Higginson, A. D. and Barnard, C. J. (2004). Accumulating wing damage affects foraging decisions in honeybees (*Apis mellifera* L.). *Ecological Entomology*, 29, 52–9.
51. Inouye, D. W. (1978). Resource partitioning in bumblebees: experimental studies of foraging behavior. *Ecology*, 59, 672–8.
52. Jarau, S., Hrncir, M., Ayasse, M., Schulz, C., Francke, W., Zucchi, R., et al. (2004). A stingless bee (*Melipona seminigra*) marks food sources with a pheromone from its claw retractor tendons. *Journal of Chemical Ecology*, 30, 793–804.
53. Jay, S. C. (1986). Spatial management of honey bees on crops. *Annual Review of Entomology*, 31, 49–65.
54. Johansen, C. A. (1977). Pesticides and pollinators. *Annual Review of Entomology*, 22, 177–92.
55. Jones, C. E. and Buchmann, S. L. (1974). Ultraviolet floral patterns as functional orientation cues in hymenopterous pollination systems. *Animal Behaviour*, 22, 481–5.
56. Jürgens, A., Witt, T., and Gottsberger, G. (2003). Flower scent composition in *Dianthus* and *Saponaria* species (Caryophyllaceae) and its relevance for pollination biology and taxonomy. *Biochemical Systematics and Ecology*, 31, 345–57.
57. Kaiser, R. (1993). *The scents of orchids. Olfactory and chemical investigations.* Elsevier, Amsterdam.
58. Kang, H. and Bawa, K. S. (2003). Effects of successional status, habit, sexual systems, and pollinators on flowering patterns in tropical rain forest trees. *American Journal of Botany*, 90, 865–76.
59. Kevan, P. G. and Baker, H. G. (1983). Insects as flower visitors and pollinators. *Annual Review of Entomology*, 28, 407–53.
60. Kevan, P. G. and Lane, M. (1985). Flower petal microtexture is a tactile cue for bees. *Proceedings of the National Academy of Sciences of the USA*, 82, 4750–2.
61. Kevan, P. G. and Phillips T. P. (2001). The economic impacts of pollinator declines: an approach to assessing the consequences. *Conservation Ecology*, 5, article no. 8.
62. Klinkhamer, P. G. L. and de Jong, T. J. (1993). Attractiveness to pollinators: a plant's dilemma. *Oikos*, 66, 180–4.

63. Knoll, F. (1956). *Die Biologie der Blüte*. Springer, Berlin.
64. Knudsen, J. T., Tollsten, L., and Bergström, G. (1993). Floral scents—a checklist of volatile compounds isolated by head-space techniques. *Phytochemistry*, 33, 253–80.
65. Kremen, C., Williams N. M., and Thorp, R. W. (2002). Crop pollination from native bees at risk from agricultural intensification. *Proceedings of the National Academy of Sciences of the USA*, 99, 16812–16.
66. Kress, W. J. and Beach, J. H. (1990). Cited in: Bawa, K. S. (1990). Plant-pollinator interactions in tropical rain forests. *Annual Review of Ecology and Systematics*, 21, 399–422.
67. Kriston, I. (1971). Zum Problem des Lernverhaltens von *Apis mellifica* L. gegenü ber verschiedenen Duftstoffen. *Zeitschrift für Vergleichende Physiologie*, 74, 169–89.
68. Kullenberg, B. (1961). Studies on *Ophrys* L. pollination. *Zoologiska Bidrag från Uppsala*, 34, 1–340.
69. Laloi, D., Bailez, O., Blight, M. M., Roger B., Pham Delègue, H. M., and Wadhams, L. J. (2000). Recognition of complex odors by restrained and free-flying honeybees, *Apis mellifera*. *Journal of Chemical Ecology*, 26, 2307–19.
70. Laverty, T. M. (1994). Bumble bee learning and flower morphology. *Animal Behaviour*, 47, 531–45.
71. Laverty, T. M. and Plowright, R. C. (1988). Flower handling by bumblebees: a comparison of specialists and generalists. *Animal Behaviour*, 36, 733–40.
72. Lennartsson, T. (2002). Extinction thresholds and disrupted plant-pollinator interactions in fragmented plant populations. *Ecology*, 83, 3060–72.
73. Leppik, E. E. (1971). Origin and evolution of bilateral symmetry in flowers. *Evolutionary Biology*, 5, 49–86.
74. Lewis, A. C. (1993). Learning and the evolution of resources: pollinators and flower morphology. In *Insect learning. Ecological and evolutionary perspectives* (ed. D. R. Papaj and A. C. Lewis), pp. 219–42. Chapman & Hall, New York.
75. Lewis, A. C. and Lipani, G. (1990). Learning and flower use in butterflies: hypotheses from honey bees. In *Insect-plant interactions*, Vol. 2 (ed. E. A. Bernays), pp. 95–110. CRC Press, Boca Raton, FL.
76. Lindauer, M. (1948). Über die Einwirkung von Duftund Geschmacksstoffen sowie anderer Faktoren auf die Tanze der Bienen. *Zeitschrift für Vergleichende Physiologie*, 31, 348–412.
77. McKone, M. J., Ostertag, R., Rausher, J. T., Heiser, D. A., and Russell, F. L. (1995). An exception to Darwin's syndrome: floral position, protogyny, and insect visitation in *Besseya bullii* (Scrophulariaceae). *Oecologia*, 101, 68–74.
78. Maynard Smith, J. (1982). *Evolution and the theory of games*. Cambridge University Press, Cambridge.
79. Menzel, R. and Shmida, A. (1993). The ecology of flower colours and the natural colour vision of insect pollinators: the Israeli flora as a study case. *Biological Reviews*, 68, 81–120.
80. Menzel, R., Erber, I., and Masuhr, T. (1974). Learning and memory in the honeybee. In

Experimental analysis of insect behaviour (ed. L. Barton Browne), pp. 195 – 217. Springer, Berlin.

81. Mevi-Schütz, J., Goverde, M., and Erhardt, A. (2003). Effects of fertilization and elevated CO_2 on larval food and butterfly nectar amino acid preference in *Coenonympha pamphilus* L. *Behavioral Ecology and Sociobiology*, 54, 36 – 43.

82. Michener, C. D. and Grimaldi, D. A. (1988). The oldest fossil bee: apoid history, evolutionary stasis, and antiquity of social behavior. *Proceedings of the National Academy of Sciences of the USA*, 85, 6424 – 6.

83. Moore, P. D., Webb, J. A., and Collinson, M. E. (1991). *Pollen analysis* (2nd edn). Blackwell, Oxford.

84. Morse, R. A. (1958). The pollination of bird's-foot trefoil. *Pceedings of the Tenth International Congress of Entomology*, 4, 951 – 3.

85. Nason, J. D., Herre, E. A., and Hamrick, J. L. (1998). The breeding structure of a tropical keystone plant resource. *Nature*, 391, 685 – 7.

86. Ne'eman, G., Dafni, A., and Potts, S. G. (1999). A new pollination probability index (PPI) for pollen load analysis as a measure for pollination effectiveness of bees. *Journal of Apicultural Research*, 38, 19 – 24.

87. Neff, J. L. and Simpson, B. B. (1993). Bees, pollination systems and plant diversity. In *Hymenoptera and biodiversity* (ed. J. LaSalle and I. D. Gauld), pp. 143 – 67. CAB International, Wallingford, UK.

88. Nepi, M., Guarnieri, M., and Pacini, E. (2003). 'Real' and feed pollen of *Lagerstroemia indica*: ecophysiological differences. *Plant Biology*, 5, 311 – 14.

89. Nieh, J. C., Ramírez, S., and Nogueira-Neto, P. (2003). Multi-source odor-marking of food by a stingless bee, *Melipona mandacaia*. *Behavioral Ecology and Sociobiology*, 54, 578 – 86.

90. Nuttman, C. and Willmer, P. (2003). How does insect visitation trigger floral colour change? *Ecological Entomology*, 28, 467 – 74.

91. Olesen, J. M. and Warncke, E. (1989). Flowering and seasonal changes in flower sex ratio and frequency of flower visitors in a population of *Saxifraga hirculus*. *Holarctic Ecology*, 12, 21 – 30.

92. O'Toole, C. (1993). Diversity of native bees and agrosystems. In *Hymenoptera and biodiversity* (ed. J. LaSalle and I. D. Gauld), pp. 169 – 96. CAB International, Wallingford, UK.

93. Paxton, R. J. and Tengö, J. (2001). Double duped males: the sweet and the sour of the orchid's bouquet. *Trends in Ecology and Evolution*, 16, 167 – 9.

94. Pham-Delègue, M. H., Etievant, P., Guichard, E., Marilleau, R., Douault, P., Chauffaille, J., et al. (1990). Chemicals involved in honeybee-sunflower relationship. *Journal of Chemical Ecology*, 16, 3053 – 65.

95. Pleasants, J. M. (1980). Competition for bumblebee pollinators in Rocky Mountain plant communities. *Ecology*, 61, 1446 – 59.

96. Proctor, M., Yeo, P., and Lack, A. (1996). *The natural history of pollination*. Harper Col-

lins, London.
97. Raguso, R. A. (2004a). Why are some floral nectars scented? *Ecology*, 85, 1486–94.
98. Raguso, R. A. (2004b). Why do flowers smell? The chemical ecology of fragrance-driven pollination. In *Advances in insect chemical ecology* (ed. R. T. Cardé and J. G. Millar), pp. 151–78. Cambridge University Press, Cambridge.
99. Ranta, E. and Lundberg, H. (1980). Resource portioning in bumblebees: the significance of differences in proboscis length. *Oikos*, 35, 298–302.
100. Reader, R. J. (1975). Competitive relationships of some bog ericads for major insect pollinators. *Canadian Journal of Botany*, 53, 1300–5.
101. Real, L., Ott, J., and Silverline, E. (1982). On the trade-off between the mean and the variance in foraging: effects of spatial distribution and color preference. *Ecology*, 63, 1617–23.
102. Regal, P. J. (1982). Pollination by wind and animals: ecology and geographic patterns. *Annual Review of Ecology and Systematics*, 13, 497–524.
103. Ricketts, T. H., Daily, G. C., Ehrlich, P. R., and Michener, C. D. (2004). Economic value of tropical forest to coffee production. *Proceedings of the National Academy of Sciences of the USA*, 101, 12579–82.
104. Rodríguez, I., Gumbert, A., Hempel de Ibarra, N., Kunze, J., and Giurfa, M. (2004). Symmetry is in the eye of the 'beeholder': innate preference for bilateral symmetry in flower-naive bumblebees. *Naturwissenschaften*, 91, 374–7.
105. Roubik, D. W. (2002). The value of bees to the coffee harvest. *Nature*, 417, 708.
106. Roulston, T. A. H., Cane, J. H., and Buchmann, S. L. (2000). What governs protein content of pollen: pollinator preferences, pollen-pistil interactions, or phylogeny? *Ecological Monographs*, 70, 617–43.
107. Sazima, M., Vogel, S., Cocucci, A., and Hausner, G. (1993). The perfume flowers of *Cyphomandra* (Solanaceae): pollination by euglossine bees, bellows mechanism, osmophores, and volatiles. *Plant Systematica and Evolution*, 187, 51–88.
108. Schiestl, F. P. and Ayasse, M. (2001). Post-pollination emission of a repellent compound in a sexually deceptive orchid: a new mechanism for maximizing reproductive success? *Oecologia*, 126, 531–4.
109. Schiestl, F. P., Ayasse, M., Paulus, H. F., Löfstedt, C., Hansson, B. S., Ibarra, F., *et al*. (1999). Orchid pollination by sexual swindle. *Nature*, 399, 421–2.
110. Schmitt, U. and Bertsch, A. (1990). Do foraging bumblebees scent-mark food sources and does it matter? *Oecologia*, 82, 137–44.
111. Schmitt, U., Lübke, G., and Francke, W. (1991). Tarsal secretion marks food sources in bumblebees (Hymenoptera: Apidae). *Chemoecology*, 2, 35–40.
112. Schnetter, B. (1972). Experiments on pattern discrimination in honey bees. In *Information processing in the visual system of arthropods* (ed. R. Wehner), pp. 195–201. Springer, Berlin.

113. Seeley, T. D. (1985). *Honeybee ecology*. Princeton University Press, Princeton, NJ.
114. Snodgrass, R. E. (1956). *Anatomy of the honey bee*. Comstock, Ithaca, NY.
115. Snow, A. A., Spira, T. P., Simpson, R., and Klips, R. A. (1996). The ecology of geitonogamous pollination. In *Floral biology. Studies on floral evolution in animalpollinated plants* (ed. G. Lloyd and S. C. H. Barrett), pp. 191–216. Chapman & Hall, New York.
116. Sprengel, C. K. (1793). *Das entdeckte Geheimniss der Natur im Bau und in der Befruchtung der Blumen*. F. Vieweg, Berlin (reprinted 1972 by J. Cramer, Lehre, and Weldon & Wesley, Codicote, UK and New York. An English translation of the general part of this book by P. Haase has been published in *Floral biology. Studies on floral evolution in animalpollinated plants* (ed. D. G. Lloyd and S. C. H. Barrett), 1996, pp. 3–43. Chapman & Hall, New York.)
117. Stead, A. D. (1992). Pollination-induced flower senescence: a review. *Plant Growth Regulation*, 11, 13–20.
118. Stout, J. C. and Goulson, D. (2001). The use of conspecific and interspecific scent marks by foraging bumblebees and honeybees. *Animal Behaviour*, 62, 183–9.
119. Stout, J. C. and Goulson, D. (2002). The influence of nectar secretion rates on the responses of bumblebees (Bombus spp.) to previously visited flowers. *Behavioral and Ecological Sociobiology*, 52, 239–46.
120. Stowe, M. K. (1988). Chemical mimicry. In *Chemical mediation of coevolution* (ed. K. C. Spencer), pp. 513–80. Academic Press, San Diego.
121. Thakar, J. D., Kunte, K., Chauhan, A. K., Watve, A. V., and Watve, M. G. (2003). Nectarless flowers: ecological correlates and evolutionary stability. *Oecologia*, 136, 565–70.
122. Thien, L. B., Azuma, H., and Kawano, S. (2000). New perspectives on the pollination biology of basal angiosperms. *International Journal of Plant Sciences*, 161, S225–35.
123. Thompson, J. N. (1994). *The coevolutionary process*. University of Chicago Press, Chicago.
124. Tollsten, L. and Bergström, G. (1989). Variation and post-pollination changes in floral odours released by *Platanthera bifolia* (Orchidaceae). *Nordic Journal of Botany*, 9, 359–62.
125. Van Dulmen, A. (2001). Pollination and phenology of flowers in the canopy of two contrasting rain forest types in Amazonia, Colombia. *Plant Ecology*, 153, 73–85.
126. Vainstein, A., Lewinsohn, E., Pichersky, E., and Weiss, D. (2001). Floral fragrance. New inroads into an old commodity. *Plant Physiology*, 127, 1383–9.
127. Velthuis, H. W. W. (1992). Pollen digestion and the evolution of sociality in bees. *Bee World*, 73, 77–89.
128. Vogel, S. (1996). Christian Konrad Sprengel's theory of the flower: the cradle of floral ecology. In *Floral biology. Studies on floral evolution in animal-pollinated plants* (ed. D. G. Lloyd and S. C. H. Barrett), pp. 44–62. Chapman & Hall, New York.
129. Waddington, K. D. (1985). Cost-intake information used in foraging. *Journal of Insect Physiology*, 31, 891–7.
130. Waddington, K. D. (1990). Foraging profits and thoracic temperature of honey bees (*Apis*

mellifera). *Journal of Comparative Physiology B*, 160, 325 –9.
131. Walther-Hellwig, K. and Frankl, R. (2000). Foraging habitats and foraging distances of bumblebees, Bombus spp. (Hym., Apidae), in an agricultural landscape. *Journal of Applied Entomology*, 124, 299 –306.
132. Waser, N. M. (1986). Flower constancy: definition, cause, and measurement. *American Naturalist*, 127, 593 –603.
133. Weiss, M. R. (1991). Floral colour changes as cues for pollinators. *Nature*, 354, 227 –9.
134. Weiss, M. R. (1995). Associative colour learning in a nymphalid butterfly. *Ecological Entomology*, 20, 298 –301.
135. Weiss, M. R. and Lamont, B. B. (1997). Floral color change and insect pollination: a dynamic relationship. *Israel Journal of Plant Sciences*, 45, 185 –99.
136. Weiss, M. R. and Papaj, D. R. (2003). Colour learning in two behavioural contexts: how much can a butterfly keep in mind? *Animal Behaviour*, 65, 425 –34.
137. Williams, I. H. (1994). The dependence of crop production within the European Union on pollination by honey bees. *Agricultural Zoology Reviews*, 6, 229 –57.
138. Wolf, T. J., Ellington, C. P., and Begley, I. S. (1999). Foraging costs in bumblebees: field conditions cause large individual differences. *Insectes Sociaux*, 46, 291 –5.

第 13 章　昆虫和植物：如何运用我们的知识

13.1　植食性昆虫会演变成害虫及其原因
　13.1.1　害虫的特征
　13.1.2　引进农作物带来的影响
　13.1.3　农业活动促进了虫害的发生
13.2　寄主植物抗性
　13.2.1　寄主植物的抗性机制
　13.2.2　部分抗性
　13.2.3　植物特性与抗性
　13.2.4　培育植物抗性的方法
13.3　混养为何能减少病虫害
　13.3.1　破坏性作物假说
　13.3.2　天敌假说
　13.3.3　诱虫作物学说
　13.3.4　多样性原则
13.4　植物源杀虫剂和拒食剂
　13.4.1　拒食剂
　13.4.2　苦楝树：印楝素
　13.4.3　拒食剂作为杀虫剂的前景
13.5　植食性昆虫可以控制杂草
　13.5.1　仙人掌和槐叶萍
　13.5.2　生物控制杂草项目的成功率
13.6　结论
13.7　参考文献

　　人类在对待昆虫与植物关系这个问题上，对二者的态度是不公平的。基本的问题是我们希望保持一种稳定的状态，即在一种动态非静态的生物环境中，能够利于某种植物（作物）生长。人们希望能够按照一种可持续的并且绿色环保合理的方法来从事农业生产。

　　在没有杀虫剂的情况下，农作物在收获期之前的损失率在10%～100%之间。使用杀虫剂后，昆虫取食造成的损失率估计为13%，然而在自然生态系统中昆虫每年造成的生物量损失约占总体的10%（第 2 章）。

　　超过25年的观察发现，对几种重要粮食作物而言，虽然可以通过单位面积生产量的提高得到一部分补偿，但害虫造成的损耗仍逐年增多[87]。但是动物害虫对几种重要食用作物的取食量正在以惊人的速度增加（图13.1）。显然地，农作物的生产量与它们对害虫，以及其他生物和非生物的限制因素的敏感性存在正相关的关系。

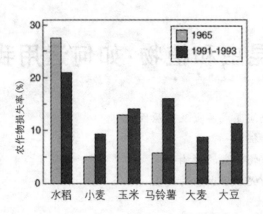

图 13.1 在 1965 年和 1991 年至 1993 年,6 种主要粮食作物受到虫害影响而造成的损失(欧克等,1994;欧克和德内,1997)[87,88]

当试图通过增加农作物产量来供养每年以 1.2% 的速率增长的世界人口时,为了减少人工杀虫剂的使用,并且让现有的农业模式变得更加合理,对昆虫—植物学的学习是达成这一目标必不可少的途径。本章节主要讨论的是昆虫—植物相互影响的一些方面,首先,将阐明一些昆虫物种演变成具有害虫特征的物种的原因,以及我们将采取什么措施来抑制这种演变;其次,我们将介绍一些能通过植食性昆虫来控制杂草的途径。

13.1 植食性昆虫会演变成害虫及其原因

13.1.1 害虫的特征

当一种农作物对昆虫而言变得可利用时,一些昆虫注定要变成害虫;而另一些昆虫,包括那些亲缘关系密切的物种,则不能很容易适应新的食物源。当一些昆虫拥有一个适宜的栖息环境时,一些物种的生理机能和行为特征,比如繁殖力、幼体的取食广泛度和多化性会促使其演变成具有害虫特征的可能。对一些昆虫群体的这些因素进行分析后发现从一些粉蝶属的生物学特征可以得出,为什么在以十字花科植物为食的虎斑狭翅蝴蝶的几十种或地理亚种中,仅仅只有两个物种欧洲粉蝶(*Pieris brassicae*)和菜粉蝶(*P. rapae*)在十字花科作物里成了害虫。相对于粉蝶其他的同属物种,这两种物种在十字花科植物中适应性较高,而且这种"广食性"似乎和其对宿主作物的预先适应有联系。这两个物种的多化性是另一种确保它们迅速扩张种群数量和繁殖数量巨大的后代的因素,这可能为它们开拓一个新的栖息地提供了证据(雌性菜粉蝶的产卵量可多达 800 个)。第三个促使这两个物种成为害虫的特性是它们在湿度适中的环境(既不极其干燥也不很湿润)中偏爱密集型的宿主植物[23]。因此,当一种昆虫暴露在对其生理和行为都可以接受的宿主植物中时,几种因素混在一起可致使这种昆虫成为潜在的害虫。

13.1.2 引进农作物带来的影响

在自然生态系统中,大多数昆虫和植物的关系是基于亿万年的演变而形成的(第 11 章)。作为进化的结果,仅就因为昆虫进攻而导致植物被淘汰的方面来说,植物和植食性昆虫之间逐步形成了一种平衡。当一种植物面对一种"未知"的昆虫物种时,情况又可能不一样。正如索思伍德[116]在一个关于昆虫和植物关系演化研究的讨论中提到:"甚至在今天,当一种植食性昆虫首先进攻一个新的宿主时,经常会造成不成比例的严重破坏。"许多作物植物,尤其是那些处于温带地区的农作物,都是引进物种。在绝大多数例子中,这些引进农作物的害虫可以从以原来的作物为食转移至以新的农作物生存。它们侵入了一个食物丰富、天敌稀少的生态位。而且,这些所谓的引进物种的抵抗力根本就无法适应当地的昆虫物种。一个很好的例子就是原产于南美后来又被引进到北美的马铃薯(*Solanum tuberosum*)。科罗拉来多马铃薯甲虫(*Leptinotarsa decemlineata*)靠当地的一种茄属植物物种(*S. rostratum*)生存,因为这种土豆不具备强有力抵抗这种昆虫的特性,这种甲虫此后便在这个具有新食物源的地方很成功地定居下来。后来这种甲虫变成了一种具有严重破坏力的害虫,并且和土豆一样迅速扩张至欧洲。

事实上,这种有害昆虫的数量,相比于庞大的潜在入侵物种而言,比例还是极少的。在世界范围内看,大约有9 000种昆虫会毁坏农作物,但在其中只有不足5%是被认为具有严重破坏力的害虫。害虫的物种数量还是很巨大的(例如,可可树的害虫有1 400种,棉花有1 360种,甘蔗有1 300种)。其实,绝大多数害虫对作物造成的伤害所导致的经济损失是微不足道的。对于任何作物来说,某一时间和某一地点的害虫数量经常是比较少的,我们只需要对最主要的害虫进行控制。因此,对于任何一种棉花作物,虽然它在世界范围内有棉花的害虫分布相当广泛,但是在一个具体的地区范围内,也仅仅是经常面临着大约5种需要控制种群数量的害虫物种[52]。

因为昆虫具有高度分散的特点,和农作物群体一样,自然界中的植物也自然而然地经常被许多害虫所侵扰。然而,在这些"访客"中,似乎只有一小部分群体想和这些植物建立一种长久的联系。例如,仅有40种昆虫物种聚居在伊利诺伊州(美国州名)的大豆牧场,而在过去的12年内,400多种植食性昆虫物种曾试图占领这个区域。尽管有超过60种蚜虫物种单独地被困在伊利诺伊州的大豆牧场内,但无论是在北美或南美,都不会是单一的,一种蚜虫物种像一个永恒的主宰者一样有能力在大豆牧场开拓出自己的生存空间[63]。即使某种新的可食用作物数量上很容易满足昆虫的需求,但在这种情况下,这种新的可食用植物的接受过程对大多数昆虫来说对也是十分困难的。

害虫来自哪里?它们属于当地的动物区系还是外来移民?在美国寄生于农作物的148种主要的昆虫物种中,仅有57种(即少于40%)属于外来引进物种。同样地,在70种主要的美国森林害虫种类中绝大多数是当地物种,只有少于30%的物种是源于欧洲或其他的地方[98]。类似地,在欧洲仅有大约20%的昆虫害虫是引进的[97]。因此,尽管在森林里有一些具有严重破坏力的害虫属于引进物种,但不管是在人为管理的生态系统中还是在自然生态系统中,主要的害虫物种还是属于当地的物种。

13.1.3 农业活动促进了虫害的发生

在西方,由于机械化耕作方式,农作物的种植方式主要以单一的作物为主。对一些昆虫而言,一旦食物是无限的,它们就会在现有的单一作物中迅速扩张。为什么这些生态系统相比于所谓的自然生态系统更加有利于昆虫数量的爆发[103]?对于这个问题现在还没有一个简单的答案,因为每一种昆虫类型、每一种宿主植物、每一种生长环境里的土壤类型和各种微气候共同形成了一个特殊的环境。一个农业生态系统的组分的多样性,导致害虫爆发的多样性。不过,在农业生态系统中,一些促使害虫爆发的最主要的因素还是十分明显的。它们主要包括两方面:一方面是,相比于原始形态或与之亲缘相近的野生物种来说,其物理和化学方面的抵抗力日益降低;另一方面,相比于自然生态系统,农业生态系统物种结构简单化。这种简单化包括急剧减少的动植物物种,农作物基因的一致性的提高,放弃作物轮作,人为的砍伐树木,填平天然水道,消除非作物生境和导致陆地景观多样性的锐减[103]。在结构上,相比于简单的陆地自然地形,复杂的陆地自然地形的天敌通常密度较高[37]。呈现在我们眼前的一个很好的例子就是,由油菜花粉传粉甲壳虫(*Meligethes aeneus*)对含油油菜籽(*Brassica napus*)的芽造成伤害的总量和景观的异质性有关。相比于占有高农业投入的简单地形而言,在复杂的地形系统中,作物的伤害程度更低,植食性昆虫的寄生虫病更多[124]。

下面将对寄主植物的抗性和混养对于增加农业生物多样性的意义进行讨论[12,50,62]。

13.2 寄主植物抗性

在自然界中,寄主植物的抗性和植食性昆虫的天敌,是两个控制昆虫种群的主要因素。因此,为了对病虫害进行控制,应该考虑培养具有抗性的寄主植物。在农业发展的初期,人们选定农作物的标准是高营养价值、高产量,以及对哺乳动物的毒害较弱和对病虫害具有一定的抗性。在我们不断地开发高收益品种的过程中,极少数保留了其野生祖先对昆虫的抗性水平,与此同时,抵抗的多样性也随之降低[97]。

培养抗性植物起源于 20 世纪,基于基本的遗传学知识以及选择、杂交植物的方法,雨果·德·弗恩斯在 1900 年重新发现了孟德尔遗传规律后,这种技术开始流行起来。虽然利用传统的育种方法开发抗虫作物品种是一个耗时长又昂贵的过程,但从利益回报和降低杀虫剂对环境的压力等方面来看,收益是巨大的。在经济方面,回报估计有投入的 120 倍,同样重要的是,最近一些含有抗虫性的棉花、水稻、蔬菜等新品种的开发,足以完全消除杀虫剂的使用[72,114]。经麦斯威尔和詹宁斯[75]、弗里切[41]、潘得和库什[90]以及斯托纳[117]广泛的验证表明,大量农作物具有对昆虫的抗性。而之所以目前报道的抗性作物种类如此稀少,不是因为缺乏抗性资源,而是因为对大量昆虫生物测定是复杂并且费力的,所以育种工作者不愿意进行这项工作。因此,对作物进行基因改造和分子标记辅助选择(MAS)[70]等新技术受到了高度的重视。

13.2.1 寄主植物的抗性机制

长期以来,研究人员在培养抗性植物过程中关注的大多是迅速识别种质库(germplasm banks)中的具有抗性的基因型的方法和监控培养过程中抗性的遗传过程,而对基础性机制少有兴趣。佩特[89]总结了此领域中的科学原理和系统的研究方法,并在最近的综述中整合了植物培养方法和抗性机制的分析[90,111,135]。

佩特认为抗性产生有 3 个"原因",强调了与抗性相关的昆虫—植物关系中的一些重要的方面:(1)无偏好性;(2)抗生性;(3)耐受性。无偏好性描述的是昆虫的反应,昆虫对于植物的选择没有偏好性,而不是植物的一种特殊性状,所以此定义逐渐被"排拒性"(antixenosis)所取代(xenosis 是希腊语中的"客人",排拒作用意思是"抵挡客人"),即为植物特性唤起昆虫的负面反应[64]。抗生性表示植物使昆虫寿命、大小、生殖力下降。抗生性与排拒性相比,显然是指那些可对植食性昆虫的生理机能产生不利影响的植物特性。耐受性是植物虽然遭受了一定的虫害,但仍然可以正常生长和繁殖。耐受性不考虑昆虫(或另一种生物)是否对组织的损伤负责的植物特性。不像排拒性和抗生性,耐受性并不代表植物对昆虫种群数量的选择压力。因此,目前耐受性不再被认为是植物"抗性"的一部分,而是作为一种仅次于抗性的植物防御机制[118]。排拒性和抗生性非常适合在特定的实验室中进行评估选择,但耐受性是一种比较难评估的植物性能,因为它需要同时观察昆虫的种群及作物[135]的生产潜力。

到目前为止,耐受性似乎是最少见的一种防御机制。超过200个关于对蔬菜中节肢害虫抗性的报告表明,耐受性约占10%,其余的情况为排拒性或抗生性[117]。然而,耐受性已至少在13种[36]作物中被记录,而其较少的研究数量也反映了对它的重视不足。

必须强调的是,虽然植物的三种类型的抗虫机制已被证明是非常有用的,但对于昆虫的攻击,通常是两种或三种类型的防御机制的结合。

13.2.2 部分抗性

虽然有一些例外存在,但通常植物只能获得部分抗性,很难达到彻底抵抗某种特定昆虫的效果。然而抗性不只有一个好处,它对昆虫的种群造成较弱的选择压力,因而作用得更持久。与各种病虫害的综合防治措施相结合,部分抗性可能是足够的,甚至是更好的,因为降低了新的昆虫生物型[46,47]的发展。

在这种背景下,需要引入两个术语:水平抗性和垂直抗性。水平抗性或多基因抗性是指抗性性状由一些微效抗性基因混合形成一种基因型所控制的;而垂直抗性是寄主植物中由一个或多个基因控制的抗性,每个基因与害虫的一种寄生能力相匹配,因此有时也被称为基因对基因抗性。在许多作物中都发现过多基因抗性,单基因抗性的例子也可以在一些文献中找到。对后一种类型,研究最深入的是小麦对小麦瘿蚊的抗性,有关抗性的26个基因都已被鉴定出来了[90]。对于每一个抗性基因来说,在小麦瘿蚊中都有相应的基因(基因对基因)与之对应,用来克服小麦的抗性。

这两种抗性的类型与稳定性存在着显著的不同。水平抗性涉及来自不同种质资源的基因的积累。这种抗性一般很难被昆虫攻克,而建立一个这种抗性需要消耗很长的时间,这种抗性一般比垂直抗性稳定。抗性的稳定性有时可能会很短暂,尤其是当植物抗性的水平很高的时候,这种抗性很容易遗传,而这种具有抗性的品种被大规模地种植。在这种情况下,昆虫可能会通过发展出能克服宿主植物抗性的新特征来破坏抗性。有这样的例子,在抗性品种广泛使用之前,需要至少从三代中筛选抗性物种,但有时候昆虫克服植物抗性的潜力巨大,以至在这种抗性品种种植之前,这种抗性品种的抗性效果完全丧失。为了增加抗性的持续性,也存在许多措施,例如减轻对害虫的选择压力[35],然而,适应新的品种通常需要很长的时间,特别是在强的选择机制[39]下。有一些长期抗性的例子被大众所熟知,例如,据报道,早在1831年,苹果品种(Winter majetin)就对苹果棉蚜具有抗性,而且至今仍保留这一特质。另一个经常被引用的例子是:自1890年以来,法国葡萄园里的部分葡萄对葡萄根瘤蚜具有有效的抗性[90]。

多基因抗性,可能并不是单纯因为自身的缘故而更持久。例如,当它涉及一种单一的化合物时就是个例外了,但是当它涉及多种化学、生理或形态机制时,一种有害生物破坏抗性的可能性要低得多。在尝试去理解为什么有些案例中的有害生物物种很容易破坏植物抗性,而在其他案例中的植物抗性却很持久时,对昆虫的适应性进行分析是一个关键因素。可以想象,昆虫可能会适应当下的生理环境,例如,昆虫去适应食物里的有害化合物比去适应新植物特性容易得多。从后者来看,一系列的变化是需要关注的,包括影响产卵行为和饲喂等各种信号的适应性。这种观点得到以下事实的认同,植物培育者选择抗虫基因品种,正是考虑到对排拒性具有抗性的品种比抗生性类型更具价值。这是因为在他们的经验里,后一种类型特性一般不持久[117]。

13.2.3 植物特性与抗性

物理(图13.2)、化学或物候因素的不同造成植物的特征会有所变化,因而室内栽培植物的抗性与野外培育植物的抗性不同,这并不奇怪。关于这方面的内容,可以参考文献综述[75,90,114,117]。

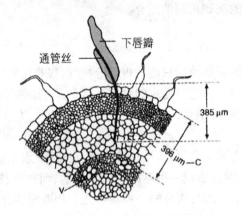

图13.2 昆虫的口器与植物特征的关系

毛蕃茄(*Lycopersicon hirsutum*)的皮质部(C)的厚度能阻止蚜虫(*Macrosiphum euphorbiae*)的口针到达维管组织(V)(奎洛斯等,1977)[100]

植物基于形态特征的抗性和化学抗性相比,通常可以提供更长效的保护。然而,物理障碍是可以被昆虫克服的,一种以无患子科(sapindaceous)植物的种子为食的姬像蝽(*Jadera haematoloma*)证明了这点。这种甲虫已在短短的50年里演替出不同长度的喙以适应新寄主的改变,使其能够食用比原来寄主更大或更小的种子(图13.3)[22]。

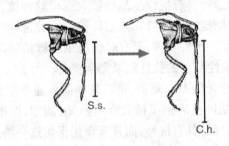

图13.3 姬像蝽的像

姬像蝽(*Jadera haematoloma*)是专食性取食无患子科(Sapindaceae)植物的昆虫,它的喙可以通过细胞壁吸食果实内的汁液,无苋子寄生在(*Sapindus saponaria*,果实半径:6.1 mm)的种群,喙的长度为6.7 mm;寄生于倒地铃(*Cardiospermum halicacabum*)(果实半径:8.5 mm)的种群,喙的长度为7.8 mm。S.s.和C.h.为按比例绘制的寄主果实半径(卡罗尔和西格尔,1996)[22]

通常情况下引入高抗性品种,往往要损失一些产量。这是因为,大多数形式的植物抗性的出现都会涉及一些资源的流动,增加了植物化感物质产量或额外的物理防御结构。在自然条件下,植物的防御系统是在植物天敌选择性压力下得以维持的,但它们也会尽量节省不必要的防御"成本"。以下的实验结果支持这种理论。野外拟南芥(Arabidopsis thaliana)有抵御害虫和病原体的能力,而与抗性相关的总糖苷浓度与腺毛密度却在逐步降低[74],这清楚地表明植物试图在减低受攻击危险的同时减少防御能量的消耗。

过去,对农作物的挑选是以提高农业产量为目的,但这种作物与抗性相关的次生物质的含量通常较低(表13.1),进而导致其容易遭受虫害。经观察发现,具最佳抗性的大豆品系产量仍然比无抗性品系的产量少。

表13.1 一些作物的次生代谢物质的含量化合物

	植物	含量%	参考文献
叮类生物碱	羽扇豆	0.5	140
葫芦素	南瓜	1	55
2-十三烷酮	西红柿	1.5	137
糖苷生物碱	马铃薯块茎	4	59
介子油苷	卷心菜	20	60
禾草碱	大麦	20	71

同样,当无虫害条件下,两种在抗虫等位基因上存在差异的大麦品种在生长环境中属于竞争关系,抗虫性低的品种产量会更高。然而当受到蚜虫(Schizaphis graminum)侵害时,结果就会被扭转:具有抗性的品种将更有竞争性[139]。然而,这样的例子很难被发现,原因之一是抗性品种和敏感品种的性能取决于多个环境条件,会很容易地掩盖抗性特征相关的损耗[13]。

不同的昆虫营养需求不同,对植物防御会产生不同的反应。只对一种昆虫具抗性的植物品种,通常很容易受其他昆虫的攻击。开发研制对大多数虫害具抗性的植物品种往往是很困难的,如开发抗三大棉花害虫抗性品种时遇到的诸多困难。较少受棉铃虫(Helicoverpa)幼虫影响的光滑叶品种却很容易受到牧草盲蝽(Lygus lineolaris)的损害。冰血(Frago-bract)品系的苞片稍作调整后使棉花花蕾显露出来,结果显示象鼻虫(Anthonomus grandis)的危害减少了,但牧草盲蝽对其危害增加。植物对植食性昆虫的抗性不仅是"直接"的,而且也可以通过三级营养水平的"间接"过程。不同品种可能产生不同的化感物质,如非洲菊(Gerbera jamesonii)曾出现过捕食性螨虫[69],如果植物培育方案中忽视了这方面,那么这样的间接性抗性因素可能在无意中被淘汰。这将导致在野外条件下,对农作物抗性的选择下降,特别是在天敌在昆虫的死亡率发挥显著作用的时候。

13.2.4 培育植物抗性的方法

利用作物品种的遗传变异和野外近缘品种,通过慎重的选择和育种方法,旨在提高抗虫害和抗疾病的能力。目前包括这些步骤:(1)利用种群增长模型探索抗性管理策略;(2)制定有效的实验步骤;(3)进一步开发排拒性技术,作为一种帮助的形式;(4)评价潜在的分子生物技术运用的可能性[109]。

1. 抗虫育种的常规方法

抗性基因的资源实际上是很丰富的,包括野生物种等分布在世界不同地区的大量品种[36,113]。1300多个包含改良农作物的重要基因库已在联合国粮农组织(FAO)的世界信息和预警系统(WIEWS)数据库中注册了。然而,在过去的100年里,重要的粮食作物品种的80%~90%已经丢失了,这表明了我们的潜在食物储层确实处于令人震惊的贫瘠状态[129]。

野生的品种为植物抗虫抗病提供了很有价值的原料[2]。例如,对于两个植物品种来说,高水平的抗性基因已经从药用野生稻转移到耕种稻中(*O. sativa*)[57]。据估计,如今植物育种家培育的种系中,本地种以及野生近缘种只占种质系(germplasm lines)的6%。

如上所述,抗虫性有时依赖于一个位点(单基因),但经常涉及几个独立的位点(寡基因,oligogenic),甚至许多位点(多基因遗传)赋予不同的抵抗方式。孟德尔遗传性状研究是相对简单的。单基因抗性常常在作物中被发现。一个经典的例子是水稻对稻飞虱的抵抗能力[90]。然而,在野生植物中,抗虫性是很少基于一个单一的抗性基因的。通常是几种防御模式(如化学、物理和营养因子的不平衡)相结合,并通过多个等位基因控制的复杂系统起作用。当抗性基因坐落在外来种质上时,将它们吸收到农艺当中,还需要有很多的工作去做。根据作物品种的生殖系统(即自花授粉或异花授粉),可用于各种育种计划,还有如潘德、库什[90]和史密斯等人[115]对抗虫育种的方法的著作。

2. 生物技术

传统的育种方法经改进后,现在可以通过巧妙的生物技术来实现快速繁育。近年来由于分子生物学和组织培养方法的发展,不仅使基因能在相关物种之间转换,而且在不相关的物种间也可以实现转换,甚至是其他亲缘关系更远的物种中,如动物、细菌和病毒[8,109]。基因工程的方法允许将新的基因引入作物,使它们具有抗虫性。例如,将昆虫病原基因引入到昆虫的寄主植物中,可以有效控制昆虫的种群。因此,一种可以控制产生毒素的昆虫病原菌苏云金芽孢杆菌的基因被引入到西红柿、水稻、棉花和云杉树的基因中。一些具有毒害作用的蛋白质,比如凝集素、淀粉酶抑制剂和蛋白酶抑制剂可以阻碍昆虫的正常生长,也可以通过转基因技术使得植物可以产生这些物质。例如,通过转移四季豆中编码α-淀粉酶抑制剂的cDNA到豌豆(*Pisum sativum*)中,可以使豌豆具有抵抗豌豆象的能力。α-淀粉酶抑制剂的抑制作用可以通过烹饪解除[108]。

转基因作物的引入正在以一个惊人的速度发展。四种商业化的转基因作物大豆、玉米、棉花和油菜,在近9年内,这些作物的覆盖面积占全球的百分比增加到29%(图13.4)。除草剂的耐受力是最显著的(72%),其次是抗虫性(20%)。最近已开始引进具有耐除草剂和抗虫性的多重基因的品种。这种组合被应用到了棉花和玉米上,到2004年这两种转基因作物已经占据全球作物面积的7%[56]。

转基因的方法可以被视为一种获得其他植物性状的特定技术,与传统的植物育种有一些重要的区别。事实上,转基因是使一个基因从一种作物的基因组上转移到另一种作物的基因组上。到基因组中的基因通常是非常详细的,因为转基因的目的是明确的。传统的植物育种一般都是植物物种或品种的交叉,一般都是通过未知基因和对应的性状的转移来增强植物的抗性。当植物增强的抗虫性是未知的时候,就很难确定其对环境会造成怎样的影响,以及对三级营养关系的影响。

另一种和传统植物育种之间的差异,例如 *Bt* 作物(含苏云金芽孢杆菌基因编码的杀虫蛋白的作物)的抗虫性是由很多的基因一起参与的。寄主植物抗性的稳定性和耐用性,在很大程度上取决于对耐药性的植物的遗传基础,也就是说这取决于一个主要的或多个微效基因。

第13章 昆虫和植物:如何运用我们的知识

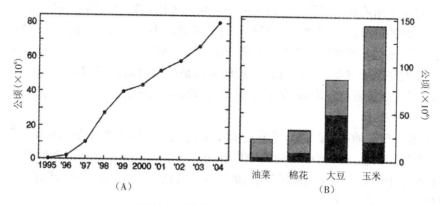

图 13.4 转基因作物的种植情况

(A)从 1996 年到 2004 年,全球转基因作物的种植情况;(B)全球范围内四种作物的种植情况(2004),其中黑色代表的是转基因的作物(詹姆斯,2004)[56]

迄今开发的 Bt 作物转移到植物基因组中的抗虫害基因都是一个基因,因此可以认为是一个垂直抗性。垂直抗性的稳定性一般小于水平抗性,因此它可能被某些昆虫克服[90,113]。

以下有几个选项可以提高垂直抗性[113]:

(1)连续释放培养,在额外的基因被连续释放之后,再释放主要的基因直到它变得无效。这种方法已经使用在水稻抗褐飞虱的基因转移上。

(2)基因聚合或者堆叠。例如,目前使用的水稻品种中的抗性基因是由 6 个曾是连续释放转移的抗性基因的聚合或堆叠而成的。

(3)基因交替,一个基因与另外一个基因交替出现。

转 Bt 基因作物做连续品种培养是不可取的,因为抗性的增强很可能会产生一些具有交叉耐药性的毒素[120]。此外,使用 Bt 喷洒有机作物,有可能对控制昆虫造成危害。

第三个主要的区别是,常规的植物育种和转基因 Bt 农作物育种在抗虫植物水平上是两种不同的方向。传统的植物育种的表达水平有它自己的特点,比如植物的化感素在常规情况下永远不可能超过自然的最高水平。在转 Bt 基因作物中,Bt 毒素基因的表达水平是由启动子[68,92,93]和插入到植物基因组中的基因拷贝数量决定的[65]。通过这种方法,转基因的植物中的 Bt 毒素水平可以高达植物体中所有蛋白质水平的 2% 到 5% 之间,这些植物是当前转基因 Bt 植物的 20 倍到 50 倍。植物如此高的表达水平,不仅比那些至今研究过及用过的剂量对非目标有机体有更大的影响,也有可能增强昆虫的抵抗力。目前 Bt 毒素技术的利与弊正在被哈奇和他的同事们研究[49]。

转基因抗虫作物是否可以更环保地对害虫防治做出贡献将取决于以下几个方面。第一,转基因对非目标生物的影响,包括传粉、生物防治和珍稀物种的保护。这涉及地上和地下的相互作用[48,67]。第二,转基因作物和害虫耐受性的适应性发展是一个在短期的或长期内需要解决的问题[121]。第三,转基因和基因渗入的野生近缘植物的异性杂交,这是需要考虑的重要方面,因为一旦这个转基因转移到野生亲缘植物中将不会再返回,而且它的特征会影响生态与非农业生态系统的相互作用[29]。大量关于转基因的潜在生态危机(和利益)的文献已在皮尔逊、普托迪维尔[95]和卡拉汉等人[86]的书中被概括出来了。

很明显,基因工程为农作物产量的提高以及抗虫性开创了一个途径[49,76,77],还被一些人当作农业生产技术应用的根本手段。与传统的杀虫剂应用相比较,建立在对昆虫毒性上的高端技术取得了明显的进步,因为毒性对非目标物种的影响减小了。然而与抗性机制相比,基于一个单独的强毒质还是有弊端的。昆虫已被证明会激发抗体去抵御杀虫剂,这种情况也可能会在基因改良的农作物中出现。抗性管理的最佳策略是所谓的高剂量避难(逃避)策略。这种做法的目的在于,通过强制栽培 Bt 农作物,来降低害虫抗性发展的风险,进而提高易感染害虫的存活率[21]。

因此,这些有价值的基因改造技术有可能是从过去的一些失败中学习而来的[48]。正如斯托内所说的那样[117]:"传统的抗虫植物的培育方法不可能被放弃。这方面的研究人员应该好好利用生物技术中的新发展所提供的机会,更应该保持在行为、生理、生态、进化等方面昆虫与其宿主植物的相互作用的关系。"

13.3 混养为何能减少病虫害

很久以前,农民就知道了在一块土地上种植多种作物可以提高农作物的产量。年轻的普利尼(公元 23–79 年)在他的《自然历史》这本书中写道当油菜(*Brassica napus*)和常见的野豌豆(*Vicia sativa*)长在一起的时候,很多本应出现在这些植物上的昆虫却不存在了。从此以后,大量研究就开始评估混种对昆虫种群动态的影响[3,54]。一项对叶甲科的条跳甲属十字花科(*Phyllotreta cruciferae*)对不同西兰花的危害的研究,可以阐明普利尼的观察发现。图 13.5 描述了被标记的叶甲从混种的花椰菜/野豌豆到单独种植花椰菜的变迁情况。首先将甲虫放于其生存的自然环境中,并将 350 个甲虫分为三组,每一组都染上不同的颜色,然后将它们释放在不同的情境中。经过 24 小时后发现混种的比单独种植的迁移率要高。特别指出的是,单一种植的植物留下的甲虫比混种的多很多,混种可以人为地使害虫数量迅速减少[42]。

共三组,每组 350 只跳甲被标记为蓝色、橘色或粉色(经加西亚和阿尔铁里,1992[42])

在热带地区,农业间作现行很普遍,作物混作比例最低的地区是印度(17%),最高的是马拉维(94%)[131]。相比而言,在西部地区,由于农业的现代集中化,在生物均一型上已达到一个很高的程度。大面积种植的单种植物往往具有非常低的遗传多样性。通过种植不同种类的农作物来增加植被多样性是一个典型的种植控制策略,可以使农业生态系统防御害虫或天敌。

在使用上,混养种植的方法是不一致的。间作指的是在某个范围里,间隔种植多种植物,间作可以分为四种类型:

(1)混作——两种或者两种以上的作物没有秩序的混合种植;
(2)行间作——多种作物成行的间隔种植;
(3)带状间作——两个或多个作物种植在相对较宽但仍然可以相互作用的范围内独立栽培;
(4)诱虫作物系统——在植物间,种植可以引诱害虫的植物使农作物免受侵害。

间作不一定要包含两种不同的植物种类。它也可以在其他层次上实行,即基因和品种水平上。复种涉及间作(即不同农作物同时生长)或连续种植(即一片土地上每年连续种植不同的作物)[131]。

不同作物之间的相互作用使得农业生态系统更加复杂,并且同时大量地减少害虫的侵害。有力的证据表明,混合栽培比单作供养的昆虫的负荷要少。已发表的 209 篇文章中,关于植

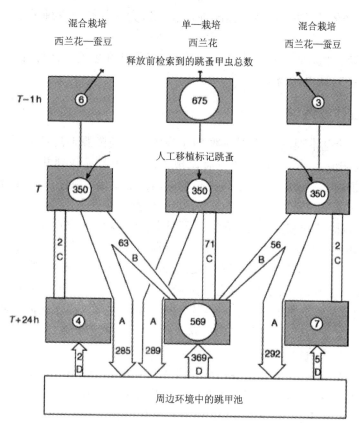

图13.5 在三组不同的作物系统中,测试被标记跳甲(*Phyllotreta cruciferae*)的选择情况

多样性在农业生态系统中对昆虫影响的调查表明,52%的植食性昆虫在混合栽培中的数量要少于在单作中的数量,然而只有15%的昆虫在混合栽培中表现出较高的总体密度(表13.2)[5]。

正如人们预测的那样,在混养中,专食性昆虫呈现低丰度,而多食性昆虫经常(但是不总是[14])生长较好并且在混合栽培中显示出更高的密度(表13.2)。当然,不是所有的农作物的混合在这方面同等有效,选择组成的农作物比单单决定实现间混作更为重要。例如,使小麦和玉米结合,实际上将会提高两者遭受共同害虫破坏的水平,比如麦长蝽(*Blissus* spp.)和线虫,然而间作小麦和土豆会减少对小麦的危害。

表13.2 混合栽培与单独栽培相比,节肢动物的相对丰度(安东尼,1991)[5]

植食性昆虫	更丰富(%)	无差异(%)	更少(%)	变化的(%)	物种总数
单食性	8	14	59	19	220
多食性	40	8	28	24	67
天敌	53	13	9	26	130

虽然许多研究已经证明了单系统与混养系统在抗虫方面的差异,但仍缺乏研究这些影响机制的准确信息。大量的生物和非生物的因素在两个系统中变化,包括植物的密度和结构复杂性,小气候的因素如温度、光照、湿度、庇护所、天敌、可选择的食物资源(花和花外的蜜腺)、掩蔽、气味和伪装[11]。

然而,发现间作的基本机制,生成预测理论和在这方面的知识管理系统的应用是至关重要的。

阐释在混合栽培中减少害虫泛滥的三个学说已经得到许多关注,包括:(1)破坏性作物假说;(2)天敌假说;(3)诱虫作物学说[3,131]。

13.3.1 破坏性作物假说

生态学中的一个基本理论是,消费者倾向于集中在可利资源丰富且易于找到的地方。其根源是"资源浓度假说"(见10.8)。通过假说预测,昆虫更容易找到并在单独的寄主植物而不是混养的寄主植物上生长。空间上,寄主植物的离散度直接影响昆虫数量,而且与寄主植物相关的植物物种对昆虫找到并利用它的寄主有直接影响。非寄主作物散发出的挥发物可能会遮蔽寄主植物的气味,从而破坏了害虫的主动寻找行为。这样的嗅觉掩蔽已经被证实,例如,土豆气味对马铃薯甲虫的影响,实验中饥饿的马铃薯甲虫对气流中马铃薯的气味表现出强大的主动搜寻反应,然而对气流中番茄气味的反应,同空气的反应没有区别。有时,寄主植物的气味,会被两种植物混合在一起的气味完全掩盖(图6.13)[123]。一个著名的例子是,将胡萝卜和洋葱混作,来阻止胡萝卜害虫侵扰。一些芳香的草本植物可以用于驱除大批滋生于蔬菜作物的昆虫[130]。球芽甘蓝和鼠尾草(*Salvia officinalis*)和百里香(*Thymus vulgaris*)混作,小菜蛾的产卵量比在单独的寄主植物上的少[33]。

在植物的混合栽培中,昆虫也表现出离开寄主植物的趋势,紧随其后的是经常移出耕作地。在黄瓜、玉米、西兰花混作中,获得的有条纹的黄瓜条叶甲的密度比在单作中要少10%到30%(表13.6)[9]。有趣的是,相关作物又对昆虫有间接影响。当在实验室条件下,提供给甲虫的两个选择是单作的叶子和黄瓜与番茄混作的叶子,昆虫更喜欢单纯的植物叶子[10],这表示植物多样性和寄主植物质量可能通过一种复杂的方式相互影响。

这种植物影响更高营养级的化感作用是非常常见的。因此,已经被证实当蚜虫的寄主植物大麦草暴露在蓟花产生的挥发物中时,大麦草对蚜虫可接受性降低[44]。在实验室测试中,还发现了一些大麦品种的气味可以降低蚜虫(*Rhopalosiphum padi*)对其的危害。这一发现表明,可以改变一些大麦品种的种植组合,来防止虫害[84]。这些化感作用的机制还有待阐明。

13.3.2 天敌假说

根据鲁特的天敌假说,混种的作物比单一种植的作物需要更多的天敌来控制虫害,因为混种通常能够提供一些额外的食物资源,比如蜜汁、花蜜以及更多的在炎热的时期可以让昆虫躲避在阴凉下与高温潮湿抗争的避难所。此外,在物种稀缺时期可为植食性昆虫和捕食者提供更多的替代食物[24]。据调查,害虫的天敌在不同植物种群的迁出率不同,但混种并没有影响天敌的迁入率[25]。一项文献调查显示,与单作相比,130种有害昆虫的天敌中的68种(53%)在混养中确实获得了更高的数量密度,而仅有9%的数量密度变得更低(表13.2)[5]。与单作相比,玉米在与蚕豆(*Vicia faba*)和南瓜(*Cucurbita moschata*)的混种中,产量有了100%的巨大提高。这都得归功于蚜虫(*Rhopalosiphum maidis*)和螨虫数量密度的降低。在混种中,蚜虫被好几种肉食性的节肢动物所袭击的可能性增大,包括两种与蚜虫有更直接的联系的瓢虫[128]。一项关于在混养中,害虫数量降低的原因的分析表明:在36项研究中的12项主要是因为天敌活动的影响(表13.3)[11]。

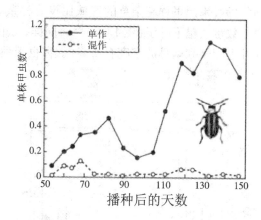

图 13.6 在单作的黄瓜,还有黄瓜、玉米和西兰花混作的植株上,黄瓜叶甲的数量变化(巴赫,1980)[9]

表 13.3 36 项关于农田间作系统的研究,报道的可以降低虫害的重要调控机制(巴利德瓦,1985)[11]

虫害控制因素	报道次数
降低作物密度	9
诱虫作物、微气候和物理阻碍物	
减少外来物种的引入	5
隐蔽和伪装	5
抗性	5
自然天敌	12

13.3.3 诱虫作物学说

诱虫作物就是种在主要农作物生长的土地边缘,或者某个特定部位来吸引有害昆虫,从而使得目标作物不被害虫侵害的作物。诱虫作物系统对热带国家的农民来说尤其有用。因为,在美国中部单作的西红柿被毛虫(*Spodoptera sunia*)破坏得非常厉害,然而西红柿与豆类的套种可以有效地防止害虫侵害。毛虫都被诱虫作物豆类吸引了。至今,诱虫作物种植仅应用在几种作物中,如棉花、大豆、马铃薯和花椰菜。在这些作物中,棉花和大豆的诱虫作物系统是最重要的。大量成功的先例也暗示这个策略可以被更广泛地应用[53]。还有一些关于保护作物的小规模实验,例如韭菜、卷心菜[7]和甜玉米[102],可以通过诱虫植物来对抗害虫,这表明这种方法值得被进一步探索。

野草也能够影响害虫的种群密度[125],它们通常能够庇护那些邻近区域的植食性昆虫的天敌。因为杂草可以给害虫的天敌提供重要的资源,例如食物和微型栖息地,这些是在无杂草的单作模式里不存在的。而这种天敌相应的害虫在这种系统中很难生存。有很多例子都说明了,杂草的存在提高了对于某种特定作物的害虫的控制力,从水果(例如苹果)到蔬菜(例如芽橄榄)、纤维植物(例如棉花)、谷类(例如高粱)以及葡萄树中都可以说明[3]这一点。在一个研究绿色控制害虫方法的实验中,摒弃了在苹果园里使用的传统杀虫剂,而挑选了一块地播种杂草来吸引害虫。在一段时间之后,果园里种有杂草的区域比没有种杂草的区域里有更多的节肢动物,因为两种有害蚜虫的数量大幅下降了(图13.7)[141]。同样,与单作相比,由于葡萄园中植物的多样性的增加,葡萄叶上叶蝉和蓟马的密度有所降低,害虫天敌的数量增加了50%,结果对两种昆虫的生物控制力都提高了[83]。

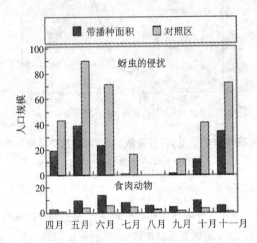

图 13.7　在一个苹果园中,捕食性节肢动物的平均数和蚜虫感染率之间的关系
注:在苹果园的一个区域播种有黑麦草,而对照区域没有黑麦草(怀斯,1995)[141]。

另外还有一种方法就是在土地的边缘种上一些可以开花的杂草,这样就能够为植食性昆虫的天敌提供花蜜。对开花植物的选择十分重要,为此,必须确定害虫是否也能够从这些制造花蜜的植物中受益。毕竟,许多害虫成虫只吃花蜜,而它们的幼虫却吃庄稼[85]。

一些田间试验也表明,在农业系统中,农作物多样性通常能有效地降低害虫的数量。更多关于通过杂草来控制昆虫数量的细节在阿列细斯和尼科尔斯的文章中被提出[3]。

13.3.4　多样性原则

农业活动本身意味着自然界生物多样性的简单化,这就导致了一个需要人不断去介入的人造生态系统的形成。在单作中多样性程度更是极其低,多样性可能是将来农业生产中控制昆虫的一个关键因素[97,119]。混作将会为迈向这个未来提供很重要的一步。混作中有一种很有趣的形式,它可以克服两种作物混养在一起的技术上的不足,那就是种植两种在基因方面完全不同的品种。当易受粉虱侵害的树薯的品种和能够抵抗粉虱的作物种植在一起时,在这种间作系统中,粉虱的数量密度就比单作时降低了60%[45]。然而,截至目前,混养在一起对某一害虫敏感或有抵抗力的作物很多仍未被发现。

混作已经被验证能提高作物产量,而且有时提高的幅度很惊人。对这种能够导致害虫数量降低的机制进行分析不是很容易,尤其是涉及的因素比较多时(表13.3)。尽管根据一些有限的研究,可以知道很多种类的控制机制,其中包括降低害虫对某一作物的关注,天敌的调控,还有各种各样的机制的转换,这些都可能对混作中的高产量做出贡献。

13.4　植物源杀虫剂和拒食剂

有确凿的证据证明,就算不是所有的,至少也是大部分的植食性昆虫在食用了植物的次生代谢产物之后,生长趋势减弱。所以,这将是保护作物的一个好办法。确实,从人类文明刚开始的时候,人们就已经学会利用植物去和害虫做斗争或者减轻害虫带来的损失。尽管早期的农业著作通常含有用植物去控制昆虫的例子,但是对于植物的描述都含混不清,所以要下的定论是确切的。然而,有据可查的记录显示,属于 16 个不同科的 185 020 种植物在欧洲和中国西部地区的农业与园艺中被用于控制害虫[82,112]。最近的一项研究表明,许多植物被用于对抗西部非洲甲虫虫害,来保护作物种子[18]。

使用植物源性化学物质来控制害虫,相比于那些最高效的合成杀虫剂,可以有效减少对于环境和健康的负面影响。尼古丁、鱼藤酮和除虫菊酯已作为十分有效的杀虫剂被广泛使用,因为它们可以迅速降解不会积聚在食物链中。然而值得注意的是,尽管许多天然杀虫剂对哺乳动物显示出较低的毒性,例如大多数有机氯化合物,但它们并不是无害的,另一种说法认为当我们寻找新的杀虫剂时,无论它是不是天然的,都应该保持谨慎,因为目标昆虫可能对其产生抗性,并且更重要的是,对于非目标无脊椎动物来说,包括自然天敌,都会存在风险。更改目标物种的行为与无毒的作用形式结合,可以提供长期的最可靠的方法[79]。具有活性的植物化学物质包括引诱剂、驱虫剂、阻碍剂,目前在文献中已经有几百个例子对比进行了讨论。

13.4.1 拒食剂

植物产生的一些化合物,可以减少摄食损伤和感染植物病原体的风险。昆虫通过减少食物摄取来响应感官检测到的拒食效应,这可能导致它要么离开该植物,要么其生长、发育、生存和繁殖会受到不利影响。与驱虫剂相比,拒食剂不引起导向运动远离刺激源[30]。在拒食化合物存在时,昆虫可能会饿死,雌性个体可能会停止产卵,直到它们找到另一个寄主。已发现一些拒食素在低剂量时,约小于百万分之一时有效,印楝素是已知的最强的拒食素之一,当给杂食性荒地蚱蜢(*Schistocerca gregaria*)喂食 0.01ppm 时,1 mg 化合物足以保护 100 m^2 被害虫残害的区域。

利用拒食素可以控制昆虫活动,成为候选化合物要有几个必要的基本特质(表 13.4),然而迄今为止只有少数测试的化合物满足条件[66]。因为所有的植物次级物质(估计有 400 000 种或更多)中只有不到 1% 被有限的昆虫物种测试过。这些化学物质作为潜在拒食素已经引起了关注(表 13.5)。

迄今为止,只有以印楝素为基础的产品上市了。蓼二醛(Polygodial)是从草本水蓼(*Polygonum hydropiper*)中提取的类倍半萜烯,在极低施用量时,可以防止蚜虫的探测行为。

表 13.4 拒食剂作为杀虫剂的标准

1	对于脊椎动物无毒或毒性很低
2	对于植物无毒或毒性很低
3	很低的浓度具有活性
4	对多种虫害都有效
5	对于有益昆虫无害(如天敌、传粉者等)
6	能渗入植物内部,被植物所吸收
7	不与其他的控制虫害方法产生冲突
8	在环境中可降解
9	原始资源足够丰富
10	可以用于商业运作(成本较低)
11	保质期较长

表 13.5　基于野外试验结果,一些有希望应用于虫害治理实践中的拒食素

化合物	植物来源	可控昆虫	参考文献
柠檬苦素类似物(Meliacins)化合物	楝科	多种昆虫	107
补身烷(Drimanes)化合物	水蓼	蚜虫	94
柠檬苦素	葡萄柚	马铃薯甲虫	80

13.4.2　苦楝树:印楝素

几个世纪以来,印度农民、家庭主妇和民间治疗师就已经知道印楝树有许多非凡的特性,包括对许多昆虫强烈的排斥性。半个多世纪前,阿尔及利亚农学家注意到,蝗灾过后,只剩下楝树没有被毁灭,表明楝树的叶提取物中含有高浓度的令沙漠蝗虫不喜欢的味道[132]。随着DDT和一系列后续的广谱合成杀虫剂的使用,楝树没有再被应用于防治昆虫方面方面,直到20世纪70年代,德国昆虫学家施穆特勒尔鼓励研究人员在世界各地开展研究印楝的可用属性[107]。

楝树(*Azadirachta indica*,楝科),可能原产于缅甸,很长一段时间已广泛种植在亚洲的热带地区和非洲,已成为这些地方的原产植物了。因为它的快速增长和优良木材,现在还在中美洲广泛种植。这种树有非常强的适应性,能够承受干旱的条件。二回羽状的叶子被损坏时有大蒜的香味,其果实呈橄榄状(图13.8)。

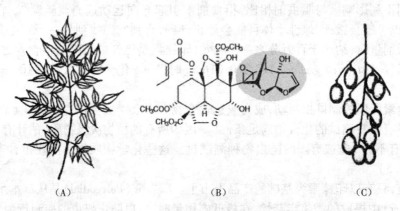

图 13.8　印楝棒

(A)印楝(*Azadirachta indica*)的二回羽状复叶合和圆锥花序;(B)印楝素的结构式;(C)印楝素可以作为昆虫拒食素(施穆特勒尔,2002)[107]

楝科与其姐妹科芸香科为近缘科,可以产生和积累具有生物活性的苦味的化合物(nortriterpenoids),根据结构特点将其称为柠檬苦素。印楝素是楝树的70多种三萜中唯一一个由许多互相接近的活性官能团高度氧化的柠檬苦素类化合物(图13.8)。它主要产生于印楝树(*A. indica*)的种子,每克干种子内浓度为大约3.5毫克,这对许多昆虫来说是一个非常强大的拒食素,尤其是鳞翅目的幼虫和直翅目的一部分。除了昆虫的拒食行动,摄入印楝素后会导致生长抑制、畸形,由于干扰昆虫的内分泌系统打乱了繁殖,严重者导致死亡[27,78,107]。有证据表明印楝素分子的左半部分是无效部分,而另一半(正确的一半,即图13.8中灰色部分的分子)对昆虫的拒食活动尤为重要[17]。到目前为止,没有大量的合成印楝素类似物的来源物质。唯一与印楝素有相似生物活性的物质是二氢印楝素(dihydroazadirachtin),这种化合物在光照下也比

印楝素更稳定[111]。

人们发现印楝素会通过抑制消化酶而降低食物利用率[126]。对于一些昆虫而言，一些相关的化合物如印楝沙兰林(salannin)也存在于籼稻种子和川楝素中，与相关的川楝(*Melia toosendan*)的树皮相隔开，甚至比印楝素更难吃[73]。而印度大陆的小农户使用传统印楝提取物，现在一些西方国家的市场上也会有商业楝产品并有一些专利配方[40]。

在目前所有已知的能阻止昆虫取食或产卵行为的植物化合物中，印楝素具有最大的潜力[81]，在许多方面实现了拒食素的作用。尤其是环境中存在有益的生物时[122]，它对哺乳动物无毒性[19]，在农作物系统中传递[6]，刺吸式口器的昆虫也不敢进食，例如几个水稻害虫飞虱[106]。然而，其对紫外线极其敏感，使得防晒过滤器的使用成为必须，如卵磷脂。目前采集的新鲜种子为印楝素的主要来源[101]。由于其复杂的分子结构，印楝素迄今为止没有被化学合成。

如施穆特勒尔所述[107]，有令人信服的证据表明：楝树有"解决全球问题"的潜在能力。此外，楝科的相关植物也可以有类似的机会。

13.4.3 拒食素作为杀虫剂的前景

正如之前所讨论的(第7章和第8章)，昆虫通过反复接触后，有可能对一些阻碍物逐渐适应，尤其在杂食性昆虫可能会出现时[58]。实际上，在许多昆虫中已经观察到，昆虫可以与低纯度的印楝素适应，包括日本甲虫[51]和亚洲黏虫[136]。有趣的是，当检测商业化高纯度的印楝素和印度棟树油时，没有发现此适应性的产生(图13.9)[20]。

在最近的一项研究中，给杂食性的粉纹夜蛾的初孵幼虫喂养带有经过单一拒食剂处理过的叶子，或者将其与其他抑制剂混合一起喂养直到三龄，然后进行选择性生物测定。单一拒食素饲养的幼虫显示出强适应性，而混合喂养的幼虫则没有表现出适应性。早些年前，曾有人发现这样一个结论，在减少昆虫取食方面，混合的化感物质比单独的纯化感物质更有效率。因此，基于两种或更多化感物质比基于一种单一化学物质的农药更适合于对农作物持久的保护。包含许多其他种类混合物的印度楝产品比基于单一一种印度楝物质能提供更好的昆虫控制效果。

另一个重要问题是，当考虑对自然物质的混合物进行修饰时，我们发现害虫的抗性也处于不断的发展之中。带有菱形斑纹的小菜蛾幼虫(*Plutella xylostella*)的长期选择实验表明，其可以产生对印楝素的抗性，尽管比使用杀虫剂溴氰菊酯时产生的抗性低。而对于印楝种子提取物，含有包括印楝素外的其他药剂，其抗性比纯印楝素产生得更慢[133]。据推测，印楝素对昆虫行为和生理产生的综合影响，使昆虫更难形成抗性。不同于基于单一活性成分的普通杀虫剂，植物的化学防御包括大量不同的，对昆虫产生行为、生理和毒理学特性的化合物。因此昆虫更难以适应。比如理论上，开发拒食性农药应该很容易，但对于寡食性昆虫是不太可能的，因为这将导致其宿主范围快速改变[58]。

另外，对于拒食剂难以识别的原因，是因为物种的敏感性与给定的拒食素化合物之间存在很大差别(表7.6、表13.6)。

表13.6 面粉中可以引起两种蝗虫取食量减半的拒食素浓度(ppm) (伯尼斯和查普曼，1978)[15]

	印楝素	马兜铃酸
沙漠蝗(*Schistocerca gregaria*)	0.1	0.1
飞蝗(*Locusta migratoria*)	100	0.01

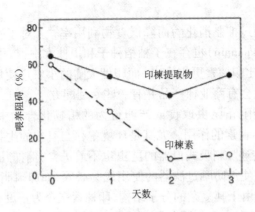

图 13.9 斜纹夜蛾(*Spodoptera litura*)5 龄诱虫,取食 1.3 ng/cm² 印楝素处理过的卷心菜叶碟,或包含相同数量的印楝树抽提物后的拒食现象(邦福德和恩斯曼,1996)[20]

大多数研究者在测试候选化合物时,只测试少数甚至只有一个昆虫,因此特定有效的拒食素很容易逃脱人们的关注。一项研究测试七种直翅目昆虫对印楝素的灵敏度,种间的差异相差了 6 个数量级[78]。关于拒食素方法的基础知识在一些论文中已被列出[40,58,66]。

大规模使用拒食素会在未来实现么?回顾目前农业活动造成的环境压力,我们不能离开未被探索的自然防御物质。事实上,许多植物在很大程度上依赖于这种化合物,这也是人们对植物王国的不断探索的动力。印楝素的应用似乎支持弗雷泽和查波[40]的观点:"使用天然产品抑制剂来控制昆虫正在迅速成为现实。"期望在许多国家的农药市场上推出楝产品,例如仅印度就有 100 多个商品制造商[91]。

13.5 植食性昆虫可以控制杂草

许多植物物种已经被人类有意或无意地转移到世界的其他区域。外来植物一旦存活于它们的自然栖息地之外,有时就会发展成为激进的入侵者,而与本地植物相互竞争,对自然生态系统造成恶劣的影响,或对农业生产造成巨大的损害。在世界的一些地区有 60%~97% 的杂草是移民外来物种(表 13.7),而全球对于除草剂的支出比杀虫剂约高出 30%[97]。

表 13.7 北美和澳大利亚杂草的起源(加斯曼,1995,皮门特尔,1986)[43,96]

	杂草的数量	杂草的起源				
		本地	欧洲	美洲	亚洲	非洲
加拿大	516	40	52	4	3	0
加拿大,常见杂草	126	2	71	25	2	0
澳大利亚	637	7	39	26	7	18
澳大利亚,维多利亚州	83		60	23	4	10
美国,种耕作物种的杂草	80	28	50	8	5	1

外来植物成为杂草,有时可以通过从植物的原产地引入特定的昆虫进行控制。有两个非常成功的案例,就是通过引入植食昆虫控制外来入侵杂草的。

13.5.1 仙人掌和槐叶萍

仙人掌是原产于北美和南美的物种。约30个种的仙人掌被引入澳大利亚做盆景或花园植物,在引入后,其中两个种——仙人掌和无刺仙人掌失控后变成了入侵物种。1839年,无刺仙人掌(*O. stricta*)作为一种盆景植物,从美国南部引入澳大利亚,并且作为一种篱笆植物种植在澳大利亚东部。它逐渐发展成为一种令人讨厌的杂草,很难控制。到1900年已经占领了澳大利亚沿海地区的400万公顷,然后迅速蔓延到内陆地区,如小麦种植区、牧场和农业用地的边缘,使其他植物的生长受到严重阻碍。到1925年,已经出现在昆士兰和新南威尔士约2 500万公顷的土地上。由于其快速的增长,这个区域大约一半的地方都覆盖着这种植物,其高度超过1米,这种密度几乎超出了人类和牲畜的承受范围,农场都被废弃了。1920年人们开始尝试通过昆虫和螨取食这些仙人掌来控制杂草,但没有重大进展,直到1927年和1930年之间,释放了大量的仙人掌螟(*Cactoblastis cactorum*),其幼虫取食仙人掌茎部。3年内,在其幼虫的取食下,仙人掌种群开始退化。生物防治项目在1939年已经取得很明显的成功。多德说:"仙人掌入侵的领土已经被神奇地改变了,它们从把一些地区弄得荒芜,到现在这些地区重新变得繁荣起来,这是我们的胜利。"目前这种幼虫仍然使仙人掌保持着低密度、稳定和平衡增长的趋势。

槐叶萍(*Salvinia molesta*)是一种漂浮生长的水生蕨类植物,约2~10厘米长,原产于巴西东南部,自1939年来,经由人类活动传播至世界上许多热带和亚热带地区。其无限增长已经造成了严重的问题,因为它的种群聚集可以达到一米的厚度,覆盖湖面和稻田,彻底阻塞所有的水道,包括移动缓慢的河流和水渠。其入侵至巴布亚新几内亚的塞皮克河漫滩后快速扩散,生长能力十分惊人。8年内,覆盖范围达250平方公里,重达200万吨,严重扰乱了当地人的生活,人们被迫迁移,并放弃了整个村庄[104]。后来,有人发现巴西的槐叶萍能被一种很小的(2毫米长)象鼻虫(*Cyrtobagous salviniae*)攻击。20世纪80年代初,这种昆虫被广泛引入,不到1年时间,象鼻虫种群数量从几千增长到100多万,破坏了30 000吨槐叶萍。在非洲和印度,象鼻虫也缩减了99%以上槐叶萍种群,实现了新的、低密度的平衡[104]。

13.5.2 生物控制杂草项目的成功率

并不是所有的昆虫在控制杂草方面都能取得上述那些惊人的成就,而事实上只在较少情况下才能达到完全控制(图13.10)[43,110],并且控制程度在不同环境下会发生变化。

通常认为,利用植食性昆虫控制的是入侵的物种,然而在某些情况下,也可以用于控制本地的杂草。因此,英国正在执行控制蕨菜(*Pteridium aquilinum*)的生物控制方案。这种植物在全球范围内种植得非常成功,已引起世界不同地区的食草动物对其的进攻。一种原产于南非的,取食叶肉组织的叶蝉(*Eupteryx maigudoi*),正在被调查是否适合在英国作为生物控制剂使用[38]。

通过使用植食性昆虫来进行杂草控制的优点,已不需要再强调了(表13.8)。然而其最薄弱的一点,就是结果的不可预测性。这是由以下事实导致的:一些必需的统计参数只有在植食性昆虫被释放后才能获取,因为它们的值主要取决于当时的条件。

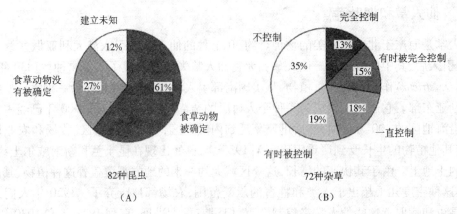

图 13.10　通过引入外来植食性动物对于控制入侵杂草的成功率
（A）成功在欧洲入侵杂草建立种群的节肢动物的比例；（B）昆虫对杂草控制的程度（朱利恩，1992）[60]

表 13.8　生物控制杂草的优势与劣势

优势	劣势
对于目标物种的合理化长久治理 对于环境安全，没有副作用 只针对特定的物种（或者只涉及很少的一部分近缘种） 通过密度制约和主动扩散，控制杂草的物种能达到自身永久存在 收益很高 不会复发	见效较慢 如果杂草与某种农作物近缘，那么可选择的昆虫种类会大幅减少

　　而一些负面的、使用昆虫控制杂草间接的和非靶标效应的问题，已经引起了人们对于这个方法可靠性的关注。由于缺乏释放后的监测，已严重阻碍对效果的评估。为了减少生态风险，加强公众对这种方法的信任，需要进行更广泛的风险和收益评估[26]。

　　一个引起关注的有趣的问题是：最有效的控制生物是否是对寄主植物最为适应的昆虫？高度专化的昆虫，面对无限的食物资源可能会迅速大量繁殖。也有人认为，相比昆虫与植物有着长期紧密关系的类型，昆虫与植物的关联不是很密切时，植物可能会更容易受到虫害[38,97]。仙人掌螟（*Cactoblastis cactorum*）在控制仙人掌（*Opuntia stricta*）和无刺仙人掌（*O. Inermis*）方面取得了非常大的成功，但是仙人掌螟是发现于南美洲，是仙人掌的另一种植物（非仙人掌和无刺仙人掌）中的植食性昆虫。

　　而象鼻虫是高度适应槐叶萍的物种。使用严格的单食性昆虫可以降低它们转移到其他宿主上的风险。而另一个复杂因素是，在本土地区，天敌对植食性昆虫种群密度的调节作用。人们普遍认为，用生物控制杂草的重要一环是解决如何将昆虫的天敌与植食性昆虫分离开来的问题[38]。

　　目前，对于杂草控制还缺乏坚实的理论基础。也许只基于一个控制因子的做法是完全错误的，应该加入多元化的控制系统，包括植物病原体等等，会比较合适。马缨丹（*Lantana camara*）可以作为一个例子，在夏威夷它是一种杂草。只有在引进数种昆虫之后，才可获得对该物种的控制。因此，更好地了解调节植物种群的因素，将有助于使用纯天然制剂（包括昆虫）去改善杂草控制方法[28]。

13.6 结论

正如卢肯比尔[72]所总结的那样："对自然界中的虫害防治只需发现并利用好我们身边的动物和植物。这样就能解决我们面临的许多严重的虫害问题。"本章指出了证明卢肯比尔看法的一些方法，阐述了利用大自然提供的工具可以尽量减少虫害带来的损失。

基于生物学原理，对于昆虫进行成功控制的共同原则是多样化的[97]。抗性育种依赖于大型基因库的可用性，并且当各种因素相结合时，其抗病虫害的能力是更为有效的。作物栽培措施的多元化，如间作轮作，往往能减少严重虫害的风险。拒食素化合物应用于混合物中是有效的，并且当它们影响各种行为和生理机制时，这种效果更强。利用生物控制杂草的成功例子的数量相当有限，有可能是由于没有引起人们对于多样化的足够重视。

本章的重点是将我们学到的昆虫与植物关系的知识应用到农作物生产中去。应该强调的是，这方面的知识与自然生态系统的管理（在自然保护方面）一样重要。当前我们对于世界各地多样性的关注正在减少，我们需要科学家以及决策者一起，通过保护自然栖息地等方式来保留现存的遗骸，正如威尔逊在《生物的多样性》书中的那些具有说服力的并且富有表现力的阐述一样[138]。

可持续农业多样化的关键因素是以自然生态系统管理为基础，这同样是一个重要概念。

13.7 参考文献

1. Akhtar, Y. and Isman, M. B. (2003). Binary mixtures of feeding deterrents mitigate the decrease in feeding deterrent response to antifeedants following prolonged exposure in the cabbage looper, Trichoplusia ni (Lepidoptera: Noctuidae). Chemoecology, 13, 177–82.

2. Alonso-Blanco, C. and Koornneef, M. (2000). Naturally occurring variation in Arabidopsis: an underexploited resource for plant genetics. Trends in Plant Science, 5, 22–9.

3. Altieri, M. A. and Nicholls, C. I. (2004). Biodiversity and pest management in agroecosystems (2nd edn). Food Products Press, New York.

4. Ames, B. N., Profet, M., and Gold, L. S. (1990). Dietary pesticides (99.99% all natural). Proceedings of the National Academy of Sciences of the USA, 87, 7777–81.

5. Andow, D. A. (1991). Vegetational diversity and arthropod population response. Annual Review of Entomology, 36, 561–86.

6. Arpaia, S. and van Loon, J. J. A. (1993). Effects of azadirachtin after systemic uptake into *Brassica oleracea* on larvae of *Pieris brassicae*. Entomologia Experimentalis et Applicata, 66, 39–45.

7. Åsman, K. (2002). Trap cropping effect on oviposition behaviour of the leek moth *Acrolepiopsis assectella* and the diamondback moth *Plutella xylostella*. Entomologia Experimentalis et Applicata, 105, 153–64.

8. Babu, R. M., Sajeena, A., Seetharaman, K., and Reddy, M. S. (2003). Advances in genetically engineered (transgenic) plants in pest management—an overview. *Crop Protection*, 22, 1071–86.
9. Bach, C. E. (1980). Effects of plant density and diversity on the population dynamics of a specialist herbivore, the striped cucumber beetle, *Acalymma vittata* (Fab.). *Ecology*, 61, 1515–30.
10. Bach, C. E. (1981). Host plant growth form and diversity: effects on abundance and feeding preference of a specialist herbivore, *Acalymma vittata* (Coleoptera: Chrysomelidae). *Oecologia*, 50, 370–5.
11. Baliddawa, C. W. (1985). Plant species diversity and crop pest control. An analytical review. *Insect Science and its Application*, 6, 479–87.
12. Barbosa, P. (1993). Lepidopteran foraging on plants in agroecosystems: constraints and consequences. In *Caterpillars. Ecological and evolutionary constraints on foraging* (ed. N. E. Stamp and T. M. Casey), pp. 523–66. Chapman & Hall, New York.
13. Bergelson, J. (1994). The effects of genotype and the environment on costs of resistance in lettuce. *American Naturalist*, 143, 349–59.
14. Bernays, E. A. (1999). When host plant choice is a problem for a generalist herbivore: experiments with the whitefly, *Bemisia tabaci*. *Ecological Entomology*, 24, 260–7.
15. Bernays, E. A. and Chapman, R. F. (1978). Plant chemistry and acridoid feeding behaviour. In *Biochemical aspects of plant-animal coevolution* (ed. J. Harborne), pp. 99–141. Academic Press, New York.
16. Bird, J. V. (1982). Multi-adversity (diseases, insects and stresses) resistance (MAR) in cotton. *Plant Diseases*, 66, 173–6.
17. Blaney, W. M., Simmonds, M. S. J., Ley, S. V., Anderson, J. C., Smith, S. C., and Wood, A. (1994). Effect of azadirachtin-derived decalin (perhydronaphthalene) and dihydrofuranacetal (furo[2,3-b] pyran) fragments on the feeding behaviour of *Spodoptera littoralis*. *Pesticide Science*, 40, 169–73.
18. Boeke, S. J., Van Loon, J. J. A., Van Huis, A., Kosu, D. K., and Dicke, M. (2001). Review of the use of plant material to protect stored seeds, with a focus on *Callosobruchus maculatus* infesting cowpea. *Wageningen University Papers*, 2001–3, 1–108.
19. Boeke, S. J., Boersma, M. G., Alink, G. M., Van Loon, J. J. A., Van Huis, A., Dicke, M., et al. (2004). Safety evaluation of neem (*Azadirachta indica*) derived pesticides. *Journal of Ethnopharmacology*, 94, 25–41.
20. Bomford, M. K. and Isman, M. B. (1996). Desensitization of fifth instar *Spodoptera litura* (Lepidoptera: Noctuidae) to azadirachtin and neem. *Entomologia Experimentalis et Applicata*, 81, 307–13.
21. Bourguet, D. (2004). Resistance to *Bacillus thuringiensis* toxins in the European corn borer: what chance for *Bt* maize? *Physiological Entomology*, 29, 251–6.
22. Carroll, S. P. and Dingle, H. (1996). The biology of post-invasion events. *Biological Conser-*

vation, 78, 207 – 14.

23. Chew, F. (1995). From weeds to crops: changing habitats of pierid butterflies (Lepidoptera: Pieridae). *Journal of the Lepidopterists' Society*, 49, 285 – 303.
24. Coll, M. (1998). Parasitoid activity and plant species composition in intercropped systems. In *Enhancing biological control: habitat management to promote natural enemies of agricultural pests* (ed. C. H. Pickett and R. L. Bugg), pp. 85 – 119. University of California Press, Berkeley, CA.
25. Coll, M. and Bottrell, D. G. (1996). Movement of an insect parasitoid in simple and diverse plant assemblages. *Ecological Entomology*, 21, 141 – 9.
26. Cory, J. S. and Myers, J. H. (2000). Direct and indirect ecological effects of biological control. *Trends in Ecology and Evolution*, 15, 137 – 9.
27. Cowles, R. S. (2004). Impact of azadirachtin on vine weevil (Coleoptera: Curculionidae) reproduction. *Agricultural and Forest Entomology*, 6, 291 – 4.
28. Crawley, M. J. (1983). *Herbivory. The dynamics of animal-plant interactions*. Blackwell Scientific, Oxford.
29. Den Nijs, H. C. M., Bartsch, D., and Sweet, J. (2004). *Introgression from genetically modified plants into wild relatives*. CAB International, Wallingford.
30. Dethier, V. G., Barton Browne, L., and Smith, C. N. (1960). The designation of chemicals in terms of the responses they elicit from insects. *Journa of Economic Entomology*, 53, 134 – 6.
31. Dicke, M. (1999). Direct and indirect effects of plants on performance of beneficial organisms. In *Handbook of pest management* (ed. J. R. Ruberson), pp. 105 – 53. Marcel Dekker, New York.
32. Dodd, A. P. (1940). *The biological campaign against prickly pear*. Government Printer, Brisbane.
33. Dover, J. W. (1986). The effect of labiate herbs and white clover on Plutella xylostella oviposition. *Entomologia Experimentalis et Applicata*, 42, 243 – 7.
34. Dunn, J. A. and Kempton, D. P. H. (1972). Resistance to attack by Brevicoryne brassicae among plants of Brussels sprouts. *Annals of Applied Biology*, 72, 1 – 11.
35. Eigenbrode, S. D. (2002). Resistance management in host-plant resistance. In *Encyclopedia of pest management* (ed. D. Pimentel), pp. 708 – 11. Dekker, New York.
36. Eigenbrode, S. D. and Clement, S. L. (1999). Plant genetic resources for the study of insect-plant interactions. In *Global plant genetic resources for insectresistant crops* (ed. S. L. Clement and S. S. Quisenberry), pp. 243 – 62. CRC Press, Boca Raton, FL.
37. Ekbom, B., Irwin, M. E., and Robert, Y. (2000). *Interchanges of insects between agricultural and surrounding landscapes*. Kluwer Academic, Dordrecht.
38. Fowler, S. V. (1993). The potential for control of bracken in the UK using introduced herbivorous insects. *Pesticide Science*, 37, 393 – 7.
39. Franc, F. H., Plaisted, R. L., Roush, R. T., Via, S., and Tingey, W. M. (1994). Selec-

tion response of the Colorado potato beetle for adaptation to the resistant potato, *Solanum berthaultii*. *Entomologia Experimentalis et Applicata*, 73, 101 –9.

40. Frazier, J. L. and Chyb, S. (1995). Use of feeding inhibitors in insect control. In *Regulatory mechanisms in insect feeding* (ed. R. F. Chapman and G. de Boer), pp. 364 –81. Chapman & Hall, London.

41. Fritzsche, R., Decker, H., Lehmann, W., Karl, E., and Geissler, K. (1988). *Resistenz von Kulturpflanzen gegen Tierische Schaderreger*. Springer, Berlin.

42. Garcia, M. A. and Altieri, M. A. (1992). Explaining differences in flea beetle *Phyllotreta cruciferae* Goeze densities in simple and mixed broccoli cropping systems as a function of individual behavior. *Entomologia Experimentalis et Applicata*, 62, 201 –9.

43. Gassmann, A. (1995). Europe as a source of biological control agents of exotic invasive weeds: status and implications. *Mitteilungen der Schweizerischen Entomologischen Gesellschaft*, 68, 313 –22.

44. Glinwood, R., Ninkovic, V., Pettersson, J., and Ahmed, A. (2004). Barley exposed to aerial allelopathy from thistles (*Cirsium* spp.) becomes less acceptable to aphids. *Ecological Entomology*, 29, 188 –95.

45. Gold, C. S., Altieri, M. A., and Bellotti, A. C. (1989). Effects of cassava varietal mixtures on the whiteflies *Aleurotrachelus socialis* and *Trialeurodes variabilis* in Colombia. *Entomologia Experimentalis et Applicata*, 53, 195 –202.

46. Gould, F. (1986). Simulation models for predicting durability of insect-resistant germ plasm: a deterministic diploid, two-locus model. *Environmental Entomology*, 15, 1 –10.

47. Gould, F. (1988). Genetics of pairwise and multispecies plant-herbivore coevolution. In *Chemical mediation of coevolution* (ed. K. C. Spencer), pp. 13 –55. Academic Press, San Diego, CA.

48. Groot, A. T. and Dicke, M. (2002). Insect-resistant transgenic plants in a multitrophic context. *Plant Journal*, 31, 387 –406.

49. Haq, S. K., Atif, S. M., and Khan, R. H. (2004). Protein proteinase inhibitor genes in combat against insects, pests, and pathogens: natural and engineered phytoprotection. *Archives of Biochemistry and Biophysics*, 431, 145 –59.

50. Hastings, A. (1999). Outbreaks of insects: a dynamic approach. In *Theoretical approaches to biological control* (ed. B. A. Hawkins and H. V. Cornell), pp. 206 –15. Cambridge University Press, Cambridge.

51. Held, D. W., Eaton, T., and Potter, D. A. (2001). Potential for habituation to a neem-based feeding deterrent in Japanese beetles, *Popillia japonica*. *Entomologia Experimentalis et Applicata*, 101, 25 –32.

52. Hill, D. E. (1997). *The economic importance of insects*. Chapman & Hall, London.

53. Hokkanen, H. M. T. (1991). Trap cropping in pest management. *Annual Review of Entomology*, 36, 119 –38.

54. Hooks, C. R. R. and Johnson, M. W. (2003). Impact of agricultural diversification on the insect community of cruciferous crops. *Crop Protection*, 22, 223-38.

55. Howe, W. L., Sanborn, J. R., and Rhodes, A. M. (1976). Western corn rootworm adult and spotted cucumber beetle associations with Cucurbita and cucurbitacins. *Environmental Entomology*, 5, 1043-8.

56. James, C. (2004). *Preview: Global status of commercialized biotech/GM crops*: 2003. ISAAA Briefs No. 32. Ithaca, NY.

57. Jena, K. K. and Khush, G. S. (1990). Introgression of genes from *Oryza officinalis* Well × Watt to cultivated rice, *O. sativa* L. *Theoretical and Applied Genetics*, 80, 737-45.

58. Jermy, T. (1990). Prospects of antifeedant approach to pest control—a critical review. *Journal of Chemical Ecology*, 16, 3151-66.

59. Johns, T. and Alonso, J. G. (1990). Glycoalkaloid change during domestication of the potato, *Solanum* section Petota. *Euphytica*, 50S, 203-10.

60. Josefsson, E. (1967). Distribution of thioglucosides in different parts of Brassica plants. *Phytochemistry*, 6, 1617-27.

60a. Julien, M. H. (1992). *Biological control of weeds* (3rd edn). CAB International, Wallingford.

61. Julien, M. H. and Griffiths, M. W. (1998). *Biological control of weeds: a world catalogue of agents and their target weeds* (4th edn). CAB International, Wallingford.

62. Kim, K. C. and McPheron, B. A. (1993). *Evolution of insect pests. Patterns of variation*. John Wiley, New York.

63. Kogan, M. (1986). Plant defense strategies and hostplant resistance. In *Ecological theory and integrated pest management practice* (ed. M. Kogan), pp. 83-134. John Wiley, New York.

64. Kogan, M. and Ortman, E. F. (1978). Antixenosis—a new term proposed to define Painter's 'nonpreference' modality of resistance. *Bulletin of the Entomological Society of America*, 24, 175-6.

65. Kota, M., Daniell, H., Varma, S., Garczynski, S. F., Gould, F., and Moar, W. J. (1999). Overexpression of the *Bacillus thuringiensis* (*Bt*) Cry2Aa2 protein in chloroplasts confers resistance to plants against susceptible and *Bt*-resistance insects. *Proceedings of the National Academy of Sciences of the USA*, 96, 1840-5.

66. Koul, O. (2004). *Insect antifeedants*. CRC Press, Boca Raton, FL.

67. Kowalchuk, G. A., Bruinsma, M., and Van Veen, J. A. (2003). Assessing responses of soil microorganisms to GM plants. *Trends in Ecology and Evolution*, 18, 403-10.

68. Koziel, M. G., Beland, G. L., Bowman, C., Carozzi, N. B., Crenshaw, R., Crossland, L., *et al.* (1993). Field performance of elite transgenic maize plants expressing an insecticidal protein derived from *Bacillus thuringiensis*. *Bio/Technology*, 11, 194-200.

69. Krips, O. E., Willems, P. E. L., Gols R., Posthumus, M. A., Gort, G., and Dicke, M.

(2001). Comparison of cultivars of ornamental crop *Gerbera jamesonii* on production of spider-mite induced volatiles, and theirattractiveness to the predator *Phytoseiulus persimilis*. *Journal of Chemical Ecology*, 27, 1355 – 72.

70. Liu, X. M., Smith, C. M., Gill, B. S., and Tolmay, V. (2001). Microsatellite markers linked to six Russian wheat aphid resistance genes in wheat. *Theoretical and Applied Genetics*, 102, 504 – 10.

71. Lovett, J. V. and Hoult, A. H. C. (1992). Gramine: the occurrence of a self defence chemical in barley, *Hordeum vulgare* L. *Proceedings of the Sixth Australian Society of Agrononomists Conferences*, Armidale, pp. 426 – 9.

72. Lugenbill, P. (1969). Developing resistant plants—the ideal method for controlling insects. *USDA, Agricultural Research Service and Product Research Report*, 111, 1 – 14.

73. Luo, L. E., van Loon, J. J. A., and Schoonhoven, L. M. (1995). Behavioural and sensory responses to some neem compounds by *Pieris brassicae* larvae. *Physiological Entomology*, 20, 134 – 40.

74. Mauricio, R. and Rausher, M. D. (1997). Experimental manipulation of putative selective agents provide evidence for the role of natural enemies in the evolution of plant defense. *Evolution*, 51, 1435 – 44.

75. Maxwell, F. G. and Jennings, P. R. (1980). *Breeding plants resistant to insects*. John Wiley, New York.

76. Metz, M. (2003). Bacillus thuringiensis: *the cornerstone of modern agriculture*. Haworth Press, Binghampton, NY.

77. Morandini, P. and Salamini, F. (2003). Plant biotechnology and breeding: allied for years to come. *Trends in Plant Science*, 8, 70 – 5.

78. Mordue (Luntz), A. J. and Blackwell, A. (1993). Azadirachtin: an update. *Journal of Insect Physiology*, 39, 903 – 24.

79. Morgan, E. A. and Mandava, N. B. (1990). *CRC handbook of natural pesticides*, Vol. 4, *Insect attractants and repellents*. CRC Press, Boca Raton, FL.

80. Murray, K. D., Alford, A. R., Groden, E., Drummond, F. A., Storch, R. H., Bentley, M. D., et al. (1993). Interactive effects of an antifeedant used with *Bacillus thuringiensis* var. *san diego* delta endotoxin on Colorado potato beetle (Coleoptera: Chrysomelidae). *Journal of Economic Entomology*, 86, 1793 – 801.

81. National Research Council (1992). *Neem: a tree for solving problems*. National Academy Press, Washington, DC.

82. Needham, J. (1986). *Science and civilization in China*, Vol. 6(1). Cambridge University Press, Cambridge.

83. Nicholls, C. I., Parrella, M. P., and Altieri, M. A. (2000). Reducing the abundance of leafhoppers and thrips in a northern California organic vineyard through maintenance of full season

floral diversity with summer cover crops. *Agricultural and Forest Entomology*, 2, 107 – 13.

84. Ninkovic, V., Olsson, U., and Pettersson, J. (2002). Mixing barley cultivars affects aphid host plant acceptance in field experiments. *Entomologia Experimentalis et Applicata*, 102, 177 – 82.

85. O'Brien, D. M., Boggs, C. L., and Fogel, M. L. (2004). Making eggs from nectar: the role of life history and dietary carbon turnover in butterfly reproductive resource allocation. *Oikos*, 105, 279 – 91.

86. O'Callaghan, M., Glare, T. R., Burgess, E. P. J., and Malone, L. A. (2005). Effects of plants genetically modified for insect resistance on nontarget organisms. *Annual Review of Entomology*, 50, 271 – 92.

87. Oerke, E. C. and Dehne, H. W. (1997). Global crop production and the efficacy of crop protection-current situation and future trends. *European Journal of Plant Pathology*, 103, 203 – 15.

88. Oerke, E. C., Dehne, H. W., Schönbeck, F., and Weber, A. (1994). *Crop production and crop protection. Estimated losses in major food and cash crops.* Elsevier, Amsterdam.

89. Painter, R. H. (1951). *Insect resistance in crop plants.* University Press of Kansas, Lawrence.

90. Panda, N. and Khush, G. S. (1995). *Host plant resistance to insects.* CAB International, Wallingford.

91. Parmar, B. S. and Sinha, S. (2002). Results with commercial neem formulations produced in India. In *The neem tree*, Azadirachta indica *A. Juss. and other meliaceous plants* (2nd edn) (ed. H. Schmutterer), pp. 562 – 92. Neem Foundation, Mumbai.

92. Perlak, F. J., Deaton, R. W., Armstrong, T. A., Fuchs, R. L., Sims, S. R., Greenplate, J. T., et al. (1990). Insect resistant cotton plants. *Bio/Technology*, 8, 939 – 43.

93. Perlak, F. J., Fuchs, R. L., Dean, D. A., McPherson, S. L., and Fischhoff, D. A. (1991). Modification of the coding sequence enhances plant expression of insect control protein genes. *Proceedings of the National Academy of Sciences of the USA*, 88, 3324 – 8.

94. Pickett, J. A., Dawson, G. W., Griffiths, D. C., Hassanali, A., Merritt, L. A., Mudd, A., et al. (1987). Development of plant antifeedants for crop protection. In *Pesticide science and biotechnology* (ed. R. Greenhalgh and T. R. Roberts), pp. 125 – 8. Blackwell, Oxford.

95. Pilson, D. and Prendeville, H. R. (2004). Ecological effects of transgenic crops and the escape of transgenes into wild populations. *Annual Review of Ecology, Evolution and Systematics*, 35, 149 – 74.

96. Pimentel, D. (1986). Biological invasions of plants and animals in agriculture and forestry. In *Ecology of biological invasions of North America and Hawaii* (ed. H. A. Mooney and J. A. Drake), pp. 149 – 62. Springer, New York.

97. Pimentel, D. (1991). Diversification of biological control strategies in agriculture. *Crop Protection*, 10, 243 – 53.

98. Pimentel, D. (1993). Habitat factors in new pest invasions. In *Evolution of insect pests. Patterns of variation* (ed. K. C. Kim and B. A. McPheron), pp. 165–81. John Wiley, New York.

99. Pimentel, D. (ed.) (1997). *Techniques for reducing pesticides: environmental and economic benefits*. John Wiley, Chichester.

100. Quiros, C. F., Stevens, M. A., Rick, C. M., and Kok-Yokomi, M. K. (1977). Resistance in tomato to the pink form of the potato aphid (*Macrosiphum euphorbiae* Thomas): the role of anatomy, epidermal hairs, and foliage composition. *Journal of the American Society of Horticultural Science*, 102, 166–71.

101. Raval, K. N., Hellwig, S., Prakash, G., Ramos-Plasencia, A., Srivastava, A., and Büchs, J. (2003). Necessity of a two-stage process for the production of azadirachtin-related limonoids in suspension cultures of *Azadirachta indica*. *Journal of Bioscience and Bioengineering*, 96, 16–22.

102. Rea, J. H., Wratten, S. D., Sedcole, R., Cameron, P. J., Davis, S. I., and Chapman, R. B. (2002). Trap cropping to manage green vegetable bug *Nezara viridula* (L.) (Heteroptera: Pentatomidae) in sweet corn in New Zealand. *Agricultural and Forest Entomology*, 4, 101–7.

103. Risch, S. J. (1987). Agricultural ecology and insect outbreaks. In *Insect outbreaks* (ed. P. Barbosa and J. C. Schultz), pp. 217–38. Academic Press, San Diego, CA.

104. Room, P. M. (1990). Ecology of a simple plant-herbivore system: biological control of *Salvinia*. *Trends in Ecology and Evolution*, 5, 74–9.

105. Root, R. (1973). Organization of a plant-arthropod association in simple and diverse habitats. The fauna of collards. *Ecological Monographs*, 43, 95–124.

106. Saxena, R. C. (1987). Antifeedants in tropical pest management. *Insect Science and its Application*, 8, 731–6.

107. Schmutterer, H. (ed.) (2002). *The neem tree*, Azadirachta indica *A. Juss. and other meliaceous plants; sources of unique natural products for integrated pest management, medicine, industry and other purposes* (2nd edn). Neem Foundation, Mumbai.

108. Schroeder, H. E., Gollasch, S., Moore, A., Tabe, L. M., Craig, S., Hardie, D. C., et al. (1995). Bean α-amylase inhibitor confers resistance to the pea weevil (*Bruchus pisorum*) in transgenic peas (*Pisum sativum* L.). *Plant Physiology*, 107, 1233–40.

109. Sharma, H. C., Sharma, K. K., and Crouch, J. H. (2004). Genetic transformation of crops for insect resistance: potential and limitations. *Critical Reviews in Plant Sciences*, 23, 47–72.

110. Sheppard, A. W. (1992). Predicting biological weed control. *Trends in Ecology and Evolution*, 7, 290–1.

111. Simmonds, M. S. J. (2000). Molecular- and chemosystematics: do they have a role in agro-

chemical discovery? *Crop Protection*, 19, 591 – 6.

112. Smith, A. E. and Secoy, D. M. (1981). Plants used for agricultural pest control in western Europe before 1850. *Chemistry and Industry*, 3 January 1981, 12 – 17.

113. Smith, C. M. (1989). *Plant resistance to insects. A fundamental approach.* John Wiley, New York.

114. Smith, C. M. and Quisenberry, S. S. (1994). The value and use of plant resistance to insects in integrated crop management. *Journal of Agricultural Entomology*, 11, 189 – 90.

115. Smith, C. M., Khan, Z. R., and Pathak, M. D. (1994). *Techniques for evaluating insect resistance in crop plants.* CRC Press, Boca Raton, FL.

116. Southwood, T. R. E. (1973). The insect/plant relationship—an evolutionary perspective. *Symposia of the Royal Entomological Society of London*, 6, 3 – 30.

117. Stoner, K. A. (1992). Bibliography of plant resistance to arthropods in vegetables, 1977 – 1991. *Phytoparasitica*, 20, 125 – 80.

118. Stowe, K. A., Marquis, R. J., Hochwender, C. G., and Simms, E. L. (2000). The evolutionary ecology of tolerance to consumer damage. *Annual Review of Ecology and Systematics*, 31, 565 – 95.

119. Sunderland, K. and Samu, F. (2000). Effects of agricultural diversification on the abundance, distribution, and pest control potential of spiders: a review. *Entomologia Experimentalis et Applicata*, 95, 1 – 13.

120. Tabashnik, B. E., Liu, Y. B., Finson, N., Masson, L., and Heckel, D. G. (1997). One gene in diamondback moth confers resistance to four *Bacillus thuringiensis* toxins. *Proceedings of the National Academy of Sciences of the USA*, 94, 1640 – 4.

121. Tabashnik, B. E., Carrière, Y., Dennehy, T. J., Morin, S., Sisterson, M. S., Roush, R. T., et al. (2003). Insect resistance to transgenic *Bt* crops: lessons from the laboratory and field. *Journal of Economic Entomology*, 96, 1031 – 8.

122. Tang, Y. Q., Weathersbee, A. A., and Mayer, R. T. (2002). Effect of neem seed extract on the brown citrus aphid (Homoptera: Aphididae) and its parasitoid *Lysiphlebus testaceipes* (Hymenoptera: Aphidiidae). *Environmental Entomology*, 31, 172 – 6.

123. Thiéry, D. and Visser, J. H. (1987). Misleading the Colorado potato beetle with an odor blend. *Journal of Chemical Ecology*, 13, 1139 – 46.

124. Thies, C. and Tscharntke, T. (1999). Landscape structure and biological control in agrosystems. *Science*, 285, 893 – 5.

125. Thresh, J. M. (1981). *Pests, pathogens and vegetation: the role of weeds and wild plants in the ecology of crop pests and diseases.* Pitman, Boston, MA.

126. Timmins, W. A. and Reynolds, S. E. (1992). Azadirachtin inhibits secretion of trypsin in midgut of *Manduca sexta* caterpillars: reduced growth due to impaired protein digestion. *Entomologia Experimentalis et Applicata*, 63, 47 – 54.

127. Tonhasca, A. and Byrne, D. A. (1994). The effects of crop diversification on herbivorous insects: a meta-analysis approach. *Ecological Entomology*, 19, 239-44.
128. Trujillo-Arriaga, J. and Altieri, M. A. (1990). A comparison of aphidophagous arthropods on maize polycultures and monocultures, in Central Mexico. *Agriculture Ecosystems & Environment*, 31, 337-49.
129. Tuxill, J. (1999). *Nature's cornucopia: our stake in plant diversity*. Worldwatch Institute, Washington, DC.
130. Uvah, I. I. I. and Coaker, T. H. (1984). Effect of mixed cropping on some insect pests of carrots and onions. *Entomologia Experimentalis et Applicata*, 36, 159-67.
131. Vandermeer, J. (1989). *The ecology of intercropping*. Cambridge University Press, Cambridge.
132. Volkonsky, M. (1937). Sur l'action acridifuge des extraits de feuilles de *Melia azedarach*. *Archives de l'Institut Pasteur d'Algérie*, 15, 427-32.
133. Völlinger, M. and Schmutterer, H. (2002) Development of resistance to azadirachtin and other neem ingredients. In *The neem tree*, Azadirachta indica *A. Juss. and other meliaceous plants; sources of unique natural products for integrated pest management, medicine, industry and other purposes* (2nd edn) (ed. H. Schmutterer), pp. 598-606. Neem Foundation, Mumbai.
134. Wapshere, A. J., Delfosse, E. S., and Cullen, J. M. (1989). Recent developments in biological control of weeds. *Crop Protection*, 8, 227-50.
135. Welter, S. C. (2001). Contrasting plant responses to herbivory in wild and domesticated habitats. In *Biotic stress and yield loss* (ed. R. K. D. Peterson and L. G. Higley), pp. 160-84. CRC Press, Boca Raton, FL.
136. Wheeler, D. A. and Isman, M. B. (2000). Effect of *Trichilia americana* extract on feeding behavior of Asian armyworm, *Spodoptera litura*. *Journal of Chemical Ecology*, 26, 2791-800.
137. Williams, W. G., Kennedy, G. G., Yamamoto, R. T., Thacker, J. D., and Bordner, J. (1980). 2-Tridecanone: a naturally occurring insecticide from the wild tomato *Lycopersicon hirsutum* f. *glabratum*. *Science*, 207, 888-9.
138. Wilson, E. O. (1992). *The diversity of life*. Penguin Books, London.
139. Windle, P. N. and Franz, E. H. (1979). The effects of insect parasitism on plant competition: greenbugs and barley. *Ecology*, 60, 521-9.
140. Wink, M. (1988). Plant breeding: importance of plant secondary metabolites for protection against pathogens and herbivores. *Theoretical and Applied Genetics*, 75, 225-33.
141. Wyss, E. (1995). The effects of weed strips on aphids and aphidophagous predators in an apple orchard. *Entomologia Experimentalis et Applicata*, 75, 43-9.

附录 A:进一步阅读建议

与昆虫—植物相互作用相关的书籍:

Abrahamson, W. G. (ed.) (1989). *Plant-animal interactions*. McGraw-Hill, New York, 480 pp.

Ahmad, S. (ed.) (1983). *Herbivorous insects. Host seeking behavior and mechanisms*. Academic Press, San Diego, 257 pp.

Ananthakrishnan, T. N. (1992). *Dimensions of insect-plant interactions*. Oxford and IBH Publishing, New Delhi, 184 pp.

Ananthakrishnan, T. N. (1994). *Functional dynamics of phytophagous insects*. Oxford and IBH Publishing, New Delhi, 304 pp.

Ananthakrishnan, T. N. (ed.) (2001). *Insects and plant defense dynamics*. Oxford and IBH Publishing, New Delhi, 256 pp.

Ananthakrishnan, T. N. and Raman, A. (eds) (1993). *Chemical ecology of phytophagous insects*. Science Pub - lishers, Lebanon, NH, 332 pp.

Barbosa, P. and Letourneau, D. K. (eds) (1988) *Novel aspects of insect-plant interactions*. John Wiley, New York, 362 pp.

Barbosa, P. and Schultz, J. C. (eds) (1987). *Insect outbreaks*. Academic Press, San Diego, 578 pp.

Barbosa, P. and Wagner, M. R. (1989). *Introduction to forest and shade tree insects*. Academic Press, San Diego, 639 pp.

昆虫—植物关系国际学术会议文集:

Proceedings of the 1st International Symposium, Insect and Foodplant, Wageningen, 1957 (ed. J. de Wilde). *Entomologia Experimentalis et Applicata*, 1, 1 - 118 (1958).

Proceedings of the 2nd International Symposium, Insect and Host Plant, Wageningen, 1969 (ed. J. de Wilde and L. M. Schoonhoven). *Entomologia Experimentalis et Applicata*, 12, 471 - 810 (1969).

Proceedings of the 3rd International Symposium, The Host-Plant in Relation to Insect Behaviour and Repro - duction, Budapest, 1974 (ed. T. Jermy). Plenum Press, New York. Also: *Symposia Biologica Hungarica*, 16, 1 - 322 (1976).

Proceedings of the 4th International Symposium, Insect and Host Plant, Slough, 1978 (eds. R. F. Chapman and E. A. Bernays). *Entomologia Experimentalis et Applicata*, 24, 201 - 766 (1978).

Proceedings of the 5th International Symposium, Insect - Plant Relationhips, Wageningen, 1982 (ed. J. H. Visser and A. K. Minks). Pudoc, Wageningen (1982).

(以下部分内容省略……)

附录 B:植物次生代谢物的分子结构

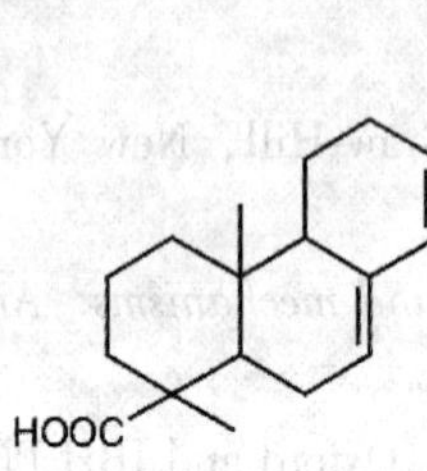

(1)松香酸

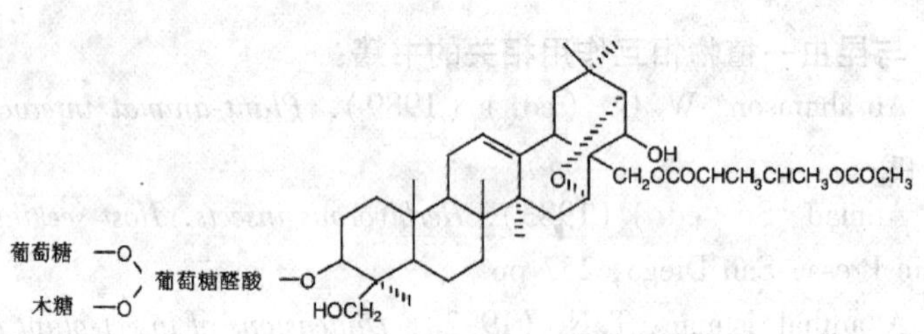

(2)七叶素
葡萄糖
木糖
葡萄糖醛酸

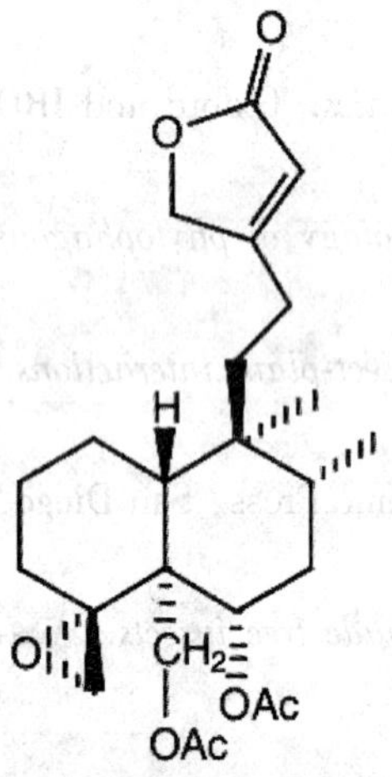

(3)散血草素 I

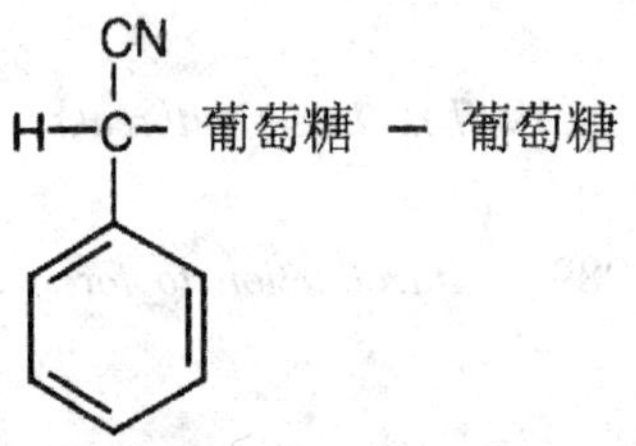

(4)苦杏仁苷
葡萄糖 — 葡萄糖

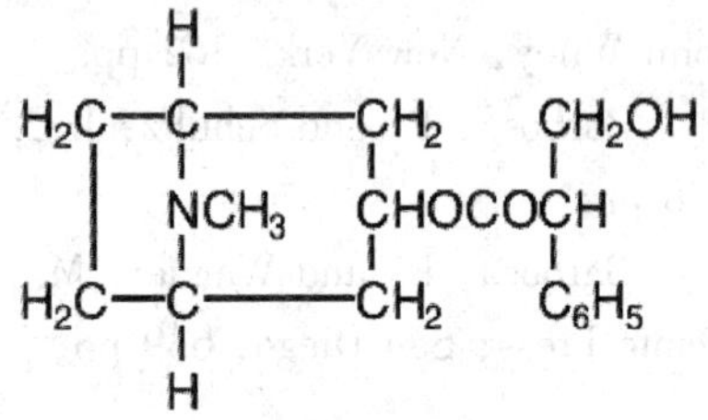

(5)阿托品

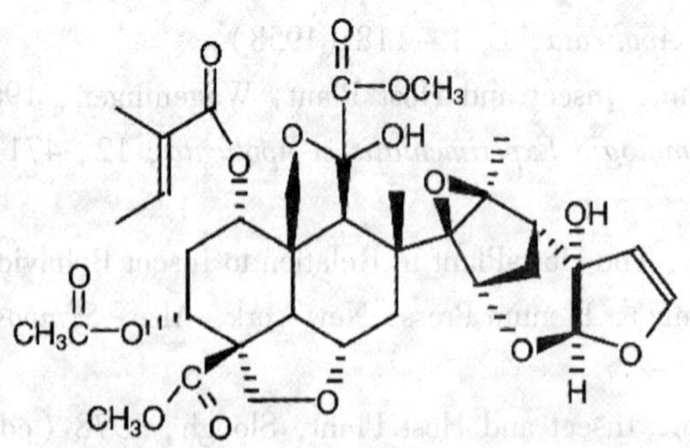

(6)印楝素

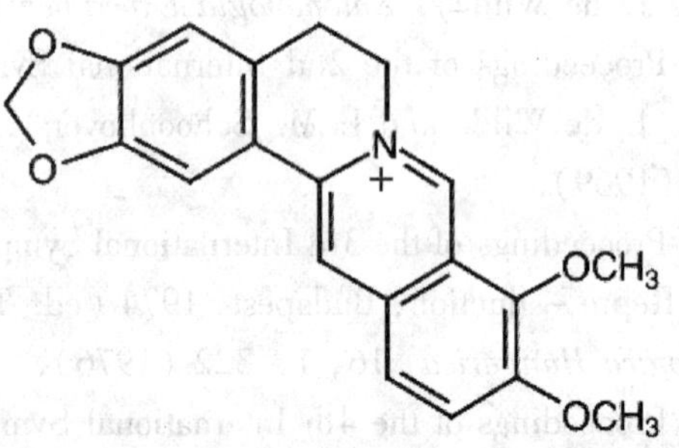

(7)黄连素

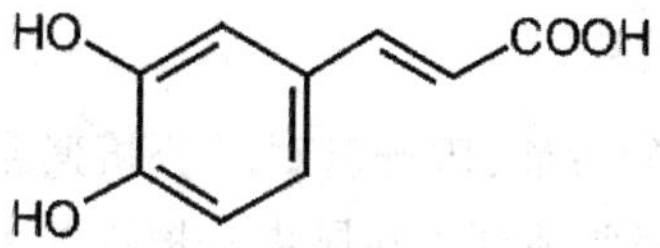

(8)咖啡酸

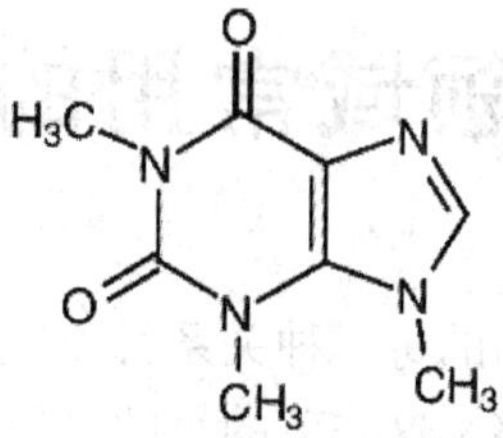

(9)咖啡因

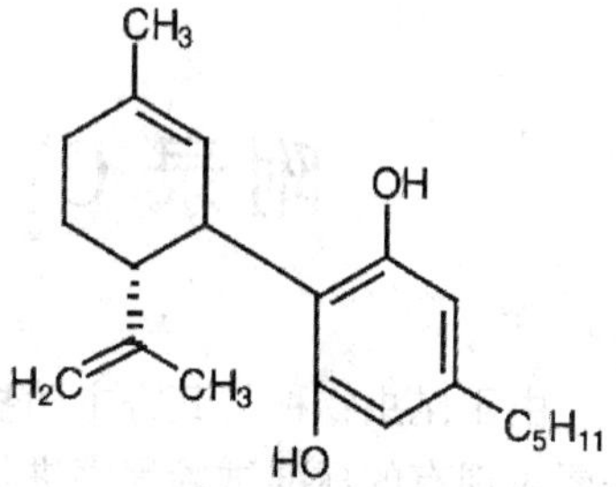

(10)大麻二醇

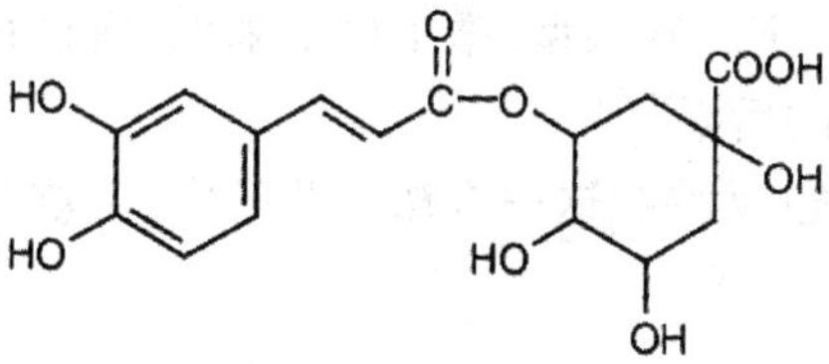

(11)绿原酸

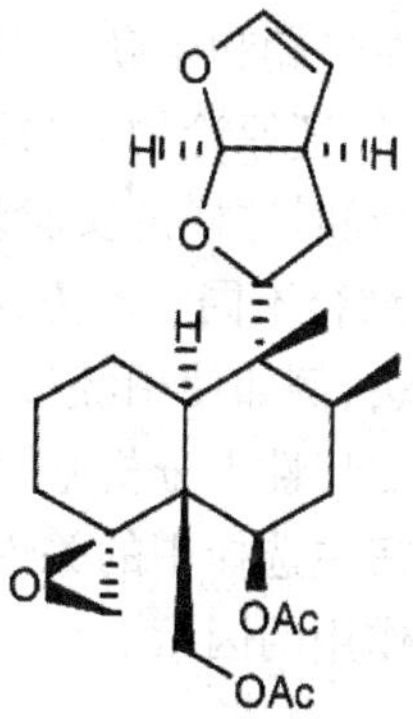

(12)桐定

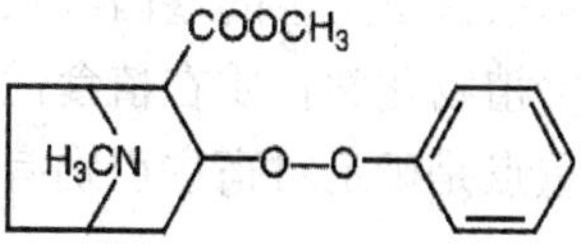

(13)可卡因

(以下部分内容省略……)

附录 C：本领域常用的实验方法

由于昆虫物种与其寄主植物之间的每一种关系都有其独特的方面，因此其科学分析通常需要对现有的标准试验程序进行调整。这一节的目的是要简明扼要地列出在昆虫—植物关系研究中经常使用的一些程序，但通常需要加以修改以适应特定的情况。要了解更多细节，请参阅米勒和米勒[78]、史密斯等人[118]和黑尔[43]的综合评论。本节讨论的大多数技术都涉及实验室研究。行为和实验研究的方法可以在米勒和米勒[78]、丹特和沃尔顿[24]的相关文献里找到。

就像在生物学的其他领域一样，还原论的方法导致了一种两难的局面：通常最清晰的一组实验条件给出了最清晰的答案，但同时这样的实验环境与人们试图理解的自然情况也是最遥远的。这个困难可以（部分）通过结合不同实验方法的结果来解决。此外，正在出现将还原论方法与功能分析结合起来的新方法，例如使用经过仔细鉴定的基因型，举个例子来说，通过在自然条件下发生的基因沉默或突变特性来分析基因变化的后果[26,58]。

植物化学物质在植食性昆虫宿主选择中的关键作用也反映在这一方法学章节中，该章节的重点是确定对宿主识别有重要作用的因素。

（以下部分内容省略……）

物种索引

注：斜体字的数字指插图和表格。

Acacia farnesiana 金合欢 73
A. pennata 羽叶金合欢 82
Acalymma vittata 黄瓜条叶甲 400
Acanthoscelides obtectus 菜豆象 279
Acer saccharum 糖槭 82
A. tegmentosum 青楷槭 45
Aconitum spp. 乌头属 363
A. henryi 川鄂乌头 367
A. napellus 欧乌头 367
Acremonium strictum 直立顶孢霉 139
Acrolepiopsis assectella 邻菜夜蛾 175
Agropyron desertorum 大麦草 47
Ajuga remota 筋骨草 62
Azadirachta indica 楝树 21
Bemisia tabaci 棉粉虱 9
Brassica oleracea 甘蓝 37
Bretschneiderea sinensis 伯乐树 71
Brunsfelsia spp. 玄参科 17
Callophrys rubi 绿细纹蝶 134
Carya 山核桃 13
Cecropia peltata 号角树 74
Chenopodium murale 藜 41
Costelytra zealandica 新西兰金龟子 63
Crataegus monogyna 山楂 58
Crocidosema plebegana 卷蛾 11
Dasineura barassicae 瘿蚊 13
Encarsia formose 丽蚜小蜂 40
Eucalyptus globulus 桉树 36
Euphydryas editha 蛱蝶 12
Euproctis cherysorrhoea 棕尾毒蛾 7
Fenusa pumila 潜叶蜂 12
Festuca arundinacea 羊茅 37

Gossypium hirsutum 棉花 11
Grammia geneura 灯蛾 18
Helicoverpa zea 玉米夜蛾 90
Ilex aquifolium 冬青树 25
Juniperus spp. 柏树 21
Leptinotarsa decemlineata 马铃薯甲虫 7
Lithocolletis ostryarella 细蛾 13
Lygus lineolaris 绣叶虫 9
Lymantria dispar 舞毒蛾 10
Malva parviflora 锦葵 11
Mamestra configurata 蓓带夜蛾 12
Matsucoccus acalyptus 松蚧 22
Myzus persicae 绿桃蚜虫 7
Oryzopsis hymenoides 落芒草 47
Osmunda regalis 王紫萁 66
Ostya sp. 角木 13
Panonychu sulmi 六点黄蜘蛛 71
Papillio machaon 凤蝶 11
Pastinaca sativa 欧洲防风草 72
Pieris brassicae 菜粉蝶 7
Piper arieianum 胡椒 23
Poplilia japonica 日本甲虫 9
Prunus spp. 李属 65
Pseudaletia unipunctata 黏虫 42
Pteridium aquilinum 欧洲蕨 82
Rhododendron spp. 杜鹃花 21
Rhopalosi phumpadi 禾缢管蚜 19
Rumex hydrolapathum 酸模 42
Schistocerca gregaria 沙漠蝗 10
Spodoptera frugiperda 行军虫 14
Tanacetum vulgare 艾菊 15
Taraxacum officinale 蒲公英 46

Therioaphis trifolii 苜蓿斑蚜 19
Troparolum 旱金莲 7
Reseda 木樨草 7

Ulmus procera 英国大叶榆 93
Urtica dioica 刺荨麻 15
Viburnum tinus 地中海荚蒾 45

（以下部分内容省略……）

作者索引

Allen, S. E. 艾伦 *117*
Andow, D. A. 安杜 *2*
Avé, D. A. 艾维 *69*
Barkman, J. J. 巴克曼 *19,20*
Basset, Y. 巴西特 *48*
Benecke, R. 贝内克 *79*
Bernays, E. A. 伯尼斯 *41,42,117*
Blaakmeer, A. 布莱克模 *70*
Chapman, R. F. 查普曼 *67*
Coley, P. D. 科利 *75*
Cox, C. B. 考克斯 *20*
Crawley, M. J. 克劳利 *7*
Croteau, R. 克罗托 *68*
Damman, H. 达曼 *23*
Davies, R. G. 戴维斯 *35*
Davis, B. N. K. 戴维斯 *16*
Eigenbrode, S. D. 艾根布罗特 *37*
Feeny, P. 非尼 *79*
Ferro, D. N. 法罗 *19*
Field, M. D. 菲尔德 *118*
Fitter, A. H. 菲特 *18*
Frost, S. W. 弗罗斯特 *13,35*
Gershenzon, J. 赫尔申宗 *68,74,78*
Gouinguené, S. 苟英谷那 *87*
Grevillius, A. Y. 格里弗柳斯 *8*
Hamilton, J. G. 哈密赖 *11*
Harborne, J. B. 哈伯纳 *59*
Hartmann, T. 哈特曼 *95*
Haukioja, E. 哈乌基奥 *89*
Hay, R. K. M. 海 *18*
Hedin, P. A. 埃丁 *92*
Heinrichs, E. A. 海因里希斯 *83*
Howlett, B. G. 豪利特 *41*
Isely, F. B. 艾斯利 *42*

Janzen, D. H. 詹曾 *42*
Jeffree, C. E. 杰弗里 *37,39*
Joern, A. 纳恩 *81*
Jones, C. G. 琼斯 *89*
Kennedy, C. E. J. 肯尼迪 *17*
Kogan, M. 科根 *39*
Marquis, R. J. 马奎斯 *21,23*
Matile, P. 马蒂 *77*
Mattson, W. J. 马特森 *9,117*
Matzinger, D. F. 马卿格 *76*
Medrano, F. G. 梅德拉诺 *83*
Moore, D. 摩尔 *43*
Nault, L. R. 纳特 *66*
Neuvonen, S. 内乌沃宁 *89*
Norris, D. M. 诺里斯 *39*
Pimentel, D. 皮门特尔 *2,3*
Reese, J. C. 里斯 *118*
Ross, J. A. M. 罗斯 *35*
Schmitz, G. 施密茨 *15*
Schoonhoven, L. M. 斯洪霍芬 *10*
Southwood, T. R. E. 塞尔本南木 *17*
Stoutjesdijk, P. 斯陶特 *19,20*
Strong, D. R. 斯特朗 *5,14*
Styer, W. E. 斯太尔 *66*
Thieme, H. 蒂姆 *79*
Van Dam, N. M. 万·戴姆 *72*
Van den Boom, C. E. M. 旺当 *70*
Van Poecke, R. M. P. 万·波克 *70*
Vandenberg, P. 范登堡 *76*
Vereijken, B. H. 维尔捷克 *22*
Visser, J. H. 维瑟 *69*
Whitham, T. G. 惠瑟姆 *73,93*
Wiklund, C. 维克隆德 *11*
Woodhead, S. 伍德黑德 *67*

（以下部分内容省略……）

主题索引

注：斜体数字指插图。

acarodomatia 虱巢 45
active space 活跃空间 167
addressing mechanisms 寻址机制 159
aerial plant 气生植物 91
alkaloids 生物碱 66
approximate digestibility 近似消化率 124
caloric value 热值 117
complexes 复合体 18
cyanogenic glycosides 生氰苷类 65
day-foraging 日行性 168
defence 防御 57
defoliation-induced 脱叶诱导 81
domatia 虫菌穴 44
drimane-type 补身烷 62
drymatter units 干物质单位 124
dynamic systems 动力系统 66
ecdysteroid 蜕化类固醇 62
egg-laying 产卵 169
energetic processing 能量处理 127
extrafloral nectaries 花外蜜腺 44
feeding reaction 摄食反应 169
flavolans 缩合单宁 63
flaush feeders 旺盛的摄食者 121
furanocoumarins 香豆素 66
gallic acid 没食子酸 64
glossy phenotypes 光滑表型 38
gravimetric method 重量分析法 124
haustellate 吸食性口器 34
hexahydroxydiphenic acid 六羟二酚酸 64
hydrogen cyanide 氰化氢 65
infochemicals 化学信息素 160
kiness 动性 162

leaf boundary layer 边界层 18
limiting nutrient 限制性营养 125
locomotion compensator 运动补偿器 173
lunularic 半月苔酸 63
maintenance costs 维持性消耗 125
malvin 锦葵花甙 63
metabolic load 代谢负荷 127
microclimate 微气候 18
modular 模块化 24
myrosinase 芥子酶 65
net assimilate 净同化物 21
nutritional indices 营养指数 124
octadecanoid 十八碳酸 88
odour-filled space 气味空间 166
palisade parenchyma 栅栏组织 12
patterns ofassociation 联系模式 159
performance 性能 124
phaseolin 菜豆素 63
phloridzin 根皮苷 66
photomenotaxis 横向趋光性 165
plant-infesting 植物滋生 6
polygodial 水蓼二醛 62
prunasin 洋李甙 65
relative growth rate 相对生长率 125
resistance 抵抗 57
respiration rates 呼吸率 125
sap-feeding 吸食汁液 20
secondary plant substances 植物次生代谢物质 57
semiochemmicals 化学信息素 160
senescence feeders 衰老的摄食者 121
sink leaves 库叶 88

source leaves 原叶 88
taxes 税金 162
taxol 紫杉醇 57
trichomes 毛状体 18

unitary 单位化 24
utilization 利用率 124
warburganal 活乐木醛 62
yield components 产量构成因素 44

（以下部分内容省略……）

译者说明

本书原版最后一部分内容为附录,包括:(1)附录 A:进一步阅读建议;(2)附录 B:植物次生代谢物的分子结构;(3)附录 C:本领域常用的实验方法;(4)物种索引;(5)作者索引;(6)主题索引。因为本书主要用于研究生教学,译者认为附录内容对教学的意义和作用不大,因此仅对原版附录内容进行了摘译。

此外,由于时间原因,本书原版有些物种的拉丁名暂时没有找到对应的中文名,因此译著中未将全部物种名称翻译成中文,敬请读者见谅。